W0107440

Artificial Intelligence

Editors: S. Amarel A. Biermann L. Bolc P. Hayes A. Joshi
D. Lenat D.W. Loveland A. Mackworth D. Nau R. Reiter
E. Sandewall S. Shafer Y. Shoham J. Siekmann W. Wahlster

Springer
Berlin
Heidelberg
New York
Barcelona
Hong Kong
London
Milan
Paris
Singapore
Tokyo

Ray Reiter

Hector J. Levesque Fiora Pirri (Eds.)

Logical Foundations for Cognitive Agents

Contributions in Honor of Ray Reiter

With 32 Figures

 Springer

Editors

Prof. Dr. Hector J. Levesque

University of Toronto
Department of Computer Science
10 King's College Road
Toronto, Ontario M5S 1A4, Canada
hector@ai.toronto.edu

Prof. Dr. Fiora Pirri

Università di Roma „La Sapienza"
Dipartimento di Informatica e Sistemistica
via Salaria 113
I-00198 Roma, Italy
pirri@dis.uniroma1.it

ISBN-13: 978-3-642-64306-4   e-ISBN-13: 978-3-642-60211-5
DOI: 10.1007/978-3-642-60211-5
Library of Congress Cataloging-in-Publication Data applied for

Die Deutsche Bibliothek – CIP-Einheitsaufnahme
Logical foundations for cognitive agents: contributions in honor of Ray Reiter /
Hector J. Levesque; Fiora Pirri (eds.). – Berlin; Heidelberg; New York; Barcelona;
Hong Kong; London; Milan; Paris; Singapore; Tokyo: Springer, 1999
(Artificial intelligence)
ISBN-13: 978-3-642-64306-4

This work is subject to copyright. All rights are reserved, whether the whole or part of
the material is concerned, specifically the rights of translation, reprinting, reuse of
illustrations, recitation, broadcasting, reproduction on microfilm or in any other way,
and storage in data banks. Duplication of this publication or parts thereof is permitted
only under the provisions of the German copyright law of September 9, 1965, in its
current version, and permission for use must always be obtained from Springer-Verlag.
Violations are liable for prosecution under the German Copyright Law.

© Springer-Verlag Berlin Heidelberg 1999
Softcover reprint of the hardcover 1st edition 1999

The use of general descriptive names, trademarks, etc. in this publication does not
imply, even in the absence of a specific statement, that such names are exempt from the
relevant protective laws and regulations and therefore free for general use.

Cover design: Künkel + Lopka Werbeagentur, Heidelberg
Typesetting: Camera-ready by the editors
SPIN: 10723537   45/3142 – 543210 – Printed on acid-free paper

Preface

It is a pleasure and an honor to be able to present this collection of papers to Ray Reiter on the occasion of his 60th birthday.

To say that Ray's research has had a deep impact on the field of Artificial Intelligence is a considerable understatement. Better to say that anyone thinking of doing work in areas like deductive databases, default reasoning, diagnosis, reasoning about action, and others should realize that they are likely to end up proving corollaries to Ray's theorems. Sometimes studying related work makes us think harder about the way we approach a problem; studying Ray's work is as likely to make us want to drop our way of doing things and take up his. This is because more than a mere visionary, Ray has always been a true *leader*. He shows us how to proceed not by pointing from his armchair, but by blazing a trail himself, setting up camp, and waiting for the rest of us to arrive. The International Joint Conference on Artificial Intelligence clearly recognized this and awarded Ray its highest honor, the Research Excellence award in 1993, before it had even finished acknowledging all the founders of the field.

The papers collected here sample from many of the areas where Ray has done pioneering work. One of his earliest areas of application was databases, and this is reflected in the chapters by Bertossi *et al.* and the survey chapter by Minker. His work in logic programming and nonmonotonic reasoning is well represented in chapters by Amati *et al.*, by Criscuolo *et al.*, by Etherington and Crawford, by Lifschitz, and by Denecker *et al.* His insights into diagnosis are evident in the chapters by de Kleer and by McIlraith. An application of his ideas to search shows up in the chapter by Lin. Ray has championed a cognitive view of robotics and high-level agent control, and we see this in the chapters by de Giacomo and Levesque, by Lakemeyer, by Lespérance *et al.*, by Shanahan, and by Zhang and Mackworth. Non-robotic agents are considered by Scherl and in the proposal by Sandewall. Ray's general approach to reasoning about action is discussed in the chapters by Bacchus, by Boutilier and Goldszmidt, and by Schubert. Finally, some basic questions regarding the logical approach to cognitive agents are taken up in chapters by McCarthy and by Gabbay. In sum, more than just representing a who's who of researchers interested in logical methods in Artificial Intelligence, the papers in this volume testify to the breadth and depth of Ray's influence. Incidentally, in the memoir by Rosemberg, we learn among other things of Ray's lepidopterological research, a branch of logic quite unlike the others.

Ray, on behalf of all of your friends and colleagues represented here, we hope you enjoy this collection, and we wish you a very happy birthday.

Toronto, Roma   *Hector Levesque*
June 1999   *Fiora Pirri*

Contents

List of Contributors

Giambattista Amati

Fahiem Bacchus

Leopoldo Bertossi

Craig Boutillier

Luigia Carlucci Aiello

John McCarthy

James M. Crawford

Giovanni Criscuolo

Giuseppe De Giacomo

Johan De Kleer

Marc Denecker

David W. Etherington

Dov M. Gabbay

Moises Goldszmidt

Sheila A. McIlraith

Michael Jenkin

Gerhard Lakemeyer

Yves Lespérance

Hector J. Levesque

Vladimir Lifshitz

Fanghzen Lin

Alan K. Mackworth

V. Wictor Marek

Eliana Minicozzi

Jack Minker

Javier Pinto

Fiora Pirri

Richard Rosemberg

Erik Sandewall

Murray Shanahan

Richard B. Scherl

Lenhart K. Schubert

Kenneth Tam

Miroslav Truszczyński

Ricardo Valdivia

Ying Zhang

Ray Reiter – A Memoir

Richard Rosemberg

Department of Computer Science
University of British Columbia
Vancouver, BC, Canada V6T 1Z4

Preface

Let me try to make a few things clear. This non-technical accounting of a friendship and two lives must of necessity be incomplete and indeed replete with errors. Since Ray was not given the opportunity to correct the manuscript prior to publication, he will have to produce his version of the past, if he wishes; otherwise, by default, this essay is the received truth. I hesitate to use such words as truth and default in such an informal manner but given the unique status of this essay in a volume of papers dedicated to logic, it may not be too far-fetched if words and concepts are somewhat fuzzy. (Oh there I go again.) Given that memory is notoriously fallible, although Ray has often claimed that mine is distinguished by its ability to remember everything, important or not, I offer this disclaimer. I have tried to produce a document that informs and perhaps even describes certain interesting aspects of Ray's wide-ranging involvement in life, both academic and otherwise. And I must reiterate that everything that follows is filtered through time and distance, as well as a concern with a friendship of almost forty-seven years.

1 The Early Years I: High school

We share much. We were born two months apart in 1939, Ray on June 12 and I on August 12. Our parents were immigrants to Canada from Poland and worked in the garment industry. We both spoke Yiddish as children and still maintain a fondness for this language of our youth. At age thirteen, in the fall of 1952, we met in high school, Harbord Collegiate Institute, an academic public school. At that time, Ontario streamed its students into three divisions: trade schools, book-keeping and accounting schools, and academic schools, intended to lead to university or other post-secondary forms of education. Although Ray and I were both from working-class families, our academic abilities convinced our families that Harbord Collegiate would be the best way to achieve future success. It is a matter of dispute whether in the class we met, I sat behind Ray or he behind me; in any case, we became friends and have remained so all these years.

We both enjoyed mathematics, not a great surprise. As with many other Jewish boys our age, we had attended religious schools to prepare for our Bar Mitzvahs, confirmations, at age thirteen. However, while Ray had stopped attending and was continuing to develop his other interests, I continued to attend a religious school five times a week. My apparent commitment to religion served to fuel an endless debate that Ray and I carried on for many years. In contrast, he had a commitment to insects, primarily butterflies and moths. Very early on, he became an entrepreneur, collecting and selling insects to biological companies. For example, Toronto was periodically plagued with June bugs, those large awkward beetles that clogged the sidewalks on hot summer days. While most of us attempted to avoid crushing them, Ray collected them in large quantities. But his real passion was butterflies and it continues to this day. He has actually published in this field, a fact not generally known.

Although Ray could never understand why I remained religious, and to tell the truth I had some difficulty articulating my own beliefs, we nevertheless belonged to a religious, Zionist youth organization. What's the logic in this, I hear some of you asking? Well, our group consisted of boys and girls, and in clear opposition to religious principles, we had dances and even danced quite slowly. In later years, I discovered that our parent organization was quite right-wing, yet another example of youthful indiscretions, given our later commitment to the peace process, a recognition of the rights of Palestinians, and an almost complete abandonment of religion. I should mention that after sharing our first year of high school together, Ray's family moved and for the next few years, we met regularly in our youth group. I must not forget to mention that we also played basketball against other youth groups. The image of Ray, in shorts, shooting hoops is one that may interest some of you.

2   The Early Years II: The University of Toronto

We both entered the University of Toronto in the fall of 1957, not in mathematics but in Engineering Physics. By this time I realized that while I enjoyed mathematics, I probably could not make a living at it, but Ray still had a more intimate kinship with the muse. Nevertheless, there we were in engineering, albeit the least engineering-like of all the sub disciplines. Still, even I was often upset by the inelegance and cook-book-like atmosphere of engineering physics. Not surprisingly therefore, that once again, Ray and I did not go the distance. He left Engineering Physics to enter that most challenging program, MPC, Mathematics, Physics, and Chemistry. Ray was not going to settle for an approach that did not derive results from basic principles in a principled matter. Although formal logic was lacking from both our programs, in retrospect, one could discover a basic component of Ray's scientific temperament, namely, a need for formal satisfaction and a rejection of facile assumptions. More later.

We saw each other quite frequently during this time although we were in different faculties, I in Applied Science and Ray in Arts. At the University of Toronto, there were a number of denominational colleges in which students could enroll. Jewish students would typically select the non-denominational University College, where they would have lunch and often hang out in the JCR, the Junior Common Room. Since in those days, engineering was notoriously lacking in women students, I would often have lunch in the JCR, to visit with Ray and other friends, and of course to encounter women. In this regard, I have to point out an instance of Ray's independence and my early conservatism and lack of initiative. A year after I began to attend university, my parents, as was quite typical then, moved about ten miles away from downtown Toronto, where the University is located. This meant a long trip every day for me. Ray lived closer but decided to get his own apartment and he did.

In the summers we both managed to get a variety of jobs, not in the same place and not particularly relevant to our academic interests, except perhaps for the summer between our third and final year. That summer I worked for the federal government measuring tides in Nova Scotia while Ray worked in Fort Churchill, Manitoba, fairly far north, investigating the Northern Lights (Aurora Borealis). We exchanged letters from our rather remote outposts describing our jobs. Recall that we both grew up in a rather large city and up to then we had not ventured very far. Upon our return, tales of our summer adventures gave us a certain status in some quarters with Ray's stories, polar bears and all, considerably more exciting.

It is probably worth noting that as we entered the final year (1960/61) as undergraduates at the University of Toronto, neither of us had any direct experience with computers. In the electrical option of Engineering Physics, I had a laboratory on the performance characteristics of a single transistor. But of course we had heard about computers and some of the work of Norbert Wiener on cybernetics. It might seem surprising that upon graduation, Ray took a job with IBM and did some programming on the 1401 in assembler. I entered the masters' program in Electrical Engineering and began to work on control theory. So the theoretician was engaged in low-level programming and the engineer was hip deep in the calculus of variations and Pontryagin's maximum principle. Oh yes, I was also introduced to analogue computers, soon to follow the engineer's slide rule into technological oblivion. By the spring of 1962, Ray was planning to leave IBM and begin a round-the-world trip by himself. I was planning to go somewhere to do a Ph.D. in Electrical Engineering even though I had not yet begin the research for my M.A.Sc. thesis.

Well, Ray took off for Southeast Asia and I was accepted by the University of Michigan to do a Ph.D. in Electrical Engineering. What I remember most from Ray's descriptions of his travels were his stories of trekking through Nepal. Now that's an adventure; it conjured up images of high peaks, snow, strange languages and people. I was envious and wondered why I hadn't gone along. Ray returned with many exotic butterflies and moths, the first of a long succession of foreign acquisitions. I have always envied his ability to venture forth alone in a determined manner to explore strange places. His self sufficiency and ease in nature may seem at odds in one so committed to the adventure of the mind but Ray is one of the very few I know who can effortlessly make the transition from academia and its varied trappings to the jungle, the forest, and the mountains in search of a rare species of butterfly and of course he does not require the comfort of tourist accommodations, arranged in advance.

When the University catalogue arrived, I discovered something called the Program in Communication Sciences (CommSci), an interdisciplinary program to explore the common theme of information in such apparently disparate disciplines as computers, mathematics, electrical engineering, linguistics, psychology, neurophysiology, and biology. Since my commitment to engineering was somewhat tenuous at best, I immediately transferred to CommSci upon my arrival in Ann Arbor. Meanwhile, Ray returned to Toronto to begun his masters degree in applied mathematics, or theoretical physics by another name. Over the next year on visits to Toronto, I tried to tell Ray what I was doing and what CommSci seemed to be about. I was also still trying to finish my masters degree, which required the use of an analogue computer to solve a pair of simultaneous nonlinear differential equations. During this effort, I became convinced that digital computers were the wave of the future and perhaps this enthusiasm was detected by Ray and led to a major decision that he made in the early fall of 1963.

In the summer of 1963, Ray decided to do a Ph.D. in physics at Queen's University in Kingston, Ontario. In fact, he went to Kingston ostensibly to start his program but just as classes started, he changed his mind, called me (I think) and arrived soon after in Ann Arbor, after having been accepted in CommSci. Furthermore, consistent with a growing reputation for good fortune, Ray managed to qualify for a fellowship awarded to University of Toronto graduates accepted by the University of Michigan for post graduate studies. I had rented an apartment with a friend, Mike Sniderman, a theatre student, that I had met the previous year while living in a co-op house. Ray joined us and we set up house on Division Street in Ann Arbor, very close to campus. After having met about eleven years previously, Ray and I were once again classmates, this time in graduate school at the University of Michigan. Interesting and exciting times were ahead in both our personal and academic lives.

3   The Middle Years I: The University of Michigan

I don't remember the courses we took that year except for one. It must be recalled that I was trained as an engineer with a background in continuous mathematics. I knew virtually nothing about set theory, probability theory, algebra, or not surprisingly, logic. The previous year I had taken a course in modern algebra and had my earlier decision not to study mathematics once again confirmed in a particularly forceful manner. In the fall of 1963, Ray and I both enrolled in a course, with the somewhat menacing title of, Introduction to Metamathematics. This was to be my first course in logic and I believe the same held for Ray, but he was a real mathematician, I thought, able to deal comfortably with abstract concepts. The course was a chastening experience for both of us in completely different ways. From the outset, I seemed to have a natural feel for logic to the point that a bright, attractive undergraduate woman decided that I was the one to be her study partner. I must admit that I was flattered and easily acquiesced and even more surprisingly, I received the highest grade in the course, A+, while Ray got an A. Nevertheless, I reluctantly acknowledged that I had been extremely lucky and that Ray seemed to have found a real and natural home in the world of axioms, inferences, and proofs. So I abandoned logic while Ray fully embraced it and the action seemed to be reciprocated. A lifelong affair had begun and I was there at the beginning.

And now for a few interesting tales and memories of graduate student life in Ann Arbor in the mid-1960s. Ray, Michael, and I moved to a new apartment in the fall of 1964, above Democratic Party headquarters. We all lived there for about a year until Ray moved into his own apartment in a house, one of whose tenants was a classmate of ours, John Seeley Brown, now Vice-President of Research and Director of Xerox PARC. Another very good friend was Abbe Mowshowitz, who introduced me to the area of the social impact of computers, but that is a later story. During this period, Ray, Abbe, and John taking the lead of another friend, Billy Ash, all bought BMW motorcycles, and of course leather jackets. There was one final touch that must be mentioned; both Billy and John rode their bikes in the warm weather wearing rubber thongs on their feet. I was impressed although both Ray and Abbe declined to adopt the style. There you have it; Ray was a member of a bike 'gang' of Ph.D. students, hardly rebels with a cause. In this regard, I was left out except to play the role of passenger on occasion. On a trip back from Toronto, after having visited our parents, Ray picked me up and we covered the nearly three hundred miles uneventfully except for the last thirty miles, in a driving rainstorm. For one year, however, I did have an Italian motor scooter, not quite the equal of a killer BMW.

We partied on weekends, listened and danced to the Beatles and the Stones, smoked but declined a more intimate contact, visited local music legends such as Washboard Willie, and of course did some research. Ray's research was on parallel computation, nominally supervised by Harvey Garner, a professor in Electrical Engineering but motivated by Dick Karp who was on leave from IBM. I was one of the first students of John Holland in what was then called adaptive systems and later, genetic algorithms. Abbe worked on graph theory under Anatol Rapaport and Frank Harari. We become very close friends, especially Ray and Abbe who met for late night coffee on a regular basis.

In the summer of 1965, Ray, his friend Cathy, Avis, my first wife, and I spent about two months in Mexico. We drove down from Ann Arbor and I recall that when we stopped at a gas station near Bogalusa, Louisiana, the attendant advised us to by-pass the town. The previous week a civil rights worker had been killed and given that we were driving a car with Michigan plates and that Ray and I had beards, it seemed like good advice, but the bravado of youth prevailed and we stupidly drove down the main street quite slowly. Fortunately we were ignored and made our way safely to Mexico. It was a glorious summer. While traveling in Mexico, Ray and I were frequently hailed with shouts of 'Fidel', quite appropriate given our black beards and somewhat revolutionary demeanour. We visited a number of famous archeological sites including Palenque and Chitzen Itza. We spent a couple of weeks roughing it on an island called Isla Mujeres (Isle of Women), a favourite stopping point of giant sea turtles. Roughing it in this context meant sleeping on large hammocks, eating rice on the beach for lunch, having barbecues at night, drinking tequila and smoking local flora. I'm not sure about Ray but thoughts of academic pursuits were not uppermost in my mind.

Avis and I had decided to get married during the trip and although a local priest, an American, offered to perform the ceremony, we actually got married in Chicago, near the end of September, with Ray as best man. However, before he was allowed to appear in the synagogue, my then mother-in-law-to-be insisted that he replace his somewhat scruffy desert boots. He agreed and someone lent him pair of standard leather shoes; the marriage then took place without further immediate problems. After a brief honeymoon, one day in a local motel, Avis and I returned to Ann Arbor to continue our studies.

We were now in the stretch run at Michigan and spent a lot of time talking about our future. It was now very clear that we were all going to be academics but when and where. It should also be remembered that during the time we were at Michigan, the U.S. was getting heavily involved in Viet Nam. So on occasion we went on marches, picketed draft boards, and took part in sit-ins. I have often thought that the U.S. was extremely generous to provide an education for foreigners, Canadians, at a major university, and take little or no notice when they exercised constitutional rights held by all Americans to criticize the government. Once Ray discovered that a U.S. Admiral would be visiting Michigan and he informed my sister, who was also a student at Michigan and was actively involved in anti-war protest. The Admiral was greeted by a sizable group of students whose morale was obviously raised, although the event certainly had no impact on the conduct of the war.

We finished in 1967 and while I remained at Michigan for another year as a Visiting Assistant Professor, Ray went off to London on a kind of postdoc at the University of London. By early 1968, I had accepted a position at the University of British Columbia for the fall and to smooth the transition, Avis and I planned a three month trip to Europe. We would be picking up a car in London and staying with Ray for a few days. He had managed to find lodgings at quite an elegant home (the Reiter luck in operation), owned I believe by a Lady Cohen, whose husband had served the British government in Uganda. I am forever grateful to Ray for having introduced me to Indian food in London. We also met up later that summer to travel together in Eastern Europe, with an Austrian friend of Ray's. Over the years, Ray has developed some very strong relationships with wonderful European women. In our travels we went to Romania, Hungary, Bulgaria, and Czechoslovakia. Our experiences were not very pleasant on several occasions. Most memorable was the abortive trip to Prague. As we drove towards Prague we frequently encountered military convoys and deciding that something was up, headed very quickly to Austria. Remember this was August 1968 and as we subsequently learned the Russian army had entered Prague in force, shortly after we departed.

So we had done Michigan. Ray was spending a second year in London and had immersed himself in logic and theorem-proving. Abbe Mowshowitz had accepted a position at the University of Toronto, and I was going to Vancouver. We were separated by many thousands of miles but one year later we found ourselves as colleagues in the Department of Computer Science at the University of British

Columbia. We were quickly labelled 'The Michigan Mafia', a designation that lasted several years.

4 The Middle Years II: The University of British Columbia

The department had been started in the spring of 1968 and only had about eight faculty members including the three of us. Upon the arrival of Ray and Abbe, we began to build an academic department, research motivated and in the tradition of the only model we had shared, Michigan's. In short order, we had a graduate program, first a master's degree, followed a year or two later with a Ph.D. program. But we made a few changes based on our personal experiences. The Michigan program required a heavy dose of courses and a series of oral comprehensive examinations. At UBC, the number of courses was reduced and tailored to a breadth requirement while the comprehensives were replaced by a defense of thesis proposal. This more humane approach reflected a shared belief that graduate students were largely self-motivated and did not require a series of ever-increasingly high hurdles to produce quality work. Or else it reflected a belief that it was not necessary for each generation to raise the bar to ensure that its accomplishments were not overlooked. In any case, we all felt committed to the creation of a department motivated by the promise of research in a new and exciting discipline.

Upon his return to North America, Ray was clearly committed to the necessity of a formal approach, based on First Order Logic, to deal with problem-solving in Artificial Intelligence (AI). Any time spent with him would inevitably involve such words as theorem proving, resolution, John McCarthy and Cordell Green. Ray was well and truly launched on his academic career. and his reputation began to spread beyond UBC.

5 A Reiter Aphorism

Don't continue to read when you encounter a passage that you do not understand; otherwise, you may not really understand subsequent material.

Ray traveled a lot for the next few years, establishing contacts and working relationships at BBN and other places. He worked with Bonnie Webber in natural language and with Jack Minker on inference in large databases. He fashioned his logic skills into a very powerful prism that illuminated many disparate areas of AI. And even when research in AI took a decidedly anti-logic turn for several years, Ray was unwavering in his belief that his way of doing business was fundamentally sound and principled. He was right.

In the mid 1970s, the AI community was highly receptive to the ideas contained in the famous paper by Marvin Minsky, 'A framework for representing knowledge'. Although primarily concerned with problems in vision, it offered an approach to capturing the structure of knowledge in situations that could not readily be dealt with in an axiomatic approach, or so Minsky argued. Ray had previously been up-set by the somewhat idiomatic and unsound principles underlying the highly ac-claimed reasoning component of Terry Winograd's renowned MIT thesis, namely, microplanner. The appearance of Minsky's paper was to set Ray on a research pro-gram that won him richly deserved international fame. Minsky had referred to cer-tain kinds of knowledge as being default, that is, to be assumed as in effect in normal situations. Ray saw that there were difficulties in this seemingly obvious assumption and the rest is history. The name Ray Reiter is forever linked with the term 'default logic', and I was there but not really aware of the significance of his efforts. That this volume exists is of course a measure of the success of those very same efforts.

Ray is well-known for his distaste of sloppy thinking and his complimentary approval of clear and careful thinking.

6   A Reiter Aphorism

To a student: Look, go away, take your ideas and apply them to an example; then show me, in detail, how it works.

Some students did not return but others did, frequently chastened but usually wiser. His uncompromising integrity about the quality research is what distinguishes Ray and has probably resulted in a certain amount of grief for him.

Have I mentioned Ray's love of classical music? He taught himself to play the recorder and learned some baroque music, which he rarely allowed anyone to hear. His musical tastes certainly influenced mine and one composer that he came to admire was Wagner especially his epic, 'Der Ring des Nibelungen'. In fact, near the end of July, 1980, Ray and I drove to Seattle to see the 4th part of the Ring, Götterdämmerung (The Twilight of the Gods). It started at 6 pm and lasted to about 2 am, with a dinner break. I think he is now over his passion for Wagner but for two Jews to actively pursue their interest in Wagner in such an active manner is unusual, perhaps.

One of the best things we did, and Ray deserves much of the credit, was con-vincing Alan Mackworth to come to UBC. Alan had finished his Ph.D. in vision at Sussex in England and was looking for a job, and a possible return to Canada. Alan was educated in England but did his undergraduate degree in Engineering Physics at the University of Toronto. (The careful reader will recall that I had also graduated in the same program and that Ray had completed the first year before switching to mathematics.) Alan decided to come to UBC in 1974 and together with Ray is re-sponsible for putting UBC on the AI map. After Ray left, Alan continued to make UBC a desirable place for students and faculty to do research. Of course, the loca-tion helped as well. Have I inserted the mandatory blurb, advertising the mountains, the sea, the temperate climate, the availability of year round sailing, golf, climbing, and winter skiing? Granted it rains a little but that just makes everyone appreciate sunshine all the more. In recognition of its growing reputation, UBC hosted IJCAI in 1981, as many readers will recall.

In the summer of 1979, Abbe and his family moved to New York, where Abbe took a position at CCNY. Since arriving at UBC, his interests had shifted from his thesis research in graph theory to computers and society. He introduced an undergraduate course and published a well-received book, ' The Conquest of Will: Information Processing in Human Affairs', in 1976. Ray had spent a leave of absence at Rutgers and there was a strong possibility that he might not return, given his appreciation of New York, where he lived. In the end, he did return and New York, while attractive was also quite expensive, a contributing factor to his decision.

Surprisingly, at least to me, I was the next to leave. The department was in a no-growth phase and seemed destined to continue in a stagnant state for the foreseeable future. I left to assume the position of Director of the Division of Computer Science in the Department of Mathematics, Statistics, and Computer Science, at Dalhousie University in Halifax, Nova Scotia. At the time, May 1984, it looked as if it was going to be a permanent move as we sold our house and cut most of our bridges with Vancouver, except for one. Instead of resigning, I took a one year leave, which I extended to two years while at Dalhousie. Unbeknownst to me, Ray was also finding the situation at UBC untenable. He rightly felt that his achievements were not adequately recognized by either the department or the University at large. So as I discovered that Dalhousie was unwilling or unable to make a commitment to Computer Science, my wife Sheryl and I decided to return to Vancouver and UBC in the spring of 1986. By this time Ray had accepted a position at the University of Toronto and a return to the city of our youth.

We arrived in Vancouver in May of 1986 and shortly after Ray left for Toronto, We had been together at UBC for about ten and sixteen years, respectively, and now Abbe, Ray, and I were separated by great distances. Fortunately, the Internet was available.

7   The Middle Years III: The University of Toronto (Again)

At UBC, a new era had begun and the University administration discovered the department. Maria Klawe become the head, accompanied by her husband Nick Pippenger. Finally, the department in the post-Reiter era began to grow. Among those hired, in the early 1990s was Craig Boutilier, who had received his Ph.D. under Ray at Toronto. Curiously, he had been a student of mine at Dalhousie, where I taught him in an introductory AI course. I hesitate to note that despite its physical size, Canada is quite a small country. Ray has acknowledged, and I certainly agree, that Craig is a bright star in the AI community. Now, Ray and I are in competing departments, an irrelevant situation with respect to our friendship.

He is now living in a house in downtown Toronto, not very far from where we both grew up. There is a nice symmetry in his return to his roots. Toronto is a far different city now than it was in our youth. It is as ethnically diversified as New York and for a world traveler like Ray it must be a great city to live in. Whenever I travel to Toronto to visit my sister and her family, once or twice a year, Ray and I meet for dinner and with very little difficulty we share our feelings, yes feelings, about recent events in our lives. I generally speak of my children, my wife, common friends in Vancouver and elsewhere, UBC, my work, and politics. He speaks of friends, close friends, his work, travels, and recent books read. Altogether civilized and very satisfying.

In 1989, Ray, Abbe, and I all celebrated our fiftieth birthdays by returning to Ann Arbor for a three-day nostalgia jag. We visited our old haunts, drank beer, and wandered the city, complaining about how much it had changed for the worse. Abbe was soon to move to Holland to pursue his business interests. Ray was continuing to develop research projects at Toronto with Hector Levesque, among others, and moving towards applying logic to real world robots, dealing with perception, errors, and multiple agents. As with all his efforts, the premium is on quality, clarity, cleanness, and elegance. He has also continued his good fortune in recognizing those areas appropriate for his techniques and temperament. He has confided to me on occasion that while many other AI researchers are more technically skilled than he is, he possesses the ability to identify those problems amenable to his approach and appropriate to his skills. A rare talent indeed.

Ray's life has been diversified and full and if the past is any guide, he will continue to surprise us with his fertility and his continuing relevance. My friend has exquisite taste in research, in butterflies and moths, and in people and relationships. Our friendship has endured all these years because we have shared much, respected each other's choices, and maintained, albeit at a distance more recently, a willingness to listen, to care, and to respond when appropriate. Aside from my family, there is no one that I have known longer than Ray. We have been involved in each other's lives for so long that I cannot imagine a future without being able to call him up to ask for advice or to share a moment of joy or even sorrow.

8   The Future

It is now 1999 and as Ray, Abbe, and I acknowledge our sixtieth birthdays, we can all look back on lives that have been eventful, with the usual distribution of fortune and misfortune. All that any of us can hope for is to live life to the fullest, given that we only get shot at it. I am reminded that in the early heyday of robot planning, the model was to take a picture of the simple world, spend some time, actually lots of time, processing the data to produce an internal representation, receive a problem command, translate it into a formal theorem to be proven, prove it if possible (more time), and finally execute the proof. In this approach, the real world is sampled minimally, most activity takes place internally, and on rare occasions, the state of the world is changed. Imagine actually living such a life, even a richly elaborated version. Admittedly, current robotics research in AI is more realistic.

If I had a glass of champagne, I would lift it and toast my friend: ' Well done, keep up the good work, and remember, *Dos mentsch tracht un Gott lacht*'.[1]

---

[1] Man plans and God laughs. (It rhymes in Yiddish.)

Default Logic and purity of reasoning

Gianni Amati[1], Luigia Carlucci Aiello[2], and Fiora Pirri[2]

[1] Fondazione Ugo Bordoni
via Baldassarre Castiglione 59,
00142 Roma, Italia
gba@fub.it
[2] Dipartimento di Informatica e Sistemistica,
Università di Roma "La Sapienza",
via Salaria 113,
00198 Roma, Italia.
aiello, pirri@dis.uniroma1.it

Foreword

Ray Reiter has worked out Default Logic during his sabbatical leave in 1979, at Imperial College in London. The paper on Default Logic then appeared in the landmark issue of the Artificial Intelligence Journal on "Commonsense Reasoning" [1].

A vast literature flourished from Ray's initial proposal, and after 20 years, it is still studied and used in many fields of Artificial Intelligence. Default Logic is one of the inventions of AI that will still be celebrated in the future.

The secret of the success of Default Logic is in its ability to represent a good deal of the dynamics of reasoning, yet giving it the status of a logic. It has become paradigmatic for the evergreen challenge: how close can logic come to meet reasoning; how far can reasoning go to reach the abstract and formal sky of logic. Even more: Default Logic has challenged scientists interested in purity of reasoning, those interested in the mechanical and algorithmic nature of proofs that handle exceptions, and those interested in the logical foundations of cognitive agents. Default Logic has disclosed them both the possibility of making explicit aspects that were implicit or hidden in the initial formulation, and provided them the intuition to make a big step towards the formalization and mechanization of reasoning.

We worked on Default Logic from 1989 till 1996. We are among those who believe in purity of reasoning. That is, reasoning mechanisms can be given a formal and symbolic apparatus independent of the instrument or tool that will make it work. This view allowed us to understand an important pattern that links the different forms of commonsense reasoning, i.e. the implicit and explicit contextual definability.

We here briefly report our recent results on the logic of defaults, and dedicate them to Ray on his 60th birthday.

1 An overview of Default Logic

The fundamental idea of Default Logic is to combine a classical theory W with a set D of special inference rules, called *default rules* and intended to capture patterns of reasoning like the following one: "Normally, if α was shown to be true, then γ can be inferred". Here, "normally" means that the applicability of the default rule itself is not granted: the presence of an exception may prevent applicability. To make exceptions explicit, rules are defined as follows: $\alpha : \beta/\gamma$, which reads as "If α is proved and β is consistent (or, equivalently, $\neg\beta$ is not provable; that is: the exception $\neg\beta$ does not occur) then infer γ". α is called the *prerequisite*, β the *justification*, and γ is called the *conclusion*.

According to Reiter [52], a default theory is a pair $\langle W, D \rangle$ where W is a set of well formed formulae of $\mathcal{L}$ (we can assume here that $\mathcal{L}$ is classical propositional logic PC) and D is a set of defaults, i.e. rules of the form $\delta = \dfrac{\alpha : \beta_1, \ldots, \beta_n}{\gamma}$, where α, β_i, $i = 1, \ldots, n$, and γ are formulae of PC and are respectively called *prerequisite, justifications*, and *conclusion* of the default rule. The intuition of Default Logic is that defaults are sort of 'meta-rules' which can be used to supply the underlying incomplete theory with supplementary assumptions [52].

The conjectures or possible explanations of a default theory are called *extensions* and are obtained by means of a fixed point equation. A default theory can have zero, one or multiple extensions. Extensions have many interesting properties. Reiter (see [52], Theorem 24) has shown that if E and F are two extensions of a default theory and $E \subseteq F$ then $E = F$. Furthermore, a default theory $\langle W, D \rangle$ has an inconsistent extension if and only if the set W of assumptions is inconsistent (see [52], Corollary 2.2). Extensions, however, do not always exist, for example the default theory:

$$(\{p \supset q, \neg q\}, \{\frac{: \neg q}{p}\})$$

does not have an extension, though its initial set of assumptions is consistent. The reader is referred to [52] for an introduction to Default Logic. Following Reiter, many different formulations of Default Logic have been proposed (see for instance [36], [37], [28], [17], [12], [54], [55], [22], [18], [24]). Each of them tries to overcome some aspect of Reiter's system regarded as counter-intuitive. One of the aspects of Reiter's formulation, which is considered a flaw, is the lack of *semi monotonicity*[3]: in some cases, when adding new defaults to the initial default theory, extensions may not be extendible to new ones or, as extreme case, consistent axioms may not have an extension (see [36], [37] for the treatment and solution of this problem).

A comparison among the different formulations of Default Logic presented in the literature can be found in [20] and [2], where a systematic approach to Default Logic is presented. A general treatment of Default Logic can be found in Besnard [10] and Lukaszewics [38].

The aim of our research has been to show that, at least in the propositional case, Default Logic has a very natural behavior both from a proof theoretical and a semantical point of view, under a suitable embedding into modal logics in which

---

[3] Default Logic is said semi monotonic if, given two default theories, say $\Delta = \langle W, D \rangle$ and $\Delta' = \langle W, D' \rangle$ with $D \subseteq D'$, and an extension E of Δ, then there is an extension E' of Δ' such that $E \subseteq E'$.

fixed points are definable. This property sheds light on two very important aspects of commonsense reasoning, namely: the connection between implicit and explicit definability of concepts relative to natural kinds, and contextual reasoning.

2 The logical nature of reasoning

Our research has initially been concerned with the logical nature of a default proof. Our intuition on the proof theory of Default Logic is that default proofs are variations of Hilbert-style proofs; namely default proofs are those Hilbert-style proofs that comply with the restrictions posed by the defaults on the applicability of inference rules. In other words, defaults are special inference rules endowed with explicit hypotheses – the justifications. The use of explicit hypotheses makes a default proof somehow puzzling and explains why classical proofs, i.e. Hilbert-style proofs, can be turned into default proofs by the introduction of suitable restrictions that take into account the explicit hypotheses.

The proof-theoretical flavor of the application of defaults is that a conclusion of a default $\delta \in D$ must be added to the set of premises W, when the prerequisite of δ has been proved and its justifications are in some relation of consistency with some set of formulae.

In general, the application of a new default δ' can invalidate the condition under which some default δ has been previously applied, hence the conclusion of default δ must be retracted. To better clarify this point, consider the definition of extension, as given in [52]:

Definition 1. (EXTENSION) Let $\langle W, D \rangle$ be a default theory in propositional logic. For any set $\mathcal{C}$ of propositional formulae, let $\Gamma(\mathcal{C})$ be the smallest set satisfying the following three properties:

D1. $W \subseteq \Gamma(\mathcal{C})$;
D2. $Cn(\Gamma(\mathcal{C})) = \Gamma(\mathcal{C})$;
D3. If $\dfrac{\alpha : \beta}{\gamma} \in D$ and $\alpha \in \Gamma(\mathcal{C})$ and $\neg\beta \notin \mathcal{C}$ then $\gamma \in \Gamma(\mathcal{C})$.

$\mathcal{C}$ is an *extension* iff $\mathcal{C} = \Gamma(\mathcal{C})$

The idea of Default Logic is thus clear: to infer conclusions one has to refer to a context $\mathcal{C}$ which represents the current state of affairs. To put it better, given a certain amount of information X, in order to gather what does follow from X, one might use both what is provable from X, through classical logic, and what it is not provable from $\mathcal{C}$. In fact, the application of default rules relies also on sentences – justifications – whose negation cannot be proved from the context $\mathcal{C}$. These sentences – justifications – may be regarded as context-dependent beliefs, therefore each application of a default resorts to the context $\mathcal{C}$. Despite this proof-theoretic flavor of Default Logic, given by the fact that defaults are sort of inference rules, the definition of an extension depends on a pseudo-inductive construction; namely an extension is the smallest deductively closed set E of formulae that coincides with the context C itself. Reiter has shown in [52] that a semi-inductive construction of an extension can be given.

The above definition of extension, though intuitive, doesn't give any explicit connection with the classical notion of proof. In fact, fixed points, in the form defined above, require that one first "guesses" a context C – a candidate belief set among the contexts – and then "checks" whether this guess is right or not (see for instance Section 3.4 in [42]). Furthermore, differently from standard logics, where the deductive closure operator Cn leads to a unique extension of the initial theory, in Default Logic mutually exclusive conclusions can be inferred by using different plausible contexts. Therefore, there might be alternative extensions.

Our basic idea is to interpret defaults as special rules that impose a restriction on the juxtaposition of monotonic Hilbert-style proofs, without resorting to a fixed point construction as in Definition 1. For the purpose of our definitions, we shall use the following notations: $\text{PREREQ}(D) = \{\alpha : \frac{\alpha : J}{\gamma} \in D\}$, $\text{CONCL}(D) = \{\gamma : \frac{\alpha : J}{\gamma} \in D\}$, $\text{JUST}(D) = \bigcup_{\frac{\alpha : J}{\gamma} \in D} J$. We assume also that the underlying logic is propositional logic[4].

Definition 2. (DEFAULT PROOF) A *default proof* from $\langle W, D \rangle$ is a sequence:

$$s = \langle \phi_1, \dots, \phi_n \rangle$$

of wffs s.t. $\phi_1, \dots, \phi_n$ is a classical propositional proof from $W \cup \text{CONCL}(D)$ and such that there is a *proof justification set* $J(s)$ satisfying the two following conditions:

(i) Either ϕ_i derives from $W \cup \{\phi_1, \dots, \phi_{i-1}\}$ or there is some default $\alpha : J/\gamma$ such that $s = \langle \phi_1, \dots, \phi_j = \alpha, \dots, \phi_i = \gamma \rangle$ with $J \subseteq J(s)$;

(ii) if $\psi \in J(s)$ then for some $\alpha : J/\gamma \in D, \psi \in J \subseteq J(s)$ and $s = \langle \phi_1, \dots, \phi_j = \alpha, \dots, \phi_i = \gamma, \dots, \phi_n \rangle$.

If the deduction relation of a logic enjoys monotonicity, then any two proofs can be freely juxtaposed. Otherwise, proofs must be juxtaposed according to some composition principle. We present Default Logic as a logic where the juxtaposition of default proofs is subordinate to a restriction condition Ψ. So Default Logic is defined as a pair $(\mathcal{L}, \Psi)$ where condition Ψ is a relation $\Psi(W, H_D, s)$ which takes into account the theory W in $\mathcal{L}$, the set of formulae H_D of $\mathcal{L}$ involved in the set D of defaults and a classical proof s itself. If relation $\Psi(W, H_D, s)$ holds, then s is called a Ψ-proof. We now postulate the condition under which the relation $\Psi(W, H_D, s)$ holds for Reiter's Default Logic (see [2]).

---

[4] Clearly the following definitions and results can be extended to any decidable fragment of first order logic – like Horn clauses, or formulae having no function symbols other than constants and no existential quantifiers.

Let DEFAULT-PROOFS(W, D) be the set of all the default proofs for a given default theory $\langle W, D \rangle$, and let $E(s)$, with $s \in$ DEFAULT-PROOFS(W, D), be defined as:

$$E(s) = \{\phi \; : \; \phi \text{ is mentioned in } s\}$$

Let $J(s)$ be the proof justification set for s and let X CONSIST Y denote that X is consistent with each element $y \in Y$, i.e. $Y \cup \{y\}$ is consistent, then:

(CONSISTENCY CONDITION)   Let $s \in$ DEFAULT-PROOFS(W, D),
$W \cup E(s)$ CONSIST $J(s)$ implies $\Psi(W, H_D, s)$.

Definition 3. (Ψ-PROOFS) A Ψ-proof is a proof $s \in$ DEFAULT-PROOFS(W, D) such that $\Psi(W, H_D, s)$, and Ψ satisfies the consistency condition.

Let $s = \langle \phi_1, \dots, \phi_n \rangle \in$ DEFAULT-PROOFS(W, D), the relation Ψ can be defined as follows:

$\Psi(W, H_D, s)$:
for all $i \leq n$, if ϕ_i does not derive from $W \cup \{\phi_1, \dots, \phi_{i-1}\}$
then $(W \cup \{\phi_1, \dots, \phi_{i-1}\})$ is consistent with J;
where J is the justification of the applied default as defined in Condition (i) of Definition 2.

Example 1. Let $\langle W, D \rangle$ be a default theory with:
$W = \emptyset$ and $D = \{\dfrac{: \beta}{\gamma}; \dfrac{: \neg\beta}{\alpha}; \dfrac{\gamma : \gamma}{\neg\gamma}\}$
then the sequences $\langle \gamma \rangle$, $\langle \gamma, \alpha \rangle$ and $\langle \gamma, \alpha, \neg\gamma \rangle$ are in $\Psi(W, H_D, s)$.

Although propositional logic is compact, as Default Logic is nonmonotonic, its closure operator is not compact. However, as the Ψ restriction affects monotonicity it affects also the compactness property of the closure operator of the logic. In fact, we say that a theory E is Ψ-compact if any element in E has a Ψ-proof with all its elements in E and such that E is "consistent" with all the justifications of D used in the Ψ-proof:

(Ψ-COMPACTNESS)   Let Ψ satisfy the consistency condition. A theory E is said to be Ψ-compact with respect to $\langle W, D \rangle$ iff for all $\alpha \in E$ there is a Ψ-proof s for α whose elements are all in E. Moreover, if $J(s) \neq \emptyset$ then $W \cup E$ is consistent with all the β in $J(s)$.

We now explain the condition "if $J(s) \neq \emptyset$ then $W \cup E$ is consistent with all the β in $J(s)$". By Reiter's results we have that a default theory $\langle W, D \rangle$ is inconsistent iff W is inconsistent. Suppose that $E \vdash \bot$, then for any proof s of $\alpha \in E$ it is not the case that $W \cup E$ is consistent, therefore $W \cup E$ cannot be consistent with all the β in $J(s)$. By the above condition we have that $J(s) = \emptyset$, so s is just a monotonic proof from W, then $E \subseteq W$, whence $W \vdash \bot$. The above condition is thus coherent with what is meant to be the set of consequences of a default theory.

Analogously, given a default theory $\langle W, D \rangle$, we say that a set of sentences E is Ψ-saturated by D if any "applicable" default of D is vacuously applicable, that is, the conclusion of any "applicable" default rule of D is already in E:

(Ψ-SATURATION) A theory E is said to be Ψ-saturated with respect to $\langle W, D \rangle$ iff for all defaults $\alpha : \beta/\gamma$, γ is in E provided all juxtapositions $\langle s; \gamma \rangle$ are Ψ-proofs – with s a Ψ-proof, $\alpha \in E(s) \subseteq E$ and $J(\langle s; \gamma \rangle) = J(s) \cup J$.

We propose, then, to define extensions of default theories $\langle W, D \rangle$ as:

(Ψ-EXTENSION) A Ψ-extension is a set of formulae Ψ-saturated and Ψ-compact with respect to $\langle W, D \rangle$.

We can now introduce the following theorem, whose proof can be found in [2], stating the soundness and completeness of Ψ-proofs, with respect to Reiter's Default Logic:

Theorem 1. (SOUNDNESS AND COMPLETENESS) *Let $\langle W, D \rangle$ be a default theory. E is a consistent Reiter extension for $\langle W, D \rangle$ iff E is a Ψ-extension.*

Example 2. Let $\langle W, D \rangle$ be the default theory of the previous example:
Then the sets $E_1 = \{\gamma, \alpha\}$ and $E_2 = \{\gamma\}$ are not Ψ-saturated. On the other hand, the set $E_3 = \{\gamma, \alpha, \neg\gamma\}$ is not Ψ-compact, since $W \cup E_3$ is obviously not consistent but $J(s) \neq \emptyset$. Observe that the set $E_4 = \{\alpha : \vdash \alpha\}$, i.e the set of all the tautologies, is both saturated and compact (compact because $J(s) = \emptyset$ and saturated because no $\gamma \in E$).

In general, one would expect that the set of consequences of a default theory is larger than that of a monotonic proof from a set of assumptions W. If we look closely to the meaning of a Ψ-proof, we can see that Ψ is, in fact, a restriction on default proofs, that is, proofs in which defaults act as rule of inferences of the form $\frac{\alpha}{\gamma}$.
The restriction Ψ sieves, among all, those default proofs fair with respect to all the conditions that make a set of defaults applicable or not. In other words Ψ-conditions interiorize the context within the proof itself, by giving up those proofs not accounting for it.

We have thus shown that, under a suitable restriction on the set of default proofs given in a Hilbert-style fashion, Reiter's Default Logic enjoys an interesting proof theoretic behaviour[5].

Our studies on the proof theoretical aspects of Default Logic have also considered the actual construction of default proofs. We have devised a tableau system for Default Logic and reported it in [5]. We omit it here, for brevity sake, and move instead to the presentation of the results on our semantic account of Reiter's Default Logic provided via the modal logic $KD4Z$.

---

[5] In [2] we have also shown that, according to a suitable definition of the restriction relation Ψ, there is a family of default logics in between Reiter and Lukaszewics' [37], that we call *linear default logics* and Reiter's Default Logic is the minimal one.

3   The modal logics of Default Logic

The approach followed to give a semantics to Default Logic are mainly based on the idea that modal logics are the good logical environment to represent a belief-set, which is in fact implicit in the notion of context used to apply defaults. These approaches resort to two paradigms, which we call the *preference paradigm* and, according to Marek and Truszczynski, the *negation as failure to prove* paradigm. In the following, we shall use Λ to denote any modal logic – we refer the reader to [27] for the basics of modal logic – and, as previously, PC to denote propositional calculus.

The preference paradigm [31], [35], [11], defines a preference relation among modal structures which are sets of classical interpretations. Preference criteria for default reasoning were introduced by Etheringthon [19]. Lin and Shoham [35] use a bimodal logic to provide a semantical characterization of Default Logic by means of preferred modal models.

The negation as failure paradigm is based on the following idea: "Find a solution (i.e. a Λ-*expansion*) to the equation: $\mathcal{C} = Cn_\Lambda(I \cup \{\Diamond\gamma : \neg\gamma \notin \mathcal{C}\}$", where I is a translation of the defaults into the modal logic Λ.

The above schema has been introduced by the early work of McDermott and Doyle (see [47]), where Λ is PC. In [46] the earlier fixed point definition – through PC – is strengthened by substituting PC with a modal logic required to be either T, $S4$ or $S5$. Actually, with the above definition, McDermott was close to find Reiter's extensions, if he had only considered I as a suitable modal translation for a default theory. It comes out that, on the basis of the McDermott definition, the characterization of stable expansions of I by a modal fixed point depends on the logic Λ and (when characterizing Reiter's Default Logic) it depends on the modal translation of default theories. This, however has not been immediately clear in the further studies of the modal approaches to nonmonotonic reasoning.

After the first modal translation of Default Logic given by Konolige [32], all the further studies focused on the connection between Moore's notion of groundedness in autoepistemic logic and suitable variations on the modal representation of Default Logic. The breakthrough in the modal study of Default Logic comes from Marek, Schwartz and Truszczynski in [56], [40], [65], [41], [42], [39], [61] who reconsider the earliest attempt of McDermott and Doyle, and of McDermott. In fact, in the McDermott and Doyle's fixed point equation it is sufficient to strengthen classical logic with the necessitation rule, that is the logic N, (which however has been done by McDermott who devised the needs of the necessitation rule) to obtain the new equation:

$$\mathcal{C} = Cn_N(I \cup \{\neg\Box\gamma : \gamma \notin \mathcal{C}\}) \tag{1}$$

The above equation, together with a suitable modal translation of defaults, namely, for $\delta \in D, tr(\delta) = \Box\alpha \wedge \Box\Diamond\beta \supset \gamma$, leads to a correct and complete characterization of Reiter's extensions.

Marek and Truszczynski [65],[42],[39] have shown that, with the negation as failure to prove paradigm, a family of modal logics can be devised to capture default reasoning. Indeed, Marek, Truszczynski and Schwartz have established more general results, by showing that Doyle and McDermott's fixed point is so powerful to yield infinitely many not equivalent nonmonotonic modal logics, by varying the choice of the underlying monotonic modal logic [59]. Among these modal logics, in particular $KD45$, $Sw5$, $S4f$, $S4.2$ and $S4.3$ have been widely studied and proposed as good candidates for representing knowledge and belief. A strong argument in favor of the above cited modal logics is that they are maximal [57],[58],[60],[59] and even the largest in their range [59] (i.e. in the class of modal logics which have the same nonmonotonic modal logic).

Recently, we have introduced boxed nonmonotonic modal logics by an alternative fixed point construction [7]. The underlying idea is to interpret the membership relation of a formula and the context $\mathcal{C}$, as in equation (1), by means of a nonmonotonic provability operator.

In this framework, all the formulae involved are *boxed*, (a boxed formula has all the propositional variables in the scope of the modal operator $\Box$). The modal expansions $\mathcal{C}$ of I (where I is the modal translation of the default theory, see equation (1), is defined according to the following paradigm:

"α is in $\mathcal{C}$ if and only if the *provability* of α can be derived by means of the modal logic Λ from the set I of formulae and the set of *unprovable* formulae of $\mathcal{C}$".

When I refers to a default theory $\langle W, D \rangle$ then I is $\Box W \cup \{\Box\alpha \land \Box\Diamond\beta \to \Box\gamma\}$. More formally:

$$\alpha \in \mathcal{C} \equiv \Box\alpha \in Cn_\Lambda(I \cup \{\neg\Box\gamma : \gamma \notin \mathcal{C}\}) \qquad (2)$$

that is:

$$Cn_\Lambda(\Box\mathcal{C}) \equiv Cn_\Lambda(I \cup \{\neg\Box\gamma : \gamma \notin \mathcal{C}\}) \qquad (3)$$

We have shown in [7] and [6] the following:

(1) Under the above translation, K, D and 4 should be considered the minimal set of axiom schemata for default reasoning.
(2) There exists a "structural" property for a logic that ensures the completeness theorem with respect to Reiter's Default Logic. In particular, the modal logic $KD4Z$, obtained by adding the axiom $Z = \Box(\Box\alpha \to \alpha) \to (\Diamond\Box\alpha \to \Box\alpha)$ to $KD4$, has this structural property.
(3) By embedding the context into the modal language, default extensions can be characterized through theorems of the modal logic $KD4Z$.

We have seen that a semantics for Default Logic is a particular class of structures verifying some correspondence between an extension of a default theory $\langle W, D \rangle$ and the set of formulae realized in that class of structures.

The previously cited study of Marek, Schwartz and Truszczynski has been devoted to the identification of some property that could clearly devise such a class of structures. Some of these properties have indeed been devised – such as, for instance, the existence of the final cluster property in a modal frame.

We have good arguments to claim that, so far, the logic $KD4Z$ is the one who exhibits the most general properties and thus its structure could be identified as the one giving the right interpretation for default reasoning. In fact, $KD4Z$-frames have the final cluster property; they have an initial world – characterizing the set of assumptions; between the first world and the last cluster, any set of worlds siblings of one another can be introduced and, finally, they have the finite depth property, that is, any irreflexive sequence of worlds is finite.

We have given in [7], [8] two embeddings of a default theory into $KD4Z$ and shown that both these embeddings, one through a meta-logical construct – namely the negation by failure paradigm – and the other through a logical one, can be used to give a representation theorem for Reiter extensions[6].

Let δ be a default in $\langle W, D \rangle$ then the translation of δ into the language $\mathcal{L}$ of modal logic is $tr^1(\delta) = \Box\alpha \wedge \Box\Diamond\beta \rightarrow \Box\gamma$ then

$$Tr^1_{\langle W,D\rangle} = I = \Box W \cup \{tr^1(\delta) : \delta \in \langle W, D \rangle\}$$

Theorem 2. (FIRST REPRESENTATION THEOREM) *Let $\langle W, D \rangle$ be a default theory and $Tr^1_{\langle W,D\rangle} = I$ its translation into the language $\mathcal{L}$ of $KD4Z$. $\mathcal{C} \cap \mathcal{L}$ is a Reiter extension iff $Cn_{KD4Z}(\Box\mathcal{C}) \equiv Cn_{KD4Z}(I \cup \{\neg\Box\gamma : \gamma \notin \mathcal{C}\})$*

The above characterization is in the same style as the negation by failure paradigm, the only difference relies in the translation of the defaults, which (see [4]) would not suit the logic devised by [40].

We have given an even more sophisticated characterization, as the ongoing presentation shall show. In [8] we have proved that fixed points are definable in $KD4Z$. In other words, given a sentence E, in which a propositional variable p occurs, there are sentences $\mathcal{C}$, not containing p, such that, by substituting $\mathcal{C}$ for p in E, one gets the following $KD4Z$ theorem:

$$\vdash_{KD4Z} E(\mathcal{C}) \equiv \mathcal{C} \qquad\qquad (4)$$

The above result clearly suggests a way of representing, through the logic, the context that a default application refers to. Even more interesting is the fact that, given E, these sentences $\mathcal{C}$, with such nice behaviour, are not unique.

In particular, if p is a propositional letter used as a variable denoting a context, we have the following translation of a default theory into $KD4Z$:

---

[6] The versatility of $KD4Z$ has been emphasized in [7] where it is shown that both Autoepistemic and Default Logic can be represented by a suitable definition of the fixed point. So far, it seems that only $KD4Z$ enjoys this property.

- for each default $\delta \in D$ the translation of the δ with respect to p is:

$$tr_\delta(p) = \Box(p \to \alpha) \wedge \Box\Diamond(p \wedge \beta) \to \Box(p \to \gamma)$$

- the translation of the default theory $\langle W, D \rangle$, with respect to p is:

$$Tr^2_{\langle W,D \rangle}(p) = \Box(\Diamond p \wedge W \wedge \bigwedge_{\delta \in D} tr_\delta(p))$$

Now, let $\Phi(p)$ be a formula of the logic containing the propositional variable p and the justification set of the defaults in D (see [3] for details), we have the following representation theorem where, we recall from Section 2, X CONSIST Y means that $X \cup y$ is consistent for each $y \in Y$, and by J_D we denote the set of justifications occurring in the set D:

Theorem 3. (SECOND REPRESENTATION THEOREM) *Let $\langle W, D \rangle$ be a default theory and $Tr^2_{\langle W,D \rangle}(p)$ its translation into the language $\mathcal{L}$ of $KD4Z$. C^* is a Reiter extension iff:*

- $\vdash_{KD4Z} C \equiv Tr^2_{\langle W,D \rangle}(C) \cup \Phi(C)$;
- $*$ *is a mapping of $\mathcal{L}$ into $\mathcal{L}_{PC}$*
- $\vdash_{KD4Z} C \equiv Tr^2_{\langle W,D' \rangle}(C) \wedge \bigwedge_{\beta \in J_{D'}} \Box\Diamond(C \wedge \beta)$, *where $D' \subseteq D$ and C^* CONSIST $J_{D'}$.*

The above representation theorem says that Default Logic can have a pure logical characterization when the logic, into which a default theory is embedded, enjoys the definability of fixed points, that is, theorems of the form (4) above.

It should be observed that, following results from Solovay and Montague (see [48], [64]), logics enjoying the definability of fixed points cannot have among their axiom schemata and rules of inferences both T and Necessitation.

Clearly, from the second representation theorem we can also get a proof theory for Default Logic in $KD4Z$ itself[7]. However, the notion of Ψ-proofs gives a more interesting insight into the logical structure of defaults. Finally, we can say that Default Logic has, among the different default formalisms presented in the literature, a very clear and interesting logical behaviour. In fact, as we have argued in the foregoing discussion, Default Logic can be characterized both from a proof-theoretical point of view, through Ψ-proofs, and from a structural point of view, through modal frames having certain general properties, such as the final cluster property, insertion of worlds and finite depth property. Furthermore, the logic up-springing from these properties has to enjoy the definability of fixed points.

---

[7] In [8] we have presented a tableau calculus for $KD4Z$; we have then discovered it to be incomplete. This is marginal with respect to the results presented there and we will soon publish a complete version.

4   Default Logic: a strategy for explicit Definability

As we have shown above, the logic $KD4Z$ reveals to be very interesting not only because it "internally" expresses external fixpoint constructions in the style of [47], [6], but – and more important – in so as it is a self referential language in which commonsense statements are expressible and in which explicit definitions from implicit ones are obtainable via fixed point equations, where a context occurs as a parameter.

We recall here some basic results on fixed points given in [8]. A modal predicate $E(c)$, with parameter c, is called an *implicit consistency predicate* whenever $\models_{KD4Z} E(c) \rightarrow \Box\Diamond c$. In the following, the subscript $KD4Z$ will be dropped.

Theorem 4. *Let $E(c)$ be an implicit consistency modal predicate. Then:*

1. $\models E(E(\top)) \equiv E(\top)$;
2. $\models E(\bot) \equiv \bot$;
3. *any fixed point $\mathcal{T}$ of $E(c)$ is of the form $E(\top) \land \Box\Diamond\gamma$, with γ any formula.*

Theorem 4 states that there exist formulae, which substituted for c, turn the fixed point equation (4) into a theorem of the logic $KD4Z$; it also specifies what has to be done in order to get these formulae. E.g let $E(c)$ be $\Box(c \rightarrow \alpha) \land \Box\Diamond(c \land \beta)$ then $\Box\alpha \land \Box\Diamond\beta$ is a *fixed point solution*, that is, $\Box\alpha \land \Box\Diamond\beta$ substituted for c satisfies (4). In this sense we can say that the formula, which is a solution to the fixed point equation, *realizes* a context, because it gives an explicit representation of the context, previously designated by a parameter.

This amounts to saying that, in the case of a nonmonotonic theory which is implicitly defining a concept, the context is its explicit definition and – more generally – in the case of a default theory, the context is the minimal set of formulae whose consequences are true in the theory.

This provides a strong connection between the definability of concepts corresponding to natural kinds, such as *bird*, *lemon*, etc, (which possess necessary but not sufficient conditions) and commonsense reasoning, via Default Logic and its semantics in $KD4Z$.

This very well reflects our intuition on the fact that concepts corresponding to natural kinds are not definable per sé, i.e. *in the vacuum*, but only relative to a specific body of knowledge, i.e. a context. Informally, birds in Italy are defined differently from Antarctica, where Penguins are an exception to the property of flying or in other parts of the world, where birds (e.g. Kiwi) happen not to have wings at all.

5   Default Logic: from a single to multiple Contexts

Default Logic gives its deep insight on purity of reasoning also into *contexts*. A context is the snapshot of a current state of affairs, to which an intelligent agent relativizes his reasoning. Agent reasoning is local and indexical, in the sense that truth depends on the actual embedded situation in which the agent is or on her specific state of mind. A typical example is when the same term denotes different objects, depending on the situation in which it is uttered. Classical instances of indexicals are words like "I", "here", "this".

Since McCarthy's Turing Award speech, in 1971, where contexts have been introduced into AI, some work has been done both to understand all the possible ways contexts are involved in commonsense reasoning (see e.g.[44], Guha's thesis [29], Shoham's discussion [62] and McCarthy's notes [45]) and to reify such a notion into a formal system (see e.g. [34],[15],[26],[25], [13],[14],[16], [21]). In particular, both in [13] and in [21], the **ist**$\langle *, * \rangle$ predicate is interpreted as a modal operator. In particular, both Smullyan [63] and Perlis[50] refer to contexts in their papers on self reference; while in [13] contexts are interpreted in a multimodal setting and in [21] via fibring. By generalizing the notion of indexicals, McCarthy argues that contexts give different interpretations to the same term [45] or sentence. For instance, "Spock is the lieutenant of the Enterprise", in the context of Star Treck stories, while it is the name of a dog in the context of the children's book "Look for Spock". Applications of contexts have been addressed in [30],[29],[23].

In [9] we extended the results on definability of concepts described in [8] to the definability of contexts, by showing that a fixed point theorem for one context can be extended to handle multiple contexts, so that we can treat contexts as parameters and arbitrarily nest them, one into the other. The extension of definability to multiple contexts enlightens an interesting point: *contextual reasoning is a generalization of Default Logic*.

As we have already pointed out, in the modal logic $KD4Z$, $\Box$ is interpreted as "it is provable in a decidable fragment of first order logic", so $\Box p$ means "p is classically provable". This interpretation is better understandable when contextual formulae are concerned. A contextual sentence is a sentence of the form $E(c)$ in which a parameter, usually denoted by c, occurs in the scope of a $\Box$. This requirement is akin to the usual reification of a context [45] and allows for substitutions of sentences for parameters. So, if ϕ is a sentence, $\Box \Diamond (c \wedge \phi)$ is a contextual sentence. Contextual sentences can be built by parameters, denoting contexts, boolean connectives and first order formulae. The modal operators are used to state that a formula ϕ is *derivable* from a given context and we write $\Box(c \rightarrow \phi)$, or that it is *consistent* with the context, and we write $\Box \Diamond (c \wedge \phi)$. Observe that the representation we give is weaker than the one proposed by McCarthy, where **ist(c,p)** stands for "p is *true* in context c". Notice that an advantage of our modal approach is that we can deal also with a consistency representation: $\Box \Diamond (c \wedge p)$ as said above means "p is consistent with context c". If $E(c)$ is a contextual sentence then $\Box E(c)$ is a *modal predicate* in the context c. A context is computed by solving a fixed point equation of the form $E(c) \equiv c$ and a solution to the equation is found by suitably substituting a formula for c. A formula $\mathcal{T}$ is a solution of the above equation, iff $\models_{KD4Z} E(\mathcal{T}) \equiv \mathcal{T}$. Reiter extensions of a default theory are characterized by one context: an extension is, in fact, a context. There are situations in which multiple contexts have to be handled: to show that different information sources are conflicting or alternative, to state that two situations are contradictory, to complete the information at hand, to specify sub-context inheritance, to indicate that a term (like "now") changes its value according to the context, etc. To express such a variety of uses of contexts, we have extended the results from [8], to manage many parameters, each designating a formula. So instead of $E(c)$ we deal with sentences of the form $E(c_1, \ldots, c_n)$. The computation of several contexts is achieved by solving a system of fixed point equations, instead of a single equation.

The novelty here is that relations among contexts have to be declared. For example "Bill" is the name of the US president, and "bill" is a bird's beak; to maintain the difference we require two contexts, say c_1 and c_2. However we also need to declare what is the relationship between these two contexts: $\Box(c_1 \equiv \neg c_2)$ means that the two contexts can never be chosen together; $\Box\Diamond(c_1 \wedge c_2)$ states that the two contexts are consistent. The language for multiple contexts is a natural extension of that for a single context: a countable number of parameters – denoting contexts – is added, and formulae are built by first order formulae, boolean connectives, parameters and modal operators. A theory with n contexts can be reduced to a system of n logical equations, which yields a set of solutions for those contexts, as follows. Let $E(c_0, \ldots, c_n)$ be a modal predicate. Let $E_i(c_0, \ldots, c_n)$ be of the form

$$E(c_0, \ldots, c_n) \wedge \Box\Diamond c_i \ 0 \leq i \leq n$$

For each i, $0 \leq i \leq n$, $E_i(c_0, \ldots, c_i)$, is an *implicit consistency modal predicate with respect to the context c_i*.

Theorem 5. *Let $E_i(c_0, \ldots, c_i)$, $0 \leq i \leq n$, be an implicit consistency modal predicate w.r.t. the context c_i. Then there exists a sentence $\mathcal{T}$ in which only those variables of $E(c_0, \ldots, c_n)$ other than $c_0, \ldots, c_n$ occur and such that, if $\mathcal{T}_i = \mathcal{T}$ or $\mathcal{T}_i = \bot$:*

$$\{\models E_i(\mathcal{T}_1, \ldots, \mathcal{T}_i, \ldots \mathcal{T}_n) \equiv \mathcal{T}\}_{1 \leq i \leq n}$$

The next characterization of the structure of fixed points gives an easy way to compute solutions of a system of fixed point equations.

Corollary 1. *All the fixed point solutions to the equations $E_i(c_0, \ldots, c_n) = c_i$ are of the form $E(\mathcal{T}_1, \ldots, \mathcal{T}_n) \wedge \Box\Diamond\gamma$, for any γ, where $\mathcal{T}_i$ is either $\top$ or $\bot$.*

We solve the system of logical equations as follows: solve one equation with respect to the given context, possibly with the solution $\bot$. A sentence not mentioning the context is obtained. Substitute the solution into the remaining equations and proceed iteratively until no equation is left. By Theorem 4 this is equivalent to simultaneously substituting $\bot$ and $\top$ in $\Box E(c_1, \ldots, c_n)$ and obtaining a solution $\mathcal{T} \wedge \Box\Diamond\gamma$ for arbitrary γ. In other words, a formula is relatively provable w.r.t. a certain context when it is a solution to the system of equations in which many contexts are considered. To recover the classical theories from the modal solutions to the contextualized theories, we simply form, for each fixed point solution $\mathcal{T}_i$, the theory $\Delta_i = \{\phi \mid \mathcal{T}_i \models \Box\phi\}$.

Sometimes axioms are described as *'implicit definitions'* of the concepts they introduce [51]. We take this view and regard a set of axioms as a system of logical equations which determines the *unknown* parameters. This system of equations characterizes certain combinations of sentences as admissible, and other as inadmissible, distinguishing them by means of substitutions. We admit only those substitutions that turn an equation into a *theorem* of the logic. By means of these equations a definite class of admissible sentences is defined, namely the class of those which satisfy the system. These sentences explicitly define the contexts.

We illustrate with an example, taken from [9], how *default inheritance* can be handled with multiple contexts. Observe that conventional nonmonotonic default inheritance mechanisms provide only for the inheritance of *properties*, not theories. This ability of contexts to inherit theories is one of the most useful features of our proposal. Consider the following set of sentences:

1. *Italians go to church on Sunday, love soccer and eat pasta for dinner;*
2. *Neapolitans eat pizza for dinner;*
3. *No one eat pasta and pizza for dinner;*
4. *Neapolitans are Italians;*

We want the context of Naples to inherit what is true in the context of Italy, except for eating pasta for dinner. To this end we introduce three contexts, namely c_I, c_N and c. Where c_I is the context for Italy, c_N for Naples and c designates their intersection. We then rewrite the above sentences so as to suitably introduce the designated contexts, along with the following rules:

i. Each sentence has to be provable from its inner context;
ii. Sentences establishing whether contexts are consistent or not have to be defined.

Let $E(c, c_I, c_N)$ be the conjunction of the following contextual sentences:

$C1.\ \Box(c \vee c_I \equiv Italians \wedge goToChurch \wedge loveSoccer)$
$C2.\ \Box(c_I \rightarrow eatPasta)$
$C3.\ \Box(c_N \rightarrow Neapolitans \wedge eatPizza)$
$C4.\ \neg\Box(c_I \wedge c_N), \Box\Diamond(c_I \vee c)$
$C5.\ \Box\Diamond c_I \rightarrow \Box(c_I \equiv Italy), \Box\Diamond c_N \rightarrow \Box(c_N \equiv Naples)$

(C1) says that (1), but for *eatPasta*, refers both to the contexts c and c_I: in the context $c \vee c_I$ people are Italians, go to church and love soccer. (C2) says that *eatPasta* is provable from c_I; (C3) says that in the context c_N people are Neapolitans and eat pizza. The two modal sentences in (C4) are consistency conditions for inheritance, the first one asserts that contexts are alternative and the second one can be interpreted as "either the context of Italy or what it can be inherited from it must be consistent". The sentences in (C5) establish the meaning of the two contexts c_I and c_N whenever they are consistent.

There are only three consistent solutions to the set of equations

$$\{\Box E(c, c_I, c_N) \wedge \Box\Diamond c_i\}_{1 \leq i \leq 3}$$

$$1. \; \Box E(\top, \top, \bot)$$
$$2. \; \Box E(\top, \bot, \top)$$
$$3. \; \Box E(\bot, \top, \bot)$$

The first and the third ones are equivalent. Observe that, in particular, $\Box E(\top, \top, \bot)$, by demodalization, yields the following:

$$Italy \equiv Italians \wedge eatPasta \wedge goToChurch \wedge loveSoccer$$

and $\Box E(\top, \bot, \top)$, by demodalization, yields:

$$Naples \equiv Neapolitans \wedge Italians \wedge eatPizza$$
$$\wedge goeToChurch \wedge loveSoccer$$

The contextualization yields what one expected: Neapolitans inherit Italian's entire theory, except for eating pasta.

References

1. *Artificial Intelligence Journal*, 13:81–132, 1980.
2. G. Amati, L. Carlucci Aiello, and F. Pirri. Default as restrictions on classical Hilbert-style proofs. *Journal of Logic Language and Information*, 3(4):303–326, 1995.
3. G. Amati, L. Carlucci Aiello, D. Gabbay, and F. Pirri. Provability logic for default reasoning. Technical Report Rap. 29.94, Universita' di Roma "La Sapienza", 1994. Paper presented at the *Logic Colloquium*, Haifa, 1995.
4. G. Amati, L. Carlucci Aiello, D. Gabbay, and F. Pirri. A structural property on modal frames characterizing Default Logic. *Journal of IGPL*, 4(1):1–24, 1996.
5. G. Amati, L. Carlucci Aiello, and F. Pirri. A proof theoretical approach to default reasoning I. Tableaux for Default Logic. *Journal of Logic and Computation*, 6(2):205–231, 1996.
6. G. Amati, L. Carlucci Aiello, and F. Pirri. Modal nonmonotonic reasoning via boxed fixed points. *6th International Workshop on nonmonotonic reasoning, Oregon, USA*, 1996.
7. G. Amati and F. Pirri. Is there a logic of provability for nonmonotonic reasoning? *Proceedings of the Fourth International Conference on the Principles of Knowledge Representation and Reasoning (KR-96)*, pages 493–506, 1996.
8. G. Amati, L. Carlucci Aiello, and F. Pirri. Definability and commonsense reasoning. *Artificial Intelligence Journal*, 93:1 – 30, 1997. Abstract in Definability and commonsense reasoning, *Third Symposium on Logical Formalization of Commonsense Reasoning*, Stanford, USA, 96.
9. G. Amati and F. Pirri. Contexts as relativized definitions: a formalization via fixed points. *Fourth Symposium on Logical Formalizations of Commonsense Reasoning*, London, 1998.
10. P. Besnard. *Deafult Logic*. Springer-Verlag, Berlin, 1989.
11. P. Besnard and T. Schaub. Possible world semantics for Default Logic. In *Working Notes of the 4th International Workshop on Nonmonotonic Reasoning*, pages 34–40, 1992.
12. G. Brewka. Cumulative Default Logic: in defence of nonmonotonic inference rules. *Artificial Intelligence Journal*, 50:183–206, 1991.

13. Saša Buvač, Vanja Buvač, and Ian A. Mason. The semantics of propositional contexts. In *Proceedings of the Eight International Symposium on Methodologies for Intelligent Systems*, volume 869 of *Lecture Notes in Artificial Intelligence*. Springer Verlag, 1994.
14. Saša Buvač and Richard Fikes. A declarative formalization of knowledge translation. In *Proceedings of the ACM CIKM: The 4th International Conference on Information and Knowledge Management*, 1995.
15. Saša Buvač and Ian A. Mason. Propositional logic of context. 1993.
16. Saša Buvač and John McCarthy. Combining planning contexts. In Austin Tate, editor, *Advanced Planning Technology–Technological Achievements of the ARPA/Rome Laboratory Planning Initiative*. AAAI Press, 1996.
17. J.P. Delgrande and W.K. Jackson. Default Logic revisited. In J. Allen, R. Fikes, and E. Sandewall, editors, *Proceedings of the Second International Conference on the Principles of Knowledge Representation and Reasoning (KR-91)*, pages 118–127, San Mateo, CA, 1991. Morgan Kaufmann.
18. P. Doherty and W. Lukaszewics. Nml3, a non-monotonic logic with explicit default. *Journal of Applied Non-Classical Logics*, 2:9–48, 1992.
19. D. Etherington. *Reasoning with incomplete information*. Morgan Kaufman, San Mateo, CA, 1988.
20. C. Froidevaux and J. Mengin. Default Logics: a unified view. *Computational Intelligence*, 10:331–369, 1994.
21. D. Gabbay and R. Nossum. Structured contexts with fibred semantics. In *Proceedings of the international and interdisciplinary Conference on Context*, 1997.
22. M. Gelfond, V. Lifschitz, H. Przimusinska, and M. Truszczynski. Disjunctive defaults. In J. Allen, R. Fikes, and E. Sandewall, editors, *Proceedings of the Second International Conference on the Principles of Knowledge Representation and Reasoning (KR-91)*, pages 230–237, San Mateo, CA, 1991. Morgan Kaufmann.
23. M.R. Geneserth and R.E. Fikes. Knowledge interchange format, version 3.0. ref.man. Technical report, University of Stanford, 1992.
24. L. Giordano and A. Martelli. On cumulative default reasoning. *Artificial Intelligence Journal*, 66:161–180, 1994.
25. F. Giunchiglia. Contextual reasoning. *Epistemologia, special issue on I Linguaggi e le Macchine*, XVI:345–364, 1993.
26. F. Giunchiglia, L. Serafini, E. Giunchiglia, and M. Frixione. Non-Omniscient Belief as Context-Based Reasoning. In *Proceedings of the Thirteenth International Joint Conference on Artificial Intelligence (IJCAI-93)*, pages 548–554, Chambery, France, 1993. Also IRST-Technical Report 9206-03, IRST, Trento, Italy.
27. R. Goldblatt. *Logics of time and computation*. CSLI, 1987.
28. R.A. Guerreiro, M.A. Casanova, and A.S. Hemerly. Contribution to a proof theory for generic defaults. In L. Carlucci Aiello, editor, *Proceedings of the Ninth European Conference on Artificial Intelligence (ECAI-90)*, pages 213–218, London, 1990. Pitman.
29. R.V Guha. *Contexts: A formalization and some applications*. PhD thesis, University of Stanford, 1991.
30. R.V. Guha and D.B Lenat. Cyc: a midterm report. *AI Magazine*, 11(3):32–59, 1990.
31. J.Y. Halpern and Y. Moses. Toward a theory of knowledge and ignorance: preliminary report. *Logics and Models of Concurrent Systems*, pages 459–476, 1988.
32. K. Konolige. On the relation between default and autoepistemic logic. *Artificial Intelligence Journal*, 35:343–382, 1988.
33. Vladimir Lifschitz. Computing circumscription. In *Proceedings of the Ninth International Joint Conference on Artificial Intelligence (IJCAI-85)*, pages 121–127, 1985.

34. Vladimir Lifschitz. On the satisfiability of circumscription. *Artificial Intelligence Journal*, 28:17–27, 1986.
35. F. Lin and Y. Shoham. Epistemic semantics for fixed-points non-monotonic logics. In *Proceedings of the Third Conference on Theoretical Aspects of Reasoning about Knowledge (TARK-90)*, pages 111–120, 1990.
36. W. Lukaszewics. Considerations on Default Logic. In *Proceedings 1984 Nonmonotonic reasoning workshop*, pages 165–193, New Paltz, N.Y., 1984.
37. W. Lukaszewics. Considerations on Default Logic: an alternative approach. *Computational Intelligence*, 4:1–16, 1988.
38. W. Lukaszewics. *Non-Monotonic Reasoning; Formalization of Commonsense Reasoning*. Ellis Horwood series in Artificial Intelligence, 1991.
39. V. W. Marek, G.F. Schwarz, and M. Truszczynski. Modal nonmonotonic logics: ranges, characterization, computation. *Journal for Association of Computing Machinery*, 40:963–990, 1993.
40. V. W. Marek and M. Truszczynski. Modal logic for default reasoning. In *Annals of Mathematics and Artificial Intelligence*, pages 275–302, 1990.
41. V. W. Marek and M. Truszczynski. More on modal aspects of Default Logic. *Fundamenta Informaticae*, 17:99–116, 1992.
42. V. W. Marek and M. Truszczynski. *Nonmonotonic Logic, context-dependent reasoning*. Springer-Verlag, 1993.
43. J. McCarthy. Circumscription – a form of nonmonotonic reasoning. *Artificial Intelligence Journal*, 13:27–39, 1980.
44. J. McCarthy. Generality in artificial intelligence. *Communications of the ACM*, 30:1030–1035, 1987.
45. J. McCarthy. Notes on formalizing contexts. In *Proceedings of the Fifth National Conference on Artificial Intelligence (AAAI-83)*, pages 555–560, San Mateo, CA, 1993. Morgan Kaufmann.
46. D. McDermott. Non-monotonic logic II: non-monotonic modal theories. *Journal of the ACM*, 29:33–57, 1982.
47. D. McDermott and J. Doyle. Non-monotonic logic I. *Artificial Intelligence Journal*, 13:41–72, 1980.
48. R. Montague. Syntactical treatments of modality, with corollaries on reflexion principles and finite axiomatizability. *Acta Philosophica Fennica*, 16:153–167, 1963.
49. R.C. Moore. Semantical considerations on nonmonotonic logics. *Artificial Intelligence Journal*, 25:75–94, 1985.
50. D. Perlis. Languages with self-reference II: Knowledge, belief, and modality. *Artificial Intelligence Journal*, 34:179–212, 1988.
51. L. Popper, Karl. *The Logic of Scientific Discovery*. Routledge, London, 1995.
52. R. Reiter. A logic for default reasoning. *Artificial Intelligence Journal*, 13:81–132, 1980.
53. R. Reiter. Circumscription implies predicate completion (sometimes). In *Proceedings of the National Conference on Artificial Intelligence (AAAI-82)*, pages 183–188, 1982.
54. R. Rychlik. Some variations on Default Logic. In *Proceedings of the Ninth National Conference on Artificial Intelligence (AAAI-91)*, pages 183–188, Cambridge, MA, 1991. MIT.
55. T. Schaub. On commitment and cumulativity in Default Logic. In *Symbolic and Quantitative Approaches for Uncertainty, European conference (ECSQAU)*, pages 305–309, Berlin, 1991. Springer-Verlag, LNCS 548.
56. G. Schwarz. Autoepistemic modal logics. In *Proceedings of the Third Conference on Theoretical Aspects of Reasoning about Knowledge (TARK-90)*, pages 97–109, 1990.

57. G. Schwarz. Minimal model semantics for nonmonotonic modal logics. In *Proc. of the 1st International workshop on Logic Programming and Non-Monotonic Reasoning*, pages 260–274. MIT Press, Cambridge, MA, 1991.
58. G. Schwarz. Minimal model semantics for nonmonotonic modal logics. In *Proc. 7th Annual IEEE Symposium on Logic in Computer Science*, pages 34–43, Santa Cruz, 92.
59. G.F. Schwarz. In search of a "true" logic of knowledge: the nonmonotonic perspective. *Artificial Intelligence Journal*, 79:39–63, 1995.
60. G.F. Schwarz and M. Truszczynski. Modal logic S4f and the minimal knowledge paradigm. In *Proceedings of the Third Conference on Theoretical Aspects of Reasoning about Knowledge (TARK-92)*, pages 97–109, San Mateo, CA, 1992. Morgan-Kaufmann.
61. G.F. Schwarz and M. Truszczynski. Minimal knowledge problem: a new approach. *Artificial Intelligence Journal*, 67:113–141, 1994.
62. Y. Shoham. Varieties of contexts. In *Artificial Intelligence and Mathematical theories of Computation*. Academic Press, London, 1991.
63. R.M. Smullyan. *Diagonalization and Self-Reference*. Oxford, 1994.
64. R.M. Solovay. Provability interpretations of modal logic. *Israel Journal Math.*, 25:287–304, 1976.
65. M. Truszczyński. Modal interpretations of Default Logic. In *Proceedings of the Twelfth International Joint Conference on Artificial Intelligence (IJCAI-91)*, pages 393–398, 1991.

Computing Domain Specific Information

Fahiem Bacchus

Department of Computer Science
University of Waterloo
Waterloo, Ontario, Canada, N2L 3G1
fbacchus@logos.uwaterloo.ca

1 Introduction

Ray Reiter has contributed much to computer science in general and to Artificial intelligence in particular. Most recently his work has been a tour de force in the area of reasoning about actions. Ray and his collaborators have succeeded in providing a unifying formalism covering a wide range of complex issues. Much of this work, done within the framework of the situation calculus is nicely explained in Ray's book on the subject [Rei99].

This article is in celebration of Ray's 60th birthday. In it I will attempt to show how Ray's approach to modeling actions in the situation calculus can be used as a unifying framework for computing domain specific information. My thoughts on this subject are still quite preliminary, and I will be posing more questions than answering them. Nevertheless, I hope to convince the reader that the situation calculus provides a nice framework in which to ask these questions. I also hope to convince the reader[1] that the framework of the situation calculus provides us with useful tools for attacking these questions.

First, what is the nature of the domain specific information under investigation? By "domain specific" I mean that we have a set of actions specific to a particular domain. The fact that these are all of the actions in this domain means that there are certain restrictions on the domain's dynamics. These restrictions are implicit in the action specifications. I am interested in developing techniques for making explicit some of these implicit restrictions. More specifically I am interested in making explicit, information that can then be utilized to speed up planning.

In my own work I have been concerned with the problem of planning, and have concluded that the main hurdle in this area is computational. Even the simplest kinds of planning problems are hard to solve. The approach I have advanced is that of "knowledge-based" planning: giving the planner extra information that it can use to help it guide its search for a plan.[2] In the approach I have developed [BK98] a temporal logic is utilized to express this information. We have found that many quite natural pieces of information can be expressed in this formalism, and that a planning

---

[1] Although perhaps with less success since I do not as yet have answers to many of these questions.

[2] Here we will take a plan to be the problem of finding a sequence of actions that will transform the initial state to a state satisfying the goal.

algorithm can be developed to utilize this information. Empirically, the results are dramatic–orders of magnitude speedups in planning time.[3]

As it currently stands, the control knowledge used in this approach must be written by the user. The user will typically examine the domain, the actions in the domain, and use their intuitions to come up with useful control knowledge. I am interested in automating this process, and in this paper I will provide some ideas in this direction.

The problem of computing information for planning domains, information designed to speed up planning, has been examined before. There have been typically two different kinds of approaches. Approaches based on learning, e.g., [Min88], [Kha97], and approaches based on reasoning, e.g., [Etz93], [PS93], [GS96], [FL98], [GS98]. I will concentrate on approaches based on reasoning here. These approaches have typically employed the solution to the frame problem that is implicit in the STRIPS formalization of planning: fluents only change when the appear on the add or delete list of an action. Using this property of STRIPS specified planning domains, these methods have typically used graph constructions to compute state invariants, operator reachability, etc.

In the situation calculus we have a logical theory that captures all of the logical consequences of the domain's dynamics, and the STRIPS assumption can be captured explicitly using the approach to the frame problem given in [Rei99]. Thus, I believe, the situation calculus it provides us a useful formalism in which much of these previous techniques can be unified and generalized.

My approach will be to illustrate some of the ideas I have had in this direction through a specific example–the logistics domain.

2 The Logistics Domain

The logistics domain consists a collection of primitive predicates:

1. $truck(x)$, x is a truck. Trucks can be used to transport objects within a city.
2. $plane(x)$, x is an airplane. Airplanes can be used to transport objects between airports which typically lie in different cities.
3. $object(x)$, x is an object. Objects can be loaded and unloaded from trucks and planes.
4. $at(x, y)$, x is at location y.
5. $in(x, y)$, x is in y.
6. $in\text{-}city(x, y)$, x is in city y.
7. $airport(x)$, x is an airport.

There are four actions in this domain, which we specify below using ADL operators [Ped89]. These operators are specified using the syntax of the TLPLAN system [BK98].

---

[3] A similar "knowledge-based" approach to planning has also been used by Ray in his work on planning in the situation calculus [Rei99], where the control knowledge is embedded in a "bad-situation" predicate. Ray has also been able to apply these ideas to planning problems that go beyond simple classical planning.

1. We may load objects into trucks or planes. Loading causes the object to stop being at some location and instead causes the object to in some vehicle.

```
(def-adl-operator (load ?o ?v ?l)
   (pre (?o ?l) (at ?o ?l)
        (?v)    (at ?v ?l)
        (and (or (truck ?v) (plane ?v)) (object ?o))))
   (add (in ?o ?v))
   (del (at ?o ?l)))
```

2. We may unload objects from a vehicle. This is the inverse of the load operator.

```
(def-adl-operator (unload ?o ?v ?l)
   (pre (?o ?v) (in ?o ?v)
        (?l)    (at ?v ?l))
   (add (at ?o ?l))
   (del (in ?o ?v)))
```

3. We can drive a truck between two locations that are in the same city. The precondition of this operator eliminates "trivial" moves by requiring that the starting and ending location be different.

```
(def-adl-operator (drive ?t ?l1 ?l2 ?c)
   (pre (?t)  (truck ?t)
        (?l1) (at ?t ?l1)
        (?c)  (in-city ?l1 ?c)
        (?l2) (in-city ?l2 ?c)
        (not (= ?l1 ?l1)))
   (add (at ?t ?l2))
   (del (at ?t ?l1)))
```

4. And finally we can fly a plane between two airports. Again the precondition eliminates "trivial" moves.

```
(def-adl-operator (fly ?p ?l1 ?l2)
   (pre  (?p)  (plane ?p)
         (?l1) (at ?p ?l1)
         (?l2) (airport ?l2)
         (not (= ?l1 ?l2)))
   (add (at ?p ?l2))
   (del (at ?p ?l1)))
```

Two things become apparent from examining these operators. First, it is easy to develop an intuitive understanding of this domain, and we have in fact used such intuitions in our descriptions above. Second, it should also be clear that many of these intuitions are unstated. For example, if there is a domain constant that satisfies both of the predicates truck and object, the load operator will happily load it onto another truck. Yet this is typically not what is intended in this domain. What can we deduce about this domain from the fact that these operators are the only way in which we can change things?

3 Conversion to Successor State Axioms

Since executing an instance of one of these operators is the only way in which changes can occur in this domain, the closure conditions required to employ Reiter's solution to the frame problem are satisfied. It is not difficult to see how we can express this domain as an equivalent theory in the situation calculus using precondition and successor state axioms (see [Rei99] for the details of such a transformation). In fact, a translation into the situation calculus can easily be automated. Here we simply compute the necessary axioms by hand. First the precondition axioms:

$$Poss(\texttt{load}(o, v, l), s) \equiv \tag{1}$$
$$\texttt{at}(o, l, s) \wedge \texttt{at}(v, l, s) \wedge (\texttt{truck}(v, s) \vee \texttt{plane}(v, s)) \wedge$$
$$\texttt{object}(o, s).$$

$$Poss(\texttt{unload}(o, v, l), s) \equiv \tag{2}$$
$$\texttt{in}(o, v, s) \wedge \texttt{at}(v, l, s).$$

$$Poss(\texttt{drive}(t, l_1, l_2, c), s) \equiv \tag{3}$$
$$\texttt{truck}(t, s) \wedge \texttt{at}(t, l_1, s) \wedge$$
$$\texttt{in-city}(l_1, c, s) \wedge \texttt{in-city}(l_2, c, s) \wedge l_1 \neq l_2$$

$$Poss(\texttt{fly}(p, l_1, l_2), s) \equiv \tag{4}$$
$$\texttt{plane}(p, s) \wedge \texttt{at}(p, l_1, s) \wedge \texttt{airport}(l_1, s) \wedge l_1 \neq l_2$$

Notice that in these axioms we have made certain predicates like $\texttt{truck}$ into fluents even though it is clear that they are situation invariant. Nevertheless, our aim is to discover such information automatically, so at this point assume that all predicates are fluents.

Now we construct the successor state axioms, one for each primitive predicate. These can be constructed automatically by examining which operators add and delete the fluent. Reiter has show that these axioms have the general form (for relation fluents F)

$$F(\boldsymbol{x}, do(a, s)) \equiv \gamma_F^+(\boldsymbol{x}, a, s) \vee \left(F(\boldsymbol{x}, s) \wedge \neg \gamma_F^-(\boldsymbol{x}, a, s)\right).$$

In this formula γ_F^+ and γ_F^- are the positive and negative effect axioms for the fluent F.

In those cases where a fluent is unaffected by any operator its positive and negative effect axioms become equivalent to FALSE (under no condition can the fluent be converted from true to false or vice versa). So for fluents that do not appear in the add or delete clauses of any operator the successor state axiom reduces to

$$F(\boldsymbol{x}, do(a, s)) \equiv F(\boldsymbol{x}, s).$$

That is, the fluent will be true after doing any action a if and only if it was true before doing the action. A number of the fluents in this domain are situation invariant in this manner.

$$\texttt{truck}(t, do(a, s)) \equiv \texttt{truck}(t, s) \tag{5}$$

$$\texttt{plane}(p, do(a, s)) \equiv \texttt{plane}(p, s) \tag{6}$$

$$\texttt{in-city}(l, c, do(a, s)) \equiv \texttt{in-city}(l, c, s) \tag{7}$$

$$\texttt{airport}(l, do(a, s)) \equiv \texttt{airport}(l, s) \tag{8}$$

$$\texttt{object}(l, do(a, s)) \equiv \texttt{object}(l, s) \tag{9}$$

Note that since these successor state axioms can be automatically derived from the operator descriptions we have an automatic mechanism for detecting that these predicates are state invariant. That is, any ground instance of these will be true in any situation if and only if it is true in the initial situation S_0. Since these predicates are not fluents we can legitimately drop their situation arguments.[4]

The successor state axioms for the other two fluents are more complex.

$$
\begin{aligned}
\texttt{at}(x, l, do(a, s)) \equiv & \tag{10} \\
& \exists v. a = \texttt{unload}(x, v, l) \lor \exists l_1, c. a = \texttt{drive}(x, l_1, l, c) \lor \\
& \exists l_1. a = \texttt{fly}(x, l_1, l) \lor \\
& (\texttt{at}(x, l, s) \land \neg \exists v. a = \texttt{load}(x, v, l) \land \\
& \quad \neg \exists l_2, c. a = \texttt{drive}(x, l, l_2, c) \land \neg \exists l_2. a = \texttt{fly}(x, l, l_2)).
\end{aligned}
$$

$$
\begin{aligned}
\texttt{in}(x, v, do(a, s)) \equiv & \tag{11} \\
& \exists v. a = \texttt{load}(x, v, l) \lor (\texttt{in}(x, v, s) \land \neg \exists v. a = \texttt{unload}(x, v, l)).
\end{aligned}
$$

---

[4] This is a very simple type of state invariant, algorithms like that described by Fox and Long [FL98] can derive more complex invariants.

4   Inner and Outer Boundaries

The main idea of this paper is the notion of inner and outer boundaries of formulas. Consider a formula $\phi(s)$ whose only free variable is the situation variable s. This formula characterizes a set of situations: those situations s in which ϕ is true. Now consider those situations s' that are one action away from $\phi(s)$. These can be characterized by the formula

$$\neg\phi(s') \wedge \exists a.\phi(s/do(a, s')) \wedge Poss(a, s').$$

Notice, that we do not allow these situations to already satisfy ϕ. We refer to this set of situations "one action" away as the outer boundary of $\phi(s)$. More generally, we can define a sequence of non-intersecting shells that characterize the situations that are exactly one, two, three, ... , ect. actions away from a situation satisfying $\phi(s)$.

$$\mathcal{O}_0[\phi, s'] \triangleq \phi(s/s') \quad \text{(where we substitute } s' \text{ for } s \text{ in } \phi)$$

$$\mathcal{O}_1[\phi, s'] \triangleq \neg\mathcal{O}_0[\phi, s'] \wedge \exists a_1.\mathcal{O}_0[\phi, do(a_1, s')] \wedge Poss(a_1, s')$$

$$\vdots$$

$$\mathcal{O}_i[\phi, s'] \triangleq \bigwedge_{j=0}^{i-1} \neg\mathcal{O}_j[\phi, s'] \wedge \exists a_i.\mathcal{O}_{i-1}[\phi, do(a_i, s')] \wedge Poss(a_i, s')$$

Here, $\mathcal{O}_i[\phi, s']$ will be a formula characterizing those situations s' that are exactly i actions (some action sequence $\langle a_i, a_{i-1}, \ldots, a_1 \rangle$) away from a situation satisfying ϕ.

Every formula $\phi(s)$ also has an inner boundary. This is the set of situations satisfying $\phi(s)$ that also have the property that they can be reached by a single action from a situation that falsifies $\phi(s)$. They can be characterized by the formula

$$\phi(s) \wedge \exists a, s^-.s = do(a, s^-) \wedge \neg\phi(s^-).$$

Or, equivalently, since s^- is clearly a member of $\mathcal{O}_1[\phi, s^-]$, and $\phi(s)$ is $\mathcal{O}_0[\phi, s]$

$$\mathcal{O}_0[\phi, s] \wedge \exists a, s^-.s = do(a, s^-) \wedge \mathcal{O}_1[\phi, s^-].$$

Note that not all situations satisfying $\phi(s)$ have this property. In particular, it is possible to execute an action that achieves $\phi(s)$, and then to execute other actions without destroying $\phi(s)$ to reach situations that cannot be reached from outside of $\phi(s)$ in only a single action.

Like the outer boundaries we can generalize this to compute the inner boundaries $\mathcal{I}_i$ of the various outer boundaries $\mathcal{O}_i$.

$$\mathcal{I}_0[\phi, s] \triangleq \mathcal{O}_0[\phi, s] \wedge \exists a, s^-.s = do(a, s^-) \wedge \mathcal{O}_1[\phi, s^-].$$

$$\vdots$$

$$\mathcal{I}_i[\phi, s] \triangleq \mathcal{O}_i[\phi, s] \wedge \exists a, s^-.s = do(a, s^-) \wedge \mathcal{O}_{i-1}[\phi, s^-]$$

The inner boundaries of $\phi(s)$ specify conditions that are implied by the fact that we have just made a transition into an outer boundary.

For example, say we are in a situation s where $\neg \text{in}(\text{object}_1, \text{truck}_1, s)$ and then we move to a situation s' where $\text{in}(\text{object}_1, \text{truck}_1, s')$ (thus s is a member of $\mathcal{O}_1[\text{in}(\text{object}_1, \text{truck}_1, s), s]$). We know from in's successor state axiom that there must exists an l such that $s' = do(load(\text{object}_1, \text{truck}_1, l), s)$, and from the effects of load that we must also have $\neg \text{at}(\text{object}_1, l, s')$, for that location l in s'. The negation of this at predicate is implied by the fact that we have just made the transition into $\mathcal{O}_0[\text{in}(\text{object}_1, \text{truck}_1, s), s]$. Of course, since the inner boundary is also contained in the outer boundary it also implies the formula characterizing that outer boundary.

We can use inner and outer boundaries in a number of ways. A wide range of planning problems involve goals that are simply lists of ground atomic formulas: we want to achieve all of these facts. Say that we have two ground fluents $F(c, s)$ and $F'(a, s)$, and that the goal is to find a situation satisfying both these facts. In Green's formalization [Gre69] of planning this would be the problem of deducing the logical formula $\exists s . F(c, s) \wedge F'(a, s)$. Furthermore, say that neither of these facts are true in the initial situation S_0, rather that S_0 is in the 3rd outer boundary of $F(c, s)$ and the 2nd outer boundary of $F'(a, s)$.

Logically we would have that $\mathcal{O}_3[F(c, s), S_0]$ and $\mathcal{O}_2[F'(a, s), S_0]$ are both true. Furthermore, we would know that to reach a situation satisfying both of these fluents we must traverse all of the inner and outer boundaries from $\mathcal{O}_3[F(c, s), s]$ to $\mathcal{O}_0[F(c, s), s]$, and from $\mathcal{O}_2[F'(a, s), s]$ to $\mathcal{O}_0[F'(a, s), s]$. It could be, e.g., that one of the boundary sets leading to $\mathcal{O}_0[F(c, s), s]$ entails $\neg F'(a, s)$, while the boundary sets leading to $\mathcal{O}_0[F'(a, s), s]$ do not mention $F(c, s)$ at all. This fact could be used inform the planner to solve $F(c, s)$ first and then work on achieving $F'(a, s)$: if it was to achieve $F'(a, s)$ first it might have to delete it again to achieve $F(c, s)$.

In the sequel I will demonstrate some other examples of the use of these boundary sets. It should also be noted that these boundary sets are clearly related to strongest postconditions and weakest preconditions (first investigated by Hoare, and also used by Rosenschein [Ros81]). This relationship deserves to be properly characterized.

5 Some Examples

The definitions given in the previous section can be used to generate the logical formulas characterizing the various boundary sets. Figure 1 shows for the two fluents at and in the different ways that these fluent can be reached with a single action. In the figure we first give the fluent that is achieved by an action a. Then below this we link the fluent to all the different actions that could achieve it. Each action is in a separate column, as these actions are disjoint. Above the action are listed the extra conditions that the action achieves (beyond the fluent we started with), and below

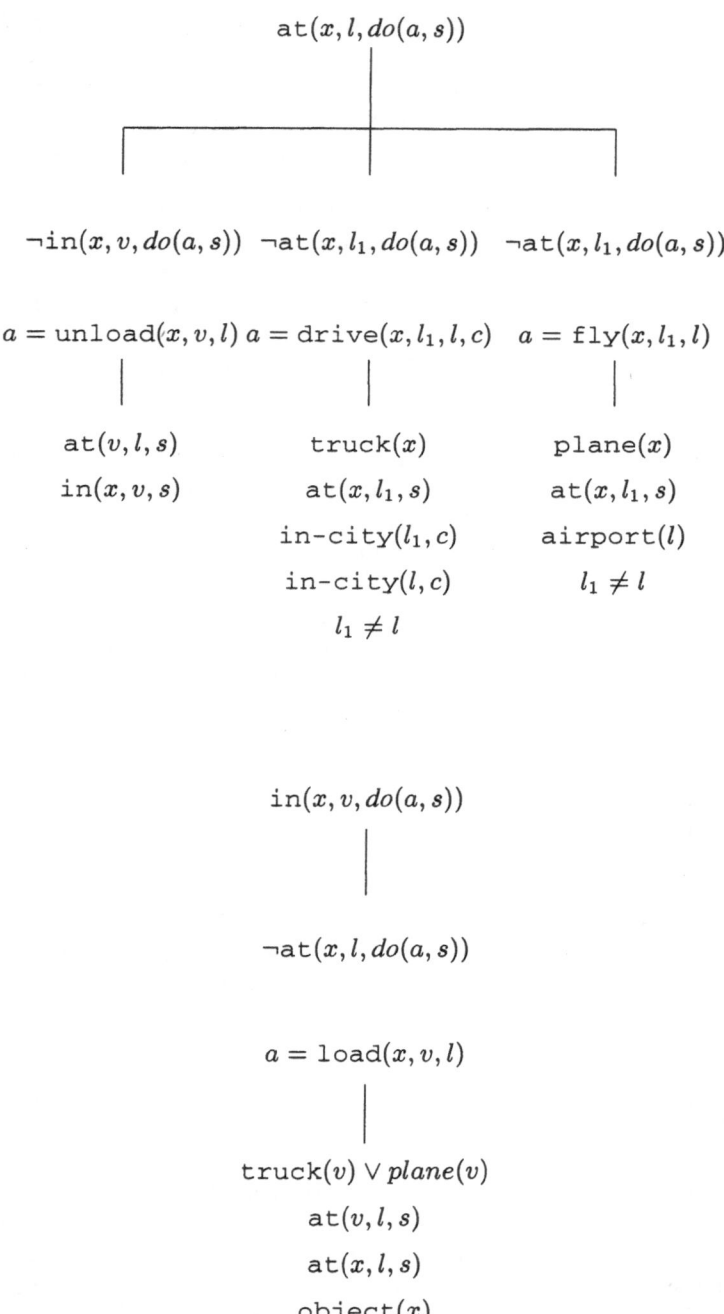

Fig. 1. Boundaries of at and in

the action are listed the various conditions that had to be previously true in order to execute the action.

Thus, using the first part of Figure 1 we can see that $\mathcal{O}_1[at(x, l, s), s]$ is characterized by the following formula

$\exists v.at(v, l, s) \wedge in(x, v, s) \vee$
$truck(x) \wedge \exists l_1.at(x, l_1, s) \wedge l_1 \neq l \wedge \exists c.\texttt{in-city}(l_1, c) \wedge \texttt{in-city}(l, c) \vee$
$plane(x) \wedge \exists l_1.at(x, l_1, s) \wedge airport(l) \wedge l_1 \neq l$

Here are some examples of conclusions that can be derived from these boundary sets.

1. Planes can only travel to airports.[5]

$$\forall x, l, s.plane(x) \wedge \neg object(x) \wedge \neg truck(x) \wedge \neg at(x, l, S_0) \Rightarrow$$
$$at(x, l, s) \Rightarrow airport(l)$$

This can be deduced from the fact that when x fails to satisfy $object(x)$ and $truck(x)$, the outer boundary of $at(x, l, s)$ entails $airport(l)$.

2. Planes can travel to any airport with a single action.

$$\forall x, l, l_1, s.plane(x) \wedge \neg object(x) \wedge \neg truck(x) \wedge airport(l) \wedge$$
$$at(x, l_1, s) \wedge \neg at(x, l, s) \Rightarrow$$
$$\mathcal{O}_1[at(x, l, s), s]$$

3. A plane is always at a unique location.

$$\forall x.plane(x) \wedge \neg object(x) \wedge \neg truck(x) \wedge \exists! l.at(x, l, S_0) \Rightarrow$$
$$\forall s.\exists! l.at(x, l, s)$$

This can be deduced from the fact that for planes the inner boundary of $at(x, l, s)$ entails the negation of a previous instance of the at fluent.

4. An object can only be in a truck or a plane.

$$\forall v, x.object(x) \wedge \neg plane(x) \wedge \neg truck(x) \wedge \neg in(x, v, S_0) \Rightarrow$$
$$\forall s.in(x, v, s) \Rightarrow truck(v) \vee plane(v)$$

This can be deduced from the fact that $truck(v) \vee plane(v)$ is entailed by the outer boundary of $in(x, v, s)$.

5. The only way an object can change location is for it to be put into a truck or a plane.

$$\forall x.object(x) \wedge \neg plane(x) \wedge \neg truck(x) \Rightarrow$$
$$\forall s, l.\mathcal{O}_1[at(x, l, s), s] \Rightarrow \exists v.in(x, v, s)$$

This can be deduced from the outer boundary of $at(x, l, s)$.

---

[5] The axioms specifying the domain's actions do not say anything about the properties of S_0. Thus for the most part any conclusions we can draw will require that certain conditions be true in S_0.

6. For any object o_1 it can be seen that traversing the boundaries of at does not entail any conditions on any other object. The only other types of individuals mentioned in these boundaries are trucks or planes. Hence, it can be concluded that the in or at status of any other object is irrelevant to achieving an at goal involving o_1. This observation could be utilized to avoid manipulating an object that is already at its goal location.

7. Consider a situation in which we have $in(o_1, v, s)$, $at(v, loc_1, s)$, and the goal requires $at(o_1, loc_1)$. This means that we are in $\mathcal{O}_1[at(o_1, loc_1, s), s]$. Furthermore, the inner boundary for this at instance only entails negating $in(o_1, v, s)$ and achieving $at(o_1, loc_1, s)$. By the previous item neither of these fluents affect the achievement of any the other goals that don't involve o_1. This observation can be utilized to construct a control rule that forces the planner to immediately unload o_1 from v in such situations. In particular, there is no need to leave the set $\mathcal{O}_1[at(o_1, loc_1, s), s]$ unless it is to enter the set $at(o_1, loc_1, s)$.

8. More generally, since moving through the outer boundaries of a literal represents progress towards achieving that literal, we can stop the planner from destroying such progress (i.e., stop it from moving back to a higher outer boundary) if none of the outer boundary conditions are relevant to any other goal literal.

This short set of examples give a very incomplete sketch of some of the things that can be accomplished using the notion of inner and outer boundaries. Some of inferences demonstrated in these examples can already be achieved by previously developed systems. However, the situation calculus offers potentially two advantages over previous approaches.

First, one can infer universally quantified assertions. Some of the previous work in this area is only able to infer information about a particular planning problem, not a general planning domain. For example, Fox and Long's TIM system [FL98] requires a fully specified initial state as input. Having a particular planning problem to deal with allows one to reduce the problem to one of propositional reasoning.

And second, one can deal with complex formulas (there are, of course, potentially difficult computational problems that still remain to be solved). For example, Etzioni's STATIC system [Etz93] works with a graph containing only literals—logical connectives are only modeled implicitly (by and-or nodes in the graph), and quantifiers are beyond the scope of his system.

6   Conclusion

This paper really has only one idea: that of boundary sets of formulas. The interesting thing is that such boundary sets can easily be logically characterized within the situation calculus by appealing to the formal framework Ray Reiter has developed over the past number of years. Furthermore, as I stated at the onset, I have posed more questions than I have answered. Nevertheless, I do believe that if the reader examines the examples I have given they will see that reasoning about such sets can generate a lot of interesting domain specific information.

Furthermore, I am fairly confident that most, if not all, of the previous work in this area can recast in terms of boundary set reasoning. In fact, my ideas about how boundary sets can be used, has to a large part been inspired by this previous work.

I am pursuing these ideas, and as I see it the main problems to be solved include the following.

1. Computational realizations. How can we generate the inferences given above automatically? Clearly we can reduce the problem to one of logical deduction, but then the problem would be to guide the search for proofs. Alternately, we can try to develop some more specialized algorithms (perhaps using graph structures like those given in Figure 1), and prove that those algorithms are generating logically sound inferences.
2. Exploring and more formally categorizing the range of inferences that can be made using boundary sets. For example, many planning domains contain fluents that are functional in one argument (like at is for planes). Another example is that there are set of deductions involving trucks that are quite analogous to the examples involving planes given above.
3. Identifying those categories of inferences that would be useful in helping to speed up a planner.

So as the reader can see, my contribution to the celebration of Ray's birthday is a research program that can be conducted within the formal framework he has been developing. It is not as yet a real research contribution. I hope to make substantive progress on this program in the near future.

References

[BK98] Fahiem Bacchus and Froduald Kabanza. Using temporal logics to express search control knowledge for planning. Under review, currently available at http://www.lpaig.uwaterloo.ca/~fbacchus/on-line.html, 1998.

[Etz93] Oren Etzioni. Acquiring search-control knowledge via static analysis. *Artificial Intelligence*, 62(2):255–302, 1993.

[FL98] M. Fox and D. Long. The automatic inference of state invariants in TIM. *JAIR*, 9:367–421, 1998.

[Gre69] Cordell Green. Application of theorem proving to problem solving. In *Proceedings of the International Joint Conference on Artificial Intelligence (IJCAI)*, pages 219–239, 1969.

[GS96] A. Gerevini and L. Schubert. Accelerating partial-order planners: Some techniques for effective search control and pruning. *Journal of Artificial Intelligence Research*, 5:95–137, 1996.

[GS98] A. Gerevini and L. K. Schubert. Inferring state constraints for domain-independent planning. In *Proceedings of the AAAI National Conference*, pages 905–912, 1998.

[Kha97] Roni Khardon. Learning action strategies for planning domains. Technical Report TR-10-97, Harvard University, 1997. Available at http://www.dcs.ed.ac.uk/home/roni/pubabs.html.

[Min88] Steve Minton. *Learning Search Control Knowledge*. Kluwer Academic Publishers, 1988.

[Ped89] E. Pednault. ADL: Exploring the middle ground between STRIPS and the situation calculus. In *Proceedings of the International Conference on Principles of Knowledge Representation and Reasoning*, pages 324–332, 1989.

[PS93]   M. Poet and D.E. Smith. Threat-removal strategies for partial-order planning. In *Proceedings of the AAAI National Conference*, pages 492–499, 1993.

[Rei99] Ray Reiter. *Knowledge In Action: Logical Foundations for Describing and Implementing Dynamical Systems*. 1999. Unpublished draft, available at http://www.cs.utoronto.ca/~cogrobo/.

[Ros81] Stanley J. Rosenschein. Plan synthesis: A logical perspective. In *Proceedings of the International Joint Conference on Artificial Intelligence (IJCAI)*, pages 115–119, 1981.

Specifying Database Transactions and Active Rules in the Situation Calculus

Leopoldo Bertossi, Javier Pinto, and Ricardo Valdivia

Departamento de Ciencia de la Computación, Escuela de Ingeniería, Pontificia Universidad Católica de Chile, Santiago 6904411, CHILE

Summary. This paper provides a predicate logic based semantics for active rules in active databases [18], [9]. Our main contribution is a proposal to integrate the specification of active rules with the specification of the dynamics of transaction based change in relational databases. Our approach extends previous work in which the situation calculus, a language of many sorted predicate logic, is used to specify updates in databases [14]. To achieve this, we first specify database transactions in the situation calculus, and then active rules using the notion of occurrence as proposed in [13].

1 Introduction

This chapter provides a predicate logic based semantics for active rules in active databases [18], [9]. Our work is based on a dialect of the situation calculus (*Sitcalc*), a language of many sorted predicate logic that was originally proposed to represent knowledge about dynamic domains [12]. An important characteristic of *Sitcalc* is its treatment of actions, primitive transactions in a database setting[1], at the object level. Furthermore, the language has sorts for domain individuals and situations. In [14], Reiter proposed a logical framework based on the *Sitcalc* to specify transaction based database updates. The only transactions involved in the original proposal are primitive in the sense that they are not decomposable. Logical specifications, in Reiter's style, include axioms to specify preconditions for actions and axioms to specify the direct effects of actions. A completeness assumption is used in order to solve the *frame problem*, allowing to succinctly represent the changes and non-changes produced by the execution of actions. With this kind of specification all possible and legal courses of evolution of the database are open to hypothetical reasoning; therefore, virtual updates can be easily considered.

In database management systems (DBMSs), changes in the state of the database result from the execution of complex transactions, which are composed of sequences of update operations. Thus, the database changes correspond to the net effects of all the operations involved in the transaction. Once the sequence of updates is successfully executed, they are committed, with a *Commit* operation. That is, the changes on the database are made visible to other transactions.

Transactions can be used as mechanisms to support recovery from failures, concurrency management and consistency maintenance. For instance, to implement

---

[1] The terms *update operation*, *primitive transaction*, and *action* are used interchangeably.

consistency maintenance, one can verify the satisfaction of the integrity constraints before committing the transaction. If the new database state is consistent, the *Commit* action is triggered, otherwise the transaction is aborted by triggering a *Rollback* operation. In this chapter, we extend Reiter's work on the formalization of database updates by formalizing the notions of complex transaction, commit, rollback, etc. as they appear in relational databases.

Traditional DBMSs are passive, that is, transactions are executed only when they are invoked explicitly. In contrast, an active DBMS (ADBMS) continuously monitors the state of the database and reacts as specified by *active rules*. These rules have also been called *Event-Condition-Action* or ECA rules [18], [19]. In this context, an ECA rule's action corresponds to a complex database transaction; to avoid confusion we will refer to a these actions as rule actions. On the other hand, events in an ECA rule correspond to changes in the database tables. In our formalization, the fact that a change in a database table has occurred will be stored in auxiliary tables; then, events are modeled as simple formulas that refer to the values stored in these auxiliary tables. Therefore, from a logical viewpoint, events and conditions will be modeled as conditions on the state of the database.

In order to formalize the notion of active rule, we use the notion of occurrence which allows us to specify that some actions must occur given specific conditions. These conditions might refer to the occurrence of some event(s), or to the actual state of the database. The applicability conditions for active rules are modeled as conditions for action occurrences. Since events and actions in active rules are expressed in terms of database transactions, it is also necessary to provide an account for (complex) transactions as they appear in real DBMSs. In part, this work has been directed towards providing a unifying criteria regarding the meaning of active rules, and to provide a framework for comparing different proposals for active database systems.

This chapter is structured as follows. Section 2 describes the execution modes of active rules, in particular of rules in Starburst [17]. Section 3 introduces *Sitcalc* and specifications of database updates. It continues with a more precise description of the task of specifying transactions and active rules. The section ends with the introduction of occurrences in *Sitcalc* [13]. Section 4 shows how to represent and specify database transactions and rollbacks. Active rules in the *Sitcalc* are specified in section 5. Section 6 shows an example of reasoning from an specification containing active rules. Finally, section 7 presents conclusions and future work.

2   Active Rules

The syntactic structure of active rules is similar in many DBMSs. Nevertheless, some fundamental aspects (for instance, the exact instant of execution of rules and the way events are considered) may differ radically from one system to another. In this section we review some issues regarding the different ways in which active rules can be handled (for more on these issues, consult [7], [18]).

An active rule is composed of three basic elements: *Events*, *Conditions* and *Actions*. The informal meaning is " Whenever the *Event* occurs, evaluate the *Condition*; if it holds, then execute the *Action*". An event is some change in the extension of database relations. In our proposal, these changes are reflected on the contents of auxiliary tables that record these changes. Thus, from a logical point of view, an event is a truth condition in a state. This truth condition refers to the auxiliary tables.

There is no standard terminology to refer to the process of rule triggering and execution. We choose the following (inspired by [8]): An ECA rule is said to be fired if its Event part is satisfied. Among all the rules that are fired, one or more may be considered for execution. If a rule is considered for execution, and its Condition part is satisfied, then its Action part is executed.

There are several possible semantics for the behavior of an active rule. In this regard, some important issues must be considered:

Granularity. Given that many DBMS operate on sets of tuples, there are several possible choices for rule firing. First, rules can be processed after the modification of individual tuples. Second, rules can be processed after all modifications specified by an update operation have been performed. Finally, rules can also be processed at the end of the execution of the transaction.

Simultaneous firing of rules. In some languages, the simultaneous firing of rules are disallowed. If this is not the case, then the execution may be sequential (using some form of conflict resolution) or concurrent. We consider the case in which rules have priorities for conflict resolution.

Rule chaining. When a rule is executed, it can lead to a state where other rules might also be applicable, chaining their execution. This possibility opens the problem of termination of the execution [1].

Scope of the transaction. In general terms, active DBMSs consider the execution of a rule inside the range of execution of the transaction where the event occurred, but there are systems with higher flexibility in the relationship between rules and transactions.

We use Starburst as a model in order to describe our approach to specify the semantics of active rules. In the next section we describe how Starburst addresses these issues.

2.1   Starburst

Active rules in Starburst [17] are associated to the notion of a *transition*. The action in a rule may be a sequence of SQL operations. When this sequence is executed, it gives rise to a transition. A transaction may involve several transitions; the first transition results from the execution of the operations explicitly specified in the user transaction. Transitions also arise from the execution of the actions that are associated to rules.

A rule may mention several alternative events. If at least one of them occurs, the rule fires. In Starburst, all rules are assigned a priority. The conditions of the fired rules are evaluated in order of priority. The first rule whose *condition* holds, is *executed*. If a rule has been fired, and its condition evaluated, it is said to have been *considered* (regardless of whether or not its *condition* holds). After a rule is executed, the process starts again and stops when no rules can be executed.

To fire an active rule in Starburst, the net effects of the transitions are considered, that is, changes caused on the tuples as a consequence of a set of SQL operations are seen as effects of a single operation. For example, if a tuple is updated and then deleted, only the deletion of the original tuple is considered.

Notice that an original user transaction can be extended by the execution of active rules. At the end of the original transaction and at the end of a rule execution, rules are evaluated for firing. A rule is fired if one or more of the events that fire it has occurred since the last time the rule was considered (see above for the difference between firing and considering a rule), or since the beginning of the original transaction if the rule has not been considered yet. The events are always checked with respect to the net effects, and only at the end of the user transaction (but before committing) or at the end of the transaction associated to an executed active rule, and not inside a transaction or rule execution. Starburst has a natural execution semantics, each rule considers each event exactly once. Thus, events that make a rule to be considered, are forgotten when the rule is evaluated for firing at a later time. According to this semantics, rules can be fired at the end of the states resulting from execution of user transactions (transitions) or after transitions generated by the execution of other rules.

3   Specifying Database Updates

In this section we briefly describe *Sitcalc* and specifications of databases updates by primitive, non decomposable transactions. In *Sitcalc* there are sorts $\mathcal{I}$, $\mathcal{A}$ and $\mathcal{S}$ for domain individuals, actions (or primitive transactions) and situations, respectively. First order variables and quantification are used for each sort. In formulas, variables are implicitly assumed to be universally quantified with maximum scope. Basic dynamic properties, called *fluents* elsewhere, are represented with predicates that take one situation argument (its last), and are of the form $F(\bar{x}, s)$. Their extensions correspond to the contents of a database table at situation s.

There is distinguished constant, S_0, to denote the initial situation, along with a successor function, do, such that the term $do(a, s)$ denotes the situation that results from executing a in situation s. The following axioms characterize the space of situations:

$$S_0 \neq do(a, s), \tag{1}$$

$$do(a_1, s_1) = do(a_2, s_2) \supset (a_1 = a_2 \wedge s_1 = s_2), \tag{2}$$

$$(\forall P)[P(S_0) \wedge (\forall a, s)(P(s) \supset P(do(a, s)))$$
$$\supset (\forall s)P(s)]. \tag{3}$$

The last axiom is a second order induction axiom on situations. With these axioms, situations form a tree whose root is the initial situation.

There is also a predicate, $Poss(a, s)$, with the intended meaning that (the execution of) a is possible at situation s. This predicate has to be axiomatized giving preconditions for all the named actions in the language. A binary predicate, $<$, between situations can be defined with the intended meaning that $s < s'$ holds when s' can be reached from s by executing a finite sequence of possible actions. This definition is based on the induction axiom plus the following constraints, where $s \leq s'$ is an abbreviation for $s < s' \vee s = s'$:

$$\neg s < S_0, \quad s < do(a, s') \equiv (Poss(a, s') \wedge s \leq s'). \tag{4}$$

A formal specification of the dynamics of a database for primitive transactions is expressed by the axioms [10]: $\Sigma = \mathcal{D}_{found} \cup \mathcal{D}_{S_0} \cup \mathcal{D}_{una} \cup \mathcal{D}_{ap} \cup \mathcal{D}_{ssa}$, where:

1. $\mathcal{D}_{found}$ contains the foundational axioms of the situation calculus (1) - (4).
2. $\mathcal{D}_{S_0}$ is the initial database as a theory; it describes the world at the initial situation S_0.
3. $\mathcal{D}_{una}$ contains unique names axioms for actions.
4. $\mathcal{D}_{ap}$ contains the action precondition axioms of the form

$$\forall s \forall \bar{x}\ Poss(A(\bar{x}), s) \equiv \Pi(\bar{x}, s)$$

, for each named parameterized action A. Here, formula $\Pi(\bar{x}, s)$ is simple in s^2, which means that its only situation term is the free variable s, and it does not mention $Poss$.
5. $\mathcal{D}_{ssa}$ contains the "successor state axioms" (SSAs). They characterize the evolution of the database by describing, for each table F, when $F(\bar{x}, do(a, s))$ becomes true, that is when the property F holds at a successor state that results from executing a possible action a in situation s^3. They have the form:

$$\begin{aligned} Poss(a, s) \supset F(\bar{x}, do(a, s)) \equiv \\ [\Psi_F(\bar{x}, a, s) \vee (F(\bar{x}, s) \wedge \neg \Psi_{\neg F}(\bar{x}, a, s))], \end{aligned} \tag{5}$$

where Ψ_F and $\Psi_{\neg F}$ are formulas simple in s. They describe the only actions and conditions that can make F respectively true or false at the situation $do(a, s)$.

In addition, we may have integrity constraints which should be satisfied after transactions are executed. These integrity constraint do not play the role of *ramification* or *qualification* constraints [11], rather they specify conditions that should be satisfied after transactions are finished. If these constraints are not satisfied, then a rollback should be performed, otherwise, the transactions are committed. Logically, these constraints could be written as:

$$(\forall s, s')\ S_0 \leq s \wedge s = do(Commit, s') \supset \gamma(s),$$

where $Commit$ is a special action, introduce later, used to perform the actual commitment of a transaction. As specified in the axiomatization of $Commit$, one of its preconditions is that the γ constraints be satisfied.

---

[2] This means that s the only situation term appearing in the formula.

[3] These axioms solve the problem of specifying both changes and persistencies in a succinct way.

Example: We consider a simple database of a company with the table $Emp(x, s)$, meaning that person x is an employee of the company when the database is in situation s, and with the primitive transactions $hire(x)$ and $fire(x)$. In this case there is one successor state axiom:

$$\forall(a, s) Poss(a, s) \supset \forall x[Emp(x, do(a, s)) \equiv$$
$$a = hire(x) \lor Emp(x, s) \land a \neq fire(x)],$$

and two action precondition axioms:

$$\forall(x, s)[Poss(hire(x), s) \equiv \neg Emp(x, s)],$$
$$\forall(x, s)[Poss(fire(x), s) \equiv Emp(x, s)].$$

3.1 Occurrences in the Situation Calculus

From the logical specification of the dynamics of change of a traditional database, it is possible to reason about all its possible legal evolutions. However, these specifications do not consider transactions that *must occur* given certain environmental conditions (i.e., the database state). In order to extend this style of specification to ADBMSs, it is necessary to extend the situation calculus with *occurrences*. This is necessary, given that the actions specified by active rules must be *forced to occur* when the associated event happens and the corresponding condition is satisfied. That is, the future is not open to all possible evolutions, but constrained by the necessary execution of actions mentioned in the rules that fire, given that their related conditions hold.

The notion of occurrence in *Sitcalc* we use here is based on [13]. The starting point is the observation that every situation s identifies a unique sequence of actions. That is, situations can be identified with histories of action executions: $s = do(a_n, \ldots (do(a_2, do(a_1, S_0)))\ldots)$. We say that the actions $a_1, a_2, \ldots, a_n$ have occurred before s. Next, the predicate *proper* is introduced to characterize those situations whose histories are considered to be proper with respect to the domain that is being axiomatized. In particular, when modeling active databases, we consider that situations are proper when their histories are consistent with the semantics given to the rule execution.

Thus, the predicate *occurs* is introduced, such that $occurs(a, s)$ represents the fact that action a occurs in situation s. If this holds, then the situation $do(a, s)$ must be in the history of every proper situation that is accessible from s:

$$proper(s_h) \equiv$$
$$\forall a, s\ (s < s_h \land occurs(a, s) \supset do(a, s) \leq s_h). \tag{6}$$

Finally, we want to consider only those situations that are contained in a proper history. For this purpose, the predicate *legal* is introduced, and defined in order to axiomatize the intuition that a situation or history s is legal if for every (posterior) situation accessible from s, there is a proper future with s in its history:

$$legal(s) \equiv \forall s'\ s < s' \supset \exists s_h\ s' \leq s_h \land proper(s_h). \tag{7}$$

In summary, a situation is legal, if it results from the execution of possible actions, and all occurrence constraints are satisfied by its history. Notice that *legal* is defined in terms of *proper*, in consequence, with *legal* we can consider a sub-tree of the situation structure, whose branches contain proper future situations.

Whenever the condition of an active rule in consideration is satisfied, the action mentioned in the rule must occur (be executed). In a database whose situations are all legal there will be only proper futures. Thus, in a situation calculus tree we only consider branches in which the given sequence of operations has occurred. Therefore, the specification of an active rule must include the presence of the predicate *occurs* associated to sequence of operations.

4 Specifying Transactions

In *Sitcalc*, transactions will be represented as finite lists of primitive actions, that is of elements of the sort $\mathcal{A}$. We introduce a new sort, $\mathcal{L}$, constructed from $\mathcal{A}$. We denote a generic variable for lists with l. We consider (a) The function ; : $\mathcal{L} \times \mathcal{A} \rightarrow \mathcal{L}$ as a list constructor; (b) An extended function do so that it includes the execution of lists of primitive transactions, that is, $do : \mathcal{L} \times \mathcal{S} \rightarrow \mathcal{S}$; (c) The empty list, $[]$, with no effects on a situation. The basic axiomatization for lists executions is:

$$do([], s) = s,$$
$$do(l; a, s) = do(a, do(l, s)).$$

Also, the usual predicates on lists can be introduced in an obvious manner: *member*, to represent the membership of an action to a list, *first*, *last*, and *tail* are predicates with their usual meanings.

Finally, the *Poss* predicate can be extended to finite sequences of transactions:

$$Poss([], s) \equiv True,$$
$$Poss([a_1, \ldots, a_n], s) \equiv \bigwedge_{k=1}^{n} Poss(a_k, do([a_1, \ldots, a_{k-1}], s)).$$

Since transactions will start with new actions *Begin* and *Commit* of the sort $\mathcal{A}$, we need to distinguish the transaction body that does not include these limit actions:

$$transaction\_body(l) \equiv$$
$$\forall a \, (member(a, l) \supset a \neq Begin \wedge a \neq Commit).$$

A transaction is a list of actions delimited by the actions *Begin* and *Commit*:

$$transaction(l) \equiv$$
$$(\exists l', l'') \, l = (l'; End) \wedge first(Begin, l') \wedge tail(l'', l') \wedge transaction\_body(l'').$$

Initially, we consider the following preconditions for the primitive transactions *Begin* and *Commit* (later on we will modify these preconditions):

$$Poss(Begin, s) \equiv True,$$
$$Poss(Commit, s) \equiv \gamma(s),$$

where $\gamma(s)$ is a condition that must hold for the transaction to be committed. Formula $\gamma(s)$ comes from a (finite conjunction of) static integrity constraint(s) of the form $\forall s\, \gamma(s)$.

In [4] a technique for treating dynamic integrity constraints as static integrity constraints is presented.

Notice that the actions *Commit* and *Rollback* do not produce any changes on the tables of the database by themselves. This is because they do not appear in the SSAs of the tables presented in section 3. Actions *Commit* and *Rollback* have a meaning only at the end of a transaction, whose effects are made visible by the *Commit* and forgotten by the *Rollback*. This is the case when, before attempting a commit, the integrity constraints are not satisfied.

During the execution of a transaction, a *Rollback* operation cancels the transaction and returns the database to the state of the database at the beginning of the transaction. Thus, a rollback produces a new situation whose state is equivalent to the state of the database at the time the transaction started.

In a first attempt at formalizing rollback, we introduce new fluents that would *copy* the fluents associated to the database relations at the situation where the transaction is started (i.e., at the time of the execution of a *Begin* action). I.e., for each fluent F, we introduce a new fluent B_F, in the following manner:

$$Poss(a, s) \supset$$
$$B_F(\bar{x}, do(a, s)) \equiv (F(\bar{x}, s) \wedge a = Begin) \vee$$
$$(B_F(\bar{x}, s) \wedge a \neq Begin).$$

Thus, every time a transaction is started, with a *Begin* action, all fluents are *backed up* in auxiliary fluents B_F. These fluents are utilized in order to restore the fluent values in the event of a rollback. Therefore, to account for this behavior, the successor state axioms for a fluent F would be rewritten as:

$$Poss(a, s) \supset$$
$$F(\bar{x}, do(a, s)) \equiv [\Psi_F(\bar{x}, a, s) \vee$$
$$(F(\bar{x}, s) \wedge \neg\Psi_{\neg F}(\bar{x}, a, s))] \wedge$$
$$a \neq Rollback \vee$$
$$B_F(\bar{x}, s) \wedge a = Rollback.$$

These SSAs can be easily rewritten in the form (5).

5   Representing Active Rules

Now, we introduce a new sort, $\mathcal{R}$, for rules (and variables for this sort $r, \dots$). We also introduce a rule name of sort $\mathcal{R}$, a constant, $\mathtt{rule}_{R_i}$, for each of the rules R_i

$(i = 1, \ldots, m)$ specified in the active database. We add unique names axioms for rules: $\texttt{rule}_{R_i} \neq \texttt{rule}_{R_j}$ for $i \neq j$; and domain closure for rules: $\forall r \; \bigvee_{i=1}^{m} r = \texttt{rule}_{R_i}$.

These new elements, allow us to distinguish rules from their associated sequences of actions. Furthermore, we introduce new, auxiliary predicates that play a role similar to the tables associated to update events in Starburst [17]. A dynamic predicate, $V_{F,R}$, which holds for the instances of table F that become true (and were false before) during a transition. Analogously, $V_{\neg F,R}$, which holds for the instances that become false (and were true before). These predicates are used to detect the occurrences of events associated to rules R; that is, events are identified with specific updates in tables.

The action *Begin*, being the beginning of the user transaction, initializes all the auxiliary predicates as empty. Furthermore, a *Considered* predicate is required. It also initializes the auxiliary predicates as empty for those rules R considered according to Starburst's specification. The fluent *Considered* is specified below.

We need successor state axioms for the auxiliary predicates below.

$$
\begin{aligned}
Poss(a, s) \supset \\
V_{F,R}(\bar{x}, do(a, s)) \equiv \; & [\Psi_F(\bar{x}, a, s) \vee \\
& (V_{F,R}(\bar{x}, s) \wedge \\
& \neg(a = Begin \vee Considered(\texttt{rule}_R, s)))].
\end{aligned}
\tag{8}
$$

$$
\begin{aligned}
Poss(a, s) \supset \\
V_{\neg F,R}(\bar{x}, do(a, s)) \equiv \; & [F(\bar{x}, s) \wedge \Psi_{\neg F}(\bar{x}, a, s) \vee \\
& (V_{\neg F,R}(\bar{x}, s) \wedge \\
& \neg(a = Begin \vee Considered(\texttt{rule}_R, s)))].
\end{aligned}
\tag{9}
$$

Here, Ψ_F and $\Psi_{\neg F}$ are the formulas appearing in (5), the SSA for base table F.

Now, we specify the basic transitions, that is, the initial list of actions corresponding to the user transaction, and the lists of actions associated to rules. The initial list is a sequence of actions delimited by actions *Begin* and *End*. The list associated to a rule R always ends with an *End* action. We do not use *Commit* instead of *End*, because it is reserved for the end of the (user) transaction. We will also use a predicate *transition_body* defined in terms of *transaction_body*, that allow us to prevent lists of actions from containing any of the distinguished auxiliary actions we introduced before:

$$
transition\_body(l) \equiv transaction\_body(l) \wedge \neg member(End, l).
$$

$$
\begin{aligned}
initial\_list(l) \equiv & first(Begin, l) \wedge last(End, l) \wedge \\
& (\forall a)\,(member(a, l) \wedge \neg first(a, l) \wedge \neg last(a, l) \supset \\
& \qquad\qquad a \neq Begin \wedge a \neq End)
\end{aligned}
$$

Also, we define the predicate *rule_list* to specify the action list associated to a rule.

$$rule\_list(\mathbf{rule}_i, l) \supset (\exists l') \, l = l'; end \wedge$$
$$transition\_body(l'), \tag{10}$$

Furthermore, the list of actions must be unique for each rule, that is, we need the axiom $rule\_list(r, l) \wedge rule\_list(r, l') \supset l = l'$.

In a particular application, the predicate *rule_list* should be axiomatized directly, by giving the lists of specific actions associated to each rule. For instance, if rule R_i is $(Event; Condition; a_1, \ldots, a_n)$, where the a_i's are the primitive actions to be executed when *Event* and *Condition* apply, then we introduce the axiom:

$$rule\_list(R_i, [a_1, \ldots, a_n, End]).$$

Now we want to specify a predicate *Poss* relative to rules, in such a way that *if a rule is possible, then the list of actions associated to it will occur.* A rule is possible if it has been considered by an event and its condition holds.

First, we specify that a rule is fired in a situation as:

$$Fired(R, s) \equiv Event_R(s) \wedge AssertionPoint(s). \tag{11}$$

where $Event_R(s)$ is a formula that describes the event that fires the rule (represented by entries in the auxiliary tables). [4] *AssertionPoint(s)* states that a rule can be executed at the end of the execution of the user transition or at the end of the sequence associated to a rule [1], i.e.:

$$AssertionPoint(s) \equiv \exists s' \, s = do(End, s'). \tag{12}$$

Second, a rule R is considered if, after it has been fired, its conditions have been evaluated. The rule is taken to be considered whether or not its condition is satisfied. Notice that for a rule to be considered there can be no other rule of higher priority that is at the same time considered and with its condition satisfied. Thus, we may write:

$$Considered(R, s) \equiv Fired(R, s) \wedge$$
$$\bigwedge_{i=1}^{m}((Fired(rule_{R_i}, s) \wedge \tag{13}$$
$$priority(\mathbf{rule}_{R_i}) > priority(R)) \supset \neg Cond_{R_i}(s).$$

Where, $Cond_R(s)$ refers to the additional conditions, expressed as a formula, which may involve both the auxiliary tables and the base tables. Also, *priority* is a function: $\mathcal{R} \to I\!N$, (we may consider $I\!N$ as a new sort with an underlying specification) which introduces a total order between rules. For this function it must hold:

$$\forall r, r' \, r \neq r' \supset priority(r) \neq priority(r'). \tag{14}$$

---

[4] That is, $Event_R(s)$ is expressed in terms of $V_{F,R}$ and $V_{\neg F,R}$. It could be expressed in terms of views defined on these predicates. As shown in [3], it is possible to derive successor state axioms for views, which can be used in the same way as successor state axioms for the base predicates.

This sentence establishes a total order between the elements of $\mathcal{R}$.

Third, a rule R is possible if:

$$Poss(R, s) \equiv Considered(R, s) \wedge Cond_R(s) \tag{15}$$
$$\wedge (rule\_list(\mathbf{rule}_R, l) \supset Poss(l, s)).$$

Notice that only one rule can be possible at each assertion point. This fact follows from the formalization presented in this section (in particular, the use of the priority assignment function).

Even when the conditions for an active rule are satisfied, we have to make sure that it will be executed, that is, the associated sequence of actions must occur. Therefore, we extend the predicate *occurs* to the case of sequences of actions.

$$occurs(l; a, s) \equiv occurs(a, do(l, s)) \wedge occurs(l, s). \tag{16}$$

In our specification we have the axiom:

$$Poss(r, s) \wedge rule\_list(r, l) \supset occurs(l, s). \tag{17}$$

Predicate *occurs* will force the execution of the sequence of actions associated to the a rule when the conditions for the execution of the rule are satisfied.

Finally, notice that for executing the *Commit*, that ends the transaction, there should be no pending rules. That is, we modify its precondition to:

$$Poss(Commit, s) \equiv \gamma(s) \wedge AssertionPoint(s) \wedge \tag{18}$$
$$\forall r \neg Poss(r, s),$$

where $\gamma(s)$ are the integrity constraints to be satisfied.

Now, we have to state that *Commit* and *Rollback* have to be executed as appropriate. In order to trigger a *Commit*, we add:

$$(\forall s)\ s \geq S_0 \wedge Poss(Commit, s) \supset occurs(Commit, s).$$

For triggering the *Rollback* action, we add:

$$(\forall s)\ s \geq S_0 \wedge AssertionPoint(s) \wedge [(\forall r) \neg Poss(r, s)] \wedge \neg\gamma(s) \supset$$
$$occurs(Rollback, s).$$

6   Reasoning from the Specification

Given the specification of the dynamics of the relational database, including a specification of the active rules in terms of the way their checking and triggering interact with the database evolution, one can perform reasoning from the specification. Among others, we may find the following reasoning problems that can be formulated as obtaining corresponding sentences as logical consequences of the specification[5].

---

[5] In an extended version of this chapter we will deal with these reasoning problems. Here we restrict ourselves to their formulation in the context of the specification

1. There are well known problems regarding the behavior of transactions; according to [1], the following are particularly relevant:

   Termination: i.e., whether one can guarantee that the transactions will end.

   Confluence: In some systems it is possible to arrive at a situation in which several rules can be executed at an assertion point. In this case, one would be interested in proving that all possible executions lead to the same state of the database.

   Given a ground user transition l_U, the termination problem can be posed in the following form:

   $$\Sigma \cup \Omega \models \exists s \; initial\_list(l_U) \wedge s = do(l_U, S_0) \supset \\ \exists s' \; s \leq s' \wedge legal(s') \wedge Poss(Commit, s'). \tag{19}$$

   Here Σ is the specification of the dynamics of the database as proposed by Reiter (without the specification of the active rules), and Ω is the specification of the active rules as presented before. The condition on the possibility of executing $Commit$ can be replaced by its definition (18).

   The confluence problem arises in systems in which there is no complete conflict resolution strategy. To model this situation, one would need to modify or eliminate the priorization of rules, leading to a modification of the specification of $Considered$ (13). To deal with this problem, one needs to consider all possible execution sequences, and compare the states that result from each execution sequence.

2. Given a ground user transition l_U, determine if the active rules preserve the integrity constraints at the end of the corresponding user transaction:

   $$\Sigma \cup \Omega \models \exists s \; initial\_list(l_U) \wedge s = do(l_U, S_0) \supset \\ (\exists s' \; s \leq s' \wedge legal(s') \wedge AssertionPoint(s') \wedge \\ \forall r \; \neg Poss(r, s') \supset \gamma(s')). \tag{20}$$

3. Given a ground user transition l_U and a rule R, determine if the rule will be executed along the execution of the corresponding transaction:

   $$\Sigma \cup \Omega \models \exists s \; initial\_list(l_U) \wedge s = do(l_U, S_0) \supset \\ \exists s' \; s \leq s' \wedge legal(s') \wedge Poss(\mathbf{rule_R}, s') \\ \wedge \neg \exists s'' \; s \leq s'' \leq s' \wedge Poss(Commit, s''). \tag{21}$$

Example: The following example, from [17], allows us to illustrate statement (20). It is based on the tables:

$$Emp(emp\_no, name, salary, dept\_no, s),$$
$$Dept(dept\_no, mgr\_no, s).$$

We want to specify a variation of the cascade elimination methodology for maintaining the referential integrity constraint.

In this case, the only existing rule is R: "If an employee is eliminated from Emp (the Event), then the department managed by that employee must be eliminated with all its employees (the Action)". There is no condition to be evaluated here. We expect that after executing this rule, and at the end of the user transaction, (20) will hold, that is the referential consistency will be restored.

In this particular example, the Σ specifies actions, del_emp and del_dept, that eliminate employees and departments, respectively., and are always possible. Successor state axioms for Emp and $Dept$ are also included in Σ. The specification is extended by Ω, which includes axioms for sequences, successor state axioms for the auxiliary tables, and axioms for active rules.

The event associated to rule R (the other elements of the specification given in previous sections are general) is specified by:

$$Event_R(s) \equiv (\exists\, emp\_no, name, salary, dept\_no)$$
$$V_{\neg Emp, R}(emp\_no, name, salary, dept\_no, s).$$

The transition, l_R, associated to the rule is:

$$del\_emp_R(emp\_no, name, salary, dept\_no); del\_dept_R(dept\_no, mng); End.$$

Here del_emp_R and del_dept_R are new primitive transactions associated to R. They are always possible; they have the intended effects, but only on the right conditions: they eliminate employees and departments which are associated to managers already eliminated by means of the action del_emp (what is recorded in the auxiliary tables). These new actions with these conditions have to be inserted in the SSAs for Emp and $Dept$.

If the user transaction is such that its associated transition, l_U, is:

$$Begin; del\_emp(E, N, S, D); End,$$

then this is the transition term that must be introduced in (20). The integrity constraints, $\gamma(s)$, is the conjunction of

$$Emp(emp\_no, name, salary, dept\_no, s) \supset \exists mng\, Dept(dept\_no, mng, s)$$

and

$$Dept(dept\_m, mng, s) \supset (\exists\, name\_m, salary\_m, dept\_no)$$
$$Emp(mng, name\_m, salary\_m, dept\_no, s).$$

We could prove (20) by means of derived induction principle on states that allows jumping from states at ends of user and rules transactions to new states of the same kind, because intermediate states, along transitions, may violate the integrity constraint momentarily.

7 Conclusions and Further Work

In this chapter, we have shown that it is possible to introduce the notion of active rule in the specification of the dynamics of a relational database. The specification ensures that rules will be executed in appropriate conditions and with the expected effects. In this way, the specification of the dynamics of active rules is integrated in a homogeneous and common formalism with the specification of the dynamics of the underlying tables. All the advantages of the original specifications are inherited by the extended specification, in particular, the possibility of integrating in a single formalism the usual extensional aspects of databases and their dynamic aspects; and the possibility of doing hypothetical reasoning with respect to to a virtual development of the database [4], [2], and the possibility of proving properties of the evolution of the database. For instance, satisfaction of integrity constraints [4], [5], or the satisfaction of a particular goal when active rules are executed (for example, termination).

We can see that in the context of our *Sitcalc* based specifications of database updates and active rules, these rules inherit a clear and well understood semantics. Namely, the semantics of predicate logic. New semantics are not needed.

Apart from the specification of active rules, in this chapter we specified some useful and common notions of databases, in particular, of the rollback operation.

Most of the specification we gave is domain independent and can be used with different specific cases of active rules. In this sense, the specification is modular.

Also, we considered the case of conditions in rules that are evaluated statically. Nevertheless, temporal conditions in active rules are also interesting [15], [16]. For example, a usual condition like non decreasing salaries for employees falls in this category. We can handle this case in our specification by accommodating history encoding [6] in the context of specifications of database updates, as done in [2].

Acknowledgments:

This research has been partially financed by FONDECYT (Grants 1971304 and 1980945), ECOS/CONICYT (Grant C97E05), DAAD and the Catholic University of Chile (DIPUC).

Afterword:

(By Leopoldo Bertossi): I became interested in doing research in databases after reading Ray Reiter's papers on foundations of databases. There I discovered that databases could be a fascinating area for scientific research. Having coauthored a couple of papers with Ray was also an extremely stimulating scientific experience for me. In my research activity, I have always had Ray as a model of conceptual depth, clarity and excellent scientific writing. I am very glad to have had the opportunity to interact with Ray, as a great person and scientist.

(By Javier Pinto): I met Ray Reiter in 1989 when I started working towards my Ph.D. degree at the University of Toronto. I feel now, as I felt then, that I was extremely privileged for having him as my thesis supervisor. So, it is a great pleasure to be able to contribute to this volume in honour of Ray. As acknowledged by everybody in the field, Ray is an outstanding researcher who has made central contributions to the foundations of AI. He has also contributed to the understanding of the close relationship between the fields of Knowledge Representation and Databases. I am forever in debt with Ray, and the more I know him, the more I admire him, not only as a scientist, but also as a human being.

References

1. A. Aiken, J. Widom, and J. Hellerstein. Behavior of database production rules: Termination, confluence, and observable determinism. In *Proc. ACM-SIGMOD Conference*, pages 59–68, 1992.
2. M. Arenas and L. Bertossi. Hypothetical Temporal Queries in Databases. In A. Borgida, V. Chaudhuri, and M. Staudt, editors, *Proceedings ACM SIGMOD/PODS 5th International Workshop on Knowledge Representation meets Databases (KRDB'98): Innovative Application Programming and Query Interfaces*, 1998. http://sunsite.informatik.rwth-aachen.de/Publications/CEUR-WS/Vol-10/.
3. M. Arenas and L. Bertossi. The Dynamics of Database Views. In Burkhard Freitag, Hendrik Decker, Michael Kifer, and Andrei Voronkov, editors, *Transactions and Change in Logic Databases*, volume 1472 of *LNCS*, pages 197–226. Springer-Verlag, 1998.
4. L. Bertossi, M. Arenas, and C. Ferretti. SCDBR: An Automated Reasoner for Specifications of Database Updates. *Journal of Intelligent Information Systems*, 10(3), 1998.
5. L. Bertossi, J. Pinto, P. Saez, D. Kapur, and M. Subramaniam. Automating Proofs of Integrity Constraints in the Situation Calculus. In Z. W. Ras and M. Michalewicz, editors, *Foundations of Intelligent Systems*, pages 212–222. Springer, LNAI 1079, 1996.
6. J. Chomicki. Efficient Checking of Temporal Integrity Constraints Using Bounded History Encoding. *ACM Transactions on Database Systems*, 20(2):149–186, June 1995.
7. U. Dayal, E. Hanson, and J. Widom. Active database systems. In W. Kim, editor, *Modern Database System: The Object Model, Interoperability, and Beyond*. ACM Press, 1995.
8. K. Dittrich, S. Gatziu, and A. Geppert. The active database management system manifesto: A rulebase of adbms fe atures. In *2nd Workshop on Rules in Databases (RIDS)*, Athens, Greece, 1995. Lecture Notes in Compure Science, Springer.
9. U. Jaeger and J.C. Freytag. An Annotated Bibliography on Active Databases. *SIGMOD Record*, 24(1):58–69, 1995.
10. F. Lin and R. Reiter. State Constraints Revisited. *Journal of Logic and Computation. Special Issue on Actions and Processes*, 4(5):655–678, 1994.
11. Fangzhen Lin and Raymond Reiter. State Constraints Revisited. *Journal of Logic and Computation*, 4(5):655–678, 1994. Special Issue on Actions and Processes.
12. John McCarthy. Situation, actions, and causal laws. Technical Report Memo 2, Stanford Artificial Intelligence Project, 1963.
13. Javier Pinto. Occurrences and Narratives as Constraints in the Branching Structure of the Situation Calculus. *Journal of Logic and Computation*, 8:777–808, 1998.
14. R. Reiter. On Specifying Database Updates. *Journal of Logic Programming*, 25(1):53–91, 1995.

15. A. P. Sistla and O. Wolfson. Temporal Conditions and Integrity Constraints in Active Database Systems. In *Proc. ACM SIGMOD'95*, pages 269–280, 1995.
16. A. P. Sistla and O. Wolfson. Temporal Triggers in Active Databases. *IEEE Transactions on Data and Knowledge Engineering*, 7(3):471–486, 1995.
17. J. Widom. The Starburst Active Database Rule System. *IEEE Transactions on Knowledge and Data Engineering*, 8(4):583–595, 1996.
18. Jennifer Widom and Stefano Ceri. *Active Database Systems, Triggers and Rules for Advanced Database Processing*. Morgan Kaufmann Publishers, Inc., 1996.
19. C. Zaniolo, S. Ceri, Ch. Faloutsos, R. T. Snodgrass, V.S. Subrahmanian, and R. Zicari. *Advanced Database Systems*. Morgan Kaufmann, 1997.

The Frame Problem and Bayesian Network Action Representations*

Craig Boutilier[1] and Moisés Goldszmidt[2]

[1] Department of Computer Science
University of British Columbia,
Vancouver, BC V6T 1Z4, CANADA
cebly@cs.ubc.ca
[2] SRI International
333 Ravenswood Ave.
Menlo Park, CA 94025, USA
moises@erg.sri.com

1 Introduction

Reasoning about action has been a central problem in artificial intelligence since its inception. Since the earliest attempts to formalize this problem, the straightforward encoding of actions and their effects has been fraught with difficulties, such as the frame, qualification and ramification problems. Representations such as the situation calculus [20] and STRIPS [10], as well as various methodologies for using these systems (e.g., [16], [28], [1], [25], [14]) have been proposed for dealing with such issues. However, such problems are exacerbated by considerations such as nondeterministic or stochastic action effects, the occurrence of exogenous events, incomplete or uncertain knowledge, imprecise observations, and so on.

Increasing interest in stochastic and decision theoretic planning [8], [9], with the objective of incorporating the above considerations into planning systems, requires that attention be paid to the natural and effective representation of actions with stochastic effects. A number of researchers have adopted for this purpose a very efficient approach to representing and reasoning with probability distributions, namely *Bayesian networks* (BNs) [21]. BNs provide a formal, graphical way of decomposing a state of belief by exploiting probabilistic independence relationships. BNs can also be augmented to represent *actions*, for instance, using the methods of *influence diagrams* (IDs) [27], [21],[3] or representations such as *two-stage* or *dynamic* BNs [7]. However, though considerable effort has been spent in characterizing the

---

* This article originally appeared in Springer Lecture Notes in Artificial Intelligence, LNAI 1081: G. McCalla (ed.), Advances in Artificial Intelligence, Proc. of 11th Biennial Conference of the Canadian Society for Computational Studies of Intelligence, AI'96, Toronto, Canada, May (21-24), 1996.

[3] IDs are representational tools used for optimal decision making in decision analysis. Actions are usually referred to as *decisions*, but for our purposes the two can be considered equivalent.

representational power of BNs in general, and developing good probabilistic inference algorithms that exploit the factored representations they provide, relatively little effort has been devoted to the study of the special features of action representations, especially with respect to classical problems such as the frame problem.

In this paper, we examine in detail the representation of actions in stochastic settings with respect to issues such as the frame and ramification problems (focusing primarily on the frame problem), providing some insight into how uncertain knowledge impacts the effort required to specify and represent actions. In particular, we provide a definition of (various aspects of) the frame problem in the context of dynamic BNs or IDs, proposing this as a standard against which future proposed solutions to the frame problem in stochastic environments can be measured. We also propose a methodology for representing actions in BNs in a very economical way, suggesting methods in which BNs and IDs can be augmented to exploit additional independencies (and better deal with ramifications) based on the rule structure that is taken for granted in nonprobabilistic representations. This bridges a wide gap between traditional probabilistic and nonprobabilistic approaches to action representation.

Our goal is to provide a detailed comparison of probabilistic and nonprobabilistic representations of actions, attempting to identify the key similarities and differences between these methods, and show the extent to which these different approaches can borrow techniques from one another. Space precludes a detailed survey and discussion of the work in this area and a number of interesting issues. We defer such discussion to a longer version of this paper [6], though we will point out some of these issues in the concluding section. In this paper, we concentrate on the issues of naturalness and compactness of action specification, and the frame problem in particular, focusing solely on the the situation calculus as the classical action representation, and the relatively elegant treatment of the frame problem proposed by Reiter [25]; from the probabilistic side, we deal exclusively with dynamic BNs and IDs. We emphasize several ways for augmenting dynamic BNs so that the size of representation and effort to specify the effects of actions in stochastic domains is essentially equivalent to that of Reiter's method.[4]

2 Actions: Semantics and Basic Representations

2.1 Semantics

Before presenting various representations of actions, we present the semantic model underlying these representations, namely that of *discrete transition systems*, a view

---

[4] There are a number of other representational methods that deserve analysis (e.g., the *event calculus* [17], the $\mathcal{A}$ language of [12] and its variants, probabilistic STRIPS rules [18],[2], probabilistic Horn rules [24]) which unfortunately we cannot provide here; but see the full paper [6].

common in dynamical systems and control theory, as well as computer science.[5] A transition system consists of a set of *states* S, a set of *actions* A, and a *transition relation* T. Intuitively, actions can occur (or be executed) at certain system states, causing the state to change as described by the transition relation. The exact nature of the transition relation varies with the type of system (or our knowledge of it).

A *deterministic transition system* is one where T is a (possibly partial) function $T : S \times A \to S$. If $T(s, a) = t$, then t is the *outcome* of action a applied at s; and if $T(s, a)$ is undefined, then we take a to be impossible at s. If T is a relation over $S \times A \times S$ then the system is *nondeterministic*: the *possible outcomes* of a at s are those states t such that $T(s, a, t)$ holds (if the set of outcomes is empty, a is not possible at s). Finally, a *stochastic transition function* is a function $T : S \times A \times S \to [0, 1]$; the probability of outcome of state t resulting when a is applied at s is $\Pr(s, a, t) \equiv_{df} T(s, a, t)$, the only requirement being that $\sum_t \Pr(s, a, t) = 1$ for each state s, and action a applicable at s. If a is not applicable at s, we take $\Pr(s, a, t) = 0$ for all t. We assume below that all actions can be applied, or attempted, at all states (perhaps with trivial effects). We note that this formulation assumes that the system is *Markovian*: the probability of moving to state t given a depends only on the current state s, not on past history.

The representation of actions in a transition system is relatively straightforward. In a deterministic system, each action a requires a tabular representation associating an outcome state with each state in S. A nondeterministic action can be represented in a 0-1 matrix of size $|S| \times |S|$, where a 1 in entry (i, j) indicates that state s_j is a possible outcome of a at s_i. A stochastic action can be represented by a similar *stochastic matrix*, where entry (i, j) is the probability of transition from s_i to s_j.[6]

One difficulty with the direct semantic view of actions, from the point of view of problem specification and representation, is that AI problems (e.g., planning problems) are rarely described in terms of an explicit state space. Rather one imagines a set of propositions, or predicates and domain objects, or random variables, that describe the system under investigation; and actions are viewed in terms of their effects on these propositions. This view underlies almost all work in action representation in AI. We assume that a set of propositional atoms **P** characterize the system. The set of states induced by this language consists of the set of truth assignments to **P**, each a possible configuration of the system.[7] A state space that can be factored in this way will often permit compact representation of actions, as we now explore.

2.2 Situation Calculus

The situation calculus (SC) was among the first logical formalisms for representing (deterministic) actions adopted in AI [20] and continues to be the focus of much

---

[5] Many of the ideas discussed here can be extended to continuous time, continuous state systems (see, e.g., [23] for continuous time extensions of the situation calculus, or [7] for continuous time action networks).

[6] Clearly if the branching factor b of nondeterministic or stochastic actions is small, sparse matrix methods can be used, requiring size $O(|S|b)$ representations.

[7] Multi-valued random variables are treated similarly.

research [16], [28], [1], [25], [14]. We adopt a somewhat simplified version of SC here. SC is a typed first-order language with two classes of domain objects, *states* and *actions*, a function symbol *do* mapping state-action pairs into states, and a set of unary predicate symbols, or *fluents* corresponding to the propositions of the underlying problem, that take state arguments.[8] We write $do(a, s)$ to denote the *successor state* of state s when action a is performed, and write $F(s)$ to denote that fluent F is true in state s.

SC can be used to describe the effects of actions quite compactly, in a way that exploits regularities in the effects actions have on particular propositions. A typical *effect axiom* is:

$$\forall s \; holding(s) \land fragile(s) \supset broken(do(drop, s)) \land \neg holding(do(drop, s)) \quad (1)$$

which states that *broken* holds (e.g., of some object of interest), and *holding* doesn't, in the state that results from performing the action *drop* if it was held in state s and is fragile. Because of the Markovian assumption in our semantics, we assume that the only state term occurring in the antecedent is a unique state variable (e.g., s) and that each state term in the consequent has the form $do(a, s)$ for some action term a. Note that Axiom (6) describes a property of a large number of state transitions quite concisely; however, it does not uniquely determine the transitions induced by the action *drop*, a point to which we return below. Furthermore, it is a natural description of (some of) the effects of dropping an object.

2.3   Dynamic Bayesian Networks and Influence Diagrams

In a stochastic setting, the effect of an action a at a given state s_i determines a probability distribution over possible resulting states. With respect to the representation discussed in Section 2.1, row i of the stochastic matrix for a is the (conditional) distribution over the resulting states when action a is executed, given that s_i was the initial state. Given that states can be factored propositionally, and this distribution is in fact a joint distribution over **P**, we would like to employ a representation that takes advantage of this factorization. Bayesian networks (BNs) [21] are one such representation.

A BN is a directed acyclic graph (dag), where nodes represent the random variables of interest (in this case the fluents) and the arcs represent direct influences or dependencies between random variables. A BN encodes the following assumption of probabilistic independence: any node in a BN is probabilistically independent of its non-descendants, given the state of its parents in the network.[9] A BN can compactly capture a distribution P by representing independence relationships in its graphical structure. To represent the distribution, we annotate each node in the

---

[8] The restriction to unary predicates means that the underlying domain is described using propositions rather than predicates itself. We adopt this merely for simplicity of exposition — rarely is the assumption made in practice.

[9] The details of algorithms for testing independence is beyond the scope of this paper. We refer to [21][Chapter 3] for details.

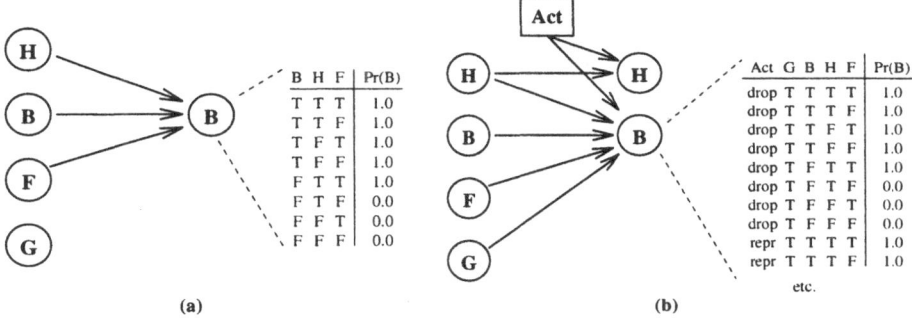

Fig. 1. Bayes Nets: (a) Without Action Node; (b) With Action Node

network with a *conditional probability table* (CPT) denoting the conditional distribution of each variable given its parents in the network. Given the assumptions of independence, any state of belief can be simply computed by multiplying entries in these tables.

The BN (or fragment of a BN) corresponding to Axiom (6) is pictured in Figure 1(a). In general, to represent the effect of action a, we have a set of variables corresponding to certain fluents in the state prior to the performance of the action, and variables corresponding to fluents after the action. In our example, the B on the left denotes the proposition *broken(s)*, while the B on the right denotes *broken(do(drop, s))*. The arcs indicate that $\Pr(broken(do(drop, s)))$ depends on the truth or falsity of *broken*, *holding* and *fragile* in state s, but does not depend on the value of a fourth fluent G (*hasglue*), nor on the values of fluents in state $do(drop, s))$ given that one knows their values in s. The CPT on the right denotes the magnitude of this influence on *broken*: for each assignment to its parents, the probability that *broken(do(drop, s))* is true must be specified (here the effect is deterministic).

Several remarks are in order at this point. First, we do not require prior probabilities on the "pre-action" nodes in the network. If they existed, we could use standard techniques to determine the probability of any post-action node. Thus, this network does not represent a complete probability distribution. The intent is to represent the stochastic matrix for a given action; therefore the network is schematic in the sense that it describes the effect of the action for *any* state, or assignment to pre-action variables. This coincides with the classical view of action representation. Second, such a network is sometimes called a *dynamic BN* or *two-stage BN* [7], since it should be viewed as schematic across time. The restriction to two stages (corresponding to states s and $do(a, s)$) is appropriate given our semantics. We also point out that, as with our description of the situation calculus, such a BN does not uniquely specify a transition matrix for action a (more in the next section).

Finally, *influence diagrams* (IDs) [13], [27] have been used in probabilistic inference and decision analysis to represent decision problems. Similar in structure to BNs, they have additional types of variables, represented by *value nodes* and *decision nodes*; we are only interested in decision nodes here. A decision node (or *action node*) is a random variable denoting the action that was executed causing the state transition. This is denoted by the square node in Figure 1(b), a variable *Act* that can take on possible actions as values (in this case, *repair, drop*). Since the state of a fluent after an action is executed will depend on the action chosen, the additional dependencies between *Act* and the other fluents must be represented. We examine the impact of these additional influences below.

3   Single Actions: Structure and the Frame Problem

As alluded to above, neither the SC nor BN description of action effects will generally provide a complete and unambiguous specification of a transition system. We focus first on the SC formulation, and compare the usual BN methods.

3.1   Frame Axioms in the Situation Calculus

There are two difficulties with Axiom (6) as a specification of the action *drop*. The first is that, while it describes the effect of *drop* on the fluents *holding* and *broken* (under some conditions), it fails to describe its effect on other fluents. For example, to completely specify the transition function for *drop*, one must also assert the effect *drop* has on other fluents in the domain (such as *hasglue*). Unfortunately, there are a typically a great many fluents that are completely unaffected by any given action, and that we do not consider part of the natural specification of an action's effects. Intuitively, we would like the user to specify how the action influences affected fluents, and assume that other fluents persist. We call this the problem of *persistence of unaffected fluents* (PUF). A second difficulty is that while Axiom (6) describes the effect of *drop* on *holding* and *broken* under the condition $holding(s) \land fragile(s)$, it fails to specify what happens when this condition is false. Once again, it is usually taken to be desirable not to force the user to have to say a fluent is unaffected in other circumstances, leaving it as a tacit assumption. We call this the problem of *persistence of affected fluents* (PAF).

The *frame problem* [20] is that of easing the user from the burden of having to specify conditions under which an action does not affect a fluent: PUF and PAF are two instances of this problem. One possible solution is to provide a means to automatically derive explicit *frame axioms* given the user's specification. For example, given the input

$$\forall s \;\; holding(s) \land fragile(s) \supset broken(do(drop, s)) \land \neg holding(do(drop, s)) \quad (2)$$

$$\forall s \;\; holding(s) \land \neg fragile(s) \supset \neg holding(do(drop, s)) \quad\quad\quad (3)$$

one could, under the assumption that this describes *all* effects of *drop*, generate axioms such as

$$\forall s \;\; (\neg holding(s) \lor \neg fragile(s)) \land broken(s) \supset broken(do(drop, s)) \quad (4)$$

$$\forall s \;\; (\neg holding(s) \lor \neg fragile(s)) \land \neg broken(s) \supset \neg broken(do(drop, s)) \quad (5)$$

$$\forall s \;\; hasglue(s) \supset hasglue(do(drop, s)) \quad\quad\quad\quad\quad\quad (6)$$

$$\forall s \;\; \neg hasglue(s) \supset \neg hasglue(do(drop, s)) \quad\quad\quad\quad\quad (7)$$

Axioms (4) and (5) deal with the PAF problem, while axioms (6) and (7) handle PUF. In general, we require $2|\mathbf{P}|$ such frame axioms, describing the lack of effect of a on each fluent in $\mathbf{P}$; or $2|\mathbf{P}||\mathcal{A}|$ such axioms for the entire set of action $\mathcal{A}$.

Other approaches deal not just with the specification problem, but also with the sheer number of axioms required. One example is the solution proposed by Reiter [25], extending the work of Pednault [22] and Schubert [26]. The aim is to directly encode the "assumption" that all conditions under which an action affects a fluent have been listed. This is accomplished by building a disjunction of all the conditions under which an action A affects a fluent F, asserting the F changes as dictated when these conditions hold, and that it retains its value otherwise. More precisely, let $\gamma_{F,A}^{+}(s)$ (resp., $\gamma_{F,A}^{-}(s)$) denote the disjunction of the antecedents of the effect axioms for action A in which F appears positively (resp., negatively) in the consequent. We know that

$$\forall s \; \gamma_{F,A}^{+}(s) \supset F(do(A, s)) \quad \text{and} \quad \forall s \; \gamma_{F,A}^{-}(s) \supset \neg F(do(A, s))$$

both follow from the action specification. Under natural conditions, we can ensure the persistence of F under all other conditions by writing:

$$\forall s \; F(do(A, s)) \equiv \gamma_{F,A}^{+}(s) \vee (F(s) \wedge \neg \gamma_{F,A}^{-}(s))$$

If we assert one such axiom for each fluent F, it is not hard to see that we uniquely determine a (deterministic) transition function for action A over the state space. In our example, these *closure axioms* for action *drop* (with some simplification) include:

$$\forall s \; holding(do(drop, s)) \equiv \bot \tag{8}$$

$$\forall s \; broken(do(drop, s)) \equiv [(holding(s) \wedge fragile(s)) \vee broken(s)] \tag{9}$$

$$\forall s \; hasglue(do(drop, s)) \equiv hasglue(s) \tag{10}$$

We require $|\mathbf{P}|$ axioms to characterize an action in this way. This is not a substantial saving over the use of explicit frame axioms, for we require $|\mathcal{A}|$ such axiom sets (one per action). However, as we see below, Reiter's method avoids the repetition of these axioms in multiple action settings. The size of the axioms is also of interest. Imagine some (presumably small) number f of fluents is affected by a, that the average action condition has c conjuncts, and that affected fluents appear in an average of e effect axioms. Then we can expect the f closure axioms for affected fluents to be roughly of size ce, while the remaining $|\mathbf{P}| - f$ axioms are of constant size.

3.2   The Frame Problem in Bayesian Networks

The BNs shown in Figure 1 do not completely characterize a transition system for the same reasons described for SC above. Unlike the work in classical reasoning about action, the use of BNs for representing dynamical systems typically assumes

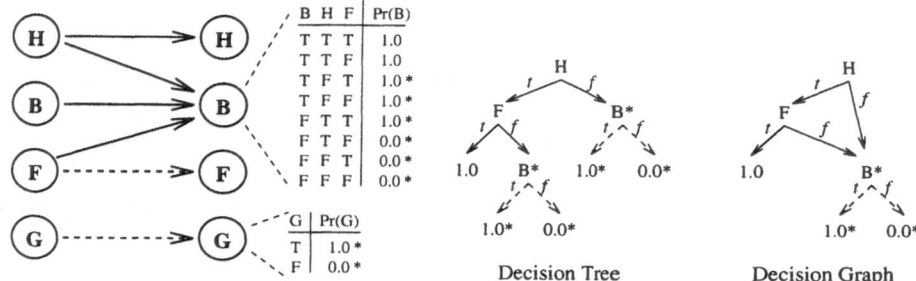

Fig. 2. (a) A Dynamic BN, and (b) Compact Representation of the CPT

that a complete specification of the action's effects, on all variables and under all conditions, is given. A (complete) dynamic BN for the action *drop* is shown in Figure 2(a) (we have left out CPTs for H, F).[10]

Most BN models of action assume this network is explicitly specified by the user; however we have used broken arcs in Figure 2(a) to indicate persistence relations among unaffected fluents (the CPT for node G shows persistence explicitly). Furthermore, the starred entries in the CPT for node B denote the persistence of fluent *broken* under the condition ¬*holding* ∨ ¬*fragile*. The fact that a user must specify these persistence relationships explicitly is the obvious counterpart of the PUF and PAF problems in the situation calculus. Therefore, we can take the *frame problem for Bayesian networks* to be exactly the need to make these relationships explicit.

As described above, there are two possible perspectives on what constitutes a solution to this problem: relieving the user of this burden, and minimizing the size of the representation of such persistence. The first type of solution is not especially hard to deal with in this representation. A rather simple idea is to have the user specify only the unbroken arcs in the network and only those unhighlighted probabilities in the CPTs. It is a simple matter to then automatically add persistence relationships.

3.3 Adding Further Structure to Bayesian Networks

The size of the dynamic BN for an action (regardless whether the persistence relations are generated automatically) is then comparable to that of Reiter's solution (with one substantial proviso). Again, assume that f fluents are affected by a, and that the average fluent is influenced by c preaction fluents. The CPTs for each of the f affected fluents will be of size 2^c, whereas the remaining $|\mathbf{P}| - f$ CPTs are of constant size. The important factor in the comparison to the Reiter's solution is the difference in the affected fluent representation, with this method requiring a representation of size roughly $f \cdot 2^k$, while SC requires a representation of size fce. Here k is the number of *relevant* preaction fluents for a typical affected fluent F, those that are part of *some* condition that influences the action's effect on F (i.e.,

---

[10] Although not used yet, arcs are allowed between post-action variables (see below, on ramifications).

the number of parents of F in the BN). Note that $k \leq ce$. The exponential term for the BN formulation is due to the fact that CPTs require a distribution for the affected fluent for *each assignment to its parents*. For instance, since B, F and H are relevant to predicting the effect of *drop* on *broken*, we must specify this effect for all eight assignments to $\{B, F, H\}$.

Axiom (2), and more significantly the closure Axiom (9), are much more compact, requiring only that the single positive effect condition be specified, with persistence under other circumstances being verified automatically. CPTs in BNs are *unstructured*, and fail to capture the regularities in the action effects that fall out naturally in SC. Only recently there is work attempting to represent regularities in BN matrices [11], and in particular, structures such as logical formulae, decision trees and rooted decision graphs have been explored as compact representations for matrices [3], [5]. Examples of representations that capture this regularity are the decision tree and decision graph shown in Figure 2(b), corresponding to the original CPT for variable B. The broken arrows indicate the persistence relationships that can be added automatically when left unspecified by the user.

In general, it will be hard to compare the relative sizes of different representations, and the logic-based method of Reiter, since this depends crucially on exact logical form of the action conditions involved. For instance, decision trees can be used to represent certain logical distinctions very compactly, but others can require trees of size exponentially greater than a corresponding set of logical formulae. However, one can also use graphs, logical formulae and other representations for CPTs–each has particular advantages and disadvantages with respect to size and speed of inference [5]. For example, Poole [24] has used Horn rules with probabilities as a way of representing Bayes nets. In general, we can see that appropriate representations of CPTs can exploit the same regularities as logical formulae; thus BNs augmented in this way are of comparable size to Reiter's model.

4   Multiple Actions: The Frame and Ramification Problems

4.1   Compact Solutions to the Frame Problem

While the proposals above for individual actions relieve the user from explicitly specifying persistence, it did little to reduce the size of the action representation (compared to having explicit frame axioms). In Reiter's method, each action requires $|\mathbf{P}|$ axioms, with f axioms having size ce and $|\mathbf{P}| - f$ axioms having constant size. To represent a transition system with action set $\mathcal{A}$ thus requires $|\mathcal{A}||\mathbf{P}|$ axioms. This is only a factor of 2 better than using explicit frame axioms.

Fortunately, Reiter's solution is designed with multiple actions in mind. Reiter exploits the fact that, since they are terms in SC, one can quantify over actions. His procedure will (under reasonable conditions) produce one axiom of the form

$$\forall s, a \ \ F(do(a, s)) \equiv \gamma_F^+(a, s) \vee (F(s) \wedge \neg\gamma_F^-(a, s))$$

for each fluent F. Here $\gamma_F^+(a, s)$ denotes the disjunction of the formulae $a = A \wedge \gamma_{F,A}^+(s)$ for each specific action A which affects fluent F positively (similar remarks

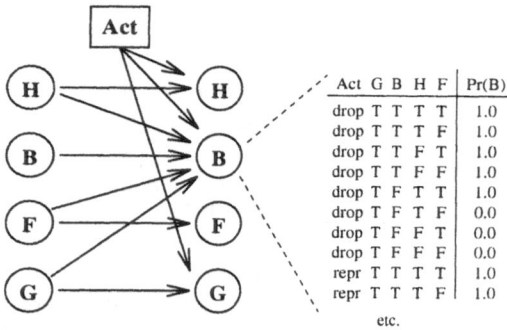

The table in the figure:

Act	G	B	H	F	Pr(B)
drop	T	T	T	T	1.0
drop	T	T	T	F	1.0
drop	T	T	F	T	1.0
drop	T	T	F	F	1.0
drop	T	F	T	T	1.0
drop	T	F	T	F	0.0
drop	T	F	F	T	0.0
drop	T	F	F	F	0.0
repr	T	T	T	T	1.0
repr	T	T	T	F	1.0

etc.

Fig. 3. A Dynamic BN with Action Node

apply to $\gamma_F^-(a, s)$). Thus, we see instead of having $|\mathcal{A}||\mathbf{P}|$ axioms, we have only $|\mathbf{P}|$ axioms, and the axiom for fluent F contains only reference to actions that influence it. If each fluent is affected by n actions (presumably n is much smaller than $|\mathcal{A}|$), each action condition has c conjuncts, and each affected fluents appear in e effect axioms for any action, then we expect this specification of the transition system to be of size $|\mathbf{P}|nce$. In our example, the axiom for fluent *broken* is:

$$\forall s \quad broken(do(a, s)) \equiv [(a = drop \land holding(s) \land fragile(s))$$
$$\lor(\neg(a = repair \land holding(s) \land fragile(s)) \land broken(s))] \tag{11}$$

Notice that actions like *paint* and *move* that have no influence (positive or negative) on *broken* are not mentioned.

Similar problems plague BNs when we consider the size of the representation of all actions in a transition system. Having a separate network for each action will require a representation of size $|\mathcal{A}|(2(|\mathbf{P}| - f) + f \cdot 2^k)$ (where f is the expected number of fluents affected by an action, k the number of conditions relevant to each affected fluent). The usual way to represent a transition system is to use an action node and condition each post-action variable on the action variable (we assume every variable can be influenced by some action). Figure 3 shows our example in this format.

The difficulty with this representation is the fact that, since (most or) every fluent is affected by some action, the action node becomes a parent of each post-action node, increasing the size of each CPT by a factor of $|\mathcal{A}|$ (as if we had a separate network for each action). Indeed, this representation, standard in decision analysis, is in fact worse, because any preaction variable that influences a post-action variable under *any* action must be a parent. Since the size of the representation is exponential in the number of parents, this will virtually always be a substantially less compact representation than that required by a set of $|\mathcal{A}|$ action networks. In general, it will have size $|\mathcal{A}||\mathbf{P}|2^m$, where m is the expected number of fluents that are relevant to a post-action node under *any* action (typically much larger than k above). In this example, G is a parent of B (because it is relevant for *repair*), and its value must

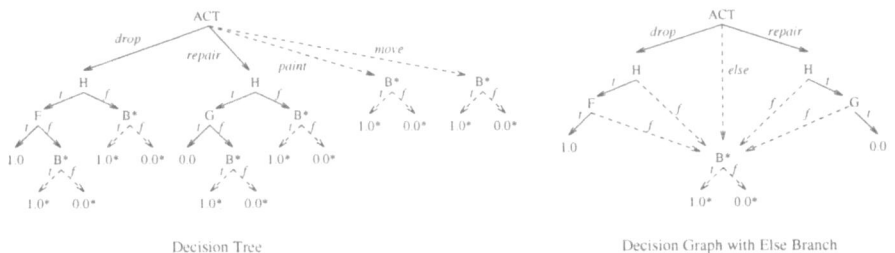

Fig. 4. Compact Representation of *broken* CPT

be considered even when we specify the effect of *drop* (for which it is actually not relevant).

In order to alleviate this problem we can use compact representations such as decision trees or decision graphs as shown in Figure 4. The broken arcs and marked nodes represent persistence relations that can be generated automatically if unspecified by the user. The tree representation provides some savings, but requires that each fluent have a separate subtree for each possible action, failing to alleviate the $|\mathbf{P}||\mathcal{A}|$ factor in the size of the representation. However, a decision graph can provide considerable savings, as shown, especially if we allow an "else" branch to exit the *Act* node (see Figure 4). This technique allows one to specify only the actions that actually influence a fluent, and only the conditions under which that effect is nontrivial, leaving the persistence relation to be filled in automatically on the "else" branch (all other actions) and for unspecified conditions for the mentioned actions. Using a graph rather than a tree means that all unfilled branches will connect to the (single) persistence subgraph. It is not hard to see that (assuming conjunctive action conditions) that the size of this representation will be exactly the same as that proposed by Reiter.

Another compact representation of multiple actions that does not require the use of else branches involves the use of proxy nodes that we called "occurrence variables". Occurrence variables are new nodes corresponding to the proposition that a given action has taken place (see Figure 5). Instead to being directly connected to the action node, each fluent has as parents the occurrence variables for actions that directly affect them. The occurrence variables signify that a particular A has "occurred" and are children of the action (or decision) node. Assuming that exactly one occurrence variable is true at any time, this representation permits CPTs to be of roughly the same size those using graphs with else branches, and more compact than graphs without else branches.

4.2 The Ramification Problem

To this point, we have ignored the issue of ramifications. We do not provide a full discussion of this problem here, but we mention the main issues. Ramifications or *domain constraints* are *synchronic* constraints on possible state configurations (as

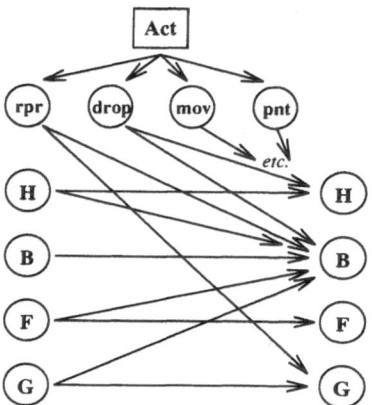

Fig. 5. The Use of Occurrence Variables

opposed to *diachronic* constraints – of the type we've discussed – that relate features of one state to features of its successor). When representing multiple actions, domain constraints allows compact specification of regular effects that are independent of the action being performed. For instance, if the location of the contents of a briefcase are always the same as the location of the briefcase itself, a statement to this effect relieves one from explicitly stating that every action a that changes the location of the briefcase also changes the location of it contents. We would like this effect (or ramification) of moving, dropping, throwing or otherwise dislocating a briefcase to be derived from the effect these actions has on the briefcase itself.

In SC, domain constraints can easily be expressed; for instance:

$$\forall s \ in(x, b, s)) \supset loc(x, s) = loc(b, x) \tag{12}$$

The ramification problem has to do with the interaction of such constraints with possible solutions to the frame problem. Solutions have been proposed by Lin and Reiter [19], Kartha and Lifschitz [15] among others.

A domain constraint is represented in a BN as an arc between two post-action variables, representing the dependency between two fluents in a single state. Note that the constraints imposed by the limited language in a BN, plus the restriction on the acyclicity of the underlying graphs limits some of the problems of including ramifications. In particular we only have to worry about modifying the specification of an action whenever a synchronic fluent becomes a new parent. Still the problem is similar to the case of SC: how to specify such domain constraints independently of a particular action, and then impose these relationships on the action network(s) in such a way that automatically derived persistence relationships account for these ramifications directly. The general ideas used for SC solutions (especially those based on "compiling" ramifications into effect axioms automatically [19]) can be applied to the BN case and are discussed elsewhere.

5 Concluding Remarks

In this paper we have taken a close look at representational issues in encoding actions in stochastic domains using Bayesian networks. In the process we defined two aspects of the frame problem for BNs, proposed possible solutions, and compared these to Reiter's method for SC. We have demonstrated that the usual models of stochastic actions have not provided the compact or natural specification methodology provided in the classical/logical setting; but that the use of "automatic" filling in of persistence relationships and the use of the compact CPT representations recently adopted in [3],[5] allow solutions to the frame problem of similar size, structure and timbre to Reiter's. In this sense, we have proposed a starting point for a methodology for the natural, compact representation of actions in BNs. The advantages of BNs as representations for probabilistic actions will be enhanced by the incorporation of such techniques.

An important issue that deserves further attention is that of nondeterminism and correlated effects. BNs are designed to take exploit independence in action effects, since this is the only way to have compact representation when effects are possibly correlated, as they can be when any sort of nondeterminism comes into play. Thus, while representing ramifications as synchronic constraints are a convenience in deterministic settings (one can do without these at the expense of additional axioms), they *must* be used in any nondeterministic representation where correlated effects are possible. Thus, the methodology reflected in BNs has an important role to play in informing the extension of classical representations such as SC to handle nondeterministic actions.[11]

Finally, we shouldn't forget that BNs are designed to facilitate efficient inference. The use of dynamic BNs in inference is very flexible–standard BN algorithms can be used to answer queries with respect to temporal projection and explanation, for action sequences of arbitrary (finite) length, and can be used for plan generation [3],[27]. In this regard, an important question is whether the compact representations of actions proposed here can enhance computation for probabilistic queries about the effects of actions, temporal projection, and planning in stochastic domains. Certain aspects of this question are investigated in [5] (in the context of BNs without actions) and [3] (with dynamic BNs).

Acknowledgements

Thanks to David Poole, Nir Friedman and especially all the participants of the 1995 AAAI Spring Symposium on Extending Theories of Action for their spirited and enjoyable discussion of these topics. Craig Boutilier acknowledges the support of NSERC Research Grant OGP0121843 and NCE IRIS-II Program IC-7. The work of Moisés Goldszmidt was funded in part by ARPA contract F30602-95-C-0251 and

---

[11] Nondeterministic actions are addressed in, e.g., [15]; the difficulties with correlations are addressed in [4].

was undertaken while the author was at the Rockwell International Science Center, Palo Alto.

References

1. Andrew B. Baker. Nonmonotonic reasoning in the framework of the situation calculus. *Artificial Intelligence*, 49:5–23, 1991.
2. Craig Boutilier and Richard Dearden. Using abstractions for decision-theoretic planning with time constraints. In *Proceedings of the Twelfth National Conference on Artificial Intelligence*, pages 1016–1022, Seattle, 1994.
3. Craig Boutilier, Richard Dearden, and Moisés Goldszmidt. Exploiting structure in policy construction. In *Proceedings of the Fourteenth International Joint Conference on Artificial Intelligence*, pages 1104–1111, Montreal, 1995.
4. Craig Boutilier and Nir Friedman. Nondeterministic actions and the frame problem. In *AAAI Spring Symposium on Extending Theories of Action: Formal Theory and Practical Applications*, pages 39–44, Stanford, 1995.
5. Craig Boutilier, Nir Friedman, Moisés Goldszmidt, and Daphne Koller. Context-specific independence in Bayesian networks. In *Proceedings of the Twelfth Conference on Uncertainty in Artificial Intelligence*, pages 115–123, Portland, OR, 1996.
6. Craig Boutilier and Moisés Goldszmidt. The frame problem and Bayesian network action representations. In *Proceedings of the Eleventh Biennial Canadian Conference on Artificial Intelligence*, pages 69–83, Toronto, 1996.
7. Thomas Dean and Keiji Kanazawa. A model for reasoning about persistence and causation. *Computational Intelligence*, 5(3):142–150, 1989.
8. Thomas Dean and Michael Wellman. *Planning and Control*. Morgan Kaufmann, San Mateo, 1991.
9. Steve Hanks (ed.). Decision theoretic planning: Proceedings of the aaai spring symposium. Technical Report SS-94-06, AAAI Press, Menlo Park, 1994.
10. Richard E. Fikes and Nils J. Nilsson. STRIPS: A new approach to the application of theorem proving to problem solving. *Artificial Intelligence*, 2:189–208, 1971.
11. Dan Geiger and David Heckerman. Advances in probabilistic reasoning. In *Proceedings of the Seventh Conference on Uncertainty in Artificial Intelligence*, pages 118–126, Los Angeles, 1991.
12. Michael Gelfond and Vladimir Lifschitz. Representing actions in extended logic programming. In K. Apt, editor, *Proceedings of the Tenth Conference on Logic Programming*, pages 559–573, 1992.
13. Ronald A. Howard and James E. Matheson, editors. *Readings on the Principles and Applications of Decision Analysis*. Strategic Decision Group, Menlo Park, CA, 1984.
14. G. Neelakantan Kartha. Two counterexamples related to Baker's approach to the frame problem. *Artificial Intelligence*, 69:379–392, 1994.
15. G. Neelakantan Kartha and Vladimir Lifschitz. Actions with indirect effects (preliminary report). In *Proceedings of the Fifth International Conference on Principles of Knowledge Representation and Reasoning*, pages 341–350, Bonn, 1994.
16. Henry A. Kautz. The logic of persistence. In *Proceedings of the Fifth National Conference on Artificial Intelligence*, pages 401–405, Philadelphia, 1986.
17. R. Kowalski and M. Sergot. A logic-based calculus of events. *New Generation Computing*, 4(1):67–95, 1986.

18. Nicholas Kushmerick, Steve Hanks, and Daniel Weld. An algorithm for probabilistic least-commitment planning. In *Proceedings of the Twelfth National Conference on Artificial Intelligence*, pages 1073–1078, Seattle, 1994.
19. Fangzhen Lin and Ray Reiter. State constraints revisited. *Journal of Logic and Computation*, 4(5):655–678, 1994.
20. John McCarthy and P.J. Hayes. Some philosophical problems from the standpoint of artificial intelligence. *Machine Intelligence*, 4:463–502, 1969.
21. Judea Pearl. *Probabilistic Reasoning in Intelligent Systems: Networks of Plausible Inference*. Morgan Kaufmann, San Mateo, 1988.
22. Edwin Pednault. ADL: Exploring the middle ground between STRIPS and the situation calculus. In *Proceedings of the First International Conference on Principles of Knowledge Representation and Reasoning*, pages 324–332, Toronto, 1989.
23. Javier A. Pinto. *Temporal Reasoning in the Situation Calculus*. PhD thesis, University of Toronto, 1994.
24. David Poole. Probabilistic Horn abduction and Bayesian networks. *Artificial Intelligence*, 64(1):81–129, 1993.
25. Raymond Reiter. The frame problem in the situation calculus: A simple solution (sometimes) and a completeness result for goal regression. In V. Lifschitz, editor, *Artificial Intelligence and Mathematical Theory of Computation (Papers in Honor of John McCarthy)*, pages 359–380. Academic Press, San Diego, 1991.
26. Lenhart K. Schubert. Monotonic solution of the frame problem in the situation calculus: An efficient method for worlds with fully specified actions. In H. E. Kyburg, R. P. Loui, and G. N. Carlson, editors, *Knowledge Representation and Defeasible Reasoning*, pages 23–67. Kluwer, Boston, 1990.
27. Ross D. Shachter. Evaluating influence diagrams. *Operations Research*, 33(6):871–882, 1986.
28. Yoav Shoham. *Reasoning About Change: Time and Causation from the Standpoint of Artificial Intelligence*. MIT Press, Cambridge, 1988.

Philosophical and Scientific Presuppositions of Logical AI

John McCarthy

Department of Computer Science, Stanford University, Stanford, CA
jmc@steam.stanford.edu

> Extinguished theologians lie about the cradle of every
> science as the strangled snakes beside that of Hercules.
> –T. H. Huxley[1]

Summary. Philosophers have done interesting work that may be useful for AI. However, some philosophical points of view make assumptions that have the effect of excluding the possibility of AI. Likewise work on AI is not neutral with regard to philosophical issues. This chapter presents what we consider the presuppositions of logical AI and also some scientific presuppositions, i.e. some results of science that are relevant. We emphasize the relation to AI rather than philosophy itself [2].

1 Philosophical Presuppositions

Q. Why bother stating philosophical presuppositions? Why not just get on with the AI?

A. AI shares many concerns with philosophy–with metaphysics, epistemology, philosophy of mind and other branches of philosophy. This is because AI concerns the creation of an artificial mind. However, AI has to treat these questions in more detail.

AI research not based on stated philosophical presuppositions usually turns out to be based on unstated philosophical presuppositions. These are often so wrong as to interfere with developing intelligent systems.

That it should be possible to make machines as intelligent as humans involves some philosophical premises, although the possibility is probably accepted by a majority of philosophers. The way we propose to build intelligent machines makes more presuppositions, some of which may be new.

This section concentrates on stating the premises without much argument. A later chapter[3] presents arguments and discusses other opinions.

---

[1] Progress in AI may extinguish some philosophies, but don't stand on one foot.

[2] This is a draft of a chapter from a forthcoming book about logical AI. The presuppositions are set forth rather baldly, but it isn't presumed that no arguments for them are required. These are projected for a more polemical chapter that will include philosophical considerations not tied to AI and will discuss other points of view.

[3] in the forthcoming book

objective world The world exists independently of humans. The facts of mathematics and physical science are independent of there being people to know them. Intelligent Martians and robots will need to know the same facts.

A robot also needs to believe that the world exists independently of itself. Science tells us that humans evolved in a world which formerly did not contain humans. Given this, it is odd to regard the world as a human construct. It is even more odd to program a robot to regard the world as its own construct. What the robot believes about the world in general doesn't arise for the limited robots of today, because the languages they are programmed to use can't express assertions about the world in general. This limits what they can learn or can be told–and hence what we can get them to do for us.

correspondence theory of truth and reference A logical robot represents what it *believes* about the world by logical sentences. Some of these beliefs we build in; others come from its observations and still others by induction from its experience. Within the sentences it uses *terms* to refer to objects in the world.

In every case, we try to design it so that what it will believe about the world is as accurate as possible, though not usually as detailed as possible. Debugging and improving the robot includes detecting false beliefs about the world and changing the way it acquires information to maximize the correspondence between what it believes and the facts of world. The terms the robot uses to refer to entities need to correspond to the entities so that the sentences will express facts about these entities. We have in mind both material objects and other entities, e.g. plans.

Already this involves a philosophical presupposition–that which is called the *correspondence theory of truth*. AI also needs a *correspondence theory of reference* , i.e. that a mental structure can refer to an external object and can be judged by the accuracy of the reference.

As with science, a robot's theories are tested experimentally, but the concepts robots use are often not defined in terms of experiments. Their properties are partially axiomatized, and some axioms relate terms to observations.

The important consequence of the correspondence theory is that when we design robots, we need to keep in mind the relation between *appearance*, the information coming through the robot's sensors, and *reality*. Only in certain simple cases, e.g. the position in a chess game, does the robot have sufficient access to reality for this distinction to be ignored.

Some robots react directly to their inputs without memory or inferences. It is our scientific (i.e. not philosophical) contention that these are inadequate for human-level intelligence, because the world contains too many important entities that cannot be observed directly.

A robot that reasons about the acquisition of information must itself be aware of these relations. In order that a robot should not always believe what it sees with its own eyes, it must distinguish between appearance and reality.

science Science is substantially correct in what it tells us about the world, and scientific activity is the best way to obtain more knowledge. 20th century corrections to scientific knowledge mostly left the old theories as good approximations to reality.

mind and brain The human mind is an activity of the human brain. This is a scientific proposition, supported by all the evidence science has discovered so far. However, the dualist intuition of separation between mind and body is related to the sometimes weak connections between thought and action. Dualism has some use as a psychological abstraction.

common sense Common sense ways of perceiving the world and common opinion are also substantially correct. When general common sense errs, it can often be corrected by science, and the results of the correction may become part of common sense if they are not too mathematical. Thus common sense has absorbed the notion of inertia. However, its mathematical generalization, the law of conservation of momentum has made its way into the common sense of only a small fraction of people–even among the people who have taken courses in physics.

From Socrates on philosophers have found many inadequacies in common sense usage, e.g. common sense notions of the meanings of words. The corrections are often elaborations, making distinctions blurred in common sense usage. Unfortunately, there is no end to philosophical elaboration, and the theories become very complex. However, some of the elaborations seem essential to avoid confusion in some circumstances. Here's a candidate for the way out of the maze.

Robots will need both the simplest common sense usages and to be able to tolerate elaborations when required. For this we have proposed two notions– contexts as formal objects [6] and [9] and *elaboration tolerance* [8][4]

science embedded in common sense Science is embedded in common sense. Galileo taught us that the distance s that a dropped body falls in time t is given by the formula

$$s = \frac{1}{2}gt^2.$$

---

[4] Hilary Putnam [12] discusses two notions concerning meaning proposed by previous philosophers which he finds inadequate. These are

(I) That knowing the meaning of a term is just a matter of being in a certain "psychological state" (in the sense of "psychological state" in which states of memory and psychological dispositions are "psychological states"; no one thought that knowing the meaning of a word was a continuous state of consciousness, of course.)

(II) That the meaning of a term (in the sense of "intension") determines its extension (in the sense that sameness of intension entails sameness of extension).

Suppose Putnam is right in his criticism of the general correctness of (I) and (II). His own ideas are more elaborate.

It may be convenient for a robot to work mostly in contexts within a larger context C_{phil1} in which (I) and (II) (or something even simpler) hold. However, the same robot, if it is to have human level intelligence, must be able to *transcend* C_{phil1} when it has to work in contexts to which Putnam's criticisms of the assumptions of C_{phil1} apply.

It is interesting, but perhaps not necessary for AI at first, to characterize those contexts in which (I) and (II) are correct.

To use this information, the English (or its logical equivalent) is just as essential as the formula, and common sense knowledge of the world is required to make the measurements required to use or verify the formula.

possibility of AI According to some philosophers' views, artificial intelligence is either a contradiction in terms [14] or intrinsically impossible [3] or [10]. The methodological basis of these arguments has to be wrong and not just the arguments themselves. We hope to deal with this elsewhere.

mental qualities treated individually AI has to treat mind in terms of components rather than regarding mind as a unit that necessarily has all the mental features that occur in humans. Thus we design some very simple systems in terms of the beliefs we want them to have and debug them by identifying erroneous beliefs. [4] treats this. Ascribing a few beliefs to thermostats has led to controversy.

third person point of view We ask "How does it (or he) know?", "What does it perceive?" rather than how do I know and what do I perceive. This presupposes the correspondence theory of truth. It applies to how we look at robots, but also to how we want robots to reason about the knowledge of people and other robots.

rich ontology Our theories involve many kinds of entity–material objects, situations, properties as objects, contexts, propositions, indivdual concepts, wishes, intentions. When one kind A of entity might be defined in terms of others, we will often prefer to treat A separately, because we may later want to change our ideas of its relation to other entities.

We often consider several related concepts, where others have tried to get by with one. Suppose a man sees a dog. Is seeing a relation between the man and the dog or a relation between the man and an appearance of a dog? Some purport to refute calling seeing a relation between the man and the dog by pointing out that the man may actually see a hologram or picture of the dog. AI needs the relation between the man and the appearance of a dog, the relation between the man and the dog and also the relation between dogs and appearances of them. None is most fundamental.

natural kinds The entities the robot must refer to often are *rich* with properties the robot cannot know all about. The best example is a *natural kind* like a lemon. A child buying a lemon at a store knows enough properties of the lemons that occur in the stores he frequents to distinguish lemons from other fruits in the store. Experts know more properties of lemons, but no-one knows all of them. AI systems also have to distinguish between sets of properties that suffice to recognize an object in particular situations and the natural kinds of some objects.

Curiously enough, many of the notions studied in philosophy are not natural kinds, e.g. proposition, meaning, necessity. When they are regarded as natural kinds, then fruitless arguments about what they really are take place. AI needs these concepts but must be able to work with limited notions of them.

approximate entities Many of the philosophical arguments purporting to show that naive common sense is hopelessly mistaken are wrong. These arguments often stem from trying to force intrinsically approximate concepts into the form of if-and-only-if definitions.

Our emphasis on the first class character of approximate entities may be new. It means that we can quantify over approximate entities and also express how an entity is approximate. An article on approximate theories and approximate entities is forthcoming.

compatibility of determinism and free will A logical robot needs to consider its choices and the consequences of them. Therefore, it must regard itself as having *free will* even though it is a deterministic device.

We discuss our choices and those of robots by considering non-determinist approximations to a determinist world–or at least a world more determinist than is needed in the approximation. The philosophical name for this view is *compatibilism*. I think compatibilism is a requisite for AI research reaching human-level intelligence.

In practice, regarding an observed system as having choices is necessary when ever a human or robot knows more about the relation of the system to the environment than about what goes on within the system. This is dicussed in [7].

mind-brain distinctions I'm not sure whether this point is philosophical or scientific. The mind corresponds to software, perhaps with an internal distinction between program and knowledge. Software won't do anything without hardware, but the hardware can be quite simple. Some hardware configurations can run many different programs concurrently, i.e. there can be many minds in the same computer body. Software can also interpret other software.

Confusion about this is the basis of the Searle Chinese room fallacy [14]. The man in the hypothetical Chinese room is interpreting the software of a Chinese personality. Interpreting a program does not require having the knowledge possessed by that program. This would be obvious if people could interpret other personalities at a practical speed, but Chinese room software interpreted by an unaided human might run at 10^{-9} the speed of an actual Chinese.

2   Scientific Presuppositions

Some of the premises of logical AI are scientific in the sense that they are subject to scientific verification. This may also be true of some of the premises listed above as philosophical.

innate knowledge The human brain has important innate knowledge, e.g. that the world includes three dimensional objects that usually persist even when not observed. This was learned by evolution. Acquiring such knowledge by learning from sense data will be quite hard. It is better to build it into AI systems.

Different animals have different innate knowledge. Dogs know about permanent objects and will look for them when they are hidden. Very likely, cockroaches don't know about objects.

Identifying human innate knowledge has been the subject of recent psychological research. See [15] and the discussion in [11] and the references Pinker gives. In particular, babies and dogs know innately that there are permanent objects and look for them when they go out of sight. We'd better build that in.

middle out Humans deal with middle-sized objects and develop our knowledge up and down from the middle. Formal theories of the world must also start from the middle where our experience informs us. Efforts to start from the most basic concepts, e.g. to make a basic ontology are unlikely to succeed as well as starting in the middle. The ontology must be compatible with the idea that the basic entities in the ontology are not the basic entities in the world. More basic entities are known less well than the middle entities.

universality of intelligence Achieving goals in the world requires that an agent with limited knowledge, computational ability and ability to observe use certain methods. This is independent of whether the agent is human, Martian or machine. For example, playing chess-like games effectively requires something like alpha-beta pruning. Perhaps this should be regarded as a scientific opinion (or bet) rather than as philosophical.

universal expressiveness of logic This is a proposition analogous to the Turing thesis that Turing machines are computationally universal–anything that can be computed by any machine can be computed by a Turing machine. The *expressiveness thesis* is that anything that can be expressed, can be expressed in first order logic. Some elaboration of the idea is required before it will be as clear as the Turing thesis.[5]

sufficient complexity yields essentially unique interpretations A robot that interacts with the world in a sufficiently complex way gives rise to an essentially unique interpretation of the part of the world with which it interacts. This is an empirical, scientific proposition, but many people, especially philosophers (see [13], [12], [2], [1]), take its negation for granted. There are often many interpretations in the world of short descriptions, but long descriptions almost always admit at most one.

The most straightforward example is that a simple substitution cipher cryptogram of an English sentence usually has multiple interpretations if the text is less than 21 letters and usually has a unique interpretation if the text is longer than 21 letters. Why 21? It's a measure of the redundancy of English. The redundancy of a person's or a robot's interaction with the world is just as real–though clearly much harder to quantify.

We expect these philosophical and scientific presuppositions to become more important as AI begins to tackle human level intelligence.

References

1. Daniel Dennett. *Brainchildren: Essays on Designing Minds*. MIT Press, 1998.
2. Daniel C. Dennett. Intentional systems. *The Journal of Philosophy*, 68(4):87–106, 1971.

---

[5] First order logic isn't the best way of expressing all that can be expressed any more than Turing machines are the best way of expressing computations. However, with set theory, what can be expressed in stronger systems can apparently also be expressed in first order logic.

3. Hubert Dreyfus. *What Computers still can't Do*. M.I.T. Press, 1992.
4. McCarthy, J.   1979.   Ascribing   mental   qualities   to   machineshttp://www-formal.stanford.edu/jmc/ascribing.html. In M. Ringle (Ed.), *Philosophical Perspectives in Artificial Intelligence*. Harvester Press. Reprinted in [5].
5. McCarthy, J. 1990. *Formalizing Common Sense: Papers by John McCarthy*. 355 Chestnut Street, Norwood, NJ 07648: Ablex Publishing Corporation.
6. McCarthy, J. 1993. Notes on Formalizing Context,
   http://www-formal.stanford.edu/jmc/context.html. In *IJCAI-93*.
7. McCarthy, J. 1996. Making Robots Conscious of their Mental States http://www-formal.stanford.edu/jmc/consciousness.html. In S. Muggleton (Ed.), *Machine Intelligence 15*. Oxford University Press.
8. McCarthy, J. 1999. Elaboration tolerance
   http://www-formal.stanford.edu/jmc/elaboration.html. *to appear*.
9. McCarthy, J., and S. Buvač. 1997. Formalizing context (expanded notes). In A. Aliseda, R. v. Glabbeek, and D. Westerståhl (Eds.), *Computing Natural Language*. Center for the Study of Language and Information, Stanford University.
10. Roger Penrose. *Shadows of the Mind: A Search for the Missing Science of Consciousness*. Oxford University Press, Oxford, 1994.
11. Pinker, S. 1997. *How the Mind Works*. Norton.
12. Hilary Putnam. The meaning of "meaning". In Keith Gunderson, editor, *Language, Mind and Knowledge*, volume VII of *Minnesota Studies in the Philosophy of Science*, pages 131–193. University of Minnesota Press, 1975.
13. Willard V. O. Quine. Propositional objects. In *Ontological Relativity and other Essays*. Columbia University Press, New York, 1969.
14. John R. Searle. *Minds, Brains, and Science*. Harvard University Press, Cambridge, Mass., 1984.
15. Spelke, E. 1994. Initial knowlege: six suggestions. *Cognition* 50:431–445.

On existence of extensions for default theories

Giovanni Criscuolo and Eliana Minicozzi

Dipartimento di Scienze Fisiche, Mostra d'Oltremare, pad. 19, 80125 Napoli, Italy
{vanni,elimi}@na.infn.it

Summary. We give necessary and sufficient conditions for the existence of extensions in a general default theory. Such conditions are rather complicated as one would expect. However, using a specific simple case, we show how to derive well-known results in the semantic of logic programs.

We also give a characterization of Reiter's Γ operator in terms of OM pairs that elucidates the "meta" component of the default rules.

1 Introduction

About thirty years ago, after completing our degree in Physics in Italy we enrolled ourselves in the master program in Computer Science at the University of British Columbia in Vancouver, where Ray, then starting his career, was assistant professor.

The atmosphere at the time was rather informal and ideal in many respects: few students and an international staff largely composed of young and therefore open-minded and enthusiastic people. We like to mention Richard Bird, Abbe Mowshowitz, the future author of *The conquest of the will* [1], Friedrich Nake, Richard Rosenberg. With some of them, including Ray of course, we established ties that lasted over the years.

Did this atmosphere and Ray in particular influence our minds? Certainly so. For example it was then the time of the hot declarative-procedural controversy in AI and we remember Ray, who was of course more on the declarative side, complaining that Computer Science students did not know enough Logic. Well, after thirty years we are still complaining that italian Computer Science students do not know enough Logic!

Jokes apart, although we had already been exposed to Logic and Recursive Function Theory by Giuseppe Trautteur in Naples, there is no doubt that Ray had a decisive impact both on our scientific formation and on our future work. See for example [2] and [3], published directly under his guidance, but also [4] and [5].

As an example on our interaction, here is the story of how transitivity in normal default theories was defeated. One of us (G.C.) went in Vancouver in 1980 to work on the problem. After a long period of frustation he suddenly rushed into Ray's office. This is, more or less, the discussion that followed:

G.C.:" Ray, there is no way to block this damned transitivity" Ray (slowly drawing on the blackboard what would have been called a seminormal default rule):" But using this kind of rules transitivity can be blocked and ..." G.C. (interrupting him):"

Yes I know, but with this kind of theories we don't have a proof theory and they may even not have extensions!" Ray:" Oh, you arrived at the same conclusions! But there is no reason to worry if this is the situation. We already have an excellent paper. The only other way out would be to change the definition of extension but I don't want to. I believe the one I gave is the best that can be done".

Of course he was right.

2 Necessary and sufficient conditions for the existence of extensions

When G.C. was back to Naples we decided to investigate the problem of the existence of extensions for default theories. The plan was the following: given a default theory $\Delta = (W, D)$ consider the normal default theory $\Delta_N = (W, D_N)$ where $D_N = \left\{ \frac{\alpha:w}{w} / \frac{\alpha:\beta_1,...,\beta_n}{w} \in D \right\}$; if E is an extension of Δ then there is an extension of Δ_N that includes it, but then the problem is to find under what conditions an extension of Δ_N includes an extension of Δ [1].

To understand the theorem we need some definitions. From now on we will consider only default theories $\Delta = (W, D)$ having a consistent W. We refer the reader to [6] for an exact definition of a default proof of a formula with respect to a closed normal default theory and for the related completeness result. What is important to remember here is that any such proof uses (in a appropriate way) a finite number of defaults.

Definition - Let $\Delta_N = (W, D_N)$ be a closed normal default theory, D_N^* a subset of D_N, E_N an extension of Δ_N and β a formula of E_N. We say that β is independent of D_N^* if there exists a default proof of β where no default of D_N^* appears, otherwise we say that β depends on D_N^*.

The following theorem is an easy generalization of a previous theorem proved in [7],where the default theories taken into consideration were those having defaults with only one justification (at the time we were interested only in seminormal theories).

Theorem 2.1 Let Δ be a closed default theory. Δ has an extension E iff there exists an extension E_N of Δ_N and a subset D^* of D, such that $D_N^* \subseteq GD(E_N, \Delta_N)$, having the following properties:

a) $\frac{\alpha:\beta_1,...,\beta_n}{w} \in D^*$ if for some j $\neg\beta_j \in E_N$ and $\neg\beta_j$ is independent of D_N^*

b) for all δ such that $\delta_N \in GD(E_N, \Delta_N) - D_N^*$, $\delta = \frac{\alpha:\beta_1,...,\beta_n}{w}$, and for each j, $1 \leq j \leq n$, if $\neg\beta_j \in E_N$ then $\neg\beta_j$ depends on D_N^*

c) for each δ in D, $\delta = \frac{\alpha:\beta_1,...,\beta_n}{w}$, if $\delta_N \in (D_N - GD(E_N, \Delta_N))$ and $\alpha \in E_N$ then either α depends on D_N^* or there exists a j such that $\neg\beta_j \in E_N$ and it is independent of D_N^*

---

[1] Notice that in general one can always obtain a one to one correspondance between D and D_N, for example changing an α to $\alpha \wedge \gamma$ where γ is a tautology.

As it can be seen, the conditions we found were so complicated (and the theorem too) that, as Ray crudely put it:"To 'save' this theorem you should find an useful application".

Let us see, after so many years, if we can put it to some use.

If we restrict ourselves to the case when $W \cup CONS(D)$ is consistent, condition c) can be dropped. More precisely we have:

Theorem 2.2 Let $\Delta = (W, D)$ a general closed default theory such that $W \cup CONS(D)$ is consistent and let E_N be the unique extension of $\Delta_N = (W, D_N)$. Then Δ han an extension E if there exists a subset D^* of D such that:

a) $\frac{\alpha : \beta_1, \ldots, \beta_n}{w} \in D^*$ if for some $j \ \neg\beta_j \in E_N$ and $\neg\beta_j$ is independent of D_N^*

b) for all δ in $D - D^*$, $\delta = \frac{\alpha : \beta_1, \ldots, \beta_n}{w}$, and for each j, $1 \leq j \leq n$, if $\neg\beta_j \in E_N$ then $\neg\beta_j$ depends on D_N^*.

Moreover $E = E_N^*$, the extension of the normal default theory $\Delta_N = (W, D_N - D_N^*)$.

Notice that this theorem must be used as a lemma to prove Theorem 2.1. It looks still rather complicated but if we make a further restriction perhaps our reader will begin to see some light.

Corollary Let $\Delta = (\Phi, D)$ be a general closed default theory whose defaults have the form: $\delta = \frac{\alpha_1 \wedge \ldots \wedge \alpha_m : \neg\beta_1, \ldots, \neg\beta_n}{w}$ with w, β and α positive literals. Let E_N the extension of $\Delta_N = (\Phi, D_N)$.

Δ has an extension E iff there exists a subset D^* of D such that:

a) $\frac{\alpha_1 \wedge \ldots \wedge \alpha_m : \neg\beta_1, \ldots, \neg\beta_n}{w} \in D^*$ if there is $j : \beta_j \in E_N$ and β_j is independent of D_N^*

b) for all $\delta \in D - D^*$, $\delta = \frac{\alpha_1 \wedge \ldots \wedge \alpha_m : \neg\beta_1, \ldots, \neg\beta_n}{w}$, and for each j, $1 \leq j \leq n$, if $\beta_j \in E_N$ then β_j depends on D_N^* ..

Moreover $E = E_N^*$, extension of $\Delta_N = (\Phi, D_N - D_N^*)$.

By now the connection with the Theory of Logic Programming and in particular with the work of Marek and Truszczynski [8] should be clear. While they exploit the relation between normal programs and particular positive programs, we exploit the relation between general default theories and particular normal default theories.

As it is well known, the relations between logic programs and default theories have been clarified and can be stated as follows:

If $c = w \leftarrow \alpha_1 \ldots \alpha_m, not\beta_1, \ldots, not\beta_n$ is a normal clause, then define $dl(c) = \frac{\alpha_1 \wedge \ldots \wedge \alpha_m : \neg\beta_1, \ldots, \neg\beta_n}{w}$. If $c = w \leftarrow \alpha_1 \ldots \alpha_m$ is a positive clause, then define $dln(c) = \frac{\alpha_1 \wedge \ldots \wedge \alpha_m : w}{w}$.

From now on we will consider only ground programs whose traslations will therefore give closed default theories. We then have: M is a stable model of the normal program P iff $Cn(M)$ is an extension of the default theory $\Delta = (\Phi, dl(P))$ [8].

Let P be a normal program, the reduct of P with respect to a model M is the positive program P^M obtained from P by removing all clauses that are irrelevant with respect to M, i.e. clauses containing in the body at least one literal $not\beta$ with $\beta \in M$, and then removing all negative literals from the remaining clauses.

Now using the previous corollary it is easy to prove the following:

Theorem 2.3 Let P be a normal program, then M is a stable model of P iff M is the minimal Herbrand model of P^M, see [8].

Indeed if M is a stable model of P then $Cn(M)$ is an extension of the general default theory $\Delta = (\Phi, dl(P))$; by the previous corollary there exists $D^* \subseteq dl(P)$ such that $Cn(M) = E_N^*$ the extension of $\Delta_N^* = (\Phi, D_N - D_N^*)$. Then there exists a positive program P' such that $dln(P') = D_N - D_N^*$ and M is the minimal Herbrand model of P'. But P' coincides with P^M because

$c = w \leftarrow \alpha_1, ..., \alpha_m, not\beta_1, ..., not\beta_n$ is an irrelevant clause with respect to M iff $dl(c) \in D^*$. Indeed c is an irrelevant clause with respect to M iff there exists a j such that $\beta_j \in M$ iff there is a default proof of β_j w.r. to Δ_N^* and then w.r. to Δ_N, which does not use any of the defaults in D_N^* iff β_j is independent of D_N^* iff $dl(C) \in D^*$.

Conversely if M is the minimal Herbrandt model of P^M then it is enough to take $D^* = \{dl(c)/c \in P$ and c irrelevant w.r.to $M\}$. Then $dln(P^M) = D_N - D_N^*$ and obviously D^* satisfyies properties a) and b) of the corollary. But then $Cn(M)$ extension of $D_N - D_N^*$ is an extension of $dl(P)$ and M is a stable model of P.

If we consider stratified normal programs we can prove:

Theorem 2.4 Let P be a locally stratified normal program and let $rank$ be any local stratification function for P then the default theory $\Delta = (\Phi, dl(P))$ has a unique extension E.

We leave the proof of Theorem 2.4 to the reader, because it is a not too difficult consequence of the following Lemma.

Lemma Let P and $rank$ be as in Theorem 2.4, and let
$$\delta_1 = \frac{\alpha_1^1 \wedge ... \wedge \alpha_m^1 : \neg\beta_1^1, ..., \neg\beta_n^1}{w^1}$$
$$\delta_2 = \frac{\alpha_1^2 \wedge ... \wedge \alpha_m^2 : \neg\beta_1^2, ..., \neg\beta_n^2}{w^2}$$
If $rank(CONS(\delta_1)) \leq rank(CONS(\delta_2))$ and there exists a j such that β_j^1 has a default proof w.r.to $(\Phi, (dl(P))_N)$ then β_j^1 is independent of δ_2

The proof is by contradiction. Suppose that β_j^1 depends on δ_2 then from the definition of default proof it follows that
$rank(CONS(\delta_2)) \leq rank(\beta_j^1)$, while we know that
$rank(\beta_j^1) < rank(CONS(\delta_1))$: a contradiction.

Of course all the above looks easy because we have at our disposal all the results so beatifully exposed in [8].

En passant we quote the fact that we did not resist the temptation of giving a new definition of extension ([9], [10]) which turned out to be equivalent to the one given by Brewka [11].(Ray forgave us, we do not know whether he also forgave Brewka).

3   An interpretation of default rules using object-meta pairs

Finally we conclude by giving another characterization of both extensions and the Γ operator in propositional default theory in terms of OM pairs, introduced in [12]. We need some definition:

Let L be a propositional language then $\bullet L$ is the propositional language whose set of atomic formulae is given by $\{\bullet A/A$ is an $L - wff\}$.

An OM pair is given by: a set O of objective axioms in a language L, a set M of axioms in the language $\bullet L$ and a set RR of reflection rules. We call $\bullet L$ the metalanguage of L.

The definition of derivation in an OM pair is straightforward. One can see such a derivation as made up by parts, each of which is either an objective subderivation or a subderivation in the metalanguage in the style of classical natural deduction, appropriately interconnected by reflection rules. $TH_{OM}(O)$ $(TH_{OM}(M))$ is the set of all the objective formulae (metaformulae) provable in the OM pair.

Given a set S of objective formulae we can interpret $\bullet A$ as $A \in S$ and $\neg(\bullet A)$ as $A \notin S$ hence we can say that S satisfies the metaxioms M if all formulae of M are true under this interpretation, For example the axiom scheme $\bullet(A \supset B) \supset (\bullet A \supset \bullet B)$ is satisfied by S iff S is closed under modus ponens.

We introduce two reflection rules: $R_{up}^r = \frac{A}{\bullet A}$ (applicable only if the premiss does not depend on any objective formula) and

$R_d = \frac{\bullet A}{A}$ We may then formulate the following theorem [12].

Theorem 3.1 Let $\Delta = (O, M, \{R_d, R_{up}^r, \})$ then $TH_{OM}(O)$ is the intersection of all theories containing O and satisfying the axioms M of the metatheory.

It follows that if the axioms of the metatheory are Horn formulae then $TH_{OM}(O)$ coincides with the minimal theory that satisfies the axioms M.

Theoren 3.2 Let $\Delta = (W, D)$ be a default theory and S be any set of formulae. Consider the OM pair $\Delta_s = (W, M_s, \{R_d, R_{up}^r\})$ with $M_s = \{\bullet\alpha \supset \bullet w/\frac{\alpha:\beta_1,\ldots,\beta_n}{w} \in D$ and for all $1 \le i \le n \ \neg\beta_i \notin S\}$.Then $\Gamma(S) = TH_{WM_S}(W)$.

Trivially $TH_{WM_S}(W)$ satisfies properties 1 and 2 of Reiter'sdefinition [6]. Morover for any default of Δ, $\frac{\alpha:\beta_1,\ldots,\beta_n}{w}$, such that $\alpha \in TH_{WM_S}(W)$ and none of the $\neg\beta_i$ is in S we have that $\bullet\alpha$ can be proved by R_{up}^r while $\bullet\alpha \supset \bullet w$ is in M_s, hence by modus ponens $\bullet w$ can be proved and, using R_d , w can be proved hence

$THW_{MS}(W)$ also satisfies property 3. But all formulae of M_S are Horn formulae and therefore $THW_{MS}(W)$ is the minimal theory that satisfies properties 2 and 3.

We can now give necessary and sufficient conditions in order that $\Gamma(S)$ be an extension.

Theorem 3.3 Let $\Delta = (W, D)$ be a default theory and S be any set of formulae. Let us define $N_S = M_S \cup \{\neg \bullet (\neg\beta_1), \dots, \neg \bullet (\neg\beta_n))/ \bullet \alpha \supset \bullet w \in M_S\}$ and $D' = \{\frac{\alpha:\beta_1,\dots,\beta_n}{w} \in D/ \text{ there is a } j : \neg\beta_j \in S\}$.Finally consider the OM pair $(W, N_S, \{R^d, R^r_{up}\})$; then $\Gamma(S)$ is an extension for Δ iff the following conditions hold:

1) $THW_{N_S}(N_S)$ is consistent,
2) for each default in D' $\frac{\alpha:\beta_1,\dots,\beta_n}{w}$, there is a $j : \neg\beta_j \in THW_{N_S}(W)$.

It follows from the previous theorem that $\Gamma(S) = \Gamma(\Gamma(S))$iff $M_S = M_{\Gamma(S)}$.

Condition 1 tells us that for each default of Δ $\frac{\alpha:\beta_1,\dots,\beta_n}{w}$ if $\bullet\alpha \supset \bullet w$ is in M_S then $\Gamma(S) = THW_{MS}(W) = THW_{N_S}(W)$ does not prove any of $\neg\beta_i$,hence $\bullet\alpha \supset \bullet w$ is in $M_{\Gamma(S)}$.

Condition 2 tell us that the converse also holds.

4   Appendix

Proof of Theorem 2.2

Because the proof of the theorem is very long we will skip some of the passages suggesting how they can be proved.

First we need the following:

Lemma If E is an extension of $\Delta = (W, D)$ then there is an extension E_N of $\Delta_N = (W, D_N)$ that contains it. Let $D' = \{\delta \in D/\delta = \frac{\alpha:\beta_1\dots\beta_n}{w}$ and there is a $j : \neg\beta_j \in E\}$ and $\Delta^*_N = (W, GD(E_N, \Delta_N) - D'_N)$. Then Δ^*_N has extension $E^*_N = E$.

The first assertion is almost obvious, the second can be proved by induction using the iterative definition of extension given by Reiter.

When $W \cup CONS(D)$ is consistent E_N is the unique extension of Δ_N and it is possible to prove that $E^*_N = E$ is the extension of $\Delta^*_N = (W, D_N - D'_N)$.

Suppose now that Δ has an extension E as in the hypothesis of theorem 2.2. Then D' is our D^* satisfying $a)$ and $b)$. Indeed if $\delta \in D'$ then there is a j such that $\neg\beta_j \in E = E^*_N \subseteq E_N$, hence there is a proof of $\neg\beta_j$ that do not uses the defaults of D'_N, hence $\neg\beta_j$ is independent of D'_N.

For all the others defaults δ if any $\neg\beta_j \in E_N$, we know that $\neg\beta_j \notin E = E^*_N$, hence if $\neg\beta_j \in E_N$, each proof of $\neg\beta_j$ must use some default of D' hence $\neg\beta_j$ depends on D'.

On the contrary suppose that a) and b) are satisfied by D^* then it is possible to prove by induction, using the iterative construction for extensions given by Reiter that E_N^* is an extension for Δ. Indeed all the defaults of D^* are blocked and the default of $D - D^*$ are blocked only if the prerequisite is not in E_N^*.

Proof of Theorem 2.1

If there exists an extension E of Δ then there exists an extension E_N of Δ_N containing it and by theorem 2.2 being $W \cup \text{CONS}(GD(E_N, \Delta_N))$ consistent, E is an extension of $\Delta' = (W, D')$, with D' such that $D' = GD(E_N, \Delta_N)$, and there exists $D^* \subseteq GD(E_N, \Delta_N)$, D^* satisfies properties a) and b).

Moreover because E is an extension for Δ then for each $\delta = \frac{\alpha : \beta_1 \ldots \beta_n}{w}$ such that $\delta_N \in D_N - GD(E_N, \Delta_N)$ $\alpha \notin E = E_N^*$ or there is a j : $\neg \beta_j \in E = E_N^*$. Now if $\alpha \in E_N$ the last assertion becomes α depends on D_N^* or there would be a j such that $\neg \beta_j$ is independent of D_N^*.

Suppose now that there exists the D^* of the theorem and that properties a), b), c) are satisfied then by theorem 2.2 the subtheory $\Delta' = (W, D')$ such that $D' = GD(E_N, \Delta_N)$ has an extension $E \subseteq E_N$ and there exists a D^* such that $D_N^* \subseteq GD(E_N, \Delta_N)$ satisfying properties a) and b). Properties c) simply says that E is an extension for all Δ.

References

1. A.Mowshowitz: *The Conquest of the Will.* Addison Wesley (1976).
2. E.Minicozzi, R.Reiter: *A Note on Linear Resolution Strategies in Consequence Finding.* Artif. Intell. **3**, 175-179.
3. R.Reiter,G.Criscuolo: *On Interacting Defaults.* Proceedings *IJCAI-81*, Vancouver (1981), 270-276.
4. G.Criscuolo, E.Minicozzi, G. Trautteur: *Limiting Recursion and the Arithmetical Hierarchy.* R.A.I.R.O. **R-3** (1975),5-12.
5. E.Minicozzi: *Some Natural Properties of Strong-Identification in Inductive Interference.* Theor. Comp. Sc. (1976) 345-360.
6. R.Reiter: *A Logic Default Reasoning.* Artif.Intell. **13** (1980) 81-132.
7. G. Criscuolo, E. Minicozzi: *On Extensions and Models for Default Theories.* Techn. Rep. Istituto di Fisica Teorica, Università di Napoli (1983).
8. V.W. Marek, M. Truszczynski: *Nonmonotonic Logic.* Springer, Berlin (1983).
9. G. Criscuolo, E. Minicozzi: *A New Definition of Extension for Default Theories.* Techn. Rep. Istituto di Fisica Reorica, Università di Napoli (1983).
10. G. Criscuolo, E. Minicozzi: *Critica al Concetto di Estensione in Teorie Default.* Atti 1 Congresso AI/IA,(1989).
11. G. Brewka: *Cumulative Default Logic: in Defense of Non Mononotonic Rules.* Artif. Intell. **50** (1991) 183-205.
12. G. Criscuolo, F. Giunchiglia, L. Serafini: *A Foundation of Metalogic Reasoning: OM pairs (Propositional Case).* Techn. Rep. 9403-02 *IRST*-Trento (1994).

An Incremental Interpreter for High-Level Programs with Sensing

Giuseppe De Giacomo[1] and Hector J. Levesque[2]

[1] Dipartimento di Informatica e Sistemistica
Università di Roma "La Sapienza"
Via Salaria 113, 00198 Rome, Italy
degiacomo@dis.uniroma1.it

[2] Department of Computer Science
University of Toronto
Toronto, Canada M5S 3H5
hector@cs.toronto.edu

Summary. Like classical planning, the execution of high-level agent programs requires a reasoner to look all the way to a final goal state before even a single action can be taken in the world. This deferral is a serious problem in practice for large programs. Furthermore, the problem is compounded in the presence of sensing actions which provide necessary information, but only after they are executed in the world. To deal with this, we propose (characterize formally in the situation calculus, and implement in Prolog) a new incremental way of interpreting such high-level programs and a new high-level language construct, which together allow much more control to be exercised over when actions can be executed. We argue that such a scheme leads to a practical way to deal with large agent programs containing both nondeterminism and sensing.

1 Introduction

The research reported in this paper is strongly based on Ray Reiter's work on the situation calculus, on the frame problem, and on cognitive robotics.

In [14] it was argued that when it comes to providing high level control to autonomous agents or robots, the notion of *high-level program execution* offers an alternative to classical planning that may be more practical in many applications. Briefly, instead of looking for a sequence of actions a such that

$$Axioms \models Legal(do(a, S_0)) \land \phi(do(a, S_0))$$

where ϕ is the goal being planned for, we look for a sequence a such that

$$Axioms \models Do(\delta, S_0, do(a, S_0))$$

where δ is a high-level program and $Do(\delta, s, s')$ is a formula stating that δ may legally terminate in state s' when started in state s. By a high-level program here, we mean one whose primitive statements are the domain-dependent actions of some agent or robot, whose tests involve domain-dependent fluents (that are caused to hold or not hold by the primitive actions), and which contains nondeterministic choice points where reasoned (non-random) choices must be made about how the execution should proceed.

What makes a high-level agent program different from a deterministic "script" is that its execution is a problem solving task, not unlike planning. An interpreter needs to use what it knows about the prerequisites and effects of actions to find a sequence with the right properties. This can involve considerable search when δ is very nondeterministic, but much less search when δ is more deterministic. The feasibility of this approach for AI purposes clearly depends on the expressive power of the programming language in question. In [14], a language called GOLOG is presented, which in addition to nondeterminism, contains facilities for sequence, iteration, and conditionals. In this paper, we extend the expressive power of this language by providing much finer control over the nondeterminism, and by making provisions for sensing actions. To do so in a way that will be practical even for very large programs requires introducing a different style of on-line program execution.

In the rest of this section, we discuss on-line and off-line execution informally, and show why sensing actions and nondeterminism together can be problematic. In the following section, we formally characterize program execution in the language of the situation calculus. Next, we describe an incremental interpreter in Prolog that is correct with respect to this specification. The final section contains discussion and conclusions.

1.1   Off-line and On-line execution

To be compatible with planning, the GOLOG interpreter presented in [14] executes in an *off-line* manner, in the sense that it must find a sequence of actions constituting an entire legal execution of a program *before* actually executing any of them in the world.[3] Consider, for example, the following program:

$$(a|b) \; ; \Delta \; ; p?$$

where a and b are primitive actions, | indicates nondeterministic choice, Δ is some very large deterministic program, and p? tests whether fluent p holds. A legal sequence of actions should start with either a or b, followed by a sequence for Δ, and end up in state where p holds. Before executing a or b, the agent or robot must wait until the interpreter considers all of Δ and determines which initial action eventually leads to p. Thus even a single nondeterministic choice occurring early in a large program can result in an unacceptable delay. We will see below that this problem is compounded in the presence of sensing actions.

If a small amount of nondeterminism in a program is to remain practical (as suggested by [14]), we need to be able to choose between a and b based on some local criterion without necessarily having to go through all of Δ. Using something

---

[3] It is assumed that once an action is taken, it need not be undoable, and so backtracking "in the world" is not an option.

like

$$(a|b) \; ; r? \; ; \Delta \; ; p?$$

here does not work, since an off-line interpreter cannot settle for a even if it leads to a state where r holds. We need to be able to *commit* to a choice that satisfies r, with the understanding that it is the responsibility of the programmer to use an appropriate local criterion, and that the program will simply fail without the option of backtracking if p does not hold at the end.

It is convenient to handle this type of commitment by changing the execution style from off-line to on-line, but including a special off-line search operator. In a *on-line* execution, nondeterministic choices are treated like random ones, and any action selected is executed immediately. So if the program

$$(a|b) \; ; \Delta \; ; p?$$

is executed on-line, one of a or b is selected and executed immediately, and the process continues with Δ; in the end, if p happens not to hold, the entire program fails. We use a new operator Σ for search, so that $\Sigma\delta$, where δ is any program, means "consider δ off-line, searching for a successful termination state". With this operator, we can control how nondeterminism will be handled. To execute

$$\Sigma\{(a|b) \; ; r?\} \; ; \Delta \; ; p?$$

on-line, we would search for an a or b that successfully leads to r, execute it immediately, and then continue boldly with Δ. In this scheme, it is left to the programmer to decide how cautious to be. If the programmer drops the search operator completely thus writing

$$(a|b) \; ; r?; \Delta \; ; p?$$

then the choice between a and b will be random. If the programmer puts the entire program within a Σ operator, thus writing

$$\Sigma\{(a|b) \; ; r?; \Delta \; ; p?\}$$

then the choice between a and b will be based on full lookahead to the end of the program, *i.e.*, the program will be executed essentially in the old way.

1.2   Sensing actions

This on-line style of execution is well-suited to programs containing sensing actions. As described in [9], [13], [18], sensing actions are actions that can be taken by the agent or robot to obtain information about the state of certain fluents, rather than to change them. The motivation for sensing actions involves applications where because the initial state of the world is incompletely specified or because of hidden exogenous actions, the agent must use sensors of some sort to determine the value of certain fluents.

Suppose, for example, that nothing is known about the state of some fluent q, but that there is a binary sensing action $read_q$ which uses a sensor to tell the robot whether or not q holds. To execute the program

$$a \; ; read_q \; ; \textbf{if } q \textbf{ then } \Delta_1 \textbf{ else } \Delta_2 \textbf{ endIf} \; ; p?$$

the interpreter would get the robot to execute a in the world, get it to execute $read_q$, then use the information returned to decide whether to continue with Δ_1 or Δ_2. But consider the program

$$(a|b) \; ; read_q \; ; \textbf{if } q \textbf{ then } \Delta_1 \textbf{ else } \Delta_2 \textbf{ endIf} \; ; p?.$$

An off-line interpreter cannot commit to a or b in advance, and because of that, cannot use $read_q$ to determine if q would hold after the action. The only option available is to see if one or a or b would lead to p for *both* values of q. This requires considering both Δ_1 and Δ_2, even though in the end, only one of them will be executed. Similarly, if we attempt to generate a low-level robot program (as suggested in [13] for planning in the presence of sensing), we end up having to consider both Δ_1 and Δ_2.

The situation is even worse with loops. Consider

$$(a|b) \; ; read_q \; ; \textbf{while } q \textbf{ do } \Delta \; ; read_q \textbf{ endWhile} \; ; p?.$$

Since an off-line interpreter has no way of knowing in advance how many iterations of the loop will be required to make q false, to decide between a and b, it would be necessary to reason about the effect of performing Δ an *arbitrary* number of times (by discovering loop invariants *etc.*). But if a commitment could be made to one of them on local grounds we could modify the program as follows

$$\Sigma\{(a|b); r?\} \; ; read_q \; ; \textbf{while } q \textbf{ do } \Delta \; ; read_q \textbf{ endWhile} \; ; p?.$$

and then use $read_q$ to determine the actual value of q, and it would not be necessary to reason about the deterministic loop. It therefore appears that an on-line execution style is often more practical for large programs containing nondeterminism and sensing actions.

2 Preliminaries

The technical machinery we use to define on-line program execution in the presence of sensing is based on that of [4], *i.e.*, we use the predicates *Trans* and *Final* to define a single step semantics of programs [10], [17]. However some adaptation is necessary to deal with on-line execution, sensing results, and the Σ operator.

2.1   Situation calculus

The starting point in the definition is the situation calculus [12]. We will not go over the language here except to note the following components: there is a special constant S_0 used to denote the *initial situation,* namely that situation in which no actions have yet occurred; there is a distinguished binary function symbol *do* where $do(a, s)$ denotes the successor situation to s resulting from performing the action a; relations whose truth values vary from situation to situation, are called (relational) *fluents,* and are denoted by predicate symbols taking a situation term as their last argument; there is a special predicate $Poss(a, s)$ used to state that action a is executable in situation s; finally, following [13], there is a special predicate $SF(a, s)$ used to state that action a would return the binary sensing result 1 in situation s.

Within this language, we can formulate domain theories which describe how the world changes as the result of the available actions. One possibility is an action theory of the following form [10]:

- Axioms describing the initial situation, S_0. Note that there can be fluents like q about which nothing is known in the initial state.
- Action precondition axioms, one for each primitive action a, characterizing $Poss(a, s)$.
- Successor state axioms, one for each fluent F,[4] stating under what conditions $F(x, do(a, s))$ holds as function of what holds in situation s. These take the place of the so-called effect axioms, but also provide a solution to the frame problem [10].
- Unique names axioms for the primitive actions.
- Some foundational, domain independent axioms.

Finally, as in [13], we include

- Sensed fluent axioms, one for each primitive action a of the form $SF(a, s) \equiv \phi_a(s)$, characterizing *SF*.

For the sensing action $read_q$ used above, we would have $[SF(read_q, s) \equiv q(s)]$, and for any ordinary action a that did not involve sensing, we would use $[SF(a, s) \equiv$ **true**].

2.2   Histories

To describe a run which includes both actions and their sensing results, we use the notion of a history. By a *history* we mean a sequence of pairs (a, x) where a is a primitive action and x is 1 or 0, a sensing result. Intuitively, the history $(a_1, x_1) \cdot \ldots \cdot (a_n, x_n)$ is one where actions $a_1, \ldots, a_n$ happen starting in some initial situation, and each action a_i returns sensing value x_i. The assumption is that

---

[4] A fluent whose current value could only be determined by sensing would normally not have a successor state axiom. However, see [7] for a proposal that overcomes this limitation.

if a_i is an ordinary action with no sensing, then $x_i = 1$. Notice that the empty sequence ϵ is a history.

Histories are not terms of the situation calculus. It is convenient, however, to use $end[\sigma]$ as an abbreviation for the situation term called the *end situation* of history σ on the initial situation S_0, and defined by: $end[\epsilon] = S_0$; and inductively, $end[\sigma \cdot (a, x)] = do(a, end[\sigma])$.

It is also useful to use $Sensed[\sigma]$ as an abbreviation for a formula of the situation calculus, the *sensing results* of a history, and defined by: $Sensed[\epsilon] = $ **true**; and inductively, $Sensed[\sigma \cdot (a, 1)] = Sensed[\sigma] \wedge SF(a, end[\sigma])$, and $Sensed[\sigma \cdot (a, 0)] = Sensed[\sigma] \wedge \neg SF(a, end[\sigma])$. This formula uses SF to tell us what must be true for the sensing to come out as specified by σ starting in the initial situation S_0.

2.3 The *Trans* and *Final* predicates

In [4] two special predicates *Trans* and *Final* were introduced. $Trans(\delta, s, \delta', s')$ was intended to say that by executing program δ starting in situation s, one can get to situation s' in one elementary step with the program δ' remaining to be executed, that is, there is a possible transition from the configuration (δ, s) to the configuration (δ', s'). $Final(\delta, s)$, instead, was intended to say that program δ may successfully terminate in situation s, that is, the configuration (δ, s) is final.

For example, the transition requirements for sequence is

$$Trans([\delta_1; \delta_2], s, \delta', s') \equiv$$
$$Final(\delta_1, s) \wedge Trans(\delta_2, s, \delta', s') \quad \vee$$
$$\exists \gamma'. Trans(\delta_1, s, \gamma', s') \wedge \delta' = (\gamma'; \delta_2).$$

This says that to single-step the program $(\delta_1; \delta_2)$, either δ_1 terminates and we single-step δ_2, or we single-step δ_1 leaving some γ', and $(\gamma'; \delta_2)$ is what is left of the sequence.

For our account here, we adopt the definitions of *Trans* and *Final* in [4] (the details of which we omit).[5]

3 Off-line lookahead

On-line executions are characterized by the fact that the robot at each step makes a transition chosen among those that are legal (we'll see later in which precise sense) and execute it in the real world. In executing it the robots commits to the transition chosen since there is no possibility of undoing it. This is to be contrasted with the off-line execution mode where we commit to a sequence of actions to be executed only after having shown that the sequence is guaranteed to terminate successfully.

Since when we execute a program on-line there is no possibility of backtracking, how to select among the possible transitions the one to execute is a critical

---

[5] For an in-depth study of *Trans* and *Final*, including a suitable treatment of procedures and constructs for concurrency, can be found in [5], [6].

step. How should this selection be done? As discussed in Section 1, one extreme possibility is to make a random choice. Obviously this selection mechanism is very efficient, however the choice made may compromise the successful termination of the program. On the other extreme we may require the robot to be very cautious and do full lookahead to the end of the program to make a transition only if it is guaranteed to lead to a successful termination of the program. Naturally, this is quite heavy computationally [6].

Here we introduce a mechanism to have a controlled form of lookahead, so that the amount of lookahead to be performed is under the control of the programmer. Namely, we introduce a *search operator* in the programming language.

We define *Final* and *Trans* for the new operator as follows. For *Final*, we simply have that $(\Sigma\delta, s)$ is a final configuration of the program if (δ, s) itself is, and so we get the requirement

$$Final(\Sigma\delta, s) \equiv Final(\delta, s).$$

For *Trans*, we have that the configuration $(\Sigma\delta, s)$ can evolve to $(\Sigma\gamma', s')$ provided that (δ, s) can evolve to (γ', s') and from (γ', s') it is possible to reach a final configuration in a finite number of transitions. Thus, we get the requirement

$$
\begin{aligned}
Trans(\Sigma\delta, s, \delta', s') \equiv \\
\exists\gamma'.\ \delta' = \Sigma\gamma' \wedge Trans(\delta, s, \gamma', s') \wedge \\
\exists\gamma'', s''.Trans^*(\gamma', s', \gamma'', s'') \wedge Final(\gamma'', s'').
\end{aligned}
$$

In this assertion, *Trans** is the reflexive transitive closure of *Trans*, defined by

$$Trans^*(\delta, s, \delta', s') \stackrel{def}{=} \forall T\,[\ldots \supset T(\delta, s, \delta', s')]$$

where the ellipsis stands for the conjunction of (the universal closure of)

$$
\begin{aligned}
T(\gamma, s, \gamma, s) \\
T(\gamma, s, \gamma', s') \wedge Trans(\gamma', s', \gamma'', s'') \supset T(\gamma, s, \gamma'', s'').
\end{aligned}
$$

The semantics of Σ can be understood as follows: (1) $(\Sigma\delta, s)$ selects from all possible transitions of (δ, s) those from which there exists a sequence of further transitions leading to a final configuration; (2) the Σ operator is propagated through the chosen transition, so that this restriction is also performed on successive transitions. In other words, within a Σ operator, we only take a transition from δ to γ', if γ' is on a path that will eventually terminate successfully, and from γ' we do the same. As desired, Σ does an off-line search before committing to even the first transition.

It is not too hard to prove that Σ has some intuitively plausible properties. In particular we have the following ones:

---

[6] In [8] on-line executions were considered to allow execution monitoring and recovery. There a brave and a cautions interpreter were defined which corresponded to exactly to these two extreme approaches.

Property 1.

$$Trans(\Sigma\delta, s, \delta', s') \supset \exists\gamma.\delta' = \Sigma\gamma$$

i.e., a program of the form $\Sigma\delta$ can evolve only to programs of the form $\Sigma\delta'$. In other words, the search operator is indeed propagated through the transition.

Property 2.

$$Trans(\Sigma\delta, s, \Sigma\delta', s') \equiv Trans(\delta, s, \delta', s') \wedge \exists s''.Do(\delta', s', s'')$$

i.e., in performing a transition step for a configuration $(\Sigma\delta, s)$ we are in fact performing a transition from (δ, s) and verifying that such transition leads to successful termination.

Property 3.

$$Trans(\Sigma\Sigma\delta, s, \Sigma\Sigma\delta', s') \equiv Trans(\Sigma\delta, s, \Sigma\delta', s')$$
$$Final(\Sigma\Sigma\delta, s) \equiv Final(\Sigma\delta, s)$$

i.e., nesting search operators is equivalent to apply the search operator only once.

4   Characterizing on-line executions

The on-line execution of a program consists of a suitable sequence of *legal* single-step transitions. We distinguish the case where we do not have sensing from the one in which we do.

4.1   Without sensing

In the absence of sensing, we say that a δ can be executed in the current situation s leading to the new situation s' with program δ' that remains to be executed, only when

$$Axioms \models Trans(\delta, s, \delta', s')$$

i.e., a transition step is legal if only if is logically implied by *Axioms*. In this way we capture the intuition that a transition is legal if on the base of our knowledge (as expressed by *Axioms*) we are certain that the transition can be executed. Analogously, we are allowed to successfully terminate a program δ in s when

$$Axioms \models Final(\delta, s)$$

i.e., δ can legally terminate in s if and only if it is logically implied by *Axioms*.

Hence, without sensing an on-line execution of a program δ starting from a situation s is a sequence $(\delta_0 = \delta, s_0 = s), \ldots, (\delta_n, s_n)$ such that for $i = 0, \ldots, n-1$:

$$Axioms \models Trans(\delta_i, s_i, \delta_{i+1}, s_{i+1}).$$

An on-line execution is successful if

$$Axioms \models Final(\delta_n, s_n).$$

It is possible to prove the following theorem.

Theorem 1. *If there exists a successful on-line execution* $(\delta_0 = \delta, s_0 = s),\ldots$ $\ldots,(\delta_n, s_n = s')$ *of a program* δ *in the situation* s *leading to* s'*, then there exists a successful off-line execution of* δ *in* s *leading to* s'*, i.e.,*

$$Axioms \models Do(\delta, s, s')$$

The converse of this theorem does not hold, since an on-line execution requires all transition steps to be logically implied by $Axioms$, while an off-line execution does not. For example, consider the program $\delta = \phi; a \mid \neg\phi; a$, where $Axioms \not\models \phi[S_0]$ and $Axioms \not\models \neg\phi[S_0]$. δ executed in S_0 has a successful off-line execution, namely $Axioms \models Do(\delta, S_0, do(a, S_0))$. But it has no successful on-line executions, since there are no transitions logically implied by $Axioms$.

We do not impose any selection criteria on on-line executions. The robot at each step makes a legal transition that is randomly chosen. Thus we cannot guarantee that the robot follows a successful on-line execution a priori. We can however make use of the search operator for giving the robot the possibility, under the control of the programmer, of doing some lookahead and avoid dead-end executions. Indeed by Property 2 above we have that:

$$Axioms \models Trans(\Sigma\delta, s, \Sigma\delta', s')$$

if and only if

$$Axioms \models Trans(\delta, s, \delta', s') \text{ and } Axioms \models \exists s''.Do(\delta', s', s'')$$

i.e., there is a legal transition from $(\Sigma\delta, s)$ to $(\Sigma\delta', s')$ if and only if (i) there is a legal transition from (δ, s) to (δ', s'), and (ii) there exists an execution of δ' in s' that successfully terminates.

By Theorem 1, if $Axioms \not\models \exists s''.Do(\delta', s', s'')$ then there are no successful on-line executions of δ' in s'. It follows that applying the search operator to a program δ we prune potential on-line executions that are bound to be unsuccessful.[7]

4.2 With sensing

First we observe that we did not require special axioms for *Trans* and *Final* in order to deal with sensing. Sensing actions are just like ordinary actions in all respects

---

[7] Note that, although we can guarantee the existence of a successful execution, it is not always the case that we can actually find an successful *on-line execution*, which in fact requires the existence of a sequence of transitions that are logically implied by $Axioms$.

except for what specified by the sensed fluent axioms involving SF. However, the existence of a given legal transition may now depend on the values sensed so far. That is, if s is $end[\sigma]$ where σ is the history of actions and sensing values starting from the initial situation S_0, then δ in s can make a legal transition leading to s' with program δ' that remains to be executed when

$$Axioms \cup \{Sensed[\sigma]\} \models Trans(\delta, s, \delta', s').$$

In other words, now we are looking for a transition that is logically implied by *Axioms* together with *the values sensed so far*.

In executing the next step we can take into account that the transition may have result in getting some new information from the sensors. Specifically, if the transition did not result in any action, *i.e.*, $s' = s,$[8] then we still consider logical implication from $Axioms \cup \{Sensed[\sigma]\}$. If, instead, an action a was performed and the value x returned, then we consider logical implication from $Axioms \cup \{Sensed[\sigma']\}$ where $\sigma' = \sigma \cdot (a, x)$, *i.e.*, we consider the value returned by action a as well.

Similarly, we are allowed to terminate the program δ successfully if

$$Axioms \cup \{Sensed[\sigma]\} \models Final(\delta, end[\sigma]),$$

where again the history σ is taken into account.

Thus, in presence of sensing, an on-line execution of a program δ starting from a situation $end[\sigma]$ is a sequence $(\delta_0 = \delta, \sigma_0 = \sigma), \ldots, (\delta_n, \sigma_n)$ such that for $i = 0, \ldots, n-1$:

$$Axioms \cup \{Sensed[\sigma_i]\} \models Trans(\delta_i, end[\sigma_i], \delta_{i+1}, end[\sigma_{i+1}])$$

$$\sigma_{i+1} = \begin{cases} \sigma_i & \text{if } end[\sigma_{i+1}] = end[\sigma_i] \\ \sigma_i \cdot (a, x) & \text{if } end[\sigma_{i+1}] = do(a, end[\sigma_i]) \text{ and } a \text{ returns } x \end{cases}$$

An on-line execution is successful if

$$Axioms \cup \{Sensed[\sigma_n]\} \models Final(\delta_n, end[\sigma_n]).$$

Note that, if no sensing action is performed then $Sensed[\sigma]$ becomes equivalent to **true**, and hence the specification correctly reduces to the specification from before.

Finally, let us focus on the meaning the search operator in the context of on-line executions in presence of sensing. By Property 2 above we have that:

$$Axioms \cup \{Sensed[\sigma]\} \models Trans(\Sigma\delta, s, \Sigma\delta', s')$$

if and only if

$$Axioms \cup \{Sensed[\sigma]\} \models Trans(\delta, end[\sigma], \delta', s') \text{ and}$$
$$Axioms \cup \{Sensed[\sigma]\} \models \exists s''.Do(\delta', s', s'')$$

---

[8] Such "null transitions" arise from tests in the program.

i.e., in looking for the existence of a successful execution of δ' in s', we obviously do not take into account how the sensing values will turn out to be (we will know these values only when we actually execute the actions in the transitions). Hence now Theorem 1 implies that if $Axioms \cup \{Sensed[\sigma]\} \not\models \exists s''.Do(\delta', s', s'')$ then there, are no successful on-line executions of δ' in s' *that do not gather new information by sensing*. It follows that applying the search operator to a program δ we prune potential on-line executions that depend on how sensing turns out in order to be successful.

5   An incremental interpreter

Next we present a simple incremental interpreter in Prolog. Although the on-line execution task characterized above no longer requires search to a final state, it remains fundamentally a theorem-proving task: does a certain *Trans* or *Final* formula follow logically from the axioms of the action theory together with assertions about sensing results?

The challenge in writing a practical interpreter is to find cases where this theorem-proving can be done using something like ordinary Prolog evaluation. The interpreter in [4] as well as in earlier work on which it was based [14] was designed to handle cases where what was known about the initial situation S_0 could be represented by a set of atomic formulas together with a closed-world assumption. In the presence of sensing, however, we cannot simply apply a closed-world assumption blindly. As we will see, we can still avoid full theorem-proving if we are willing to assume that a program executes appropriate sensing actions prior to any testing it performs. In other words, our interpreter depends on a *just-in-time history assumption*[9] where it is assumed that *whenever a test is required, the on-line interpreter* at that point *has complete knowledge of the fluents in question to evaluate the test without having to reason by cases* etc.

5.1   The main loop

As it turns out, most of the subtlety in writing such an interpreter concerns the evaluation of tests in a program. The rest of the interpreter derives almost directly from the axioms for *Final*, and *Trans* described above. It is convenient, however, to use an implementation of these predicates defined over encodings of histories (with most recent actions first) rather than situations. We get

```
/*   P is a program                  */
/*   H is a history, initially []     */
/*     H ::= [] | [(Act,1/0)|H]      */

incrInterpret(P,H) :- final(P,H).
incrInterpret(P,H) :-
```

---

[9] The notion of just-in-time histories is investigated further in [7].

```
trans(P,H,P1,[(Act,_)|H]), !, execute(Act,Sv),
   incrInterpret(P1,[(Act,Sv)|H]).
incrInterpret(P,H) :-
   trans(P,H,P1,H), !,
   incrInterpret(P1,H).
```

So to incrementally interpret a program on-line, we either terminate successfully, or we find a transition involving some action, commit to that action, execute it in the world to obtain a sensing result, and then continue the interpretation with the remaining program and the updated history.[10] In looking for the next action, we skip over transitions involving successful tests where no action is required and the history does not change. To execute an action in the world, we connect to the sensors and effectors of the robot or agent. Here for simplicity, we just write the action, and read back a sensing result.

```
execute(Act,Sv) :-
   write(Act),
   (senses(Act,_) ->
      (write(':'), read(Sv)) ; (nl, Sv=1)).
```

We assume the user has declared using `senses` (described below) which actions are used for sensing, and for any action with no such declaration, we immediately return the value 1.

5.2   Implementing *Trans* and *Final*

Clauses for `trans` and `final` are needed for each of the program constructs. For example, for sequence, we have

```
trans(seq(P1,P2),H,P,H1) :-
   final(P1,H), trans(P2,H,P,H1).
trans(seq(P1,P2),H,seq(P3,P2),H1) :-
   trans(P1,H,P3,H1).
```

which corresponds to the axiom given earlier except for the use of histories instead of situations. We omit the details for the other constructs, except for Σ (search):

```
final(search(P),H) :- final(P,H).

trans(search(P),H,search(P1),H1) :-
   trans(P,H,P1,H1), ok(P1,H1).

ok(P,H) :- final(P,H).
ok(P,H) :- trans(P,H,P1,H), ok(P1,H).
ok(P,H) :- trans(P,H,P1,[(Act,_)|H]),
```

---

[10] In practice, we would not want the history list to get too long, and would use some form of "rolling forward" [15].

```
(senses(Act,_) ->
    ( ok(P1,[(Act,0)|H]) ,
      ok(P1,[(Act,1)|H]) ) ;
    ok(P1,[(Act,1)|H])).
```

The auxiliary predicate ok here is used to handle the *Trans** and *Final* part of the axiom by searching forward for a final configuration.[11] Note that when a future transition involves an action that has a sensing result, we need the program to terminate successfully for *both* sensing values. This is clearly explosive in general: sensing and off-line search do not mix well. It is precisely to deal with this issue in a flexible way that we have taken an on-line approach, putting the control in the hands of the programmer.

5.3   Handling test conditions

The rest of the interpreter is concerned with the evaluation of test conditions involving fluents, given some history of actions and sensing results. We assume the programmer provides the following clauses:

- poss ($Act, Cond$) : the action is possible when the condition holds;
- senses ($Act, Fluent$) : the action can be used to determine the truth of the fluent;[12]
- initially ($Fluent$) : the fluent holds in the initial situation S_0;
- causesTrue ($Act, Fluent, Cond$) : if the condition holds, performing the action causes the fluent to hold;
- causesFalse ($Act, Fluent, Cond$) : if the condition holds, performing the action causes the fluent to not hold.

In the absence of sensing, the last two clauses provide a convenient specification of a successor state axiom for a fluent F, as if we had (very roughly)

$$F(do(a, s)) \equiv$$
$$\exists \phi(causesTrue(a, F, \phi) \wedge \phi[s]) \quad \vee$$
$$F(s) \wedge \neg \exists \phi(causesFalse(a, F, \phi) \wedge \phi[s]).$$

In other words, F holds after a if a causes it to hold, or it held before and a did not cause it not to hold. With sensing, we have some additional possibilities. We can handle fluents that are completely unaffected by the given primitive actions by leaving out these two clauses, and just using sensing. We can also handle fluents that are partially affected. For example, in an elevator controller, it may be necessary to use sensing to determine if a button has been pushed, but once it has been pushed, we can assume the corresponding light stays on until we perform a reset action

---

[11] In practice, a breadth-first search may be preferable. Also, we would want to cache the results of the search to possibly avoid repeating it at the next transition.

[12] The specification allows a sensor to be linked to an arbitrary formula using *SF*; the implementation insists it be a fluent.

causing it to go off. We can also handle cases where some initial value of the fluent needs to be determined by sensing, but from then on, the value only changes as the result of actions, *etc*. Note that an action can provide information for one fluent and also cause another fluent to change values.

With these clauses, the transitions for primitive actions and tests would be specified as follows:

```
trans(prim(Act),H,nil,[(Act,_)|H])  :-
    poss(Act,Cond), holds(Cond,H).

trans(test(Cond),H,nil,H)  :- holds(Cond,H).
```

where nil is the empty program. The holds predicate is used to evaluate arbitrary conditions. Because of the just-in-time histories assumption, the problem reduces to holdsf for fluents (we omit the reduction). For fluents, we have the following:

```
holdsf(F,[])  :- initially(F).

holdsf(F,[(Act,X)|H])  :-
    senses(Act,F),!, X=1. /* mind the cut */

holdsf(F,[(Act,X)|H])  :-
    causesTrue(Act,F,Cond), holds(Cond,H).

holdsf(F,[(Act,X)|H])  :-
    not ( causesFalse(Act,F,Cond), holds(Cond,H) ),
    holdsf(F,H).
```

Observe that if the final action in the history is not a sensing action, and not an action that causes the fluent to hold or not hold, we regress the test to the previous situation. This is where the just-in-time histories assumption: for this scheme to work properly, the programmer must ensure that a sensing action and its result appear in the history as necessary to establish the current value of a fluent.

5.4   Correctness

This completes the incremental interpreter. The interpreter presented above is correct under suitable hypotheses. In particular, apart from the usual assumption required when we encoding an action theory in Prolog (see [19]), we make the hypothesis that the predicate **holds** satisfies the following properties.[13] Let δ and σ contain free variables only on objects and actions:

1. If a goal **holds**(ϕ, σ) succeeds with computed answer θ, then (by $\forall \psi$, we mean the universal closure of ψ)

$$Axioms \cup \{Sensed[\sigma]\} \models \forall \phi(end[\sigma])\theta.$$

---

[13] We keep implicit the translation between Prolog terms and the programs, histories, and terms of the situation calculus

2. If a goal $\mathtt{holds}(\phi, \sigma)$ finitely fails, then

$$Axioms \cup \{Sensed[\sigma]\} \models \forall \neg \phi(end[\sigma]).$$

Although we do not attempt to show it formally here, it should be intuitively clear that our definition for $\mathtt{holds}$ under the just-in-time histories assumption does actually satisfy the requirements above.

We can formally state the correctness of the incremental interpreter as follows.

Theorem 2. *Let δ and σ contain free variables only on objects and actions. Then under the hypotheses above:*

1. *If a goal $\mathtt{trans}(\delta, \sigma, \delta', \sigma')$ succeeds with computed answer θ, then*

$$Axioms \cup \{Sensed[\sigma]\} \models \forall Trans(\delta, end[\sigma]\delta', end[\sigma'])\theta$$

   moreover $\delta'\theta$ and $\sigma'\theta$ contain free variables only on objects and actions.
2. *If a goal $\mathtt{trans}(\delta, \sigma, \delta', \sigma')$ finitely fails, then*

$$Axioms \cup \{Sensed[\sigma]\} \models \forall \neg Trans(\delta, end[\sigma], \delta', end[\sigma']).$$

3. *If a goal $\mathtt{final}(\delta, \sigma)$ succeeds with computed answer θ, then*

$$Axioms \cup \{Sensed[\sigma]\} \models \forall Final(\delta, \sigma)\theta.$$

4. *If a goal $\mathtt{final}(\delta, \sigma)$ finitely fails, then*

$$Axioms \cup \{Sensed[\sigma]\} \models \forall \neg Final(\delta, end[\sigma]).$$

Notably, because of the assumption above on $\mathtt{holds}$ —and hence because of the just-in-time histories assumption— we have that if $\mathtt{trans}$ succeeds for a program of the form $(\Sigma\delta, s)$ then an on-line successful execution exists indeed.

6   Discussion

The framework presented here has a number of limitations beyond those already noted: it only deals with sensors that are binary and noise-free; no explicit mention is made of how the sensing influences the knowledge of the agent, as in [18]; the interaction between off-line search and concurrency is left unexplored; finally, the implementation has no finite way of dealing with search over a program with loops.

One of the main advantages of a high-level agent language containing nondeterminism is that it allows limited versions of (runtime) planning to be included within a program. Indeed, a simple planner can be written directly:[14]

$$\mathbf{while} \ \neg\phi \ \mathbf{do} \ \pi a. \ (Acceptable(a)? \ ; \ a) \ \mathbf{endWhile}.$$

---

[14] The π operator is used for a nondeterministic choice of value.

Ignoring *Acceptable*, this program says to repeatedly perform some nondeterministically selected action until condition ϕ holds. An off-line execution would search for a legal sequence of actions leading to a situation where ϕ holds. This is precisely the planning problem, with *Acceptable* being used as a forward filter, in the style of [2].

However, in the presence of sensing, it is not clear how even limited forms of planning like this can be handled by an off-line interpreter, since a *single* nondeterministic choice can cause problems, as we saw earlier. The formalism presented here has more chances of being practical for large programs containing both nondeterministic action selection and sensing.

One concern one might have is that once we move to on-line execution where nondeterministic choice defaults to being random, we have given up reasoning about courses of action, and that our programs are now just like the pre-packaged "plans" found in RAP [3] or PRS [11]. Indeed in those systems, one normally does not search off-line for a sequence of actions that would eventually lead to some future goal; execution relies instead on a user-supplied "plan library" to achieve goals. In our case, with Σ, we get the advantages of both worlds: we can write agent programs that span the spectrum from scripts where no lookahead search is done and little needs to be known about the properties of the primitive actions being executed, all the way to full planners like the above. Moreover, our formal framework allows considerable generality in the formulation of the action theory itself, allowing disjunctions, existential quantifiers, *etc*. Even the Prolog implementation described here is considerably more general than many STRIPS-like systems, in allowing the value of fluents to be determined by sensing intermingled with the context-dependent effects of actions.

A more serious concern, perhaps, involves how to build an effective program to be executed on-line. There is a difficult tradeoff here that also shows up in the work on so-called *incremental planning* [1], [12]. Even if we have an important goal that needs to be achieved in some distant place or time, we want to make choices here and now without worrying about it. How should I decide what travel agent to use given that I have to pick up a car at an airport in Amsterdam a month from now? The answer in practice is clear: decide locally and cross other bridges when you get to them, exactly the motivation for the approach presented here. It pays large dividends to assume by default that routine choices will not have distant consequences, chaos and the flapping of butterfly wings notwithstanding. But as far as we know, it remains an open problem to characterize formally what an agent would have to know (test/sense) to be able to quickly confirm that some action can be used immediately as a first step towards some challenging but distant goal.

References

1. J. A. Ambros-Ingerson and S. Steel. Integrating planning, execution and monitoring. In *Proc. AAAI-88*, 1988.
2. F. Bacchus and F. Kabanza. Planning for temporally extended goals. In *Proc. AAAI-96*, 1996.

3. R. J. Firby. An investigation in reactive planning in complex domains. In *Proc. AAAI-87*, 1987.

4. G. De Giacomo, Y. Lespérance, and H. Levesque. Reasoning about concurrent execution, prioritized interrupts, and exogenous actions in the situation calculus. In *Proc. IJCAI-97*, 1997.

5. G. De Giacomo, Y. Lespérance, and H. Levesque. CONGOLOG, a concurrent programming language based on the situation calculus: language and implementation. submitted, 1999.

6. G. De Giacomo, Y. Lespérance, and H. Levesque. CONGOLOG, a concurrent programming language based on the situation calculus: foundations. submitted, 1999.

7. G. De Giacomo and H. Levesque. Progression and regression using sensors. In *Proc. IJCAI-99*, 1999.

8. G. De Giacomo, R. Reiter, and M. Soutchanski. Execution monitoring of high-level robot programs. In *Proc. of KR-98*, pages 453–465, 1998.

9. K. Golden and D. Weld. Representing sensing actions: the middle ground revisited. In *Proc. KR-96*, 1996.

10. M. Hennessy. *The Semantics of Programming Languages*. John Wiley & Sons, 1990.

11. F. F. Ingrand, M. P. Georgeff, and A. S. Rao. An architecture for real-time reasoning and system control. *IEEE Expert*, 7(6), 1992.

12. P. Jonsson and C. Backstrom. Incremental planning. In *Proc. 3rd European Workshop on Planning*, 1995.

13. H. Levesque. What is planning in the presence of sensing? In *Proc. AAAI-96*, 1996.

14. H. Levesque, R. Reiter, Y. Lespérance, F. Lin, and R. Scherl. GOLOG: A logic programming language for dynamic domains. *Journal of Logic Programming: special issue on actions*, 31(1–3):59–83, 1997.

15. F. Lin and R. Reiter. How to progress a database. *Artificial Intelligence*, 92:131–167, 1997.

16. J. McCarthy and P. Hayes. Some philosophical problems from the standpoint of artificial intelligence. *Machine Intelligence*, 4, 1969.

17. G. Plotkin. A structural approach to operational semantics. Technical Report DAIMI-FN-19, Computer Science Department Aarhus University Denmark, 1981.

18. R. Reiter. The frame problem in the situation calculus: A simple solution (sometimes) and a completeness result for goal regression. In *Artificial Intelligence and Mathematical Theory of Computation: Papers in Honor of John McCarthy*, pages 359–380. Academic Press, 1991.

19. R. Reiter. Knowledge in action: Logical foundation for describing and implementing dynamical systems. In preparation., 1999.

20. R. Scherl and H. Levesque. The frame problem and knowledge producing actions. In *Proc. of AAAI-93*, pages 689–695, 1993.

An Improved Incremental Algorithm for Generating Prime Implicates

Johan de Kleer

Xerox Palo Alto Research Center, 3333 Coyote Hill Road, Palo Alto CA 94304 USA
e-mail:dekleer@parc.xerox.com

In 1987 Ray Reiter and I wrote a paper entitled "Foundations of assumption-based truth maintenance systems: Preliminary report" [9] which showed how the behavior of the Assumption-Based Truth Maintenance System [1] can be defined using the notions of prime implicate and prime implicant . This definition of the ATMS immediately suggests generalizing the ATMS to operate on arbitrary clauses. This generalization raises two immediate computational challenges (to me, not Ray who seems immune to such challenges). First, computing prime implicates/implicants is very expensive. Second, since ATMS's are used incrementally we need to exploit previous computation. This paper describes an improved and incremental algorithm to compute prime implicates/implicants. This algorithm allows us to experiment with the ideas Ray and I laid out in our paper. Unfortunately, the task is inherently NP-complete and all this paper can accomplish is present a more clever incremental algorithm.

Summary. Prime implicates have become a widely used tool in AI. The prime implicates of a set of clauses can be computed by repeatedly resolving pairs of clauses, adding the resulting resolvents to the set and removing subsumed clauses. Unfortunately, this brute-force approach performs far more resolution steps than necessary. Tison provided a method to avoid many of the resolution steps and Kean and Tsiknis developed an optimized incremental version. Unfortunately, both these algorithms focus only on reducing the number of resolution steps required to compute the prime implicates. The actual running time of the algorithms depends critically on the number and expense of the subsumption checks they require. This paper[1] describes a method based on a simplification of Kean and Tsiknis' algorithm using an entirely different data structure to represent the data base of clauses. The new algorithm uses a form of discrimination net called tries to represent the clausal data base which produces an improvement in running time on all known examples with a dramatic improvement in running time on larger examples.

1 Introduction

Prime implicates have become a widely used tool in AI. They can be used to implement ATMSs [9], to characterize diagnoses [2], to compile formulas for TMSs [3],

---

[1] This is a revision of a paper which appeared in *Proceedings of the National Conference on Artificial Intelligence,* San Jose, CA (1992), 780-785.

to implement circumscription [5], [8], to give but a few examples. The prime implicates of a set of clauses can be computed by repeatedly resolving pairs of clauses, adding the resulting resolvents to the set and removing subsumed clauses. Unfortunately, this brute-force approach performs far more resolution steps than necessary. Tison [10] provided a method to avoid many of the resolution steps and Kean and Tsiknis [6] give an optimized incremental version. Both algorithms provide a significant advance as they substantially reduce the number of resolution steps required to compute the prime implicates of a set of clauses.

Unfortunately, both algorithms focus only on reducing the number of resolution steps required to compute the prime implicates. The actual running time of the algorithm also depends critically on the number and expense of the subsumption checks they requires. Reducing the number of resolutions certainly reduces the number of subsumption checks required. However, the number of subsumption checks required grows (see analysis in Section 3.2 and data in Section 4) faster than the square of the final number of prime implicates. As a result both algorithms are impractical on all but the tiniest examples.

Neither algorithm [6], [10] indicates how subsumption is to be performed, and any prime implicate algorithm is incomplete without such a specification. This paper proposes a method based on a simplification of Kean and Tsiknis' algorithms using an entirely different data structure to represent the data base of clauses in order to facilitate subsumption checking. This data structure is called trie [7] which has been explored extensively for representing dictionaries of words. (Tries are also used in ATMS implementations [1] to store the nogood data base.) Using tries to represent the data base dramatically improves the performance of prime implicate algorithms. The data in Section 4 shows that we can now construct the prime implicates for a much larger set of tasks.

Admittedly, no amount of algorithmic improvement can avoid the complexity produced by the sheer number of prime implicates. The number of prime implicates for many tasks grows relatively quickly so the approach is impractical for many applications. However, we can now conceive of computing prime implicates for applications impossible to before.

2   A brute-force algorithm

The key step in computing prime implicates consists of a resolution rule called consensus [6], [9], [10]. Given two clauses:

$$x \vee \beta,$$

$$\neg x \vee \gamma,$$

where x is a symbol and β and γ are (possibly empty) disjunctions of literals, the consensus of these two clauses with respect to x is the clause,

$$\beta \vee \gamma,$$

with duplicate literals removed. If the two clauses have more than one pair of complementary literals, then the consensus would contain complementary literals and is discarded (since it is a logical tautology). The prime implicates of a set of clauses can be computed by repeatedly adding the consensus of any pair of clauses to the set and continually removing all subsumed clauses (until no further consensus and subsumption is possible).

The following algorithm finds the prime implicates of a set of clauses Q:

Algorithm **Brute-Force**(Q)

1. Let $\mathcal{P}$ (the result) be $\{\}$.
2. Take the first clause q off of Q. If none we are done.
3. If q is subsumed by any clause of $\mathcal{P}$, then go back to step 2.
4. Remove all clauses of $\mathcal{P}$ which are subsumed by q.
5. Try to compute the consensus of q and every clause in $\mathcal{P}$. Whenever the consensus exists, add it to Q.
6. Add q to $\mathcal{P}$.
7. Go to step 2.

3 Improving efficiency

The algorithm presented in Section 2 is intuitively appealing but quite inefficient in practice. Constructing prime implicates is known to be NP-complete and therefore it is unlikely any really good algorithm exists. Nevertheless we can do dramatically better than the algorithm of Section 2. Through logical analysis we can eliminate a large number of the redundant consensus calculations. We can redesign the data structures to support addition and deletion of clauses relatively efficiently.

3.1 A more efficient consensus algorithm

When one observes the behavior of **Brute-Force**, almost all the clauses produced by the consensus calculation are subsumed by others. One reason for this is somewhat obvious: the consensus operation is commutative and associative when the consensi all exist (usually the case). For example, if we have three clauses α, β and γ, $consensus(\alpha, consensus(\beta, \gamma)) = consensus(consensus(\alpha, \beta), \gamma)$. For 3 clauses **Brute-Force** finds the same result in 3 distinct ways. Unfortunately, the number of ways to derive a result grows exponentially in the number of clauses used to produce the final result.

Tison [6], [10] introduced the following key intuition which suppresses the majority of the consensus calculations. To compute the prime implicates of a set of clauses, place an ordering on the symbols, then iterate over these symbols in order, doing all consensus calculations with respect to that symbol only. Once all the consensus calculations for a symbol have been made *it is never necessary to do another consensus calculation with respect to that symbol even for all the new consensus results which are produced later* (of course, when the user incrementally supplies the next clause the symbols must be reconsidered). Figure 3.1 provides a simple example. Suppose we are given clauses $A \lor B$, $\neg B \lor C$ and $\neg C \lor D$. The symbols are ordered: A, B and C. There are no consensus calculations available for A. There is only one (at 1 in the figure) consensus calculation available for B (on the first and second clauses). Finally, when processing C there are two (at 2 and 3 in the figure) consensus calculations available. One of those consensus calculations produces $\neg B \lor D$. Although this resolves with the first original clause, Tison's method tells us the result will be irrelevant because all the useful consensus calculations with respect to B have already been made.

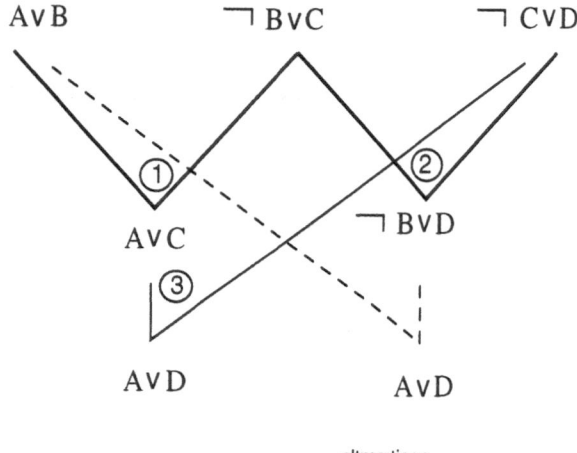

cltms:tison

Fig. 1. Tison's Method

The following algorithm incorporates this ideas. The algorithm **IPIA** (this derives from the incremental Tison method presented and proved correct in [6]) takes a current set of prime implicates N and a set S of new clauses to add.

Algorithm **IPIA(N,S)**

1. Delete any $D \in N \cup S$ that is subsumed by another $D' \in N \cup S$.
2. Remove a smallest C clause from S. If none, return.
3. For each literal l of C, construct Π_l which contains all clauses of N which resolve successfully with C.
4. Let Σ be the set containing C.
5. Perform the following steps for each literal l of C.
   (a) For each clause in Σ which is still in N compute the consensus of it and every clause in Π_l which is still in N.
   (b) For every new consensus, discard it if it has been subsumed by $N \cup S$. Otherwise, remove any clauses in $N \cup S$ subsumed by it. Add the clause to N and Σ.

Comparing this algorithm to the previous one we see that a great many consensus computations are avoided:

- Consensus calculations with respect to a literal earlier in the order are ignored.
- Two clauses produced in the same main step (choice of C) are never resolved with each other.
- Consensus calculations with a D in the original N are ignored unless the consensus of D and C exists.

3.2   Implementing subsumption checking efficiently

Thus far we have been analyzing the logic of the prime implicate algorithms in order to improve their efficiency. However, all the algorithms we know of depend critically on subsumption checking and unless that is properly implemented all the CPU time will be spent checking subsumption.

A key observation is that we are maintaining a data base of unsubsumed clauses. We need to implement 3 transactions with this data base.

1. Check whether clause x is subsumed by some clause of the data base.
2. Add clause x to the data base.
3. Remove all clauses from the data base which are subsumed by x.

To understand some of the complexities, consider the most obvious implementation: We could implement the subsumption check by a subset test and maintain the data base as a simple list. Using lists makes checking for subsumption of order the number of clauses, and thus the complexity of generating k prime implicates at least k^2 which is unacceptable.

Our implementation is based on an integration of two ideas. First, each clause is always represented in a canonical form. Second, the clause data base is represented as a discrimination tree. To achieve a canonical form for clauses we assign a unique integer id to each symbol. We order the literals of every clause in ascending order of their ids. (Complementary literals have the same id, but they can never appear in the same clause as this would produce a tautology.) This means that two sets of literals refer to the same clause if their ordered lists of literals are identical. For example, given symbols A, B and C with id's 1, 2 and 3, the clause,

$$A \vee B \vee C$$

is represented by the list,

$$[A, B, C].$$

The representation of clauses is very sensitive on the choice of id's. If the id's were 3, 1, 2 respectively, then the clause would be represented by the list,

$$[B, C, A].$$

This also makes it possible to test whether a clause of length n subsumes a clause of length m in at most $n + m$ comparisons. However, our algorithm never checks whether one clause subsumes another. Instead, our algorithm stores the canonical forms of clauses in a discrimination tree (or trie [7]). By storing canonical forms in a trie, a single clause can be checked against all existing clauses in a single operation.

The trie for clauses is relatively simple. Conceptually, it is a tree, all of whose edges are literals and whose leaves are clauses. The edges below each node in the trie are ordered by the id of the literal. Suppose that A, B, C, D and E are a sequence of nodes with ascending id's and the data base contains the three clauses:

$$A \vee B \vee C,$$

$$B \vee D,$$

$$A \vee D.$$

The resulting tree is illustrated in Figure 3.2. Because clauses are canonically ordered, our tries have the additional important property that the id of any edge is less than the id of any edge appearing below it at any depth. This property is heavily exploited in the update algorithms which follow.

Fig. 2. Data base with 3 clauses.

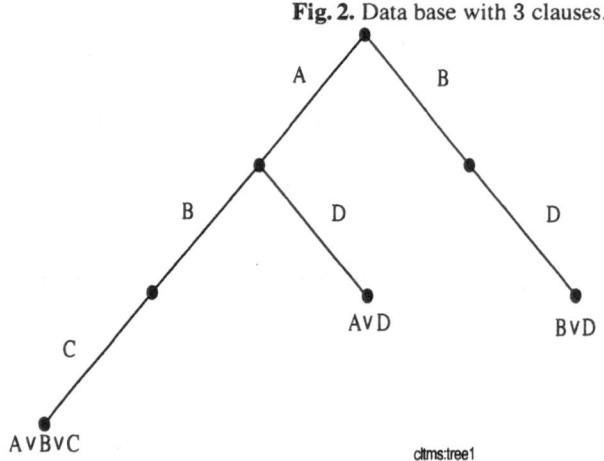

The most commonly called procedure checks whether a clause is subsumed by one in the data base. Given an ordered set of literals, the recursive function **Subsumed?** checks whether the set of literals L is subsumed by trie N. The ordered literals are represented as an ordered list, and the trie by an ordered list of edges.

Algorithm **Subsumed?**(L, N)

1. If N is a terminal clause, return success.

2. Remove literals from the front of L until the id of the first edge of N is greater than that of the first literal of L.
3. If no literals remain (L is empty), return failure.
4. For each literal l of L do the following until success or the id of the first edge of N is no longer greater than that of l.
   (a) If the first literal of N is l, recursively invoke **Subsumed?** on the remaining literals and the edges below the first element of N.
   (b) If the recursive call returns success, return success.
5. Remove the first element of N.
6. Go to Step 2.

Suppose we want to check whether $D \vee E$ is subsumed by the data base of Figure 3.2. The root of the trie has 2 outgoing edges, A and B. D has a larger id than the top two edges of the trie (A and B), therefore **Subsumed?** immediately reports failure. Suppose we want to check whether $A \vee B \vee D$ is subsumed by the trie. The first edge from the root matches the first literal, so the recursive call tries to determine whether the remaining subclause $B \vee D$ is subsumed by the trie rooted from the edge below A. Again B matches, but D does not match C so the two recursive calls to **Subsumed?** fail. Finally, the top-level invocation of **Subsumed?** again recursively calls itself and finds a successful match.

Adding a clause to the data base is very simple. Our algorithm exploits the fact that the clause to be added is not itself subsumed by some other clause, and that any clause it subsumes has been removed from the data base.

Allgorithm **Add-To-Trie**(L, N)

1. Remove edges of the front of N, until the id of the first edge of N is greater or equal to the first literal of L.
2. If the label of the first edge of N is the same as that of L, recursively call **Add-To-Trie** with the remainder of L, the edges underneath the first edge of N and return.
3. Construct the edges to represent the literals of L and return, and side-effect the trie such that it appears just before the current position N.

The potentially most expensive operation and the one which requires the greatest care is the third basic update on the trie. Here we are given a clause, not subsumed by the trie and we must remove from the trie all clauses subsumed by it.

Algorithm **Remove-Subsumed**(L, N)

1. If there are no literals, delete the entire trie represented by N and return.
2. If we are at a leaf of the trie, return.
3. For each edge e of N, do the following.
   (a) If the label of e is the first literal of L, then recursively call **Remove-Subsumed** with the rest of the literals and the edges below e.

(b) If the label of e is lower than that of the first literal of L, then recursively call **Remove-Subsumed** on the *same* literals but the edges below e.

As an extreme case suppose we want to remove all clauses subsumed by D from Figure 3.2. Because D is last in the ordering, the algorithm simply searches in left-to-right depth-first order removing all clauses containing D. After adding the clause D, the resulting trie is illustrated by Figure 3.2.

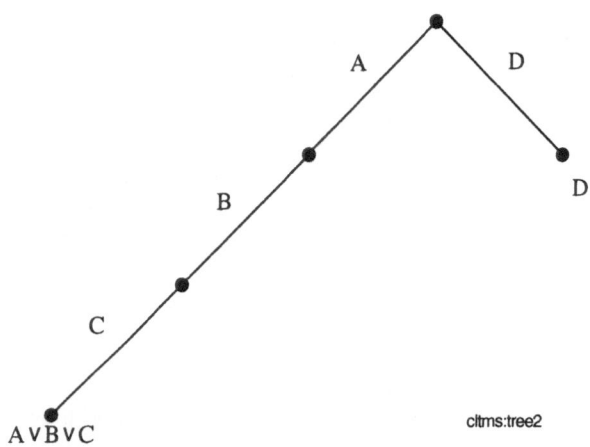

Fig. 3. Trie with 2 clauses.

4   Results

Table 1 table summarizes the performance improvement of **IPIA** produced using a trie. Each line of the table lists the name of the task, the number of clauses which specify the problem, the number of prime implicates of these clauses, the number of subsumption checks the non-trie version of **IPIA** requires, the running times of the two algorithms, the average fraction of the trie searched during subsumption checks, the fraction of resolvents which are not subsumed by the trie, and the number of non-terminal nodes in the final trie. The timings are obtained on a Symbolics XL1200. The Lisp code is not particularly optimized as it is designed for a text book [4].

The non-trie version of **IPIA** is so slow that it is not possible to actually time its performance on larger examples. We instrumented the trie version to estimate the number of subsumption checks that would have to be made by assuming that, on average, each subsumption check would check half of the current clauses. In all the cases we have tried, this estimate is within 25% of the actual number of subsumption checks so this estimate seems fairly reliable. We estimate the timing of the non-trie

Table 1. Comparison of the old and improved algorithms. " * " indicates estimates.

Task	Cls	PIs	Checks	$t_O(s)$	$t_N(s)$	Srchd	New	Size
2Ppes	54	88	13124	.22	.06	.081	.50	54
Adder	50	8400	$3.8 \times 10^{10}*$	$3.8 \times 10^5*$	721	.009		16207
K33	10	73	5166	.12	.03	.123	1	24
K44	17	641	506472	5.5	.46	.031	1	160
K55	26	7801	$9.1 \times 10^7*$	$900*$	5.7	.003	1	1560
K54	21	1316	1730120	22.6	1.1		1	263
K66	37	117685	$2.4 \times 10^{10}*$	$2.4 \times 10^5*$	224		1	
K67	43	823585	$1.3 \times 10^{15}*$	$1.4 \times 10^{10}*$	2541		1	137264
Reg	106	2814	$3.7 \times 10^9*$	$3.7 \times 10^4*$	85		.06	1365
BD	151	1907	$1.1 \times 10^9*$	$1.1 \times 10^4*$	31		.067	2493

version using 10 microseconds per subsumption check which is the fastest we've ever seen the non-trie algorithm perform. The implementation is written in basic Common Lisp and runs in Lucid and Franz as well. Both the code and the examples are available from the author.

Table 1 clearly indicates the dramatic performance achievement produced using a trie data structure to represent the clausal data base. The actual **IPIA** algorithm in [6] includes additional optimizations to the version we have presented in this paper. Those optimizations reduce the number of resolutions and subsumption checks. However, these optimizations entail additional bookkeeping and are known to produce incorrect results in some cases. Therefore, we restrict our comparison to the basic version of **IPIA**.

The tasks labeled "2Ppes" and "Reg" come from qualitative physics. The tasks labeled "Adder" and "BD" come are diagnosis problems. The tasks labeled "K"nm are taken from [6].

The analysis clearly shows that the trie-based algorithm yields substantial improvement in all cases. The final columns in the table provide some insight into why performance improves so dramatically. One hypothesis for the good performance is that most new clauses are subsumed by others and therefore immediately found in the trie and that the data structure would perform poorly if there were few subsumptions. The data does not substantiate this intuition. The column labeled "Fraction New" indicates the fraction of the new clauses which are not subsumed by the current trie. We see even in the worst case where the new clause is never subsumed, that the trie-based algorithm is far superior. The column labeled "Fraction Searched" shows the average fraction of the non-terminal nodes in the trie actually searched. This data suggests that the central advantage of using tries is that subsumption checks need only examine a small fraction of trie.

We have attempted to invent tasks which would defeat the advantage of using tries and have not been able to find any. Tries would perform poorly for a task in which all subsumption checks failed and all the nodes in the trie have to be scanned for each failing subsumption check. To achieve this the trie would have to contain a very high density of clauses. It is difficult to devise an initial set of clauses whose prime implicates densely populate the space. Moreover, the fact that clauses resolve with each to produce new ones means that if the clause set becomes too dense the clause set is likely to be reduced by subsumptions (a dense clause set is also likely to be inconsistent).

The performance of the trie-based **IPIA** is relatively sensitive to the choice of ids (although performance is always much better than the non-trie version). The choice of ids affects the canonicalized forms of the clauses and hence the size of the trie. To lower the size of the trie, the most common symbols should have lower id. Although it is impossible to tell initially which symbols will occur more commonly in the prime implicates, we use the number of occurences in the initial clause set as a guide. **IPIA** performs noticably better if less common symbols are processed first (i.e., in steps 2 and 5).

References

1. de Kleer, J., An assumption-based truth maintenance system, *Artificial Intelligence* **28** (1986) 127–162. Also in *Readings in NonMonotonic Reasoning*, edited by Matthew L. Ginsberg, (Morgan Kaufmann, 1987), 280–297.
2. de Kleer, J., Mackworth A., and Reiter R., Characterizing diagnoses and systems, *Artificial Intelligence* **59** (1993) 63–67.
3. de Kleer, J., Exploiting locality in a TMS, AAAI-90, Boston, MA (1990) 254–271.
4. Forbus, K., and de Kleer, J., *Building problem solvers* (MIT Press, 1992).
5. Ginsberg, M.L., A circumscriptive theorem prover, *Proceedings of the second international workshop on Non-Monotonic Reasoning*. Springer, LNCS 346, 100–114, (1988).
6. Kean, A. and Tsiknis, G., An incremental method for generating prime implicants/implicates, *Journal of Symbolics Computation* **9** (1990) 185–206.
7. Knuth, D.E., *The art of computer programming* (Addison-Wesley, Reading, MA, 1972).
8. Raiman, O., and de Kleer, J., A minimality maintenance system, *Proceedings of the international cconference on Knowledge Representation and Reasoning*, Boston, MA, 532-538.
9. Reiter, R. and de Kleer, J., Foundations of assumption-based truth maintenance systems: Preliminary report, *Proceedings of the National Conference on Artificial Intelligence*, Seattle, WA (July, 1987), 183–188.
10. Tison, P., Generalized consensus theory and application to the minimization of boolean functions, *IEEE transactions on electronic computers* **4** (August 1967) 446-456.

Fixpoint 3-valued semantics for autoepistemic logic

Marc Denecker[1], V. Wiktor Marek[2], and Mirosław Truszczyński[2]

[1] Department of Computer Science, K.U.Leuven, Celestijnenlaan 200A, B-3001 Heverlee, Belgium
[2] Department of Computer Science, University of Kentucky, Lexington, KY 40506-0046, USA

*Dedicated to Ray Reiter
on his 60th birthday*

Summary. The paper presents a constructive 3-valued semantics for autoepistemic logic (AEL). We introduce a derivation operator and define the semantics as its least fixpoint. The semantics is 3-valued in the sense that, for some formulas, the least fixpoint does not specify whether they are believed or not. We show that complete fixpoints of the derivation operator correspond to Moore's stable expansions. In the case of modal representations of logic programs our least fixpoint semantics expresses both well-founded semantics and 3-valued Fitting-Kunen semantics (depending on the embedding used). We show that, computationally, our semantics is simpler than the semantics proposed by Moore (assuming that the polynomial hierarchy does not collapse).

1 Introduction

We describe a 3-valued semantics for modal theories that approximates skeptical mode of reasoning in the autoepistemic logic introduced in [12], [13]. We present results demonstrating that our approach is, indeed, appropriate for modeling autoepistemic reasoning. We discuss computational properties of our semantics and connections to logic programming.

Autoepistemic logic is among the most extensively studied nonmonotonic formal systems. It is closely related to default logic introduced by Reiter in [17]. It can handle default reasonings under a simple and modular translation in the case of prerequisite-free defaults [10]. In the case of arbitrary default theories, a somewhat more complex non-modular translation provides a one-to-one correspondence between default extensions and stable (autoepistemic) expansions [5]. Further, under the so called Gelfond translation, autoepistemic logic captures the semantics of stable models for logic programs [3]. Under the Konolige encoding [6] of logic programs as modal theories, stable expansions generalize the concept of the supported model semantics [10]. Autoepistemic logic is also known to be equivalent to several other modal nonmonotonic reasoning systems including the *only-knowing* logic of Levesque [8] and the reflexive autoepistemic logic of Schwarz [18].

The semantics for autoepistemic logic [13] assigns to a modal theory T a collection of its *stable expansions*. This collection may be empty, may consist of exactly one expansion, or may consist of several different expansions. Intuitively, consistent stable expansions are designed to model belief states of agents with *perfect* introspection powers: for every formula F, either the formula KF (expressing a belief in F) or the formula $\neg KF$ (expressing that F is not believed) belongs to an expansion. We will say that expansions contain no *meta-ignorance*.

In many applications, the phenomenon of multiple expansions is desirable. There are situations where we are not interested in answers to queries concerning a single atom or formula, but in a *collection* of atoms or formulas that satisfy some constraints. Planning and diagnosis in artificial intelligence, and a range of combinatorial optimization problems, such as computing hamilton cycles or k-colorings in graphs, are of this type. These problems may be solved by means of autoepistemic logic precisely due to the fact that multiple expansions are possible. The idea is to represent a problem as an autoepistemic theory so that solutions to the problem are in one-to-one correspondence with stable expansions. While conceptually elegant, this approach has its problems. Determining whether expansions exist is a Σ_2^P-complete problem [4], [14], and all known algorithms for computing expansions are highly inefficient.

In a more standard setting of knowledge representation, the goal is to model the knowledge about a domain as a theory in some formal system and, then, to use some inference mechanism to resolve queries against the theory or, in other words, establish whether particular formulas are entailed by this theory. Autoepistemic logic (as well as other nonmonotonic systems) can be used in this mode, too. Namely, under the so called *skeptical* model, a formula is entailed by a modal theory, if it belongs to all stable expansions of this theory. The problem is, again, with the computational complexity of determining whether a formula belongs to all expansions; this decision problem is Π_2^P-complete [4].

We propose an alternative semantics for autoepistemic reasoning that, in particular, allows us to *approximate* the skeptical approach described above (as well as the dual, *brave* mode of reasoning). Our semantics has the property that if it assigns to a formula the truth value **t**, then this formula belongs to all stable expansions and, dually, if it assigns to a formula the truth value **f**, then this formula does not belong to any expansion. Our semantics is 3-valued and some formulas are assigned the truth value **u** (unknown). While only approximating the skeptical mode of reasoning, it has one important advantage. Its computational complexity is lower (assuming that the polynomial hierarchy does not collapse on some low level). Namely, the problem to determine the truth value of a formula under our semantics is in the class Δ_2^P.

As mentioned above, the semantics we propose can be applied to approximate the skeptical mode of autoepistemic reasoning. However, it has also another important application. It can be used as a pruning mechanism in algorithms that compute expansions. While searching for expansions, one can compute our 3-valued semantics for a modal theory under consideration (as mentioned, it is a simpler task computationally than the task of computing an expansion). Formulas true under this semantics are guaranteed to belong to all expansions and those that are false belong to none. This information can be used to simplify the current theory and limit the search space. As a consequence, significant speedups may be achieved.

There are parallels between our semantics and the well-founded semantics in logic programming. The well-founded semantics approximates the stable model semantics (atoms true under the well-founded semantics are in all stable models and atoms that are false under the well-founded semantics belong to none). Moreover, computing well-founded semantics is polynomial while deciding whether an atom belongs to all stable models is a co-NP-complete problem. As a result, the well-founded semantics is used as a search space pruning mechanism by some algorithms to compute stable model semantics [15]. We will show in the paper that there is, indeed, a close formal connection between our 3-valued semantics of modal theories and the well-founded semantics of logic programs.

The 3-valued semantics for autoepistemic logic introduced in this paper is based on the notion of a *belief pair*, that is, a pair (P, S), where P and S are sets of 2-valued interpretations of the underlying first-order language, and $S \subseteq P$. The motivation to consider belief pairs comes from Moore's possible-world characterization of stable expansions [12]. Moore characterized expansions in terms of *possible-world structures*, that is, sets of 2-valued interpretations. A belief pair (P, S) can be viewed as an approximation to a possible-world structure W such that $S \subseteq W \subseteq P$: interpretations not in P are known not to be in W, and those in S are known to be in W. It turns out that while expansions (or the corresponding possible-world structures) do not contain meta-ignorance, belief pairs, in general, do.

There is a natural ordering of belief pairs. We say that (P_1, S_1) "better approximates" than (P, S) the agent's beliefs entailed by the agent's initial assumptions if $S \subseteq S_1 \subseteq P_1 \subseteq P$. We will denote the corresponding ordering relation in the set $\mathcal{B}$ of all belief pairs by $\leq_p$. Our semantics of modal theories is defined in terms of an operator on the set of belief pairs. This operator, $\mathcal{D}_T$, is determined by a modal theory T (the set of initial assumptions of the agent). Intuitively, it attempts to simulate a constructive process a rational agent might use to produce an "elementary" improvement on this agent's current set of beliefs and disbeliefs: given a belief pair $B = (P, S)$, $\mathcal{D}_T(B)$ is a belief pair that provides another, under some assumptions better, approximation to the agent's beliefs.

An important property is that $\mathcal{D}_T$ is monotone with respect to $\leq_p$. Hence, it has the least fixpoint. This least fixpoint can be constructed by starting with the least informative belief pair (approximating every possible-world structure) and then iterating the operator $\mathcal{D}_T$, in each step improving on the previous belief pair until no further improvement is possible. We propose this fixpoint as a *constructive* approximation to the semantics of stable expansions.

A fundamental property that makes the above approach meaningful is that *complete* belief pairs (those with P equal to S) that are fixpoints of $\mathcal{D}_T$ are (under an obvious one-to-one correspondence) precisely Moore's autoepistemic models characterizing expansions. Thus, by the general properties of fixpoints of monotone operators over partially ordered sets, the least fixpoint described above indeed approximates the skeptical and brave reasoning based on expansions. Moreover, as mentioned above, the problem of computing the least fixpoint of the operator $\mathcal{D}_T$ requires only polynomially many calls to the satisfiability testing engine, that is, it is in Δ_2^P. Another property substantiating our approach is that under some natural encodings of logic programs as modal theories, our semantics yields both well-founded semantics [20] and the 3-valued Fitting-Kunen semantics [2], [7].

Our paper is structured as follows. The next section reviews the basics of autoepistemic logic including both syntactic and semantic definitions of expansions. We then investigate the properties of the partial ordering of belief pairs and study the operator $\mathcal{D}_T$. Subsequently, we show how the purely semantic approach can be described in proof-theoretic terms and use this proof-theoretic approach to study algorithmic issues of the least fixpoint of the operator $\mathcal{D}_T$. Next, we discuss connections between fixpoints of $\mathcal{D}_T$ and several semantics of logic programs with negation. Section 6 contains conclusions and a discussion of future work. The appendix that concludes the paper gives a proof of Theorem 6.

2   Autoepistemic logic – preliminaries

The language of autoepistemic logic is the standard language of propositional modal logic over a set of atoms At and with a single modal operator K. We will refer to this language as $\mathcal{L}_K$. The modal-free fragment of $\mathcal{L}_K$ will be denoted by $\mathcal{L}$.

The notion of a 2-*valued interpretation* of the language $\mathcal{L}$ is defined as usual: it is a mapping from At to $\{\mathbf{t}, \mathbf{f}\}$. Throughout the paper $\mathcal{A}_{At}$ (or $\mathcal{A}$, if At is clear from the context) will always denote the set of all interpretations of the set At of atoms of $\mathcal{L}$.

Autoepistemic logic was first introduced by Moore in [12] and later studied in [13]. In [13], the semantics of an autoepistemic theory T is defined in terms of *stable expansions*. For every two sets T and E of modal formulas, E is said to be a *stable expansion* of T if it satisfies the equation:

$$E = \{\varphi \colon T \cup \{\neg K\psi \colon \psi \notin E\} \cup \{K\psi \colon \psi \in E\} \models_{FOL} \varphi\}$$

(the symbol $\models_{FOL}$ stands for classical entailment, where all formulas $K\varphi$ are interpreted as propositional literals).

A possible-world treatment of autoepistemic logic was described by Moore [12]. A possible-world structure W (over At) is a set of 2-valued interpretations of At. Alternatively, it can be seen as a Kripke structure with a total accessibility relation. Given a pair (W, I), where W is a possible-world structure and I is an interpretation (not necessarily from W), one defines a truth assignment function $\mathcal{H}_{W,I}$ inductively as follows:

i. For an atom A, we define $\mathcal{H}_{W,I}(A) = I(A)$
ii. The boolean connectives are handled in the usual way
iii. For every formula F, we define $\mathcal{H}_{W,I}(KF) = \mathbf{t}$ if for every interpretation $J \in W, \mathcal{H}_{W,J}(F) = \mathbf{t}$, and $\mathcal{H}_{W,I}(KF) = \mathbf{f}$, otherwise.

We write $(W, I) \models_{ael} F$ to denote that $\mathcal{H}_{W,I}(F) = \mathbf{t}$. Further, for a modal theory T, we write $(W, I) \models_{ael} T$ if $\mathcal{H}_{W,I}(F) = \mathbf{t}$ for every $F \in T$. Finally, for a possible world structure W we define the *theory* of W, $Th(W)$, by: $Th(W) = \{F: (W, I) \models_{ael} F$, for all $I \in W\}$.

It is well known that for every formula F, either $KF \in Th(W)$ or $\neg KF \in Th(W)$ (since $\mathcal{H}_{W,I}(KF)$ is the same for all interpretations $I \in \mathcal{A}$). Thus, possible-world structures have no meta-ignorance and, as such, are suitable for modeling belief sets of agents with *perfect* introspection capabilities. It is precisely this property that made possible-world structures fundamental objects in the study of modal nonmonotonic logics [12], [10].

Definition 1. An *autoepistemic model* of a modal theory T is a possible-world structure W which satisfies the following fixpoint equation[3]:

$$W = \{I: (W, I) \models_{ael} T\}.$$

The following theorem, relating stable expansions of [13] and autoepistemic models, was proved in [8] and was discussed in detail in [19].

Theorem 1. *For any two modal theories T and E, E is a stable expansion of T if and only if $E = Th(W)$ for some autoepistemic model W of T.*

3   A fixpoint 3-valued semantics for autoepistemic logic

Our semantics for autoepistemic logic is defined in terms of possible-world structures and fixpoint conditions. The key difference with the semantics proposed by Moore is that we consider *approximations* of possible-world structures by *pairs* of possible-world structures. Recall from the previous section, that $\mathcal{A}$ denotes the set of all interpretations of a fixed propositional language $\mathcal{L}$.

Definition 2. A *belief pair* is a pair (P, S) of sets of interpretations $P, S \subseteq At$ such that $S \subseteq P$. When $B = (P, S)$, $S(B)$ denotes S and $P(B)$ denotes P. The belief pair $(\mathcal{A}, \emptyset)$ is denoted $\perp$. The set $\{(P, S): P, S \subseteq \mathcal{A}$ and $P \supseteq S\}$ of all belief pairs is denoted by $\mathcal{B}$. The belief pair $(\emptyset, \emptyset)$ is called *inconsistent* and is denoted by $\top$.

A belief pair B can be seen as an approximation of a possible-world structure W such that $S(B) \subseteq W \subseteq P(B)$. The interpretations in $S(B)$ can be viewed as states of the world which are known to be possible (belong to W). The set of these interpretations forms a lower approximation to W. The set $P(B)$ of interpretations can be viewed as an upper approximation to W: interpretations not in $P(B)$ are known not to be in W.

We will now extend the concept of an interpretation to the case of belief pairs and consider the question of meta-ignorance and meta-knowledge of belief pairs. We

---

[3] Observe that empty models are allowed. This assumption allows us to treat consistent and inconsistent expansions in a uniform manner.

will see that, being only approximations to possible-world structures, belief pairs may contain meta-ignorance. We will use three logical values, $\mathbf{f}$, $\mathbf{u}$ and $\mathbf{t}$. In the definition, we will use the *truth* ordering: $\mathbf{f} \leq_{tr} \mathbf{u} \leq_{tr} \mathbf{t}$ and define $\mathbf{f}^{-1} = \mathbf{t}, \mathbf{t}^{-1} = \mathbf{f}, \mathbf{u}^{-1} = \mathbf{u}$.

Definition 3. Let $B = (P, S)$ be a belief pair and let I be an interpretation. The truth function $\mathcal{H}_{B,I}$ is defined inductively (min and max are evaluated with respect to the ordering $\leq_{tr}$):

(a) $\mathcal{H}_{B,I}(A) = I(A)$, if A is an atom
(b) $\mathcal{H}_{B,I}(\neg F) = \mathcal{H}_{B,I}(F)^{-1}$
(c) $\mathcal{H}_{B,I}(F_1 \vee F_2) = \max\{\mathcal{H}_{B,I}(F_1), \mathcal{H}_{B,I}(F_2)\}$
(d) $\mathcal{H}_{B,I}(F_1 \wedge F_2) = \min\{\mathcal{H}_{B,I}(F_1), \mathcal{H}_{B,I}(F_2)\}$
(e) $\mathcal{H}_{B,I}(F_2 \supset F_1) = \max\{\mathcal{H}_{B,I}(F_1), \mathcal{H}_{B,I}(F_2)^{-1}\}$

The formula KF is evaluated as follows:

$$\mathcal{H}_{B,I}(KF) = \begin{cases} \mathbf{t} & \text{if for every } J \in P, \mathcal{H}_{B,J}(F) = \mathbf{t} \\ \mathbf{f} & \text{if there is } J \in S \text{ such that } \mathcal{H}_{B,J}(F) = \mathbf{f} \\ \mathbf{u} & \text{otherwise} \end{cases}$$

The truth value of a modal atom KF, $\mathcal{H}_{B,I}(KF)$, does not depend on the choice of I. Consequently, for a modal atom KF we will write $\mathcal{H}_B(KF)$ to denote this, common to all interpretations from $\mathcal{A}$, truth value of KF.

The interpretation of modal formulas given by the function $\mathcal{H}_{B,I}$ is 3-valued. Let us define the *meta-knowledge* of a belief pair B as the set of formulas $F \in \mathcal{L}_K$ such that $\mathcal{H}_B(KF) = \mathbf{t}$ or $\mathcal{H}_B(KF) = \mathbf{f}$. The *meta-ignorance* is formed by all other formulas, that is, those formulas $F \in \mathcal{L}_K$ for which $\mathcal{H}_B(KF) = \mathbf{u}$.

Clearly, a belief pair $B = (W, W)$ naturally corresponds to a possible-world structure W. Such a belief pair is called *complete*. We will denote it by (W). The following straightforward result indicates that $\mathcal{H}_{B,I}$ is a generalization of $\mathcal{H}_{W,I}$ to the case of belief pairs. It also states that a complete belief pair contains no meta-ignorance.

Proposition 1. *If B is a complete belief pair (W), then $\mathcal{H}_{B,I}$ is 2-valued. Moreover, for every formula $F \in \mathcal{L}_K$, $\mathcal{H}_{B,I}(F) = \mathcal{H}_{W,I}(F)$.*

We will now define two satisfaction relations: *weak*, denoted by $\models_w$, and *strong*, denoted by $\models$. Namely, for a belief pair B, an interpretation I and a modal formula F we define:

i. $(B, I) \models_w F$ if $\mathcal{H}_{B,I}(F) \neq \mathbf{f}$ (that is, if $\mathcal{H}_{B,I}(F) \geq_{tr} \mathbf{u}$), and
ii. $(B, I) \models F$ if $\mathcal{H}_{B,I}(F) = \mathbf{t}$

Let T be a modal theory and let B be a belief pair. We define the *derivation operator* as follows:

$$\mathcal{D}_T(B) = (\{I : (B, I) \models_w T\}, \{I : (B, I) \models T\}). \tag{1}$$

Clearly, if $(B, I) \models F$ then $(B, I) \models_w F$. Hence, $\mathcal{D}_T(B)$ is a belief pair or, in other words, $\mathcal{D}_T$ is an operator on $(\mathcal{B}, \leq_p)$. In addition, $P(\mathcal{D}_T(B))$ consists of the interpretations which weakly satisfy T according to B, while $S(\mathcal{D}_T(B))$ consists of those interpretations which strongly satisfy T according to B. The subscript T in $\mathcal{D}_T$ is often omitted when T is clear from the context.

Example 3. Consider $T = \{Kp \supset q\}$. Then $\mathcal{D}(\bot) = (\mathcal{A}, \{pq, \overline{p}q\})$ (here, by pq we mean an interpretation that assigns $\mathbf{t}$ to both p and q while $\overline{p}q$ denotes an interpretation assigning $\mathbf{f}$ to p and $\mathbf{t}$ to q). Indeed, $\mathcal{H}_\bot(Kp) = \mathbf{u}$. Consequently, for every I, $\mathcal{H}_{\bot, I}(Kp \supset q) \neq \mathbf{f}$, that is, $(\bot, I) \models_w Kp \supset q$. For the same reason, $\mathcal{H}_{\bot, I}(Kp \supset q)) = \mathbf{t}$ if and only if $I(q) = \mathbf{t}$.
   To compute $\mathcal{D}^2(\bot)$, observe that $\mathcal{H}_{\mathcal{D}(\bot)}(Kp) = \mathbf{f}$. Consequently, for every I, $\mathcal{H}_{\mathcal{D}(\bot), I}(Kp \supset q) = \mathbf{t}$. It follows that $\mathcal{D}^2(\bot) = (\mathcal{A}, \mathcal{A})$. It is also easy to see now that $(\mathcal{A}, \mathcal{A})$ is the fixpoint of $\mathcal{D}$, that is, $\mathcal{D}(\mathcal{A}, \mathcal{A}) = (\mathcal{A}, \mathcal{A})$.

The next result relates complete fixpoints of $\mathcal{D}$ to Moore's semantics of autoepistemic logic.

Theorem 2. *Let $T \subseteq \mathcal{L}_K$. Then:*

(a) *For every $W \subseteq \mathcal{A}$, (W) is a fixpoint of $\mathcal{D}_T$ if and only if W satisfies the following equation: $W = \{I : (W, I) \models_{ael} T\}$*
(b) *A possible-world structure W is an autoepistemic model of T if and only if (W) is a fixpoint of $\mathcal{D}_T$*
(c) *A modal theory E is a stable expansion of T if and only if $E = Th(S)$ for some complete fixpoint (S) of $\mathcal{D}_T$*

Proof: (a) Observe that for every $W \subseteq \mathcal{A}$, for every $I \in \mathcal{A}$ and for every $F \in T$, $\mathcal{H}_{(W), I}(F) = \mathcal{H}_{W, I}(F)$. Hence, $((W), I) \models F$ if and only if $(W, I) \models_{ael} F$. In addition, by Proposition 1, $(W, I) \models F$ if and only if $((W), I) \models F$. Thus, (W) is a fixpoint of $\mathcal{D}_T$ if and only if $W = \{I : (W, I) \models_{ael} T\}$. The assertion (b) follows directly from (a). The assertion (c) follows from (b) by Theorem 1.   □
   Theorem 2 demonstrates that complete fixpoints of the operator $\mathcal{D}_T$ describe stable expansions of T. However, in general, the operator $\mathcal{D}_T$ may also have fixpoints that are not complete. Such fixpoints provide 3-valued interpretations to modal formulas and can serve as approximations to complete fixpoints of $\mathcal{D}_T$.
   The approach to autoepistemic reasoning that we present in this paper exploits the concept of a *least* fixpoint of $\mathcal{D}_T$. Namely, we show the existence of this least fixpoint and demonstrate that it can be constructed by iterating the operator $\mathcal{D}_T$ starting with the belief pair $\bot$. Intuitively, this iterative construction models the agent who, given an initial theory T, starts with the belief pair $\bot$ (with the smallest meta-knowledge content) and, then, iteratively constructs a sequence of belief pairs with increasing meta-knowledge (decreasing meta-ignorance) until no further improvement is possible.
   Next, we demonstrate that the semantics implied by the least fixpoint of $\mathcal{D}_T$ approximates the semantics of Moore and that it coincides with the semantics of

Moore on stratified modal theories. We show that the task to compute the least fixpoint of the operator $\mathcal{D}_T$ is simpler than computing autoepistemic expansions (unless the polynomial hierarchy collapses). Finally, we study connections of our semantics to several semantics used for logic programs with negation.

Our approach relies on an observation that there is a natural partial ordering of the set $\mathcal{B}$ of belief pairs. Recall that for two belief pairs B_1 and B_2, we defined

$$B_1 \leq_p B_2 \text{ if } P(B_1) \supseteq P(B_2) \text{ and } S(B_1) \subseteq S(B_2). \tag{2}$$

This ordering is consistent with the ordering defined by the "amount" of meta-knowledge contained in a belief pair: the "higher" a belief pair in the ordering $\leq_p$, the more meta-knowledge it contains (and the less meta-ignorance). Clearly, the relation $\leq_p$ is reflexive, antisymmetric and transitive. Hence, $(\mathcal{B}, \leq_p)$ is a poset. The following two results gather some basic properties of the poset $(\mathcal{B}, \leq_p)$, truth assignment function $\mathcal{H}_{B,I}$ and the operator $\mathcal{D}$. The first one shows that the ordering $\leq_p$ is consistent with the concept of the *knowledge ordering* (also referred to as *information* ordering in the literature) of the truth values: $\mathbf{u} \leq_{kn} \mathbf{f}, \mathbf{u} \leq_{kn} \mathbf{t}, \mathbf{f} \not\leq_{kn} \mathbf{t}$ and $\mathbf{t} \not\leq_{kn} \mathbf{f}$. It also relates the ordering $\leq_p$ to the weak and strong entailment relations $\models_w$ and $\models$. The second result states that $\mathcal{D}$ is monotone with respect to $\leq_p$.

Proposition 2. *Let B_1 and B_2 be belief pairs such that $B_1 \leq_p B_2$. For every interpretation $I \in \mathcal{A}$ and every formula $F \in \mathcal{L}_K$:*

(a) $\mathcal{H}_{B_1,I}(F) \leq_{kn} \mathcal{H}_{B_2,I}(F)$.
(b) If $(B_2, I) \models_w F$ then $(B_1, I) \models_w F$
(c) If $(B_1, I) \models F$, then $(B_2, I) \models F$.

Proof: (a) We proceed by induction on the length of F. Thus, let us consider a modal formula F and let us assume that the assertion of the proposition holds for every modal formula G of length smaller than the length of F. There are three cases to consider.

First, assume that F is an atom. Then for every $I \in \mathcal{A}$, $\mathcal{H}_{B_1,I}(F) = I(F) = \mathcal{H}_{B_2,I}(F)$. In particular, $\mathcal{H}_{B_1,I}(F) \leq_{kn} \mathcal{H}_{B_2,I}(F)$ (this argument establishes the basis for the induction).

Next, assume that F is of the form $G \wedge G'$, $G \vee G'$, $G \supset G'$ or $\neg G$. In this case, the assertion follows immediately from the induction hypothesis and from the following observation: if a, b, a' and b' are truth values such that $a \leq_{kn} a'$ and $b \leq_{kn} b'$ then:

i. $(a \wedge b) \leq_{kn} (a' \wedge b')$
ii. $(a \vee b) \leq_{kn} (a' \vee b')$
iii. $(a \supset b) \leq_{kn} (a' \supset b')$
iv. $(\neg a) \leq_{kn} \neg(a')$.

Finally, let us assume $F = KG$ for some modal formula G. Take any $I \in \mathcal{A}$. Assume that $\mathcal{H}_{B_1,I}(KG) = \mathbf{t}$. It follows that for every $J \in P(B_1)$, $\mathcal{H}_{B_1,J}(G) = \mathbf{t}$.

Since $B_1 \leq_{kn} B_2$, $P(B_2) \subseteq P(B_1)$. Hence, by the induction hypothesis, for every $J \in P(B_2)$, $\mathcal{H}_{B_2,J}(G) = \mathbf{t}$. Consequently, $\mathcal{H}_{B_2,I}(KG) = \mathbf{t}$ and $\mathcal{H}_{B_1,I}(F) \leq_{kn} \mathcal{H}_{B_2,I}(F)$.

The argument in the case when $\mathcal{H}_{B_1,I}(KG) = \mathbf{f}$ is similar. Since $\mathbf{u} \leq_{kn} \mathbf{t}$ and $\mathbf{u} \leq_{kn} \mathbf{f}$, the assertion follows in the case when $\mathcal{H}_{B_1,I}(KG) = \mathbf{u}$, too.

(b) Assume that $(B_1, I) \not\models_w T$. Then, there is a formula $F \in T$ such that $\mathcal{H}_{B_1,I}(F) = \mathbf{f}$. By the assertion (a), $\mathcal{H}_{B_2,I}(F) = \mathbf{f}$. Consequently, $(B_2, I) \not\models_w T$.

(c) Assume that $(B_1, I) \models T$. Then $\mathcal{H}_{B_1,I}(F) = \mathbf{t}$ for every $F \in T$. By the assertion (a), $\mathcal{H}_{B_2,I}(F) = \mathbf{t}$ for every $F \in T$. Hence, $(B_2, I) \models T$.   □

Proposition 3. *Let B_1 and B_2 be belief pairs such that $B_1 \leq_p B_2$. For every theory $T \subseteq \mathcal{L}_K$, $\mathcal{D}_T(B_1) \leq_p \mathcal{D}_T(B_2)$, that is, the operator $\mathcal{D}_T$ is monotone on $(\mathcal{B}, \leq_p)$.*

Proof: Assume that $B_1 \leq_p B_2$. By Proposition 2(b), $\{I : (B_2, I) \models_w T\} \subseteq \{I : (B_1, I) \models_w T\}$. Hence, $P(\mathcal{D}(B_2)) \subseteq P(\mathcal{D}(B_1))$. Similarly (by Proposition 2(c)), $S(\mathcal{D}(B_1)) \subseteq S(\mathcal{D}(B_2))$. Thus, $\mathcal{D}(B_1) \leq_p \mathcal{D}(B_2)$.   □

Proposition 3 is especially important. The monotonicity of the operator $\mathcal{D}$ will allow us to assert the existence of a least fixpoint of $\mathcal{D}$. However, let us note that the poset $(\mathcal{B}, \leq_p)$ is not a lattice (and, hence, not a complete lattice). Indeed, for every $W \subseteq \mathcal{A}$, (W) is a maximal element in $(\mathcal{B}, \leq_p)$. If $W_1 \neq W_2$, then (W_1) and (W_2) have no least upper bound (l.u.b.) in $(\mathcal{B}, \leq_p)$. Thus, we will not be able to use the theorem Tarski-Knaster in its classic form. Instead, we will use its generalization (see [11]) developed for the case of posets that are *chain complete*. Let us recall that a poset is chain complete if its every *chain* (that is, a totally ordered subposet) has a l.u.b. [1], [11]. Note also that every chain complete partially ordered set has a least element. It follows from the observation that the empty set is a chain.

Theorem 3 ([11]). *Let $(P, \leq)$ be a chain-complete poset. Let D be a monotone operator on $(P, \leq)$. Then, D has a least fixpoint. This fixpoint is the limit of the sequence of iterations of D starting with the least element of $(P, \leq)$.*

To use Theorem 3, we will now show that the poset $(\mathcal{B}, \leq_p)$ is chain complete.

Proposition 4. *The poset $(\mathcal{B}, \leq_p)$ is chain complete.*

Proof: For a nonempty set C of belief pairs define $P_C = \bigcap\{P(B) : B \in C\}$ and $S_C = \bigcup\{S(B) : B \in C\}$. Consider now a chain $C \subseteq \mathcal{B}$ of belief pairs. Assume that $I \in S_C$. There exists a belief pair $(P, S) \in C$ such that $I \in S$. Since (P, S) is a belief pair, $I \in P$. Let $(P', S') \in C$. Then we have $(P', S') \leq_p (P, S)$ or $(P, S) \leq_p (P', S')$. In the first case, $P \subseteq P'$. Hence, $I \in P'$. In the second case we have $S \subseteq S' \subseteq P'$ and, again, $I \in P'$. It follows that $I \in P_C$ and, consequently, that $S_C \subseteq P_C$.

We have just proved that (P_C, S_C) is a belief pair. It is easy to see that for every $(P, S) \in C$, $(P, S) \leq_p (P_C, S_C)$. Moreover, any other upper bound B of C satisfies $(P_C, S_C) \leq_p B$. Hence, (P_C, S_C) is the l.u.b. of C.

Finally, it is evident that the belief pair $\perp = (\mathcal{A}, \emptyset)$ is a least element of the poset $(\mathcal{B}, \leq_p)$. Thus, the empty chain also has its least upper bound (the least element of $(P, \leq)$). □

As an immediate consequence of Theorem 3 and Proposition 4 we obtain the following crucial corollary.

Corollary 1. *For every theory $T \subseteq \mathcal{L}_K$, the operator $\mathcal{D}_T$ has a least fixpoint.*

The least fixpoint of the operator $\mathcal{D}_T$ will be denoted by $\mathcal{D}_T\uparrow$. We propose this fixpoint as the semantics of modal theory T. This semantics reflects the reasoning process of an agent who gradually constructs belief pairs with increasing knowledge (information) content.

The following three results provide justification for our least fixpoint semantics. The first of these results shows that the least fixpoint semantics provides a lower approximation to the skeptical semantics based on expansions and an upper approximation to the brave reasoning based on expansions.

Theorem 4. *Let T be a modal theory. If $\mathcal{H}_{\mathcal{D}_T\uparrow}(KF) = \mathbf{t}$ then F belongs to all expansions of T. If $\mathcal{H}_{\mathcal{D}_T\uparrow}(KF) = \mathbf{f}$ then F does not belong to any expansion of T.*

Proof: Assume that $\mathcal{H}_{\mathcal{D}_T\uparrow}(KF) = \mathbf{t}$. Let E be a stable expansion of T. Then $E = Th(W)$ for some autoepistemic model W of T (Theorem 1). Clearly, (W) is then a fixpoint of $\mathcal{D}_T$ (Theorem 2). Since $\mathcal{D}_T\uparrow \leq_p (W)$, by Proposition 2 it follows that $\mathcal{H}_{(W)}(KF) = \mathbf{t}$. Hence, for every $I \in W$, $\mathcal{H}_{(W),I}(F) = \mathbf{t}$ or, equivalently (Proposition 1), $\mathcal{H}_{W,I}(F) = \mathbf{t}$. Consequently, $F \in Th(W) = E$. A similar argument can be used to prove the second part of the assertion. □

The second result shows that if the least fixpoint is complete (that is, no further improvement in meta-knowledge is possible) than the least fixpoint semantics coincides with the semantics of Moore.

Theorem 5. *If $\mathcal{D}_T\uparrow$ is complete then $\mathcal{D}_T\uparrow$ is the unique autoepistemic model of T.*

Proof: For every complete fixpoint (W) of $\mathcal{D}_T$, $\mathcal{D}_T\uparrow \leq_p (W)$. Moreover, complete elements of $(\mathcal{B}, \leq_p)$ are maximal. Hence, if $\mathcal{D}_T\uparrow$ is complete, it is a unique complete fixpoint of $\mathcal{D}_T$. Thus, by Theorem 2(b), $\mathcal{D}_T\uparrow$ is the unique autoepistemic model of T. □

In the last result of this section we will show that the least fixpoint semantics is complete for the class of stratified theories, introduced by Gelfond [3] and further generalized in [9]. This property, in combination with Theorem 5, implies that for stratified theories the least fixpoint semantics coincides with the skeptical (and brave) autoepistemic semantics of Moore. This is an important property since the semantics of Moore is commonly accepted for the class of stratified theories and the agreement with this semantics is regarded as a test of "correctness" of a semantics for a modal nonmonotonic logic. Let us note that a similar test of agreement with the perfect model semantics on stratified programs is used in logic programming to justify semantics for logic programs with negation. In particular, the well-founded and stable model semantics both coincide with the perfect model semantics on stratified logic programs. This property is not quite coincidental as connections between autoepistemic logic and logic programming are well known [10] and are also discussed below in Section 5.

Theorem 6. *If T is a stratified autoepistemic theory then:*

(a) $\mathcal{D}_T\!\uparrow$ is complete
(b) T has a unique stable expansion
(c) $\mathcal{D}_T\!\uparrow$ is consistent if and only if the lowest stratum T_0 is consistent.

The proof of this theorem (as well as a precise definition of a stratified modal theory) can be found in the appendix.

To conclude this section let us observe that the semantics defined by the least fixpoint of the operator $\mathcal{D}$ has several attractive features. It is defined for every modal theory T. It coincides with the semantics of autoepistemic logic on stratified theories. In the general case, it provides a lower approximation to the intersection of all stable expansions (skeptical autoepistemic reasoning) and upper approximation to the union of all stable expansions (brave autoepistemic reasoning).

4   An effective implementation of $\mathcal{D}$

The approach proposed and discussed in the previous section does not directly yield itself to fast implementations. The definition of the operator $\mathcal{D}$ refers to all interpretations of the language $\mathcal{L}$. Thus, computing $\mathcal{D}(B)$ by following the definition is exponential even for modal theories of a very simple syntactic form. Moreover, representing belief pairs is costly. Each of the sets $P(B)$ and $S(B)$ may contain exponentially many elements. In this section, we describe a characterization of the operator $\mathcal{D}$ that is much more suitable for investigations of algorithmic issues associated with our semantics.

To this end, in addition to the propositional language $\mathcal{L}$ (generated, recall, by the set of atoms At), we will also consider the extension of $\mathcal{L}$ by three new constants $\mathbf{t}$, $\mathbf{f}$ and $\mathbf{u}$. We will call this language *3-FOL*. Formulas and theories in this language will be called *3-FOL formulas* and *3-FOL theories*, respectively. Our strategy is now as follows. First, we will show that a wide class of belief pairs can be represented by 3-FOL theories. Next, using this representation, we will describe a method to compute fixpoints of the operator $\mathcal{D}$ that is algorithmically more feasible than the direct approach implied by the definition of $\mathcal{D}$.

We start by discussing a class of 3-valued truth assignments on the language 3-FOL that are generated by 2-valued interpretations from $\mathcal{A}$ under the assumption that the new constants $\mathbf{t}$, $\mathbf{f}$ and $\mathbf{u}$ are always interpreted by the logical values they represent. Formally, given an interpretation $I \in \mathcal{A}$, we define a valuation I^e on the language 3-FOL inductively as follows (minima and maxima are computed with respect to the truth ordering of $\mathbf{t}$, $\mathbf{f}$ and $\mathbf{u}$):

   i. $I^e(A) = I(A)$, if $A \in At$

ii. $I^e(a) = a$, for $a \in \{\mathbf{t}, \mathbf{f}, \mathbf{u}\}$
iii. $I^e(\neg F) = (I^e(F))^{-1}$
iv. $I^e(F_1 \vee F_2) = \max\{I^e(F_1), I^e(F_2)\}$
 v. $I^e(F_1 \wedge F_2) = \min\{I^e(F_1), I^e(F_2)\}$
vi. $I^e(F_2 \supset F_1) = \max\{I^e(F_1), (I^e(F_2))^{-1}\}$.

Let F be a 3-FOL formula. By F^{wk} we denote the formula obtained by substituting $\mathbf{t}$ for all positive occurrences of $\mathbf{u}$ and $\mathbf{f}$ for all negative occurrences of $\mathbf{u}$. Similarly, by F^{str} we denote the formula obtained by substituting $\mathbf{t}$ for all negative occurrences of $\mathbf{u}$ and $\mathbf{f}$ for all positive occurrences of $\mathbf{u}$. Given a 3-FOL theory Y, we define Y^{str} and Y^{wk} by the standard setwise extension. Before we proceed let us note the following useful identities (the proof is straightforward and is omitted):

$$(\neg F)^{str} = \neg(F^{wk}) \quad \text{and} \quad (\neg F)^{wk} = \neg(F^{str}). \tag{3}$$

Clearly, F^{str} and F^{wk} do not contain $\mathbf{u}$. Consequently, they can be regarded as formulas in the propositional language generated by the atoms in At and the two constants $\mathbf{t}$ and $\mathbf{f}$. We will call this language 2-FOL. Formulas F^{wk} and F^{str} can be viewed as lower and upper approximations to the formula F.

It is clear that for every 2-FOL formula F, and for every interpretation $I \in \mathcal{A}$, $I^e(F) \in \{\mathbf{t}, \mathbf{f}\}$. We say that an interpretation $I \in \mathcal{A}$ is a *model* of a 2-FOL theory T if $I^e(F) = \mathbf{t}$. We will write $I \models F$ in such case. An interpretation $I \in \mathcal{A}$ is a *model* of a 2-FOL theory T ($I \models T$) if I is a model of every formula from T. The set of interpretations from $\mathcal{A}$ that are models of a 2-FOL formula F will be denoted by $\mathcal{M}od(T)$. The entailment relation in the language 2-FOL is now defined in the standard way: for two 2-FOL theories T_1 and T_2, $T_1 \models T_2$ if $\mathcal{M}od(T_2) \subseteq \mathcal{M}od(T_1)$. We have the following technical lemma.

Lemma 1. *For every interpretation $I \in A$ and for every 3-FOL formula F:*

(a) $I^e(F) = \mathbf{t}$ if and only if $I \models F^{str}$
(b) $I^e(F) = \mathbf{f}$ if and only if $I \not\models F^{wk}$.

Proof: We will prove both (a) and (b) simultaneously by induction. Clearly, both (a) and (b) are true for every atom At and for the constants $\mathbf{t}$, $\mathbf{f}$ and $\mathbf{u}$.

Consider a 3-FOL formula G and assume both (a) and (b) hold for all 3-FOL formulas with length smaller than the length of G. Assume first that $G = F_1 \vee F_2$. Clearly, $I^e(F_1 \vee F_2) = \mathbf{t}$ if and only if $I^e(F_1) = \mathbf{t}$ or $I^e(F_2) = \mathbf{t}$. Similarly, $I \models (F_1 \vee F_2)^{str}$ if and only if $I \models F_1^{str}$ or $I \models F_2^{str}$. By the induction hypothesis, $I^e(F_i) = \mathbf{t}$ if and only if $I \models F_i^{str}$, $i = 1, 2$. Hence, the assertion (a) holds for $G = F_1 \vee F_2$.

Analogous arguments can be used to show that the assertion (b) holds for $G = F_1 \vee F_2$ and that both assertions (a) and (b) hold for $G = F_1 \wedge F_2$. Thus, to complete the proof, consider the case when $G = \neg F$. Then, $I^e(\neg F) = \mathbf{t}$ if and only if $I^e(F) = \mathbf{f}$. By the induction hypothesis, $I^e(F) = \mathbf{f}$ if and only if $I \not\models F^{wk}$. Moreover, by (3), $I \not\models F^{wk}$ if and only if $I \models (\neg F)^{str}$. By the induction hypothesis,

$I^e(F) = \mathbf{f}$ if and only if $I \not\models F^{str}$. Hence, (a) holds for $G = \neg F$. The proof of (b) for $G = \neg F$ is similar. □

Lemma 1 has an important consequence. It implies that each 3-FOL theory generates a belief pair.

Corollary 2. *Let Y be a 3-FOL theory. Then, $\mathcal{M}od(Y^{str}) \subseteq \mathcal{M}od(Y^{wk})$. That is, equivalently, $(Mod(Y^{wk}), Mod(Y^{str}))$ is a belief pair.*

Proof: Let $I \in Mod(Y^{str})$ and let $F \in Y$. Then, $I \models F^{str}$. By Lemma 1(a), $I^e(F) = \mathbf{t}$. Hence, $I^e(F) \neq \mathbf{f}$ and, by Lemma 1(b), $I \models F^{wk}$. Consequently, $I \in Mod(Y^{wk})$ and $\mathcal{M}od(Y^{str}) \subseteq \mathcal{M}od(Y^{wk})$ follows. □

Let Y be a 3-FOL theory. The belief pair $(Mod(Y^{wk}), Mod(Y^{str}))$ will be denoted by $Bel(Y)$. We say that a belief pair B is *represented by a 3-FOL theory Y* if $B = Bel(Y)$. Clearly, the belief pair $\perp = (\mathcal{A}, \emptyset)$ is represented by the 3-FOL theory $\{\mathbf{u}\}$. We will now show that every belief pair in the range of the operator $\mathcal{D}_T$ is representable by a 3-FOL theory.

Let B be a belief pair and let F be a modal formula. By F_B we will denote a 3-FOL formula that is obtained from F by replacing each top level modal atom KG in F by the constant corresponding to the logical value $\mathcal{H}_B(KG)$. For a modal theory T, we define $T_B = \{F_B : F \in T\}$. We have the following result.

Theorem 7. *For every modal theory $T \subseteq \mathcal{L}_K$ and every belief pair B we have $\mathcal{D}_T(B) = Bel(T_B)$.*

Proof: First, observe that directly from the definitions of the truth assignment I^e and a 3-FOL formula F_B it follows that

$$I^e(F_B) = \mathcal{H}_{B,I}(F). \tag{4}$$

Now, $\mathcal{D}_T(B)$ is the following belief pair:

$$(\{I : \mathcal{H}_{B,I}(F) \geq_{tr} \mathbf{u}, \text{ for all } F \in T\}, \{I : \mathcal{H}_{B,I}(F) = \mathbf{t}, \text{ for all } F \in T\}).$$

Hence, by (4),

$$\mathcal{D}_T(B) = (\{I : I^e(F_B) \neq \mathbf{f}, \text{ for all } F \in T\}, \{I : I^e(F_B) = \mathbf{t}, \text{ for all } F \in T\}).$$

Finally, by Lemma 1,

$$\mathcal{D}_T(B) = (Mod(T_B^{wk}), Mod(T_B^{str})).$$

That is, $\mathcal{D}_T(B) = Bel(T_B)$. □

We will now show that, similarly to belief pairs, 3-FOL theories can be used to assign truth values to *modal* atoms (and, hence, to all modal formulas). We will then exhibit (Theorem 8) the relationship between this truth assignment and the truth assignment $\mathcal{H}_{B,I}$ introduced in Section 3. In order for the inductive argument in the proof of Theorem 8 to work, we need to extend the modal language $\mathcal{L}_K$ by the constants $\mathbf{t}$, $\mathbf{f}$ and $\mathbf{u}$. We call the resulting language *3-AEL*. We call formulas and theories in this language *3-AEL formulas* and *3-AEL theories*, respectively. Observe that the definition of the truth assignment $\mathcal{H}_{B,I}$ from Section 3 naturally extends to 3-AEL formulas.

Definition 4. Let Y be a 3-FOL theory, and let F be a 3-AEL formula. We define $\mathcal{H}_Y(KF)$ as follows. If F is a modal-free formula (that is a 3-FOL formula), then define:

$$\mathcal{H}_Y(KF) = \begin{cases} \mathbf{t} & \text{if } Y^{wk} \models F^{str} \\ \mathbf{f} & \text{if } Y^{str} \not\models F^{wk} \\ \mathbf{u} & \text{otherwise.} \end{cases}$$

If F is not modal free, then replace every modal atom KG in F, not under the scope of any other occurrence of the modal operator, by the constant corresponding to the value of $\mathcal{H}_Y(KG)$. Call the resulting formula F'. Define $\mathcal{H}_Y(KF) = \mathcal{H}_Y(KF')$ (notice that F' is a modal-free formula and the first part of the definition applies).

Let T be a modal theory and let Y be a 3-FOL theory. By the Y-*instance* of T, T_Y, we mean the 3-FOL theory obtained by substituting in each formula from T all modal atoms KF (not appearing under the scope of any other occurrence of the modal operator K) by the constant corresponding to $\mathcal{H}_Y(KF)$.

The following theorem shows that the truth values of modal atoms evaluated according to a 3-FOL theory Y and according to the corresponding belief pair $Bel(Y)$ coincide.

Theorem 8. *Let Y be a 3-FOL theory. Then, for every 3-AEL formula F,*

$$\mathcal{H}_{Bel(Y)}(KF) = \mathcal{H}_Y(KF).$$

Proof: The proof is by induction on the length of the formula F. In what follows we denote $Bel(Y)$ by B. Thus, we also have $P(B) = \mathcal{M}od(Y^{wk})$ and $S(B) = \mathcal{M}od(Y^{str})$.

First consider the case when F is a 3-FOL formula (the argument in this case will establish the basis of the induction). By the definition, $\mathcal{H}_B(KF) = \mathbf{t}$ if and only if

$$\text{for every } I \in P(B), \mathcal{H}_{B,I}(F) = \mathbf{t}. \tag{5}$$

Since F is a 3-FOL formula, $\mathcal{H}_{B,I}(F) = I^e(F)$. Hence, by Lemma 1 and by the equality $P(B) = \mathcal{M}od(Y^{wk})$, the statement (5) is equivalent to:

$$\text{for every } I \in \mathcal{M}od(Y^{wk}), I \in \mathcal{M}od(F^{str}). \tag{6}$$

The statement (6), in turn, is equivalent to $Y^{wk} \models F^{str}$. Thus, $\mathcal{H}_B(KF) = \mathbf{t}$ if and only if $\mathcal{H}_Y(KF) = \mathbf{t}$. In a similar way one can prove that $\mathcal{H}_B(KF) = \mathbf{f}$ if and only if $\mathcal{H}_Y(KF) = \mathbf{f}$. Consequently, $\mathcal{H}_B(KF) = \mathcal{H}_Y(KF)$.

Second, consider the case when F is a modal 3-AEL formula. Let F' be a formula obtained from F by replacing each modal atom KG (not in the scope of any other occurrence of K in F) by the constant corresponding to the truth value $\mathcal{H}_B(KG)$. By the induction hypothesis $\mathcal{H}_B(KG) = \mathcal{H}_Y(KG)$. Hence, by the definition of $\mathcal{H}_Y(KF)$, $\mathcal{H}_Y(KF) = \mathcal{H}_Y(KF')$.

Since F' is modal-free, $\mathcal{H}_B(KF') = \mathcal{H}_Y(KF')$. In addition, it is easy to see that $\mathcal{H}_B(KF) = \mathcal{H}_B(KF')$. Thus, $\mathcal{H}_B(KF) = \mathcal{H}_Y(KF)$.   □

Let T be a modal theory. We will now define an operator $\mathcal{SD}_T$ on 3-FOL theories. We will then show that this new operator is closely related to the operator $\mathcal{D}_T$. Let Y be a 3-FOL theory. Define $\mathcal{SD}_T(Y) = T_Y$.

The key property of the operator $\mathcal{SD}_T$ is that, for a finite modal theory T and for a finite 3-FOL theory Y, $\mathcal{SD}_T(Y)$ can be computed by means of polynomially many calls to the propositional provability procedure. The number of such calls is bounded by the number of occurrences of the modal operator K in the theory T. In each call we verify whether some 2-FOL theory X_1 entails another 2-FOL theory X_2, where the sizes of X_1 and X_2 are bounded by the sizes of the theories T and Y.

Theorems 8 and 7 imply the main result of this section.

Theorem 9. *Let T be a modal theory and let Y be a 3-FOL theory. Then,*

(a) $T_Y = T_{\mathcal{B}el(Y)}$ and $\mathcal{SD}_T(Y) = T_{\mathcal{B}el(Y)}$
(b) $\mathcal{B}el(\mathcal{SD}_T(Y)) = \mathcal{D}_T(\mathcal{B}el(Y))$.
(c) If a belief pair B is a fixpoint of $\mathcal{D}_T$, then T_B is a fixpoint of $\mathcal{SD}_T$.
(d) If Y is a fixpoint of $\mathcal{SD}_T$ then $\mathcal{B}el(Y)$ is a fixpoint of $\mathcal{D}_T$.

Proof: (a) This statement follows directly from Theorem 8.
(b) By Theorem 7, $\mathcal{D}_T(\mathcal{B}el(Y)) = \mathcal{B}el(T_{\mathcal{B}el(Y)})$.
By (a), $\mathcal{D}_T(\mathcal{B}el(Y)) = \mathcal{B}el(\mathcal{SD}_T(Y))$ follows.
(c) If a belief pair B is a fixpoint of $\mathcal{D}_T$, then $B = \mathcal{D}_T(B) = \mathcal{B}el(T_B)$ (the last equality follows by Theorem 7). Hence, $T_B = T_{\mathcal{B}el(T_B)} = \mathcal{SD}_T(T_B)$ (the last equality follows by (a)).
(d) This statement follows directly from (b).   □

Let us denote $B_\alpha = \mathcal{D}_T^\alpha(\bot)$ and $Y_\alpha = \mathcal{SD}_T^\alpha(\mathbf{u})$. It follows directly from Theorem 9(b) (by an easy induction) that for every ordinal number α, $B_\alpha = \mathcal{B}el(Y_\alpha)$. Hence, if $Y_\alpha = Y_{\alpha+1}$ (that is, if Y_α is a fixpoint of the operator $\mathcal{SD}_T$) $B_\alpha = B_{\alpha+1}$ (that is, B_α is a fixpoint of the operator $\mathcal{D}_T$).

Next, by (a), for every ordinal α, $T_{Y_\alpha} = T_{B_\alpha}$. Hence, if $B_\alpha = B_{\alpha+1}$ (that is, if B_α is a fixpoint of the operator $\mathcal{D}_T$) then

$$Y_{\alpha+2} = \mathcal{SD}_T(Y_{\alpha+1}) = T_{Y_{\alpha+1}} = T_{B_{\alpha+1}} = T_{B_\alpha} = T_{Y_\alpha} = \mathcal{SD}_T(Y_\alpha) = Y_{\alpha+1}.$$

That is, $Y_{\alpha+1}$ is a fixpoint o the operator $\mathcal{SD}_T$.
It follows that

$$\mathcal{D}_T\!\uparrow = \mathcal{B}el(\mathcal{SD}_T\!\uparrow).$$

In the case when T is finite, the number of iterations needed to compute $\mathcal{SD}_T\!\uparrow$ is limited by the number of top level (unnested) modal literals in T. Originally, they may all be evaluated to $\mathbf{u}$. However, at each step, at least one $\mathbf{u}$ changes to either $\mathbf{t}$ or $\mathbf{f}$ and this value is preserved in the subsequent evaluations.

Once $\mathcal{SD}_T\!\uparrow$ is computed, one can evaluate the truth value $\mathcal{H}_{\mathcal{SD}_T\uparrow}(KG)$ for any modal atom of the language $\mathcal{L}_K$. This task again requires polynomially many calls to a propositional provability procedure. A key point is that the logical value so computed is exactly the logical value of the modal atom KG with respect to the belief pair $\mathcal{D}_T\!\uparrow$. In other words, determining the logical value of a modal formula with respect to the semantics defined by the least fixpoint of the operator $\mathcal{D}_T$ takes a polynomial number of calls to a propositional provability procedure. Consequently, the problem to decide whether a logical value of a modal atom under this semantics is $\mathbf{t}$ is in the class Δ_P^2 (the same is true for two other decision problems of deciding whether the logical value of a modal atom is $\mathbf{u}$ and $\mathbf{f}$, respectively). Since deciding whether a modal atom is in all (some of the) expansions of a modal theory is Π_P^2-complete (Σ_P^2-complete), our 3-valued semantics is computationally simpler (unless the polynomial hierarchy collapses at some low level). These considerations yield the following formal result.

Theorem 10. *The problems to decide whether* $\mathcal{H}_{\mathcal{D}_T\!\uparrow}(KF) = \mathbf{t}$, $\mathcal{H}_{\mathcal{D}_T\!\uparrow}(KF) = \mathbf{f}$ *and* $\mathcal{H}_{\mathcal{D}_T\!\uparrow}(KF) = \mathbf{u}$ *are in the class* Δ_P^2.

5   Relationship to logic programming

Autoepistemic logic is closely related to several semantics for logic programs with negation. It is well-known that both stable and supported models of logic programs can be described as expansions of appropriate translations of programs into modal theories (see, for instance, [10]). In this section, we discuss connections of the semantics defined by the least fixpoint of the operator $\mathcal{D}$ to some 3-valued semantics of logic programs.

We will be interested in propositional logic programs over a set of atoms At. However, to prove the main results of the section and to state some auxiliary facts, we will also consider a wider class of programs. These programs, called 3-FOL programs, will play a similar role as 3-FOL theories in Section 4. Formally, a *3-FOL program clause* is an expression of the form

$$a \leftarrow b_1, \dots, b_k, \mathbf{not}(c_1), \dots, \mathbf{not}(c_m), l_1, \dots, l_n,$$

where a, each b_i, $1 \le i \le k$, and each c_i, $1 \le i \le m$ are atoms from At, and each l_i, $1 \le i \le n$, is one of $\mathbf{t}, \mathbf{f}, \mathbf{u}$ or their negation. The literals l_i will be referred to as *truth-value literals*. A 3-FOL clause in which $m = 0$ (no literals of the form $\mathbf{not}(c)$ in the body) is called a *definite* 3-FOL clause. A collection of 3-FOL clauses (definite 3-FOL clauses, respectively) is a *3-FOL program* (*definite 3-FOL program*).

We will often interpret a definite 3-FOL logic program P as a 3-FOL theory (by regarding program clauses as implications). This allows us to use for definite 3-FOL programs concepts introduced in Section 4 for 3-FOL theories. In particular, with every definite 3-FOL program we will associate 2-FOL theories P^{str} and P^{wk}, as well as the belief pair $\mathcal{B}el(P) = (Mod(P^{wk}), Mod(P^{str}))$.

Consider a 3-FOL definite logic program P. We say that a 3-valued interpretation I strongly satisfies P if for each rule

$$a \leftarrow b_1, \dots, b_k, l_1, \dots, l_n$$

from P, $I^e(a) \geq_{tr} I^e(b_i)$, for some i, $1 \leq i \leq k$ or $I^e(a) \geq_{tr} I^e(l_i)$, for some i, $1 \leq i \leq n$ (I^e is obtained from I by extending I naturally to the constants $\mathbf{t}$, $\mathbf{f}$ and $\mathbf{u}$)[4].

It is easy to see that every definite 3-FOL program has a least 3-valued model with respect to the truth ordering (see [16]). We will denote this model by $LM_3(P)$.

Since P is a definite 3-FOL program, theories P^{wk} and P^{str} are both definite 3-FOL programs. Thus, each has a least model (with respect to the truth ordering). Moreover, since $\mathbf{u}$ does not occur in P^{wk} and P^{str}, these least models are two valued. We will denote them by $LM(P^{wk})$ and $LM(P^{str})$, respectively. Let $I, J \in \mathcal{A}$. Define $I \leq_{tr} J$ if for all atoms A, $I(A) \leq_{tr} J(A)$. We have the following simple technical lemma connecting the three interpretations $LM_3(P)$, $LM(P^{wk})$ and $LM(P^{str})$. The proof is easy and is left to the reader.

Lemma 2. *Let P be a definite 3-FOL program. Then:*

(a) $LM(P^{wk}) \leq_{tr} LM(P^{str})$
(b) $LM_3(P) = \mathbf{t}$ *if and only if* $LM(P^{wk}) = \mathbf{t}$, *and* $LM_3(P) = \mathbf{f}$ *if and only if* $LM(P^{str}) = \mathbf{f}$.

Let B be a belief pair. Define the *projection*, *Proj(B)*, as the 3-valued interpretation I such that $I(p) = \mathcal{H}_B(Kp)$. We have the following theorem relating, for definite 3-FOL program P its belief pair $\mathcal{B}el(P)$ with its least model $LM_3(P)$.

Theorem 11. *For any 3-FOL definite program P, $Proj(\mathcal{B}el(P)) = LM_3(P)$.*

Proof: By the definition of $Proj(\mathcal{B}el(P))$ and by Theorem 8 we have

$$Proj(\mathcal{B}el(P))(p) = \mathcal{H}_{\mathcal{B}el(P)}(Kp) = \mathcal{H}_P(Kp) \tag{7}$$

(slightly abusing the notation, we use the same symbol P to denote both a 3-FOL program and the corresponding 3-FOL theory).

Hence, by (7) and by Definition 4, $Proj(\mathcal{B}el(P))(p) = \mathbf{t}$ if and only if $P^{wk} \models p$. The entailment $P^{wk} \models p$ is, in turn, equivalent to $LM(P^{wk})(p) = \mathbf{t}$. By Lemma 2(b), it follows then that $Proj(\mathcal{B}el(P))(p) = \mathbf{t}$ if and only if $LM_3(P)(p) = \mathbf{t}$.

Similarly, by (7) and by Definition 4, $Proj(\mathcal{B}el(P))(p) = \mathbf{f}$ if and only if $P^{str} \not\models p$ or, equivalently, if and only if $LM(P^{str})(p) = \mathbf{f}$. Hence, by Lemma 2(b), $Proj(\mathcal{B}el(P))(p) = \mathbf{f}$ if and only if $LM_3(P)(p) = \mathbf{f}$.   □

We will now study the relationship between logic programming and autoepistemic logic. Given a logic programming clause (over the alphabet At)

$$r = a \leftarrow b_1, \dots, b_k, \mathbf{not}(c_1), \dots, \mathbf{not}(c_m)$$

define:

$$ael_1(r) = Kb_1 \wedge \dots \wedge Kb_k \wedge \neg Kc_1 \wedge \dots \wedge \neg Kc_m \supset a$$

---

[4] This means that I satisfies the rule $p \leftarrow B$ in the strong Kleene truth table.

and

$$ael_2(r) = b_1 \wedge \ldots \wedge b_k \wedge \neg K c_1 \wedge \ldots \wedge \neg K c_m \supset a$$

Embeddings $ael_1(\cdot)$ and $ael_2(\cdot)$ naturally extend to logic programs P.

In the remainder of this paper we show that fixpoints of the operator $\mathcal{D}_{ael_1(P)}$ ($\mathcal{D}_{ael_2(P)}$, respectively) precisely correspond to 3-valued supported (stable, respectively) models of P (the projection function $Proj(\cdot)$ establishes the correspondence). Moreover, complete fixpoints of $\mathcal{D}_{ael_1(P)}$ ($\mathcal{D}_{ael_2(P)}$) describe 2-valued supported (stable, respectively) models of P. Finally, the least fixpoint of $\mathcal{D}_{ael_1(P)}$ captures the Fitting-Kunen 3-valued semantics of a program P, and the least fixpoint of $\mathcal{D}_{ael_2(P)}$ captures the well-founded semantics of P.

We will focus first on the embedding $ael_1(\cdot)$. It establishes the relationship between stable expansions and supported models and between the least fixpoint of the operator $\mathcal{D}_{ael_1(P)}$ and the Fitting-Kunen 3-valued semantics of a program P.

Let P be a logic program. Let us recall a definition of the 3-valued stepwise inference operator $\mathcal{T}_P$ [2]:

$$\mathcal{T}_P(I) = I' \text{ where } I'(p) = \max(\{I^e(body) : p \leftarrow body \in P\})$$

(here we treat $body$ as the conjunction of its literals). It is well known that fixpoints of the operator $\mathcal{T}_P$ are models of the program P. These models are called *3-valued supported models*. It is also known [2] that every logic program P has a least (with respect to the knowledge ordering) 3-valued supported model. This model determines a semantics of P known as *Fitting-Kunen semantics*.

For a given 3-valued interpretation I, and a logic program P, define P_I^{sp} as the *definite* 3-FOL program obtained from P by substituting, in the bodies of rules in P $I(p)$, for each atom p occurring positively and $\neg I(p)$ for each literal **not**(p). We have the following lemma (its proof is straightforward and is omitted).

Lemma 3. *Let P be a logic program over the set of atoms At.*

(a) Let I be a 3-valued interpretation of At. $\mathcal{T}_P(I) = LM_3(P_I^{sp})$
(b) If B is a belief pair then $P_{Proj(B)}^{sp} = (ael_1(P))_B$.

Equipped with Lemma 3 we are ready to prove the first of the two main results of this section.

Theorem 12. *Let P be a logic program over the set of atoms At.*

(a) For every belief pair B, $\mathcal{T}_P(Proj(B)) = Proj(\mathcal{D}_{ael_1(P)}(B))$
(b) If a belief pair B is a fixpoint of $\mathcal{D}_{ael_1(P)}$ then $Proj(B)$ is a 3-valued supported model of P
(c) If I is a 3-valued supported model of P, then $B = \mathcal{B}el(P_I^{sp})$ is a fixpoint of $\mathcal{D}_{ael_1(P)}$ and $Proj(B) = I$
(d) $Proj(\mathcal{D}_{ael_1(P)}\!\uparrow)$ is the $\leq_{kn}$-least 3-valued supported model of P (the model defining the Fitting-Kunen semantics)

(e) If a belief pair B is a complete fixpoint of $\mathcal{D}_{ael_1(P)}$, then $Proj(B)$ is a 2-valued supported model of P. Moreover, each 2-valued supported model of P is of this form.

Proof: (a) Clearly,

$$\mathcal{T}_P(Proj(B)) = LM_3(P^{sp}_{Proj(B)}) = LM_3(ael_1(P)_B) =$$
$$= Proj(\mathcal{B}el(ael_1(P)_B)) = Proj(\mathcal{D}_{ael_1(P)}(B))$$

(the first two equalities follow by Lemma 3, the third one follows by Theorem 11 and the last one by Theorem 7).
(b) If B is a fixpoint of $\mathcal{D}_{ael_1(P)}$ then by (a):

$$\mathcal{T}_P(Proj(B)) = Proj(\mathcal{D}_{ael_1(P)}(B)) = Proj(B).$$

(c) Since I is a fixpoint of $\mathcal{T}_P$, $I = \mathcal{T}_P(I) = LM_3(P^{sp}_I)$ (Lemma 3(a)). Let $B = \mathcal{B}el(P^{sp}_I)$. By Theorem 11, $Proj(B) = LM_3(P^{sp}_I) = I$. Hence, by Lemma 3(b), $(ael_1(P))_B = P^{sp}_I$. Consequently, by Theorem 7, $\mathcal{D}_{ael_1(P)}(B) = \mathcal{B}el(ael_1(P)_B) = \mathcal{B}el(P^{sp}_I) = B$.
(d) Let $B = \mathcal{D}_{ael_1(P)} \uparrow$. By (b), $Proj(B)$ is a supported model of P. Consider another supported model I of P. It follows that $B' = \mathcal{B}el(P^{sp}_I)$ is a fixpoint of $\mathcal{D}_{ael_1(P)}$ and that $Proj(\mathcal{B}el(P^{sp}_I)) = I$.
    Clearly, $B \leq_p B'$. Proposition 2 entails that for each atom p, $\mathcal{H}_B(Kp) \leq_{kn} \mathcal{H}_{B'}(Kp)$. By the definition of $Proj(\cdot)$, $Proj(B) \leq_{kn} Proj(B') = I$, that is, $Proj(B)$ is the $\leq_{kn}$-least 3-valued supported model of P.
(e) This assertion follows from the observation that if B is complete then $Proj(B)$ is 2-valued (Proposition 1).   □
    We will now discuss the second embedding, $ael_2(\cdot)$, of logic programs into autoepistemic logic.
    Recall the definition of the 3-valued version $\mathcal{GLP}_P$ of the Gelfond and Lifschitz operator (see, for instance, [16]). Given a logic program P and a 3-valued interpretation I, P_I is the program where negative body literals **not**(p) are replaced by $\neg I(p)$[5]. Then, $\mathcal{GLP}_P(I)$ is defined as $LM_3(P_I)$. Fixpoints of the operator $\mathcal{GLP}_P$ are known to be 3-valued models of P. These 3-valued models are called *stable*. The *well-founded model* of P is the $\leq_{kn}$-least fixpoint of $\mathcal{GLP}_P$ [16].
    We have now the following technical lemma and the second main result of this section on the relationship between fixpoints of the operator $\mathcal{D}_{ael_2(P)}$ and 3-valued stable models of P.

Lemma 4. *If $I = Proj(B)$, then $P_I = (ael_2(P))_B$.*

Theorem 13. *Let P be a logic program over the set of atoms At.*

(a) $\mathcal{GLP}_P(Proj(B)) = Proj(\mathcal{D}_{ael_2(P)}(B))$

---

[5] Normally, P_I is further simplified, by deleting rules with $\neg t$ in the body and deleting literals $\neg f$ in the body of rules.

(b) *If a belief pair B is a fixpoint of $\mathcal{D}_{ael_2(P)}$ then $Proj(B)$ is a 3-valued stable model of P*

(c) *If I is a 3-valued stable model of P, then $B = Bel(P_I)$ is a fixpoint of $\mathcal{D}_{ael_2(P)}$ and $Proj(B) = I$*

(d) *$Proj(\mathcal{D}_{ael_2(P)}\!\uparrow)$ is the well-founded model of P.*

(e) *If a belief pair B is a complete fixpoint of $\mathcal{D}_{ael_2(P)}$, then $Proj(B)$ is a 2-valued stable model of P. Moreover, all 2-valued stable models of P are of this form.*

6   Conclusions and future work

In this paper we investigated the conctructive approximation scheme for Moore's autoepistemic logic. We introduced the notion of a belief pair – a Kripke-style 3-valued structure for the modal language. The set of belief pairs $\mathcal{B}$ is endowed with a natural ordering $\leq_p$. This ordering is chain complete, which guarantees that every monotone operator on $(\mathcal{B}, \leq_p)$ has a least fixpoint. With every modal theory T we associated a monotone *derivation* operator $\mathcal{D}_T$ on $(\mathcal{B}, \leq_p)$. We proposed the least fixpoint of the operator $\mathcal{D}_T$ as the intended constructive 3-valued semantics of modal theory T. We proved that the complete fixpoints of the operator $\mathcal{D}_T$ coincide with Moore's autoepistemic models of T. Thus, the semantics specified by the least fixpoint of $\mathcal{D}_T$ approximates Moore's semantics. Under appropriate embeddings of a logic program P as a modal theory T ($T = ael_1(P)$ or $T = ael_2(P)$), the least fixpoint of the operator $\mathcal{D}_T$ generalizes Kunen-Fitting semantics and Van Gelder-Ross-Schlipf well-founded semantics. These results provide further evidence of the correctness of our approach.

It is natural to ask how general is the technique proposed in our paper. In the forthcoming work we show that the scheme proposed in this paper can be generalized and that one can develop a theory of approximating operators. Specifically, we elucidate the abstract content of the well-founded semantics in terms of a suitably chosen approximation operator in a chain-complete poset.

Acknowledgments

This paper is a full version of an extended abstract that appeared in the Proceedings of AAAI-98, pp. 840 – 845, MIT Press, 1998. The second and third author were partially supported by the NSF grants IRI-9400568 and IRI-9619233.

Appendix: Stratified autoepistemic theories

We will present here a proof of Theorem 6. We will start by recalling the concept of stratification. We will use the original definition by Gelfond [3]. However, our argument can easily be extended to a slightly wider class of theories considered in [9].

A modal formula is called a *modal clause* if it is of the form

$$l_1 \wedge \ldots \wedge l_k \wedge K F_1 \wedge \ldots \wedge K F_m \neg K G_1 \wedge \ldots \wedge \neg K G_r \supset p_1 \vee \ldots \vee p_s$$

where $l_1, \ldots, l_k$ are literals of $\mathcal{L}$, $p_1, \ldots, p_s$ are atoms of $\mathcal{L}$, and $F_1, \ldots, G_r$ are formulas of $\mathcal{L}$.

A theory consisting of modal clauses is called *stratified* if there are pairwise disjoint theories $T_0, \ldots, T_n$ such that

i. $\bigcup_{i=0}^n T_i = T$
ii. T_0 is modal-free
iii. For every m, $0 < m \leq n$, all clauses in T_m have nonempty conclusions (that is, $s > 0$)
iv. Whenever p appears in a conclusion of a clause in T_j, $j > 0$, then p does not appear in T_i, $i < j$ and p does not appear within the scope of the modal operator K in T_i, $i \leq j$.

We call the list $\langle T_0, \ldots, T_n \rangle$ a *stratification* of T. In the remainder of this section, we write $T = T_0 \cup \ldots \cup T_n$ to indicate that $\langle T_0, \ldots, T_n \rangle$ is a stratification of T.

A stratification $T = T_0 \cup \ldots \cup T_n$ generates an increasing family of subsets of the set of atoms At. Namely, At_0 is the set of those atoms in At that do not occur in the conclusions of modal clauses from T_i, where $i > 0$, and

$$At_i = At_{i-1} \cup \{p: p \text{ occurs in the conclusion of a clause in } T_i\},$$

for $i = 1, \ldots, n$.

For an interpretation I and a set $Z \subseteq At$, by $I|Z$ we denote the restriction of I to Z. This concept is naturally extended to sets of interpretations and to belief pairs. For a set R of interpretations, we define $R|Z = \{I|Z : I \in R\}$ and, for a belief pair B, we define $B|Z = (P(B)|Z, S(B)|Z)$.

We say that a formula F is *based* on set of atoms Z if all atoms occurring in F belong to Z. The following simple lemma (we leave it without proof) gathers several facts on restrictions.

Lemma 5. *Let $Z \subseteq At$ and let F be a formula based on Z. Then for every belief pair B and interpretation I:*

(a) $(B, I) \models F$ *if and only if* $(B|Z, I|Z) \models F$
(b) $(B, I) \models_w F$ *if and only if* $(B|Z, I|Z) \models_w F$.

Consider a stratified theory $T = T_0 \cup \ldots \cup T_n$. We will now construct a sequence of belief pairs $B_0, \ldots, B_{n+1}$. Namely, we set $B_0 = \bot$ and for every i, $0 \leq i \leq n$, $B_{i+1} = (P_{i+1}, S_{i+1})$ where:

$$P_{i+1} = \{I \in P_i : (B_i, I) \models T_i\}$$

and

$$S_{i+1} = \{I \in P_i : (B_i, I) \models T_i \text{ and for every } p \in At \setminus At_i, I(p) = \mathbf{t}\}.$$

Lemma 6. *For every* i, $1 \leq i \leq n+1$, $B_i|At_{i-1}$ *is complete. Furthermore, for every interpretation* $I \in \mathcal{A}$, $(B_i, I) \models T_i$ *if and only if* $(B_i, I) \models_w T_i$.

Proof: Clearly, $S_i \subseteq P_i$. In particular, it follows that $S_i|At_{i-1} \subseteq P_i|At_{i-1}$. Consider now a valuation $I' \in P_i|At_{i-1}$. Then, there is a valuation $I \in P_i$ such that $I|At_{i-1} = I'$. Denote by J a valuation obtained from I by setting:

$$J(p) = \begin{cases} I(p) & \text{if } p \in At_{i-1} \\ \mathbf{t} & \text{if } p \in At \setminus At_{i-1}. \end{cases}$$

Since $I \in P_i$, $(B_{i-1}, I) \models T_{i-1}$. By Lemma 5, $(B_{i-1}, J) \models T_{i-1}$. Thus, by the definition of J, $J \in S_i$ and, consequently, $I' = I|At_{i-1} = J|At_{i-1} \in S_i|At_{i-1}$. Hence, for every i, $1 \leq i \leq n+1$, $P_i|At_{i-1} = S_i|At_{i-1}$. In other words, $B_i|At_{i-1}$ is complete.

By the definition of stratification, every modal atom KF occurring in a modal clause from T_i is based on the set of atoms At_{i-1}. Thus, the second part of the assertion follows from the completeness of the belief pair $B_i|At_{i-1}$ and from Lemma 5. □

The following lemma plays the key role in the proof of Theorem 6.

Lemma 7. *Let* $T = T_0 \cup \ldots \cup T_n$ *be a stratified theory. Then for every* i, $0 < i \leq n+1$, $B_i \leq_p \mathcal{D}^i(\perp)$.

Proof: We will proceed by induction on i. Let $i = 1$. Clearly, $P(B_1) = \{I : (\perp, I) \models T_0\}$ and $P(\mathcal{D}_T(\perp)) = \{I : (\perp, I) \models_w T\}$. Since T_0 is modal-free,

$$(\perp, I) \models T_0 \text{ if and only if } (\perp, I) \models_w T_0$$

Thus $P(\mathcal{D}_T(\perp)) \subseteq P(B_1)$ follows.

Consider now $I \in S(B_1)$. Then $(\perp, I) \models T_0$ and for every $p \in At \setminus At_0$, $I(p) = \mathbf{t}$. Since every clause in $T \setminus T_0$ has at least one positive atom in the conclusion, $(\perp, I) \models T$. Thus, $I \in S(\mathcal{D}_T(\perp))$. Consequently, $B_0 \leq_p \mathcal{D}^0(\perp)$. That is, the basis for the induction is established.

For the inductive step, we need to prove that $P(B_{i+1}) \supseteq P(\mathcal{D}_T^{i+1}(\perp))$ and $S(B_{i+1}) \subseteq S(\mathcal{D}_T^{i+1}(\perp))$. Consider an interpretation $I \notin P_{i+1}$. Then, either $I \notin P_i$ or $(B_i, I) \not\models T_i$. In the first case, since $B_i \leq_p \mathcal{D}_T^i(\perp) \leq_p \mathcal{D}_T^{i+1}(\perp)$, $I \notin P(\mathcal{D}_T^{i+1}(\perp))$. In the second case, by Lemma 6, $(B_i, I) \not\models_w T_i$. Consequently, by Proposition 2, $(\mathcal{D}_T^i(\perp), I) \not\models_w T_i$ and, hence also in this case, $I \notin P(\mathcal{D}_T^{i+1}(\perp))$. Thus, $P(B_{i+1}) \supseteq P(\mathcal{D}_T^{i+1}(\perp))$ follows.

Next, consider $I \in S_{i+1}$. By the definition, $I \in P_i$, $(B_i, I) \models T_i$ and for every $p \in At \setminus At_i$, $I(p) = \mathbf{t}$. We will show that $(\mathcal{D}_T^i(\perp), I) \models T$ (or, equivalently, that $I \in S(\mathcal{D}_T^{i+1}(\perp))$).

Consider stratum T_j with $j < i$. Then $P_i \subseteq P_{j+1}$. Since $I \in S_{i+1}$, $I \in P_i$ and, hence, $I \in P_{j+1}$. By the definition of P_{j+1}, $(B_j, I) \models T_j$. By the induction hypothesis, $B_j \leq_p \mathcal{D}_T^j(\perp)$. Thus, $B_j \leq_p \mathcal{D}_T^i(\perp)$. It now follows from Proposition 2 that $(\mathcal{D}_T^i(\perp), I) \models T_j$.

Next, consider stratum T_i. Since $I \in S_{i+1}$, $(B_i, I) \models T_i$. By the induction hypothesis, $B_i \leq_p \mathcal{D}_T^i(\bot)$. Hence, by Proposition 2, $(\mathcal{D}_T^i(\bot), I) \models T_i$.

Finally, consider stratum T_j with $j > i$. Since the conclusion of every modal clause in T_j contains a positive occurrence of an atom in $At \setminus At_i$ and since $I(p) = \mathbf{t}$ for every atom $p \in At \setminus At_i$, $(\mathcal{D}_T^i(\bot), I) \models T_j$.

To summarize, it follows that $(\mathcal{D}_T^i(\bot), I) \models T$. Consequently, the interpretation I belongs to $S(\mathcal{D}_T^{i+1}(\bot))$. $\qquad\qquad\square$

We now prove Theorem 6 from Section 3.

Theorem 6 *If T is a stratified autoepistemic theory then:*

(a) $\mathcal{D}_T\uparrow$ is complete
(b) T has a unique stable expansion
(c) $\mathcal{D}_T\uparrow$ is consistent if and only if the lowest stratum T_0 is consistent.

Proof: (a) Clearly, Lemma 6 implies that B_{n+1} is a complete belief pair. By Lemma 7, $B_{n+1} \leq_p \mathcal{D}_T^{n+1}(\bot)$. Hence, it follows that $B_{n+1} = \mathcal{D}_T^{n+1}(\bot)$. Thus, $\mathcal{D}_T^{n+1}(\bot)$ is a fixpoint. Hence, it is a least fixpoint and, since it coincides with B_{n+1}, it is complete.

(b) The assertion follows directly from (a) by Theorem 1.

(c) Clearly, if T_0 is inconsistent, $B_1 = (\emptyset, \emptyset)$ and it is a least fixpoint of $\mathcal{D}_T$. On the other hand, if T_0 is consistent, it is easy to see that $S_1 \neq \emptyset$. Hence, $\mathcal{D}_T$ is consistent. $\square$

References

1. N. Bourbaki. *Elements of Mathematics Theory of Sets.* Hermann, 1968.
2. M. C. Fitting. A Kripke-Kleene semantics for logic programs. *Journal of Logic Programming*, 2(4):295–312, 1985.
3. M. Gelfond. On stratified autoepistemic theories. In *Proceedings of AAAI-87*, pages 207–211. Morgan Kaufmann, 1987.
4. G. Gottlob. Complexity results for nonmonotonic logics. *Journal of Logic and Computation*, 2(3):397–425, 1992.
5. G. Gottlob. Translating default logic into standard autoepistemic logic. *Journal of the ACM*, 42(4):711–740, 1995.
6. K. Konolige. On the relation between default and autoepistemic logic. *Artificial Intelligence*, 35(3):343–382, 1988.
7. K. Kunen. Negation in logic programming. *Journal of Logic Programming*, 4(4):289–308, 1987.
8. H. J. Levesque. All I know: a study in autoepistemic logic. *Artificial Intelligence*, 42(2-3):263–309, 1990.
9. W. Marek and M. Truszczyński. Autoepistemic logic. *Journal of the ACM*, 38(3):588–619, 1991.
10. W. Marek and M. Truszczyński. *Nonmonotonic logics; context-dependent reasoning.* Springer-Verlag, Berlin, 1993.
11. G. Markowsky. Chain-complete posets and directed sets with applications. *Algebra Universalis*, 6(1):53–68, 1976.

12. R.C. Moore. Possible-world semantics for autoepistemic logic. In *Proceedings of the Workshop on Non-Monotonic Reasoning*, pages 344–354, 1984. Reprinted in: M. Ginsberg, ed., *Readings on nonmonotonic reasoning*, pp. 137–142, Morgan Kaufmann, 1990.
13. R.C. Moore. Semantical considerations on nonmonotonic logic. *Artificial Intelligence*, 25(1):75–94, 1985.
14. I. Niemelä. On the decidability and complexity of autoepistemic reasoning. *Fundamenta Informaticae*, 17(1-2):117–155, 1992.
15. I. Niemelä and P. Simons. Evaluating an algorithm for default reasoning. In *Proceedings of the IJCAI-95 Workshop on Applications and Implementations of Nonmonotomic Reasoning Systems*, 1995.
16. T.C. Przymusinski. The well-founded semantics coincides with the three-valued stable semantics. *Fundamenta Informaticae*, 13(4):445–464, 1990.
17. R. Reiter. A logic for default reasoning. *Artificial Intelligence*, 13(1-2):81–132, 1980.
18. G.F. Schwarz. Autoepistemic logic of knowledge. In A. Nerode, W. Marek, and V.S. Subrahmanian, editors, *Logic programming and nonmonotonic reasoning (Washington, DC, 1991)*, pages 260–274, Cambridge, MA, 1991. MIT Press.
19. G.F. Schwarz. Minimal model semantics for nonmonotonic modal logics. In *Proceedings of LICS-92*, pages 34–43, 1992.
20. A. Van Gelder, K.A. Ross, and J.S. Schlipf. The well-founded semantics for general logic programs. *Journal of the ACM*, 38(3):620–650, 1991.

Toward Efficient Default Reasoning[*]

David W. Etherington[1] and James M. Crawford[2]

[1] Computational Intelligence Research Laboratory
1269 University of Oregon
Eugene, OR   97403-1269
[2] i2 Technologies
909 E. Las Colinas Blvd, 12th floor
Irving, TX   75039

Summary. Reiter's pioneering work on default reasoning aimed to formalize the notion of quickly "jumping to conclusions". Unfortunately, his resulting formalism, default logic, proved more computationally complex than classical logics. The same has been true for the other major formalisms for nonmonotonic reasoning, circumscription [20], nonmonotonic logic [22], [21], and autoepistemic logic [25]. This has dramatically limited the applicability of formal methods vis à vis real problems involving defaults. The complexity of consistency checking is one of the two problems that must be addressed to reduce the complexity of default reasoning. We propose an approach to default reasoning based on approximate consistency checking, using a novel synthesis of limited contexts and fast incomplete checks, and argue that this combination overcomes the limitations of its component parts. Our approach trades correctness for speed, but we argue that the nature of default reasoning makes this trade relatively inexpensive and intuitively plausible. This approach not only accords well with Reiter's original motivations, but converges to Default Logic in the computational limit. We describe a prototype implementation of a default reasoner based on these ideas, and a preliminary empirical evaluation.

1   Computation and Nonmonotonicity

Early work on nonmonotonic reasoning (NMR) was often motivated by the idea that defaults should make reasoning easier. For example, Reiter [28] says " [closed-world reasoning] leads to a significant reduction in the complexity of both the representation and processing of knowledge". Winograd [32] observes that agents must make assumptions to act in real time: "A robot with common sense would [go] to the place where it expects the car to be, rather than sitting immobilized, thinking about the infinite variety of ways in which circumstances may have conspired for it not to be there."

Paradoxically, formal theories of NMR have been consistently characterized by their intractability. For example, first-order default logic [29] is not semi-decidable and its inference rules are not effective. In the propositional case, most NMR prob-

---

[*] A slightly modified version of this paper appeared in *Proc. AAAI-96* [11], ©1996, American Association for Artificial Intelligence, reprinted with permission.

lems are Σ_2^P or Π_2^P-complete [15], [31].[3] Even very restricted sublanguages based on propositional languages with linear decision procedures remain NP-complete [17]. Convincing examples of broadly useful theories within demonstrably tractable languages for NMR have yet to appear.

A nonmonotonic formalism sanctions a default conclusion only if certain facts can be shown to be consistent with the rest of the system's beliefs–i.e., only if it can be shown that the default is not a known exceptional case. Unfortunately, consistency is generally even harder to determine than logical consequence. The need to prove consistency before drawing default conclusions is the first source of the intractability of nonmonotonic formalisms.

The second source of intractability is that the order in which default rules are applied can effect the extension generated. It is these two sources of intractability together that produce the Σ_2^P (or Π_2^P) time complexity of most problems in default reasoning. However, given an oracle for consistency checking, some interesting problems, such as finding an extension for a normal default theory, could be solved tractably. Conversely, an oracle for default ordering would produce tractability only for languages with very limited expressive power. Furthermore, the ability to check consistency quickly is interesting in its own right for many propositional reasoning tasks. Therefore, we believe that a first step toward developing practicable nonmonotonic reasoners is to reduce their dependency on intractable consistency checking.

Rather than limiting the expressive power of the language to the point that consistency becomes tractably testable (and, arguably, it is impossible to say anything interesting), our approach to tractable, though approximate, consistency checking is ultimately based on limiting the search for exceptions. This approach has the intuitive appeal that a default can be applied without first having to discount every possible reason this case might be exceptional. We hope to recapture the intuition that a default should be applied unless its inapplicability is readily apparent (i.e., "at the top of your mind"). Our approach trades accuracy for speed: "inappropriate" conclusions may be reached that must be retracted solely due to additional thought, but this tradeoff accords with the original arguments for default reasoning. More importantly, we argue, defaults generally seem to be used in ways that minimize the cost of this tradeoff.

We restrict our discussion to Reiter's default logic [29], which appears to have become the most successful of the various nonmonotonic reasoning frameworks proposed to date, but it is important to note that our ideas apply directly to other nonmonotonic formalisms. A *default* has the form

$$\frac{P(\bar{x}) : J(\bar{x})}{C(\bar{x})} ,$$

----

[3] Arguably, this complexity is the price of increased expressivity, allowing NMR formalisms to represent knowledge that can't be concisely expressed in monotonic logics [3], [14] but this observation is little help in building practical systems.

where P, J, and C are formulae whose free variables are among $\bar{x} = x_1, ..., x_n$; they are called the *prerequisite, justification,* and *consequent* of the default, respectively. The default can be read as saying that if P is believed, and J is consistent with what is believed, then C should be believed: informally, things satisfying P typically satisfy C unless known not to satisfy J.

This paper is organized as follows. First, we outline tractable mechanisms for quickly proving consistency in some cases, and show why they are not adequate in general. We then detail the idea of context-limited consistency checking, and present several ideas on context selection. Next, we explain how these two ideas can be combined to achieve a tractable approximation of default reasoning, and finally we present preliminary experimental results on the efficacy of the approach.

2   Sufficient Tests for Consistency

While consistency checking is intractable, the good news is that there are ways to sometimes check quickly. Beyond simply computing until a fixed time bound runs out, there are fast sufficient (but not necessary) tests for consistency, and algorithms that reason from bounds that limit what follows from the knowledge base (KB) without deciding every fact. The complexity of some of these tests is no worse than linear in the size of the KB.

For example, consider testing whether β is consistent with a KB. Provided the theory and β are each self-consistent, it suffices (but is not necessary) that no literal in $\neg\beta$ occurs in the clausal representation of the theory. This can be tested in at worst linear time even for non-clausal theories. Similarly, if $\neg\beta$ occurs only in clauses with pure literals, β is consistent with the KB. More complicated tests derive from techniques such as vivid reasoning [18], [10], [1], knowledge compilation [30] and multi-valued entailment [4].

Unfortunately, there are two serious obstacles to using such fast tests. Those fast enough to check the whole KB in real time can be expected to fail in realistic applications. It would be a peculiar KB that, for example, had the default "Birds usually fly" with *no* information about non-flying birds! Representing a rule as a default seems to presume knowledge (or at least the strong expectation) of exceptions. Hence, when the time comes to test the consistency of $Flies(Tweety)$, we expect to have facts like $\forall x.\ \neg Penguin(x) \lor \neg Flies(x)$. This will cause the fast tests described above to fail, giving no useful information. The more complicated tests, such as knowledge compilation, are too expensive to do on the whole KB before each default is applied. Moreover, since applying defaults expressly changes what is believed, compilation cannot be done once in advance.

3   Context-Limited Consistency Checking

To address both of these difficulties, it seems plausible to try to to restrict the system's "focus of attention" to a *context*–a subset of the KB "near" the default being considered–in such a way that the limited computational resources available might be expected to produce the highest return. If computational resources are limited, it makes sense to focus our search for inconsistency on the relevant parts of the KB. For example, the default that you can get a babysitter might fail for prom night, but is unlikely to be affected by the stock market; a limited reasoner that devotes much effort to seeing if market fluctuations prevent hiring a sitter seems doomed.

Focusing on limited contexts provides two benefits. First, in the propositional case, consistency checking can be exponential in the size of the theory (if $P \neq NP$). Clearly, if we need only check a small subset, efficiency will improve significantly. Second, one can use fast consistency checks and limited contexts together to help gain efficiency even in first-order logic, where full consistency checking is undecidable.

Ideally, the context should contain exactly the formulae relevant to determining consistency. Then all necessary information is available, and irrelevant search is curtailed: consistency checking is no harder than it must be for correctness. Of course, this ideal solves the problem by reducing it to the arguably harder problem of determining relevance. Conversely, using a randomly-chosen context for consistency checking could be expected to produce very cheap consistency checks (since the fastest sufficient tests will be likely to succeed), and still have (marginally) better-than-random accuracy (applicable defaults won't be contradicted, and inapplicable default *might* be detected). Naturally, any realistic context-selection mechanism will fall between these extremes. Additional effort spent on context building can reduce the accuracy lost in focusing on the context: like most approximation schemes, practical context selection involves balance.

Just what a context should contain is an open question, but a rudimentary notion suffices to illustrate the idea (c.f. [9]). Facts come into the context as they are attended to (e.g., from perception or memory), and exit as they become stale. The context should include ground facts known about the objects under discussion (e.g., $Tweety$) as well as rules whose antecedents and consequents are instantiated by either the context or the negation of the justification to be checked (e.g., if $Penguin(Tweety)$ is in the context, checking the consistency of $Flies(Tweety)$, should draw in $\forall x. \ Penguin(x) \supset \neg Flies(x)$). Such a context can be built quickly using good indexing techniques.

This simple notion of context can be elaborated in many ways. Limited forms of rule chaining can be provided if chaining can be tightly bounded. For example, if the KB has a terminological component (c.f. [2]), chains through the type hierarchy might be brought in by treating deductions from terminological knowledge as single 'rule' applications. Also, "obvious" related items can be retrieved using Crawford's [7] notion of the accessible portion of the KB, Levesque's [19] notion of limited inference, or other mechanisms that guarantee cheap retrieval.

The significant feature of our approach is the synergy between the two components: context focuses the consistency check on the part of the KB most likely to contain an inconsistency and, often, can be expected to allow fast sufficient checks to succeed where they would fail in the full KB. Such fast tests can allow context-limited consistency testing to be efficient even in large first-order KBs.

A Simple Example: Consider the canonical default reasoning example:

$$Robin(Tweety), \ Penguin(Opus), \ Emu(Edna) \ \cdots$$

$$\forall x. \ Canary(x) \supset Bird(x) \qquad (1)$$
$$\forall x. \ Penguin(x) \supset Bird(x) \qquad (2)$$
$$\forall x. \ Penguin(x) \supset \neg Flies(x) \qquad (3)$$
$$\forall x. \ Emu(x) \supset Bird(x)$$
$$\forall x. \ Emu(x) \supset \neg Flies(x) \qquad \cdots$$
$$\frac{Bird(x) \ : \ Flies(x)}{Flies(x)}$$

where the ellipses indicate axioms about many other kinds of birds and many other individual birds. To conjecture that Tweety flies, one must prove $Flies(Tweety)$ is consistent with the above theory–i.e., that $Penguin(Tweety)$, $Emu(Tweety)$, etc. aren't provable. This amounts to explicitly considering all the ways that Tweety might be exceptional, which seems unlike the way people use defaults.

On the other hand, if recent history hasn't brought exceptional types of birds to mind, the context might contain just $Robin(Tweety)$ and (1). A fast test for consistency of $Flies(Tweety)$ would succeed, and so $Flies(Tweety)$ could be assumed. Deciding if Opus can fly, however, brings $Penguin(Opus)$ into the context and hence (2) and (3), so the consistency test fails. Similarly, after a long discussion about various forms of flightless birds, facts about exceptional classes should still be in the context. Fast consistency tests for $Flies(Tweety)$ would thus probably fail, and one would have to explicitly rule out exceptions.

4   The Mitigating Nature of Defaults

Clearly context selection is difficult. Fortunately, the nature of defaults makes selection of a useful context less problematic than might be expected. For a default to be reasonable, we contend, (at least) two factors must combine favorably: the likelihood that the consequent holds given that the prerequisite holds and the likelihood that if the prerequisite holds but the justifications are not consistent (so the default is not applicable), the agent will be aware of this fact. If the default is extremely likely to apply, one can tolerate overlooking the odd exception. Similarly, if exceptions are easy to spot, it may be useful to have a default that rarely applies. However, if exceptions are common but difficult to detect, one is ill-advised to make assumptions.[4] Now, if we characterize a " good default" as one for which the probability is low that the prerequisite holds, the justification is inconsistent, and the inconsistency will not be noticed, we are guaranteed that a context-based system will produce accuracy as good as its defaults.

---

[4] We ignore the obvious third factor: the cost of errors.

Gricean principles of cooperative communication seem to enforce the second property above: if the speaker believes the hearer may draw an inappropriate default conclusion from her utterance, she must explicitly block it [16], ensuring the appropriate contradiction is in the hearer's context when the default is considered. Similarly, in many non-conversational applications of default reasoning, exceptions are perceptible, and hence can be expected to be in the current focus of attention if they occur. For example, when planning to take one's default (but flood-prone) route home, one can easily see whether it is raining heavily and, if so, block the default.

Since this is an approximate approach, it can fail in several ways. The context may fail to contain the knowledge necessary to detect an inconsistency. If this happens, any consistency check we use will fail to realize that the default is contradicted. Moreover, even if the context is built perfectly, fast consistency checks may fail, making it necessary to perform a full consistency check within the context, or to guess. Finally, if an application requires defaults for which ensuring high expected correctness requires an unmanageable context, context-limited reasoning won't help.

This being said, we can make several arguments in favor of our approach. First, as argued above, the nature of the default reasoning task minimizes the impact of the loss of correctness while maximizing the gain in efficiency. Second, the kinds of errors induced by approximate consistency checking are intuitively plausible. People, lacking logical omniscience, frequently apply defaults that they "know" are inappropriate. For example, many of us can recall driving our default route between two points only to recall–on encountering a major traffic jam–having heard that the road would be under construction. Such examples seem to indicate that human default reasoning does take place in a limited context that does not include everything we know. Third, we achieve "asymptotic correctness": if the agent has time to retrieve more formulae and reason with them, the probability of correctness (measured, for example, against the specification of standard default logic) increases. Thus, we can achieve a favorable trade of correctness for efficiency, without abandoning the semantic foundation provided by default logic. Finally, since by their very nature, defaults may be wrong despite being consistent with all one knows, agents should be prepared to accept errors in default conclusions, and deal with the resulting inconsistencies, as Perlis [26] and many others have argued. An increase in the error rate should thus be less problematic to a default reasoner than it might be to a purely deductive system.

5   Experimental Evaluation

We now turn to the results of preliminary experiments beginning the validation of our approach. A completely convincing test would involve extensive experiments on large, first-order, real-world nonmonotonic KBs, showing significant computational gains and acceptable error rates. Sadly, the intractability of nonmonotonic formalisms seems to have stifled construction of large KBs with defaults; we hope that the this work will be a step toward their construction.

Meanwhile, the goal of these experiments is more modest: preliminary determination of the effect of context limitations on the accuracy and cost of consistency checking for randomly-generated propositional theories. Such theories are generally characterized by two parameters: the number of variables and the number of clauses (the length of all clauses is generally taken to be three). For low clause-to-variable ratios, almost all problems are satisfiable, and most problems are computationally easy. At high ratios, almost all problems are unsatisfiable and most problems are easy. In between, in the so-called "transition region", lies a mixture of satisfiable and unsatisfiable problems, and many quite hard problems (Cheeseman, Kanefsky & Taylor 1991; Mitchell, Selman & Levesque 1992; Crawford & Auton 1993).

Our experiments are primarily in the underconstrained region. In the overconstrained region, almost all theories are inconsistent, so no defaults are applicable. Solving problems in the transition region generally seems to require intricate case-splitting of the kind found more in logic puzzles than in commonsense reasoning. Also, an agent's world knowledge is likely to be fairly underconstrained—we generally know sufficiently little about the world that there are many models that are consistent with what we know.[5]

Working in the underconstrained region, we face a problem: it is likely that a random literal chosen to be our "default" will be consistent with a random theory, and consistency checking in any limited context (even the empty context!) will give the right answer. To solve this problem, we add a randomly generated set of literals to our theories. Intuitively, the clauses correspond to general knowledge about the world and these literals correspond to a set of facts.

The experiments presented below investigate the success of context-limited consistency checking as problem size (number of variables, V), degree of constraint (number of clauses, C), and number of facts (number of literals, L) vary. We find that context limitations are useful in much of the underconstrained region, but their utility drops sharply as we approach the transition region. This is consistent with the generally held belief that, in the transition region, clauses throughout the theory interact with each other in complex ways. Context-limited consistency checking also becomes less useful as L becomes more than about $V/2$; in these cases unit propagation makes full consistency checking so easy that context-limitations become superfluous.

5.1 Experimental Setup

We generate random 3-SAT theories using Mitchell *et al*'s (1992) method—each clause is generated by picking three different variables at random and negating each with probability 0.5. There is no check for repeated clauses. Inconsistent theories are discarded. We then randomly select a series of L literals consistent with the theory built so far. Consistency checks are done using TABLEAU [6].

---

[5] Of course, commonsense knowledge no doubt clusters and some of these clusters may be locally quite constraining. Ideally in these cases one would want to choose the context to include the entire cluster, but this goes beyond the scope of the current experiments.

We select a random literal d to be the "default", and construct a series of concentric contexts around d. $C_{r,d}$ denotes the context around d with *radius r*. Intuitively, the radius measures how many clauses the context extends out from d. More formally, the context is the subset of the input theory, T, defined as follows: $C_{l,0}$ is $\neg l$ if $\neg l \in T$, and $\{\,\}$ otherwise. For $r > 0$,

$$C_{r,l} = C_{r-1,l} \cup \bigcup_{x \vee y \vee \neg l \in T} (\{x \vee y \vee \neg l\} \cup C_{\neg x, r-1} \cup C_{\neg y, r-1})$$

(e.g., the $r = 3$ context around l contains the $r = 2$ contexts around l, and around l's neighbors). TABLEAU is used to test satisfiability. For these tests we modified it to halt the search whenever the current partial assignment satisfies all the clauses in the theory.

5.2   Experimental Results

Experiment 1: The first experiment tests how the efficacy of context limitation varies with problem size. We varied V from 100 to 600 incrementing by 100. We set C to $2V$ (roughly centered in the underconstrained region), and L to $0.4V$, generated 200 theories, and tested 10 defaults per theory. Each check was done against contexts with radius 0 to 3, and then against the entire theory. The results appear in Figure 1.

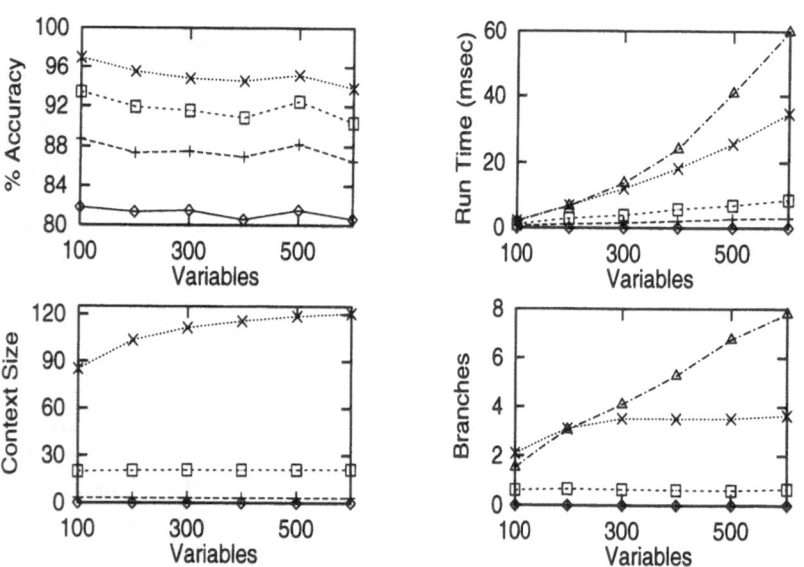

Fig. 1. Accuracy, run time), context size, and search tree size, *vs* V, at $L = 0.4V$. $r = 0, 1, 2, 3,$ and ∞ are marked $\diamond$, $+$, $\square$, $\times$ and $\triangle$, respectively.

The limited growth of the context size is not surprising. A simple probabilistic argument shows that for large problems the expected context size depends on r and C/V (not V). Further, since the number of branches in the search tree depends primarily on the size of the context, it makes sense that the number of branches does not increase appreciably with problem size. Run time depends on the number of branches and on the time spent at each node. However, the time TABLEAU spends at each node depends linearly on V even when reasoning in a restricted context. This is an artifact of the design of TABLEAU (and the context-building mechanism) that could be removed with some recoding. If this artifact were removed, run time for reasoning within the limited contexts would presumably not increase appreciably with problem size. In any case, run time in the contexts increases more slowly than run time for consistency checking in the entire theory. Accuracy (the percentage of correct answers from the consistency check) also seems relatively unaffected by problem size. We conjecture that, for large problems, accuracy is a function of r, C/V and L/V.

Combining these effects, we conclude that the effectiveness of context-limited consistency checking increases with problem size. The size of a radius r context, and thus the complexity of the consistency check, is essentially unchanged as V increases, but the accuracy of the consistency check does not seem to fall. This attractive property is due to the fact that, at least for underconstrained, random theories, the average length of the inference chains that might lead us to conclude $\neg d$ depends on C/V rather than on V. If this same effect occurs in realistic KBs then context limitation should be quite effective for large problems.

Experiment 2: The second experiment measures the effect of changing L on the effectiveness of context limitation. We fixed V at 200 and C at 400. We varied L from 20 to 180 by 20, generating 100 theories and testing 10 defaults per theory. Results are shown in Figure 2.

The most interesting result is the accuracy, which generally falls to a minimum at around $L = 0.4V$, and increases on either side of this point. We believe that the rise in accuracy below $0.4V$ is due not to any real increase in the effectiveness of the context-limited consistency check; below this point more defaults are consistent, and context-limited checks only make mistakes when defaults are inconsistent. However, note that as L falls, the difference in run time (and search tree size) between the context-limited check and the full check rises dramatically. Thus the results in Experiment 1 would have been even more favorable had we chosen a lower L/V ratio. Above $L = 0.5V$, the accuracy of the context-limited checks rises again. However, this region is not particularly interesting because so many literals are set by the input theory that full consistency checking becomes trivial.

Experiment 3: The final experiment measures how the effectiveness of context limitation changes with C. We fixed V at 200 and L at 20, and varied C from 200 to 800 by 100. This takes us from quite underconstrained to the edge of the transition region. We generated 100 theories and tested 10 defaults per theory, at each point. The results appear in Figure 3.

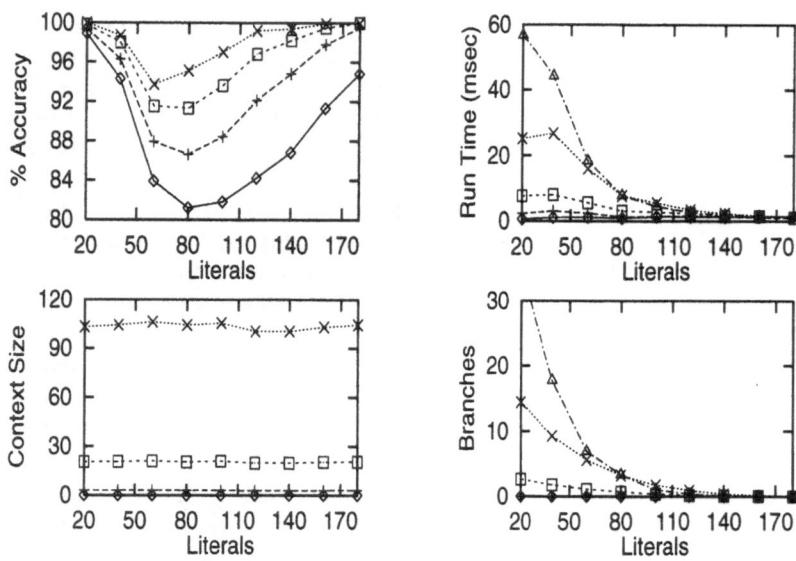

Fig. 2. Accuracy, run time, search tree size, and context size, *vs L*, at $V = 200$, $C = 400$. $r = 0, 1, 2, 3$, and ∞ are marked $\diamond$, $+$, $\square$, $\times$ and $\triangle$, respectively.

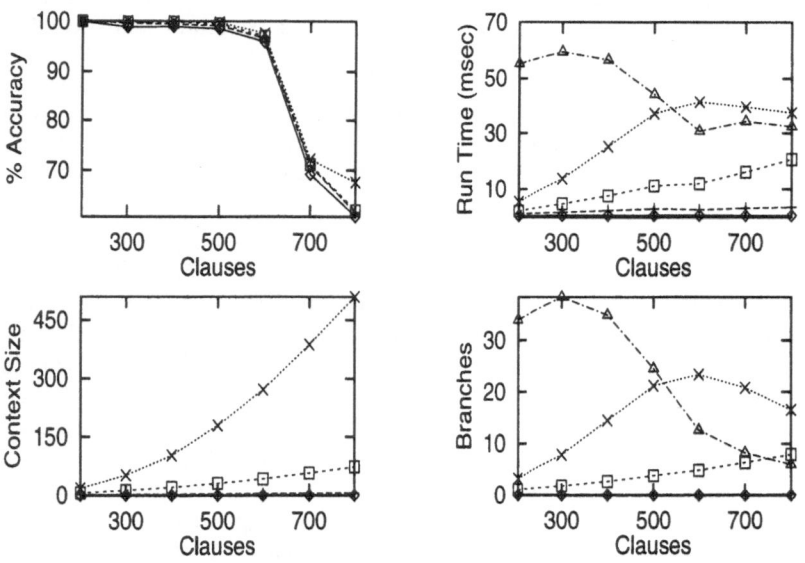

Fig. 3. Accuracy, run time, context size, and search tree size, *vs C*, at $V = 200$, $L = 20$. $r = 0, 1, 2, 3$, and ∞ are marked $\diamond$, $+$, $\square$, $\times$ and $\triangle$, respectively.

Here again *Accuracy* is the most interesting graph. Starting at about 600 clauses, or C/V about 3, the accuracy falls dramatically. We believe this is because near the transition region the interactions between the clauses in the theory become more global and any limited context is likely to miss some of them and so fail to detect inconsistencies. Our hope, of course, is that realistic theories of commonsense knowledge do not interact in this way (or do so only within local clusters that can be entirely included within the context).

One surprise is that starting at about 500 clauses, or a ratio of about 2.5, the cost of the consistency check in the radius 3 context rises *above* the cost of the full check. We believe this is due to of a kind of "edge effect" in the context. Consider a clause $x \vee y \vee z$ in the context. In some cases, there may be sets of clauses and literals in the full theory (e.g., $\neg x \vee a \vee b$, $\neg a$, and $\neg b$) but not in the context, that force the value of x. If this happens, the full check may actually be easier due to unit resolution. One way to test this hypothesis would be to unit resolve the input theory (this can be done in linear time) before any other reasoning is done.

6   Related Work

The idea of restricting the scope of consistency checking to a subset of the KB is not new. Our ideas are the logical result of a long tradition of context-limited AI reasoning systems dating back to CONNIVER (c.f. [23], [13]). This line of work limits deductive effort, resulting in incompleteness. Limiting consistency checking in default reasoning, however, results in unsoundness—unwarranted conclusions may be reached due to lack of deliberation.

More directly related is Perlis' suggestion to limit consistency checking to about seven formulae determined by immediate experience. Perlis argues that anything more is too expensive [26], [9]. He suggests that agents will have to simply adopt default conclusions and retract them later when further reasoning reveals contradictions. There are problems with Perlis' approach, however. First, consistency-checking can be undecidable even in such tiny theories. More importantly, though, the errors this approach produces do not seem justifiable, since defaults are applied with essentially *no* reflection. Our analysis can be seen as explaining *why* (and when) such context-limited consistency checking can be expected to have a high probability of correctness. Furthermore, we believe that the notion of applying fast consistency tests in limited contexts provides significant leverage, allowing contexts to be larger while still achieving tractability.

THEORIST [27] is also related in that it uses limited consistency checking to determine default applicability. However, THEORIST does not maintain a notion of context, so its errors are based on the use of incomplete reasoning mechanisms, rather than restricted focus of attention. Also, THEORIST has no notion of fast sufficient consistency checking.

7 Conclusions and Open Problems

We have described, and presented a preliminary experimental evaluation of, a practical way to trade accuracy for speed in consistency checking, that we expect to have applications to Reiter's default logic and to nonmonotonic reasoning in general (as well as to other commonsense reasoning problems that involve verifying consistency). We argue that restricting the consistency check to a focused context, combined with fast tests for consistency, can improve expected efficiency, at an acceptable and motivatable cost in accuracy.

The techniques we have outlined are not universally applicable–any gains from our approach hinge on the nature of the theories and defaults involved. It is easy to construct pathological theories in which *any* restriction of the context will falsely indicate consistency. (Such theories occur near the transition region in our experimental results.) In general, our approach will suffer if there are too many exceptions and those exceptions are hard to detect. We conjecture, however, that commonsense reasoning in general, and default reasoning in particular, is well behaved, in that complex interactions between distant parts of the KB are rare, and inconsistent defaults are generally readily apparent. In addition, we achieve "asymptotic correctness": as time allows, broadening the context increases the probability of correctness This allows a favorable trade of correctness for efficiency, without abandoning semantic foundations. Also, since by their very nature, defaults may be wrong despite being consistent with all one knows, default reasoners should be prepared to accept errors in default conclusions, and deal with the resulting inconsistencies.

The efficacy of our approach depends on the design of both the context-selection and consistency-checking mechanisms. These choices can only be based on, and ultimately verified by, extensive experiments with realistic commonsense KBs. Here we offered, and experimentally examined, only some simple first-cut mechanisms. In particular, our experiments use complete consistency checking in the context. In the propositional case, this appears to be sufficient; we believe that sufficient tests for consistency will be important primarily for first-order theories.

We are continuing to explore the application of limited contexts and fast consistency checking to nonmonotonic reasoning. Our recent work has focused on sufficient algorithms for checking consistency that are guaranteed to be fast (taking polynomial time in the size of the context) [8], and on extensions of those algorithms to first-order representations [12], where limited contexts appear to play a key role in obtaining tractability.

We reiterate that consistency checking is only one source of combinatorial complexity in default reasoning; for many problems of interest, default ordering presents another. We conjecture that limited contexts can also serve to limit the search of default orderings (e.g., by considering only defaults in the context), which may allow a tractable approximation of the overall default reasoning problem, but have not yet explored this possibility.

8   Acknowledgments

This work was supported in part by NSF grant IRI-94-12205, and by the Defense Advanced Research Projects Agency (DARPA) and US Air Force Research Laboratory (AFRL) contracts F30602-93-C-0031 and F30602-95-1-0023.

These ideas were refined over several years in discussions with many people, especially Alex Borgida, Bob Mercer, and Ray Reiter. We are indebted to them and to Matt Ginsberg, Henry Kautz, David Poole, Bart Selman and anonymous referees, for fruitful discussions and criticisms. The views and conclusions contained herein are those of the authors and should not be interpreted as necessarily representing the official policies or endorsements, either expressed or implied, of the above-named individuals, DARPA, AFRL, or the U.S. Government.

References

1. A. Borgida and D.W. Etherington. Hierarchical knowledge bases and tractable disjunction. In *Proceedings of the First International Conference on Principles of Knowledge Representation and Reasoning (KR'89)*, pages 33–43, Toronto, Canada, May 1989. Morgan Kaufmann.
2. R.J. Brachman and J. Schmolze. An overview of the KL-ONE knowledge representation system. *Cognitive Science*, 9(2):171–216, 1985.
3. M. Cadoli, F.M. Donini, and M. Schaerf. Is intractability of non-monotonic reasoning a real drawback? In *Proceedings of the Twelfth National Conference on Artificial Intelligence (AAAI-94)*, pages 946–951, 1994.
4. M. Cadoli and M. Schaerf. Approximate inference in default logic and circumscription. In *Working Notes of the 4th International Workshop on Nonmonotonic Reasoning*, 1992.
5. P. Cheeseman, B. Kanefsky, and W.M. Taylor. Where the really hard problems are. In *Proceedings of the Twelveth International Joint Conference on Artificial Intelligence (IJCAI-91)*, pages 163–169, 1991.
6. James M. Crawford and Larry D. Auton. Experimental results on the crossover point in satisfiability problems. In *Proceedings of the Eleventh National Conference on Artificial Intelligence (AAAI-93)*, pages 21–27, 1993.
7. J.M. Crawford. *Access-Limited Logic–A Language for Knowledge Representation*. PhD thesis, University of Texas at Austin, Austin, Texas, September 1990.
8. J.M. Crawford and D.W. Etherington. A non-deterministic semantics for tractable inference. In *Proceedings of the Fourteenth National Conference on Artificial Intelligence (AAAI-98)*, 1998.
9. J. Elgot-Drapkin, M. Miller, and D. Perlis. Life on a desert island: Ongoing work on real-time reasoning. In *Proceedings of the 1987 Workshop on the Frame Problem in Artificial Intelligence*, pages 349–357, Lawrence, KS, April 1987.
10. D.W. Etherington, A. Borgida, R.J. Brachman, and H.A. Kautz. Vivid knowledge and tractable reasoning. In *Proceedings of the Eleventh International Joint Conference on Artificial Intelligence (IJCAI-89)*, pages 1146–1152, 1989.
11. D.W. Etherington and J.M. Crawford. Toward efficient default reasoning. In *Proceedings of the Thirteenth National Conference on Artificial Intelligence (AAAI-96)*, 1996.
12. D.W. Etherington and J.M. Crawford. Limited contexts and tractable first-order reasoning. in preparation, 1999.

13. S.E. Fahlman. *NETL: a System for Representing and Using Real-World Knowledge*. MIT Press, 1979.
14. G. Gogic, H. Kautz, C. Papadimitriou, and B. Selman. The comparative linguistics of knowledge representation. In *Proceedings of the Fourteenth International Joint Conference on Artificial Intelligence (IJCAI-95)*, pages 862–869, 1995.
15. G. Gottlob. Complexity results for nonmonotonic logics. In *Working Notes of the 4th International Workshop on Nonmonotonic Reasoning*, 1992.
16. H.P. Grice. Logic and conversation. In P. Cole and J.L. Morgan, editors, *Syntax and Semantics, Vol 3: Speech Acts*. Academic Press, 1975.
17. H. Kautz and B. Selman. Hard problems for simple default logics. In *Proceedings of the First International Conference on Principles of Knowledge Representation and Reasoning (KR'89)*, Toronto, Canada, May 1989. Morgan Kaufmann.
18. Hector J. Levesque. Making believers out of computers. *Artificial Intelligence*, 30:81–108, 1986.
19. H.J. Levesque. A logic for implicit and explicit belief. Technical Report FLAIR TR 32, Fairchild Laboratory for Artificial Intelligence Research, 1984.
20. John McCarthy. Circumscription – a form of non-monotonic reasoning. *Artificial Intelligence*, 13:27–39, 1980.
21. Drew McDermott. Non-monotonic logic II. *Journal ACM*, 29:33–57, 1982.
22. Drew McDermott and Jon Doyle. Non-monotonic logic I. *Artificial Intelligence*, 13:41–72, 1980.
23. D.V. McDermott and G.J. Sussman. The CONNIVER reference manual. Technical Report AI Memo 259, MIT Artificial Intelligence Laboratory, 1972.
24. David Mitchell, Bart Selman, and Hector Levesque. Hard and easy distributions of SAT problems. In *Proceedings of the Tenth National Conference on Artificial Intelligence (AAAI-92)*, pages 459–465, 1992.
25. R. Moore. Semantical considerations on nonmonotonic logic. *Artificial Intelligence*, 25:75–94, 1985.
26. D. Perlis. Nonmonotonicity and real-time reasoning. In *Proceedings of the First International Workshop on Nonmonotonic Reasoning*, New Paltz, NY, October 1984.
27. D. Poole. Explanation and prediction: An architecture for default reasoning. *Computational Intelligence*, 5(2):97–110, 1989.
28. R. Reiter. On reasoning by default. In *Proceedings of the Second Conference on Theoretical Issues in Natural Language Processing*, pages 210–218, Urbanna, IL, 1978.
29. R. Reiter. A logic for default reasoning. *Artificial Intelligence*, 13:81–132, 1980.
30. B. Selman and H. Kautz. Knowledge compilation using Horn approximations. In *Proceedings of the Ninth National Conference on Artificial Intelligence (AAAI-91)*, pages 904–909, 1991.
31. J. Stillman. The complexity of propositional default logics. In *Proceedings of the Tenth National Conference on Artificial Intelligence (AAAI-92)*, pages 794–799, 1992.
32. T. Winograd. Extended inference modes in reasoning by computer systems. *Artificial Intelligence*, 13:5–26, 1980.

Action, Time and Default

D. M. Gabbay

Department of Computer Science,
King's College,
London

I am very happy to write this note, honouring Ray Reiter. Ray has recently visited me at King's College for some months and we had the opportunity to talk and see how compatible our views are on the subject. We also have a common religious (this also means a certain kind of logic) background which affects one's scientific approach (I can only speak for myself here). So here is how I see things![1]

The story is set almost twenty-five years ago. I undertook to teach a logic class in the Humanities and Social Sciences with about 100 mature students (i.e. between 40–50 years old) who were in the middle of their successful careers (judges, police superintendents, army generals, top executives, school heads, local dignitaries, etc.). They were given a sabbatical year to take a course at university. One of the classes was *logic* and that is what I was to teach them. I planned to teach mainly propositional logic, truth tables, natural deduction rules and some fallacies and paradoxes. In those days, non-monotonic logic was just about to start and logic and computation as an area was in the process of being born.

It became very clear to me right from the beginning that formal logic and natural deduction was completely disconnected from real everyday life. My students all dealt with practical reasoning situations on a day-to-day basis; some of their activities were reported in the press (we had to turn the class into a 'closed confidential session' on more than one occasion, in order to discuss some case studies) and it became apparent that the most useful and potent topics for the students were, in fact, the fallacies and the paradoxes (considered relatively marginal in traditional logic courses). I learnt more from them than they did from me!

That experience left me wondering what 'real' logic is in the human mind, and how we can model it and what kind of a course would be suitable for such people.

The emergence of non-monotonic reasoning helped understand real practical reasoning a bit better, but I believe the area went in the 'wrong' direction and became more and more technical, on many occasions 'feeding upon itself' and trying to gain 'respectability' by being formal.

I don't want to criticise research in this area: I am sure that it has enriched pure logic by expanding our understanding of what a logical system is, what proof theory is, and has given new horizons for logic. However, I don't think my class of 25 years ago would have been much happier with the new logics. They may have liked them a little better, but not that much. They would still have homed in on the fallacies and the paradoxes, as the most relevant to their day-to-day situations.

---

[1] This note is based on an editorial 'What is on my mind', *Journal of Logic and Computation*, **9**, 1–6, 1999.

Today I think that in real situations we do not really reason but we *act* . We need systems where the 'database' is mainly a list of available actions (in particular we need notation for actions as first class residents in the database) and the 'proof' is mainly a sequence of threats and actions and where the 'argument' is mostly only a cover-up to mislead everyone else and present yourself in a better light. It is a reality where assumptions constantly disappear and where a simple deduction can take up to two weeks! I am excited about these ideas in fact I have had many recent disucssions with Ray about them.[2]

Let me see what such a logical system looks like, through an example.

My name is John. I have to take an exam next week and I'm not prepared. If I show up I will fail. If I don't show up without good cause I'll also fail. So I check what a good cause might be. Illness is good cause. So I go to the university nurse and say I am sick. This might not work because she may examine me. What is consistent with my general state of health at the time? One thing which cannot be checked is depression and nervous breakdown. So I pretend to be depressed. I go to the nurse and try to get a letter to satisfy the examiners. My tutor, however, suspects this, but what can he do? He is determined not to let me get away with it. So he confides in one of the girl students, explains the situation and asks her to offer me a date to go to the nightclub. If I accept, then that will disprove my 'depression'.

What we have here is a sequence of action/counter-actions intended to *create* data so that certain goals can be obtained. The reasoning is secondary. It is only used to licence actions that will give the right results.

To explain what is happening, let us model the above argument. We start with a vocabulary:

E = general data about exams, students, etc.
$A(x)$ = x is depressed
$B(x)$ = x gets a special postponed exam
$N(x)$ = x goes to a nightclub
$D(x)$ = x records ill health

Action $\mathbf{a}(x)$:
x goes to the college nurse, looks nervous and claims depression.
preconditions: $A(x)$
postconditions: $D(x)$

Action $\mathbf{b}(x,y)$:
x asks someone to lure y to go to a nightclub.
preconditions: none
postconditions: $N(y)$
This postcondition is non-deterministic. It is probable, but not certain.

Action $\mathbf{s}(x,y)$:
x suspends y pending investigation.

---

[2] See my book, *Dynamics of Practical Reasoning*, in preparation. Postscript file available from Jane Spurr, jane@dcs.kcl.ac.uk and Ray's book *Knowledge in Action*, to appear.

preconditions: contradiction involving y
postconditions: $\neg B(y)$

Action default $\mathbf{d}(x)$:
Assume $A(x)$
preconditions: It is consistent to add $A(x)$ to the now current database.
postconditions: $A(x)$

The players in this 'deduction' are John and the Tutor. The initial database is Δ, containing the following:

1. E, (data)
2. $\forall x(D(x) \rightarrow B(x))$, (data)
3. $\forall x(N(x) \rightarrow \neg D(x))$, (data)
4. Actions available to any x, y, at any stage:
   $\mathbf{a}('x), \mathbf{b}(x,y), \mathbf{s}(x,y), \mathbf{d}(x)$.

The 'proof'/'argument'/'dialogue'/'game' goes as follows, between John and the Tutor.

5. Activate $\mathbf{d}(J)$ and add $A(J)$ to database[3]
6. Activate $\mathbf{a}(J)$ and add $D(J)$ to the database.
7. $B(J)$ from (6) and (2).
8. Activate $\mathbf{b}(T, J)$. There is a good chance that we may be able to add $N(J)$ in which case we continue.
9. Activate $\mathbf{s}(T, J)$ and add (revise!) the data with $\sim B(J)$.

The above procedures may resemble very much what is done in some planning systems or in some agents models in artificial intelligence. I am not denying that. In those systems, however, the purpose and the uses are different. I claim what we have just been doing is *logic*. To strengthen my case recall the dialogue semantics or game semantics for intuitionistic logic. See [Felscher, 1985].[4] To show that $a \wedge b \rightarrow a \wedge b$ is a theorem of intuitionistic logic, we need two players, the proponent (say John) and the opponent, say the Tutor. The proponent puts forward $a \wedge b \rightarrow a \wedge b$ and the opponent attacks and the game continues according to rules. The proponent has a winning strategy exactly for theorems of the logic.

Here is the game for this case as presented in [Felscher, 1985. p. 345][5] using my notation.

---

[3] Note how we are using the default rule. We can be, in principle, quite happy with a pair of defaults of the form $\frac{q}{q}$ and $\frac{\neg q}{\neg q}$. We just choose the one that suits us. None of the usual worries of the default formal machinery apply here.

[4] W. Felscher. Dialogues as a foundation for intuitionistic logic. In *Handbook of Philosophical Logic*, Vol. 3. D. Gabbay and F. Guenthner, eds. Kluwer, 1985.

[5] ibid.

0. $a \wedge b \rightarrow a \wedge b$, (John)
1. $a \wedge b$, attack of Tutor
2. a, counterattack by John on first conjunct of 1.
3. a, defence by Tutor against 2.
4. b, counterattack by John on second conjunct of 1.
5. b, defence by Tutor against 4.
6. $a \wedge b$, defence against 1 by John.

There are now two ways of continuing, either attacking a:

7. a, Tutor attacks first conjunct of 6
8. a, John defends against 7 in view of 3

or attacking b

7. b, Tutor attacks second conjunct of 6
8. b, John defends against 7 in view of 5.

The moves of John and Tutor are regulated. It is very easy and natural to allow additional actions in the database and allow additional moves, for example a player x activates $\mathbf{d}(x)$ and adds $A(x)$, etc.

It is clear that the new system can be viewed as just another dialogue logic with a twist.

One point to clarify. In the dialogue interaction, the 'time' involved is the sequence of moves. In our exam example the time involved is 'real' time. Does this make a difference? No, it does not. We all know that intuitionistic logic is complete for the Kripke future-time semantics, where at any time t, $t \vDash A \rightarrow B$ means 'whenever A becomes true, so does B'. Dialogue moves correspond to semantic tableaux moves against Kripke time model and so can be interpreted as real time.

So in our new kind of logic, modus ponens can take real time and actions can be involved. I could have taught this kind of logic to my class of 25 years ago and it would have made practical sense to them.

Explanatory Diagnosis:
Conjecturing Actions to Explain Observations*

Sheila A. McIlraith

Knowledge Systems Laboratory
Department of Computer Science
Stanford University
Stanford, CA 94305

Summary. In this paper we present contributions towards a logical theory of diagnosis for systems that can be affected by the actions of agents. Specifically, we examine the task of conjecturing diagnoses to explain *what happened* to a system, given a theory of system behaviour and some observed (aberrant) behaviour. We characterize what happened by introducing the notion of explanatory diagnosis in the language of the situation calculus. Explanatory diagnoses conjecture sequences of actions to account for a change in system behaviour. As such, we show that determining an explanatory diagnosis is analogous to classical AI planning with state constraints and incomplete knowledge. The representation scheme we employ provides an axiomatic solution to the frame, ramification and qualification problems for a syntactically restricted class of state constraints. Exploiting this representation, we show that determining an explanatory diagnosis can be achieved by regression followed by theorem proving in the database describing what is known of the initial state of our system. Further, we show that by exploiting features inherent to diagnosis problems, we can simplify the diagnosis task.

Foreword

It is a great pleasure to contribute a paper to this book in honour of Ray Reiter. Ray was my Ph.D. supervisor at the University of Toronto. Inspired by Ray's work on diagnosis from first principles, my early work proposed an augmentation of his characterization of diagnosis to include a formal account of testing and test generation. The work on testing convinced me that a comprehensive account of diagnostic problem solving must involve reasoning about action and change. Again, Ray's work on the situation calculus had a tremendous influence. The problem of intergrating actions into diagnostic problem solving posed many interesting challenges, which led to a solution to the ramification problem (sometimes), and to several new characterizations of diagnosis, one of which is described in this paper.

So, Happy Birthday Ray! Here's to the pleasures of life. There are many more theorems to prove, mountains to climb, and butterflies to chase. You've had a long and impressive research career. The field, and the many researchers you've touched with your work, owe you a debt of gratitude.

---

* A version of this paper originally appeared in [23]

1   Introduction

Given a theory of system behaviour and some observed aberrant behaviour, the traditional objective of diagnosis is to conjecture *what is wrong* with the system, (e.g., which components of the device are behaving abnormally, what diseases the patient is suffering from, etc.). Each candidate diagnosis consists of a subset of distinguished literals that are conjectured to be true or false in order to account for the observation in some way. Different criteria have been proposed for determining the space of such candidate diagnoses. Within formal accounts of diagnosis, two widely accepted definitions of diagnosis are consistency-based diagnosis (e.g., [8], [26]), and abductive explanation (e.g., [4], [8], [20], [25]). Such research has historically focussed on static systems. Recently, some researchers have advocated extending diagnostic problem solving (DPS) to enable reasoning about actions, under the argument that DPS is purposive in nature and that systems operate within and are affected by agents[1].

In this paper we focus upon one aspect of diagnosing such dynamic systems. In particular, given a theory of system behaviour and some observation of (aberrant) behaviour, our concern is with the task of conjecturing diagnoses to explain *what happened* to the system (i.e., what actions or events occurred to result in the observed behaviour). Knowing or conjecturing what happened is interesting in its own right, but it can also help to further constrain the space of possible states of the system. In so doing, conjecturing what happened facilitates conjecturing of what is wrong with a system, as well as predicting other relevant system behaviour. Compared to our traditional notion of *what is wrong* diagnoses, knowing *what happened* can more accurately capture the root cause of system malfunction rather than its manifestations, thus providing for the identification of future preventative as well as prescriptive actions.

In the spirit of previous foundational work in model-based diagnosis (MBD) (e.g., Reiter [26], Console and Torasso [4], de Kleer et al. [8]), this paper presents a logical characterization of the diagnosis task. We take as our starting point the existing MBD research on characterizing diagnoses for static systems *without* a representation of actions (e.g., [8], [4], [26]). Next, we exploit a situation calculus representation scheme previously proposed by the author [21] that enables the integration of a representation of action with the representation of the behaviour of a static system. With this representation in hand, we provide a logical characterization for the task of determining *what happened* to a system. The characterization is presented in the guise of *explanatory diagnosis*.

The distinguishing features of our characterization are afforded in great part by the richness of our representation scheme which provides a comprehensive and semantically justified representation of action and change. In particular, our representation provides an axiomatic closed-form solution to the frame and ramification problems for a syntactically restricted class of state constraints that is common to diagnosis. This representation captures the direct and indirect effects of actions in

---

[1] An agent could be another system, a robot, a human, or nature.

a compiled representation, which is critical to our ability to generate explanatory diagnoses efficiently. Further, our representation provides a closed-form solution to the qualification problem, thus identifying the conditions underwhich an action is possible. It is interesting to note that when we are dealing with incomplete knowledge of our initial state, conjecturing an action or sequence of actions also requires conjecturing that its preconditions are satisfied, which in many instances serves to further constrain our search.

As we show in the sections to follow, our characterization establishes a direct link between explanatory diagnosis and planning, deductive plan synthesis, and abductive planning. As a consequence of a completeness assumption embedded in our representation, we can exploit goal-directed reasoning in the form of regression [34] in order to generate diagnoses. This completeness assumption also provides for an easy mapping of our situation calculus representation to Prolog. While explanatory diagnoses can be mathematically characterized in an analogous fashion to plans, an important distinction of explanatory diagnoses is that they can be refined to exploit features of our diagnosis problem that have no meaning in the context of planning. Indeed, we use diagnosis-specific attributes to define variants of explanatory diagnosis to deal with the challenges of incomplete knowledge and large search spaces.

2   Representation Scheme

2.1   Situation Calculus Language

The situation calculus language in which we axiomatize our domains is a sorted first-order language with equality. The sorts are of type $\mathcal{A}$ for primitive *actions*, $\mathcal{S}$ for *situations*, and $\mathcal{D}$ for everything else, including domain objects [17]. We represent each action as a (possibly parameterized) first-class object within the language. Situations are simply sequences of actions. The evolution of the world can be viewed as a tree rooted at the distinguished initial situation S_0. The branches of the tree are determined by the possible future situations that could arise from the realization of particular sequences of actions. As such, each situation along the tree is simply a history of the sequence of actions performed to reach it. The function symbol *do* maps an action term and a situation term into a new situation term. For example, $do(turn\_on\_pump, S_0)$ is the situation resulting from performing the action of turning on the pump in situation S_0. The distinguished predicate $Poss(a, s)$ denotes that an action a is possible to perform in situation s (e.g., $Poss(turn\_on\_pump, S_0)$). Thus, $Poss$ determines the subset of the situation tree consisting of situations that are possible in the world. Finally, those properties or relations whose truth value can change from situation to situation are referred to as *fluents*. For example, the fluent $on(Pump, s)$ expresses that the pump is on in situation s.

The dialect of the situation calculus that we use in this paper is restricted to primitive, determinate actions. Our language does not include functional fluents, nor does it include a representation of time, concurrency, or complex actions, but we believe the results presented herein can be extended to more expressive dialects of the situation calculus (e.g., [24], [29]).

2.2 Domain Representation

In this section we overview the representation scheme we use to characterize the system we will be diagnosing. The scheme, proposed in [21], integrates a situation calculus theory of action with a MBD system description, SD [8]. The resulting representation of a system comprises both domain-independent and domain-specific axioms. The domain-independent axioms are the foundational axioms of the discrete situation calculus, Σ_{found} [17]. They are analogous to the axioms of Peano arithmetic, modified to define the branching structure of our situation tree, rather than the number line. The domain-specific axioms, T specify both the *behaviour of the static system*, and the *actions*[2] that can affect the state of the system, as well as those actions required to achieve testing and repair. Together they define our situation calculus representation $\Sigma = \Sigma_{found} \wedge T$.

The domain-specific axioms composing T result from application of a procedure proposed in [21] that compiles a typical MBD system description, SD and a set of axioms relating to the preconditions and effects of actions into a representation that provides a closed-form solution to the frame, ramification and qualification problems for a syntactically restricted class of domain axiomatizations. The resulting domain axiomatization $T = T_{SC}^{S_0} \wedge T_{domain} \wedge T_{SS} \wedge T_{AP} \wedge T_{UNA} \wedge T_{DCA} \wedge T_{S_0}$ is described below. The representation is predicated on an explicit causal ordering of fluents and a completeness assumption. The assumption states that all the conditions underwhich an action a can lead, directly or indirectly, to fluent F becoming true or false in the successor state are captured in the axiomatization of our system.

We illustrate the representation scheme in terms of a small portion of a power plant feedwater system [22] derived from the APACS project [14]. This simplified example models the filling of a vessel either by the operation of an electrically powered ($Power$) pump ($Pump$), by manual filling, or by a siphon that was started by the pump or by manual filling. For notational convenience, all formulae are understood to be universally quantified with respect to their free variables, unless explicitly indicated otherwise. For a more thorough description of this representation scheme, and for a more extensive example please see [21], [22].

Every domain axiomatization, T comprises the following sets of axioms.

$$T_{SC}^{S_0} \wedge T_{domain} \wedge T_{SS} \wedge T_{AP} \wedge T_{UNA} \wedge T_{DCA} \wedge T_{S_0}$$

The set of state constraints relativized to situation S_0, $T_{SC}^{S_0}$ capture what is implicitly true about the initial database. They can be acquired from a typical MBD system description, SD as described in [21]. Note that these state constraints hold for all situations s. We represent them explicitly for the initial situation S_0, and they are compiled into the successor state axioms, T_{SS} for all subsequent situations. In our simple example, $T_{SC}^{S_0}$ is as follows.

$$\neg AB(Power, S_0) \wedge \neg AB(Pump, S_0) \wedge on(Pump, S_0) \supset filling(S_0) \quad (1)$$

$$manual\_fill(S_0) \supset filling(S_0) \quad (2)$$

$$\neg(on(Pump, S_0) \wedge manual\_fill(S_0)) \quad (3)$$

---

[2] Actions can be performed by agents: a human, another system, or nature.

Axiom (1) states that if the pump and power are operating normally and the pump is *on* in S_0, then *filling* will be true in S_0. The set of domain constraints, T_{domain} is as follows.

$$Power \neq Pump \tag{4}$$

The set of successor state axioms, T_{SS} is composed of axioms of the following general form, one for each fluent F.

$$Poss(a, s) \supset [F(do(a, s)) \equiv \Phi_F] \tag{5}$$

where Φ_F is a simple formula[3] of a particular syntactic form. Intuitively, a successor state axiom says the following:

$Poss(a, s) \supset [fluent(do(a, s)) \equiv$
  an action made it true
  $\vee$ *a state constraint made it true*
  $\vee$ *it was already true*
    $\wedge$ *neither an action nor a state constraint*
      made it false].

The following axioms compose T_{SS} for our example.

$$Poss(a, s) \supset [on(Pump, do(a, s)) \equiv a = turn\_on\_pump$$
$$\vee (on(Pump, s) \wedge a \neq turn\_off\_pump)] \tag{6}$$

$$Poss(a, s) \supset [AB(Power, do(a, s)) \equiv a = power\_failure$$
$$\vee (AB(Power, s) \wedge a \neq aux\_power \wedge a \neq power\_fix)] \tag{7}$$

$$Poss(a, s) \supset [AB(Pump, do(a, s)) \equiv a = pump\_burn\_out$$
$$\vee (AB(Pump, s) \wedge a \neq pump\_fix)] \tag{8}$$

$$Poss(a, s) \supset [manual\_fill(do(a, s)) \equiv a = turn\_on\_manual\_fill$$
$$\vee (manual\_fill(s) \wedge a \neq turn\_off\_manual\_fill)] \tag{9}$$

---

[3] A *simple formula* only mentions domain-specific predicate symbols, fluents do not include the function symbol *do*, there is no quantification over sort *situation*, and there is at most one free *situation* variable.

$$Poss(a, s) \supset [filling(do(a, s)) \equiv a = turn\_on\_manual\_fill$$
$$\lor \; (manual\_fill(s) \land a \neq turn\_off\_manual\_fill)$$
$$\lor \; [(a \neq power\_failure$$
$$\land \; (\neg AB(Power, s) \lor a = aux\_power$$
$$\lor \; a = power\_fix))$$
$$\land \; (a \neq pump\_burn\_out$$
$$\land \; (\neg AB(Pump, s) \lor a = pump\_fix))$$
$$\land \; (a = turn\_on\_pump$$
$$\lor \; (on(Pump, s) \land a \neq turn\_off\_pump))]$$
$$\lor \; (filling(s) \land a \neq stop\_siphon)] \tag{10}$$

Axiom (6) states that if action a is possible in situation s, then the pump is on in the situation resulting from performing action a in situation s (i.e., $on(Pump, do(a, s))$) if and only if the action a is $turn\_on\_pump$, or the pump was already on in s and a is not $turn\_off\_pump$. Note that the successor state axioms presented here, and in particular axiom (10), are fully compiled. They can be expressed more compactly as intermediate successor state axioms [22].

The set of action precondition axioms, T_{AP} is composed of axioms of the following general form, one for each action prototype A in the domain.

$$Poss(A(x), s) \equiv \Pi_A \tag{11}$$

where Π_A is a simple formula with respect to situation variable s, capturing the necessary and sufficient conditions for prototype action $A(x)$ to be executable in situation s.

$$Poss(stop\_siphon, s) \equiv (\neg manual\_fill(s) \land \; \neg on(Pump, s)) \tag{12}$$
$$Poss(pump\_fix, s) \equiv \neg on(Pump, s) \tag{13}$$
$$Poss(pump\_burn\_out, s) \equiv on(Pump, s) \tag{14}$$
$$Poss(turn\_on\_manual\_fill, s) \equiv \neg on(Pump, s) \tag{15}$$
$$Poss(turn\_on\_pump, s) \equiv \neg manual\_fill(s) \tag{16}$$
$$Poss(turn\_off\_pump, s) \equiv on(Pump, s) \tag{17}$$
$$Poss(turn\_off\_manual\_fill, s) \equiv manual\_fill(s) \tag{18}$$
$$Poss(power\_failure, s) \equiv \text{true} \tag{19}$$
$$Poss(power\_fix, s) \equiv \text{true} \tag{20}$$
$$Poss(aux\_power, s) \equiv \text{true} \tag{21}$$

Finally, we provide a possible set of initial conditions for our system, T_{S_0}. These constitute the explicit aspect of the initial database. Note that in general we do not have complete knowledge of the initial state of our system. This makes the task of diagnosis all the more challenging. In this example, we do not know initially whether the pump and power are operating normally. We also do not know whether the vessel was filling in the initial state.

$$on(Pump, S_0) \land \neg manual\_fill(S_0) \tag{22}$$

T_{UNA} and T_{DCA} are the unique names axioms for actions and the domain closure axiom for actions, respectively.

This concludes the description of our representation scheme. Before advancing to issues of diagnosis, we note [30] that our proposed situation calculus representation can be viewed as an executable specification because it is easily realized in Prolog by exploiting Prolog's completion semantics and simply replacing the equivalence connectives characteristic of axioms in T_{SS} and T_{AP} by implication connectives. The Lloyd-Topor transformation [19] must then be applied to convert this theory into Prolog clausal form. Later in this paper, we will advocate using Waldinger's notion of regression to rewrite axioms of our representation and simplify computation. This type of regression rewriting is precisely achieved by Prolog's backwards chaining mechanism.

3   Preliminaries

With our representation in hand, we turn our attention to the task of diagnosis. In this section we introduce the framework for performing diagnosis relative to our representation. For our purposes we adopt the ontological and notational convention of the MBD literature and view the systems we are diagnosing as comprising a number of interacting *components*, $COMPS$. These components have the property of being either abnormal or normal in a situation. We express this property in our situation calculus language using the fluent AB. For example, $AB(Pump, s)$ denotes that the pump component is abnormal in situation s. Note that the use of AB is not mandatory to the contributions of this paper. Once again, following the convention in the MBD literature, we define our diagnoses relative to the domain-independent concept of a *system* [8], adapted to our situation calculus framework.

Definition 1 (System).
A system is a quadruple $(\Sigma, HIST, COMPS, OBS)$ where:

- Σ, the background theory, is a set of situation calculus sentences describing the behaviour of our system and the actions that can affect it.
- $HIST$, the history, is a sequence of ground actions $[a_1, \ldots, a_k]$ that were performed starting in S_0.
- $COMPS$, the components, is a finite set of constants.
- OBS_F, the observation, is a simple formula composed of fluents whose only free variable is the situation variable s, and which are otherwise ground.

Example 1.
In our power plant example above, Σ is our axiomatization $\Sigma_{found} \wedge T$ and $COMPS$ = $\{Pump, Power\}$. The observation, OBS_F could be $filling(s)$, for example. $HIST$ could be empty, i.e., [ ], or perhaps $[turn\_on\_pump]$.

4   Explanatory Diagnosis

In this section we introduce and formally characterize the notion of an explanatory diagnosis which conjectures *what happened* to result in some observed (aberrant) behaviour. Given a system, $(\Sigma, HIST, COMPS, OBS_F)$, the objective of explanatory diagnosis is to conjecture a sequence of actions, $[\alpha_1, \ldots, \alpha_n]$ such that our observation is true in the situation resulting from performing that sequence of actions in $do(HIST, S_0)$.

Definition 2 (Explanatory Diagnosis).
 An explanatory diagnosis for system $(\Sigma, HIST, COMPS, OBS_F)$ is a sequence of actions $E = [\alpha_1, \ldots, \alpha_n]$ such that,

$$\Sigma \models Poss(HIST \cdot E, S_0)^3 \wedge OBS_F(do(HIST \cdot E, S_0)).$$

Thus, E is an explanatory diagnosis when the observation is true in the situation resulting from performing the sequence of actions E in situation $do(HIST, S_0)$, and further that the preconditions for each action of the action sequence $HIST \cdot E$ are true in the appropriate situations, commencing at S_0.

The problem of determining explanatory diagnoses is an instance of temporal explanation or postdiction (e.g., [31]), and is related to the classical AI planning problem. In particular, identifying the sequence of actions composing an explanatory diagnosis, E is analogous to the plan synthesis problem, and thus is realizable using deduction on the situation calculus axioms. According to [12], a plan to achieve a goal $G(s)$ is obtained as a side effect of proving $Axioms \models \exists s.G(s)$. The binding for the situation variable s represent the sequence of actions. In our case, $Axioms \models \exists s.G(s)$ is analogous to $\Sigma \models \exists s.OBS_F(s)$. As such, our representation enables us to generate explanatory diagnoses deductively, just as we could deductively generate a plan in the situation calculus. Note that the task of generating explanatory diagnoses is analogous to plan synthesis in the presence of state constraints – a challenging problem. Our representation scheme eliminates the additional challenges presented by state constraints by providing a domain axiomatization that a priori solves the frame, ramification and qualification problems.

---

[3] **Notation:**
  $HIST \cdot E$ is an abbreviation for $[a_1, \ldots, a_k, \alpha_1, \ldots, \alpha_n]$.
  $do([a_1, \ldots, a_m], s)$ is an abbreviation for
      $do(a_m, (do(a_{m-1}, (do(a_{m-2}, (\ldots, (do(a_1, s)))))))).$
  Finally, $Poss([a_1, \ldots, a_n], s)$ is an abbreviation for
      $Poss(a_1, s) \wedge Poss(a_2, do(a_1, s)) \wedge \ldots \wedge Poss(a_n, do([a_1, \ldots, a_{n-1}], s)).$

Example 2.
Given the power plant example system $(\Sigma, [\,], \{Power, Pump\}, \neg filling(s))$, the sequence of actions $[power\_failure]$ constitutes *one* example of an explanatory diagnoses for the system. Another explanatory diagnosis for our system is $[turn\_off\_pump]$.

Observe that for certain problems there can be an infinite number of sequences of actions that constitute explanatory diagnoses. For example, the following sequences of actions also constitute valid explanatory diagnoses for our example system:

$[power\_failure, power\_fix, power\_failure]$,
$[power\_failure, aux\_power, power\_failure]$,
$[turn\_off\_pump, power\_failure, turn\_on\_pump]$,

and so on.

Definition 2 is not sufficiently discriminating to eliminate these, clearly suboptimal explanatory diagnoses. We must define a preference criterion. Probability measures, even simple order of magnitude probabilities have provided an effective preference criterion for many applications of MBD [7]. Likewise, in the case of determining explanatory diagnoses in the context of the situation calculus, probabilities will serve us well in identifying preferred explanatory diagnoses. Unfortunately, probability measures are not always available and the correct treatment of probabilities in our situation calculus framework is only now being developed. In this paper, we limit our discussion to what we refer to as a chronologically simple preference criterion.

In our chronologically simple preference criterion, we prefer diagnoses that are relativized to situations reached without performing any extraneous actions. Note that this preference criterion is syntactic in nature, relying on the notion of a primitive action as a unit measure.

Definition 3 (Simpler).
Given a sequence of actions $HIST = [\alpha_1, \ldots, \alpha_n]$, define $ACTS(HIST)$ to be the set $\{\alpha_1, \ldots, \alpha_n\}$, and $LEN(HIST)$ to be the the length of the sequence of actions composing $HIST$.

Thus, given $HISTA = [a_1, \ldots, a_n]$ and $HISTB = [b_1, \ldots, b_n]$, situation $S_A = do(HISTA, S_0)$ is simpler than situation $S_B = do(HISTB, S_0)$ iff $ACTS(HISTA) \subseteq ACTS(HISTB)$ and $LEN(HISTA) < LEN(HISTB)$.

Definition 4 (Chronologically Simple Diagnosis).
E is a chronologically simple explanatory diagnosis for system $(\Sigma, HIST, COMPS, OBS_F)$ iff E is an explanatory diagnosis for the system, and there is no explanatory diagnosis E' such that situation $S' = do(HIST \cdot E', S_0)$ is simpler than situation $S = do(HIST \cdot E, S_0)$.

We might further distinguish this criterion to prefer chronologically simple explanatory diagnoses comprised solely of actions performed by nature.

Finally, observe that the characterization of explanatory diagnosis just presented assumes that E and OBS_F occur *after* $HIST$. While this assumption is not critical to characterizing explanatory diagnoses, it acts as a form of preference, facilitating computation of E.

5   Exploiting Regression

In the previous section, we provided a characterization of explanatory diagnosis. We observed that computing an explanatory diagnosis for system (Σ, $HIST$, $COMPS$, OBS_F) is analogous to generating a plan to achieve a goal $OBS_F(s)$ starting with axioms Σ and situation $do(HIST, S_0)$. At first glance, the general problem of computing explanatory diagnoses does not look very promising for at least three reasons: the second-order induction axiom in Σ_{found}, the potential incompleteness of the initial database, and the potentially large size of the situation search space. In this section, we show how diagnoses can be computed by exploiting regression [34]. We are able to exploit regression as a direct result of the embedded completeness assumption inherent in our representation scheme.   In this context, regression is a recursive rewriting procedure that we use to reduce the nesting of the *do* function in situation terms, or to eliminate the *Poss* predicate. We show that generating explanatory diagnoses reduces to regression followed by entailment with respect to the initial database. Computationally, the merit of regression is that it searches backwards through the situation space from the observation rather than searching forward from the initial database. Under the assumption that the observation consists of fewer literals than the initial database, regression will make for more efficient search. Observe that Prolog's backwards chaining mechanism achieves the substitution performed by regression.

Following directly in the spirit of previous work by Reiter, [27],[28] and more recently [30], on the exploitation of regression for planning and query answering, we first define two regression operators, $\mathcal{R}^*$ and $\mathcal{R}_{Poss}$.

Definition 5 (Regression Operator $\mathcal{R}^*$).
Given a set of successor state axioms, T_{SS} composed of axioms of the form $Poss(a, s) \supset [F(do(a, s)) \equiv \Phi_F]$, $\mathcal{R}^*[\Psi]$, the repeated regression of formula Ψ with respect to successor state axioms T_{SS} is the formula that is obtained from Ψ by repeatedly replacing each fluent $F(do(a, s))$ in Ψ by Φ_F, until the resulting formula makes no mention of the function symbol *do*.

For example,
$$\mathcal{R}^*[on(Pump, do(turn\_on\_manual\_fill, do(turn\_on\_pump, S_0)))]$$
$$= \mathcal{R}^*[(turn\_on\_manual\_fill = turn\_on\_pump)$$
$$\vee (on(Pump, do(turn\_on\_pump, S_0))$$
$$\wedge (turn\_on\_manual\_fill \neq turn\_off\_pump))]$$
$$= \mathcal{R}^*[\text{false} \vee (on, (Pump, do(turn\_on\_pump, S_0)) \wedge \text{true})]$$
$$= \mathcal{R}^*[(turn\_on\_pump = turn\_on\_pump)$$
$$\vee (on(Pump, S_0)) \wedge (turn\_on\_pump \neq turn\_off\_pump))]$$
$$= \text{true}$$

We can similarly define a *Poss* regression operator over the set of action precondition axioms, T_{AP}. This regression operation rewrites each occurrence of the literal $Poss(a, s)$ by Π_A as defined in the action precondition axioms.

Definition 6 (Regression Operator $\mathcal{R}_{Poss}$).
Given a set of action precondition axioms, T_{AP} composed of axioms of the form $Poss(A(x), s) \equiv \Pi_A$, $\mathcal{R}_{Poss}[W]$ is the formula obtained by replacing each occurrence of predicate $Poss(A(x), s)$ by Π_A. All other literals of W remain the same.

Reiter proved soundness and completeness results for regression applied to a theory with no state constraints [28, Theorem 1, Theorem 2]. In the theorem below, we prove soundness and completeness results for our representation scheme which includes state constraints. The theory Σ_{init} mentioned in the theorem below is a subset of Σ containing only information about the initial situation, and no information about successor situations. It also excludes the induction axiom of Σ_{found}.

Theorem 1 (Soundness and Completeness).
Given

- Σ_{init}, *a subset of the situation calculus theory* Σ, *such that*
  $$\Sigma_{init} = \Sigma_{UNS} \wedge T_{S_0} \wedge T_{SC}^{S_0} \wedge T_{domain} \wedge T_{UNA},$$
  where Σ_{UNS} *is a subset of* Σ_{found} *containing the set of unique names axioms for situations.*
- *a sequence of ground actions, s_HIST such that*
  $$\Sigma_{init} \wedge \mathcal{R}^* [\mathcal{R}_{Poss} [Poss(s\_HIST, S_0)]] \text{ is satisfiable.}$$
- $Q(s)$, *a simple formula whose only free variable is situation variable s.*

Further, suppose $S = do(s\_HIST, S_0)$, *then*

- $\Sigma \models Q(do(s\_HIST, S_0))$ *iff* $\Sigma_{init} \models \mathcal{R}^*[Q(do(s\_HIST, S_0))]$,
- $\Sigma \models Poss(s\_HIST, S_0)$ *iff* $\Sigma_{init} \models \mathcal{R}^*[\mathcal{R}_{Poss}[Poss(s\_HIST, S_0)]]$,
- $\Sigma \wedge Poss(s\_HIST, S_0) \wedge Q(do(s\_HIST, S_0))$ *is satisfiable iff*
  $$\Sigma_{init} \wedge \mathcal{R}^*[\mathcal{R}_{Poss}[Poss(s\_HIST, S_0)]] \wedge \mathcal{R}^*[Q(do(s\_HIST, S_0))]$$
  is satisfiable.

Thus, assuming situation s is a possible situation and exploiting regression, $Q(s)$ holds at situation s iff its regression is entailed in the initial database. The beauty of Theorem 1 is that it enables us to generate explanatory diagnoses via regression followed by theorem proving in the initial database, without the need for the second-order induction axiom in Σ_{found}. From these results, we can characterize explanatory diagnosis with respect to regression.

Corollary 1 (Explanatory Diagnosis with Regression).
The sequence of actions $E = [\alpha_1, \dots, \alpha_k]$ *is an explanatory diagnosis for system* $(\Sigma, HIST, COMPS, OBS_F)$ *iff*

$$\Sigma_{init} \models \mathcal{R}^*[\mathcal{R}_{Poss}[Poss(HIST \cdot E, S_0)]] \wedge \mathcal{R}^*[OBS_F(do(HIST \cdot E, S_0))].$$

6   Exploiting the Task

In the previous section we showed that we could exploit regression to simplify the computation of explanatory diagnoses. A remaining source of difficulty in generating explanatory diagnoses is that our search space may be large and our initial database may be incomplete. An incomplete initial database both underconstrains our search problem, and precludes us from using certain planning machinery, such as STRIPS [10], that assumes a complete or near-complete initial database [18]. In this section we show how to exploit features of diagnosis problems to further assist in the generation of explanatory diagnoses. In particular, we propose to 1) make assumptions regarding our domain, 2) relax our criteria for explanatory diagnoses, and 3) verify rather than generate diagnoses. An additional means of simplifying our computational task is to use likelihoods of actions and action sequences to focus search for explanatory diagnoses. Detailed discussion of this option is beyond the scope of this paper.

6.1   Assumption-Based Diagnoses

In diagnostic problem solving it is common to make further assumptions that are consistent with what we know of the world. For example, we may assume that in the absence of information to the contrary, components are operating normally, or certain properties hold of the world. To support such assumption-based reasoning, we define the notion of an assumption-based explanatory diagnosis.

Definition 7 (Assumption-Based Explanatory Diagnosis).
Given an assumption $H(S)$ relativized to ground situation S such that
- $S_0 \leq {}^4 S \leq do(HIST \cdot E, S_0)$,
- $\Sigma \wedge H(S)$ is satisfiable, and
- $\Sigma \wedge H(S) \models Poss(HIST, S_0)$.

An assumption-based explanatory diagnosis for system $(\Sigma, HIST, COMPS, OBS_F)$ under assumption $H(S)$ is a sequence of actions $E = [\alpha_1, \ldots, \alpha_k]$ such that,

$$\Sigma \wedge H(S) \models Poss(HIST \cdot E, S_0) \wedge OBS_F(do(HIST \cdot E, S_0)).$$

In Example 2 of the previous section, we did not have complete information about the initial state of our system. It could actually have been the case that observation $\neg filling$ was true in S_0, i.e., $\neg filling(S_0)$, but since it was not entailed by Σ, the empty action sequence was not proposed as a valid explanatory diagnosis,

---

[4] **Notation:** The transitive binary relation $<$ defined in Σ_{found} further limits our situation tree by restricting the actions that are applied to a situation to those whose preconditions are satisfied in the situation. Intuitively, if $s < s'$, then s and s' are on the same branch of the tree with s closer to S_0 than s'. Further, s' can be obtained from s by applying a sequence of actions whose preconditions are satisfied by the truth of the $Poss$ predicate.

and we were forced to conjecture a sequence of actions to account for our observation. If we assume $\neg filling(S_0)$, then the empty sequence of actions is indeed an assumption-based explanatory diagnosis.

In generating explanatory diagnoses, we may want to make a priori assumptions about the world, conjoin these assumptions to our theory and then try to compute our explanatory diagnoses. For example, we may wish to assume that all components are operating normally in S_0. This would be achieved by making $H(S)$ in our definition equal to $\bigwedge_{c \in COMPS} \neg AB(c, S_0)$ (i.e., $\neg AB(Pump, S_0) \wedge \neg AB(Power, S_0)$). Similarly, we may wish to assume that the observation, OBS_F is true in $do(HIST, S_0)$. In our example above, this would mean assuming $H(S) = \neg filling(S_0)$. In other instances, we might want $H(S)$ to equal a *what is wrong* diagnosis that we are currently entertaining, relativized to a previous situation. For example, we might want to assume that the pump was abnormal in the initial situation, i.e. $H(S) = AB(Pump, S_0)$.

In still other instances, we may not want to fix our assumptions a priori but rather make the minimum number of assumptions necessary to generate a chronologically simple explanatory diagnosis. Such assumptions might be limited to a distinguished set of literals which the domain axiomatizer considers to be legitimately assumable (e.g., AB fluents). There might also be a partial ordering, e.g. a likelihood ordering, on such assumables fluents.

The distinction regarding when we make our assumptions affects the machinery by which we compute assumption-based explanatory diagnoses. If assumptions are made prior to computation we simply conjoin the regression of the assumption to the initial database and use regression and theorem proving as we would for generating normal explanatory diagnoses. When assumptions are interleaved with computation, generating an assumption-based explanatory diagnosis requires abduction.

In a theorem prover, abduction is generally implemented as proof-tree completion, i.e., by resolving dead-ends of proof trees with abducible literals. To generate an explanatory diagnosis, an abductive theorem prover would attempt to prove $OBS_F(s)$. If an attempted proof failed because it dead-ended on a literal or literals that were assumable, then these would be abduced and the proof continued. We have not yet implemented such an abduction engine for our situation calculus system. It is interesting to note that in the context of planning, [32] has used abduction to implement a partial order planner for the event calculus. In contrast to the deductive approach of the situation calculus, this planner abduces actions and the relative order of certain actions.

6.2   Potential Diagnoses

In addition to making assumptions to help complete our theory, we can also facilitate computation by relaxing the criteria for defining an explanatory diagnosis. To this end, we observe that the requirement in Definitions 2 and 7 that $\Sigma \models Poss(HIST \cdot E, S_0)$ may be too stringent in the case of an incomplete initial database. That is, it may not be reasonable to require that we know that an action is possible in a situation that is incompletely specified. We may prefer to consider explanatory diagnoses, where the theory allows us to consistently assume that the preconditions for $HIST$ or for $HIST \cdot E$ hold, but not necessarily that they are entailed by our theory. To this end, we propose the following alteration to our definition of explanatory diagnosis, Definition 2. A comparable refinement can be made to our definition of assumption-based explanatory diagnosis, Definition 7.

Definition 8 (Potential Explanatory Diagnosis).
A potential explanatory diagnosis for system $(\Sigma, HIST, COMPS, OBS_F)$ is a sequence of actions $E = [\alpha_1, \ldots, \alpha_k]$ such that,

$$\Sigma \wedge Poss(HIST \cdot E, S_0) \text{ is satisfiable, and}$$
$$\Sigma \wedge Poss(HIST \cdot E, S_0) \models OBS_F(do(HIST \cdot E, S_0)).$$

Note that $HIST$ is a sequence of actions that we know to have been performed starting in situation S_0. Thus, the preconditions for each of the actions in $HIST$ are true in the corresponding situations. This provides us with further information concerning the truth values of fluents at various situations, helping to constrain our search.

6.3 Verifying Likely Diagnoses

To further address the problem of generating explanatory diagnoses, we propose exploiting domain information and maintaining a library of most likely (assumption-based) explanatory diagnoses, indexed by observations and/or situation histories. With candidate diagnoses in hand, the problem of computing explanatory diagnoses reduces to a verification problem, rather than a generation problem. Given a system $(\Sigma, HIST, COMPS, OBS_F)$, and a candidate diagnosis E, such that $S = do(HIST \cdot E, S_0)$, we are interested in verifying that E is indeed a diagnosis of the system. Verifying a candidate diagnosis is simply a query evaluation problem. It can be accomplished by regression and theorem proving in the initial database, as per Theorem 1 above.

Example 3.
Given the system $(\Sigma, [\ ], \{Power, Pump\}, \neg filling(s))$, and the candidate diagnosis $E=[power\_failure]$, E can be verified to be an explanatory diagnosis with respect to the system by evaluating the query
$$\mathcal{R}^*[\mathcal{R}_{Poss}[Poss(do(power\_failure, S_0))]]$$
$$\wedge \mathcal{R}^*[\neg filling(do(power\_failure, S_0))]$$
with respect to the initial database, Σ_{init}.

7 Related Work

This work has been influenced by formal characterizations of diagnosis for systems without an explicit representation of actions (e.g., [8],[26],[4]) and by Reiter's work on the frame problem and the problem of temporal projection [28] and [30]. Aside from previous work by the author (e.g., [22],[21]), research to date has not explicitly addressed the problem of integrating a rich representation of actions into diagnostic reasoning. As such, there is little related work that exploits a comprehensive representation of action.

A recent notable exception is Thielscher's work on dynamic diagnosis [33]. Thielscher's basic representation is similar in spirit to ours, though he does not use the situation calculus and he does not exploit an axiomatic solution to the frame, ramification and qualification problems. Like us, he adopts the basic MBD ontology, employs an action theory to represent actions, and represents certain state constraints causally to capture the indirect effects of actions as dictated by the system description of the device. Where he differs is in the actual task he is performing, and the assumptions he makes. To use terminology discussed here, he is computing *what is wrong* diagnoses. Given an action history and some observed aberrant behaviour, Thielscher conjectures that certain components must be abnormal to account for the observed behavior. He is not conjecturing actions. To simplify computation, he assumes that all components are initially normal, and he uses a priori likelihood of failure to select most likely candidate diagnoses. While his paper examines *what is wrong* diagnoses, rather than *what happened* diagnoses, clearly the two are intimately related. [22] discusses the inter-relationship between these two types of diagnosis in the context of the situation calculus framework described here.

The research on temporal diagnosis and diagnosis of dynamic systems originating in the diagnosis research community (e.g., [2],[3],[13], [11],[15],[9]) and in particular [5] is also loosely related. Brusoni et al. [2] recently provided a characterization of temporal abductive diagnosis together with algorithms for computing these diagnoses under certain restrictions. Building on earlier work by some of the authors [3], they decouple atemporal and temporal diagnoses, using SD to represent the behaviour of the atemporal components and transition graphs to represent the temporal components. The later work uses temporal constraints to represent the temporal components. Also related is the work on event-based diagnosis by Cordier and Thiébaux [5]. Their work is similar in motivation to our work on explanatory diagnosis, viewing the diagnosis task as the determination of the event-history of a system between successive observations. While this work is related, the representation of action is impoverished. It does not provide a comprehensive representation of the preconditions for and the effects of actions, nor does it address the frame, ramification and qualification problems. Their transition system is sufficiently expressive for their application but the lack of a compact representation proves problematic.

In the area of reasoning about action, research on temporal explanation and postdiction has an interesting relationship to this work (e.g., [6],[1]). Of particular note is Shanahan's research [31]. While Shanahan also proposes the situation calculus as a representation language for axiomatizing his domain, he does so without an axiomatic solution to the frame and ramification problems. As such these problems must be addressed coincidentally with generating explanatory diagnoses. In contrast, our characterization of explanatory diagnosis, with its axiomatic solution to the frame and ramification problems, enables simpler characterization and computation of temporal explanation.

8 Summary

The results in this paper provide contributions to model-based diagnosis and to reasoning about actions. Our concern in this paper was, given a system that affects and can be affected by the actions of agents, and given some observed (aberrant) behaviour, how do we capture the notion of *what happened*, i.e., how do we go about conjecturing a sequence of actions that account for the behaviour we have observed. As we discussed at the beginning of this paper, not only is conjecturing candidate explanatory diagnoses interesting in its own right, but it facilitates determining the current state of the system.

We addressed this problem by providing a mathematical characterization of the notion of explanatory diagnosis in the context of a rich situation calculus representation, proposed in [21]. Our characterization made apparent the direct relationship of explanatory diagnosis to the planning task, and in particular to Green's notion of deductive plan synthesis. However, generating explanatory diagnoses is actually akin to planning in the face of state constraints and a potentially incomplete initial database. Our representation scheme addressed the challenges presented by state constraints by providing an axiomatic a priori solution to the frame, ramification and qualification problems. This enabled us to extend Reiter's results on the soundness and completeness of regression and to show that we can generate explanatory diagnoses by regression followed by theorem proving in the initial database. A remaining difficulty was that our initial database is often incomplete. Exploiting features of diagnosis problems, we proposed the notions of assumption-based and potential explanatory diagnosis, to allow for the conjectured sequences of actions that constitute a diagnosis to be predicated on some other assumptions we choose to make about the world. Finally, we proposed exploiting a library of precomputed likely diagnoses, indexed by context and observations. This enabled us to verify, rather than generate explanatory diagnoses.

In our dissertation work [22], we have also addressed the complementary problem of conjecturing *what is wrong* diagnoses. In future work we examine the application of Golog procedures [16] to diagnostic problem solving.

Acknowledgements I would like to thank Ray Reiter, Yves Lespérance, Hector Levesque and Allan Jepson for helpful comments on the work presented in this paper. I would also like to thank Oskar Dressler, for engaging me in an interesting discussion regarding this work that helped me look at the material in a different light. Finally, thanks to the KR'98 reviewers for their conscientious reviews.

References

1. Baker, A. (1991) Nonmonotonic Reasoning in the Framework of the Situation Calculus. *Artificial Intelligence* **49**:5–23.

2. Brusoni, V., Console, L., Terenziani, P., and Theseider Dupré, D. (1995) An Efficient Algorithm for Computing Temporal Abductive Diagnoses. In *Proceedings of the Sixth International Workshop on Principles of Diagnosis*, 41–48.
3. Console, L., Portinale, L., Theseider Dupré, D. and Torasso, P. (1994) Diagnosing Time-Varying Misbehavior: An Approach Based on Model Decomposition. *Annals of Mathematics and Artificial Intelligence* 11(1–4):381–398.
4. Console, L. and Torasso, P. (1991) A Spectrum of Logical Definitions of Model-Based Diagnosis. *Computational Intelligence* 7(3):133–141.
5. Cordier, M. and Thiébaux, S. (1994) Event-Based Diagnosis for Evolutive Systems. In *Proceedings of the Fifth International Workshop on Principles of Diagnosis*, 64–69.
6. Crawford, J. and Etherington, D. (1992) Formalizing Reasoning About Change: A Qualitative Reasoning Approach. In *Proceedings of the Tenth National Conference on Artificial Intelligence (AAAI-92)*, 577–582.
7. de Kleer, J. (1991) Focusing on Probable Diagnoses. In *Proceedings of the Ninth National Conference on Artificial Intelligence (AAAI-91)*, 842–848.
8. de Kleer, J., Mackworth, A. and Reiter, R. (1992) Characterizing Diagnoses and Systems. *Artificial Intelligence* 56(2–3):197–222.
9. Dressler, O. (1994) Model-based diagnosis on board: Magellan-MT inside. In *Proceedings of the Fifth International Workshop on Principles of Diagnosis*, 87–92.
10. Fikes, R. and Nilsson, N. (1971) STRIPS: A New Approach to Theorem Proving in Problem Solving. *Artificial Intelligence* 2:189–208.
11. Friedrich, G. and Lackinger, F. 1991. Diagnosing Temporal Misbehaviour. In *Proceedings of the Twelfth International Joint Conference on Artificial Intelligence (IJCAI-91)*, 1116–1122.
12. Green, C. C. (1969) Theorem Proving by Resolution as a Basis for Question-Answering systems. In Meltzer, B., and Michie, D., eds., *Machine Intelligence 4*. New York: American Elsevier. 183–205.
13. Hamscher, W. (1991) Modeling Digital Circuits for Troubleshooting. *Artificial Intelligence* 51(1–3):223–271.
14. Kramer, B. and Mylopolous, J. et al. (1996) Developing an Expert System Technology for Industrial Process Control: An Experience Report. In *Proceedings of the Conference of the Canadian Society for Computational Studies of Intelligence (CSCSI'96)*, 172–186.
15. Lackinger, F. and Nejdl, W. (1991) Integrating Model-Based Monitoring and Diagnosis of Complex Dynamic Systems. In *Proceedings of the Twelfth International Joint Conference on Artificial Intelligence (IJCAI-91)*, 1123–1128.
16. Levesque, H., Reiter, R., Lespérance, Y., Lin, F. and Scherl, R. (1997) Golog: A Logic Programming Language for Dynamic Domains. In *Journal of Logic Programming, Special Issue on Actions* 31(1–3):59–83.
17. Lin, F. and Reiter, R. (1994) State Constraints Revisited. *Journal of Logic and Computation* 4(5):655–678. Special Issue on Action and Processes.
18. Lin, F. and Reiter, R. (1995) How to Progress a Database II: The STRIPS Connection. In *Proceedings of the Fourteenth International Joint Conference on Artificial Intelligence (IJCAI-95)*, 2001–2007.
19. Lloyd, J. (1987) *Foundations of Logic Programming*. Springer Verlag, second edition.
20. McIlraith, S. (1994) Further Contributions to Characterizing Diagnosis. *Annals of Mathematics and Artificial Intelligence* 11(1–4):137–167.
21. McIlraith, S. (1997) Representing Actions and State Constraints in Model-Based Diagnosis. In *Proceedings of the Fourteenth National Conference on Artificial Intelligence (AAAI-97)*, 43–49.

22. McIlraith, S. (1997) *Towards a Formal Account of Diagnostic Problem Solving*. Ph.D. Dissertation, Department of Computer Science, University of Toronto, Toronto, Ontario, Canada.

23. McIlraith, S. (1998) Explanatory Diagnosis: Conjecturing Actions to Explain Observations. In Cohn, A. G., Schubert, L., Shapiro, S. S., eds., *Proceedings of the Sixth International Conference on Principles of Knowledge Representation and Reasoning (KR'98)*, 167–177. Morgan Kaufmann.

24. Pinto, J. (1994) *Temporal Reasoning in the Situation Calculus*. Ph.D. Dissertation, Department of Computer Science, University of Toronto, Toronto, Ontario, Canada.

25. Poole, D. (1988) Representing Knowledge for Logic-Based Diagnosis. In *Proceedings of the Fifth Generation Computer Systems Conference (FGCS-88)*, 1282–1290.

26. Reiter, R. (1987) A Theory of Diagnosis from First Principles. *Artificial Intelligence* **32**:57–95.

27. Reiter, R. (1991) *The Frame Problem in the Situation Calculus: A Simple Solution (sometimes) and a completeness result for goal regression*. Artificial Intelligence and Mathematical Theory of Computation: Papers in Honor of J. McCarthy. San Diego, CA: Academic Press. 359–380.

28. Reiter, R. (1992) The Projection Problem in the Situation Calculus: a Soundness and Completeness Result, with an Application to Database Updates. In *Proceedings First International Conference on AI Planning Systems*, 198–203.

29. Reiter, R. (1996) Natural Actions, Concurrency and Continuous Time in the Situation Calculus. In Aiello, L.; Doyle, J.; and Shapiro, S., eds., *Proceedings of the Fifth International Conference on Principles of Knowledge Representation and Reasoning (KR'96)*, 2–13. Cambridge, Massachusetts, USA.: Morgan Kaufmann.

30. Reiter, R. *Knowledge in Action: Logical Foundations for Describing and Implementing Dynamical Systems*. In preparation. Draft available at http://www.cs.toronto.edu/ cogrobo/.

31. Shanahan, M. (1993) Explanation in the Situation Calculus. In *Proceedings of the Thirteenth International Joint Conference on Artificial Intelligence (IJCAI-93)*, 160–165.

32. Shanahan, M. (1997) Event Calculus Planning Revisited. In *Proceedings 1997 European Conference on Planning (ECP 97), Springer-Verlag Lecture Notes in Artificial Intelligence (no. 1348)*, 390—402.

33. Thielscher, M. (1997) A Theory of Dynamic Diagnosis. In *Linköping Electronic Articles in Computer and Information Science* **2**(11). Research article received for discussion October, 1997. Revised March, 1998. http://www.ep.liu.se/ea/cis/1997/011/.

34. Waldinger, R. (1977) Achieving Several Goals Simultaneously. In Elcock, E., and Michie, D., eds., *Machine Intelligence 8*. Edinburgh, Scotland: Ellis Horwood. 94–136.

On sensing and off-line interpreting in GOLOG

Gerhard Lakemeyer

Department of Computer Science, Aachen University of Technology,
D-52056 Aachen, Germany, Email: gerhard@cs.uni-bonn.de

Summary. GOLOG is a high-level programming language for the specification of complex actions. It combines the situation calculus with control structures known from conventional programming languages. Given a suitable axiomatization of what the world is like initially and how the primitive actions change the world, the GOLOG interpreter derives for each program a corresponding linear sequence of legally executable primitive actions, if one exists. Despite its expressive power, GOLOG's applicability is severely limited because the derivation of a linear sequence of actions requires that the outcome of each action is known beforehand. Sensing actions do not meet this requirement since their outcome can only be determined by executing them and not by reasoning about them. In this paper we extend GOLOG by incorporating sensing actions. Instead of producing a linear sequence of actions, the new interpreter yields a tree of actions. The idea is that a particular path in the tree represents a legal execution of primitive actions conditioned on the possible outcome of sensing actions along the way.

Prologue

When I left Toronto in late 1990, Ray had just begun to revive the situation calculus as a serious contender among the various logics of action.[1] To be honest, I myself was rather skeptical at first whether his approach and, in particular, GOLOG would ever be more than just a specification language for dynamic domains. When Ray gave a talk at my then department at Bonn in 1994, he was met with even more skeptical questions regarding GOLOG's practicability by the "real" roboticists at Bonn. As there were no conclusive answers at the time, I decided it was time to put GOLOG to the test and, lo and behold, within a year and with the invaluable support of our robotics group, we conducted the first experiments controlling a real robot using GOLOG. Again, Ray's vision proved to be right, and I have since joined his quest to explore cognitive robotics. This paper[2] is a small contribution in this regard. Needless to say, none of this would have been possible without Ray's efforts and that of the other members of the Cognitive Robotics Group at Toronto.

---

[1] It is very fitting that Ray's first paper on the subject appeared in the Festschrift in honor of John McCarthy [17].

[2] An earlier version of this paper appeared in [3].

1   Introduction

When reasoning about action one is often faced with incomplete knowledge. For example, when trying to achieve a goal such as catching an airplane, there usually is not enough information at the outset for an agent to come up with a single course of action which would satisfy the goal. For instance, I may not know the departure gate of the plane until I actually reach the airport (or, to be more modern, until I check my airline's web-site.) What is needed are sensing actions which, when executed at the appropriate time, gather relevant information about the world and whose outcome determines what other actions need to be performed later.

Despite this obvious observation, dealing with sensing in both a principled and practical way has been surprisingly difficult. On the principled side, there has been substantial progress in understanding the connection between knowledge, sensing, and action, see for example [15], [16], [18], [10], [9]. There have also been several proposals to incorporate sensing actions into planning systems such as [5], [6], [1]. While planning may be workable in limited domains, we support the view of Levesque and Reiter [11] that general purpose planning is not sufficient as the main means for agents such as robots to decide how to achieve a task. The argument here is mainly one of complexity. The planning problem without sensing is already highly intractable, and adding sensing only compounds the problem.

Rather than leaving it completely up to the robot to construct a plan from a set of primitive actions, an alternative strategy would be to devise a suitable high-level programming language in which the user specifies not just a goal but also how it is to be achieved, perhaps leaving small subtasks to be handled by an automatic planner. An example of such a language is GOLOG [12], which combines the expressive power of the situation calculus with control structures known from conventional programming languages. A key property of GOLOG (or, more precisely, the GOLOG interpreter) is that it takes a program and verifies off-line whether it is legally executable. In case the verification succeeds, it also produces a plan in terms of a linear sequence of primitive actions which can then be immediately executed.

While GOLOG comes with an efficient Prolog implementation, its applicability in real world domains is severely limited because sensing actions are not handled properly. The problem is that in order to come up with a sequence of actions GOLOG needs to have all the relevant information beforehand to decide on a course of actions to achieve a goal, whereas the whole point of sensing is that some information becomes available only at run-time. This deficiency became very clear in a recent robotics application [2], where our group employed GOLOG to specify the actions of a robot who gives guided tours in a museum. While GOLOG provided more than enough flexibility in terms of the available control structures, not being able to deal with sensing proved to be rather cumbersome.[3]

---

[3] Here we do not mean sensing as it is needed for safe navigation, which was not handled at the logical level at all, but was left to lower level components of the robot. What we do mean is sensing at the abstract task level, which, in this application, involved mainly the interaction with a visitor during a guided tour.

Despite recent arguments against it [4],[4] we believe that off-line verification of a plan is a valuable feature of GOLOG, in particular, during program development where mistakes are bound to happen, and it seems desirable to be able to take into account sensing actions as well. This paper provides a step in this direction.

To illustrate the problem and our proposal, let us consider the airport example in somewhat more detail. Suppose that the agent is already at the airport, but she does not know the gate yet. Before boarding the plane, she wants to buy a newspaper and a coffee. In case the gate number is 90 or up, it is preferable to buy coffee at the gate, otherwise it is better to buy coffee before going to the gate. Let us assume, we have the following primitive actions: *buy_paper,buy_coffee*, *goto_gate*, *board_plane*, and *sense_gate*, which senses the value of *gate* (perhaps by glancing at the departure information monitor).

In GOLOG one might be tempted to write the following (grossly oversimplified) procedure to catch a plane.

proc *catch_plane*
    sense_gate;
    buy_paper;
    if *gate* $\geq$ 90 **then** *goto_gate*;*buy_coffee* **else** *buy_coffee*;*goto_gate*
    endif;
    board_plane
endproc

Let us assume also that we have a set of axioms which suitably characterize what the world is like initially, what the action preconditions are, and how actions change the world and the agent's knowledge about the world (see the next section for hints about how all this is done). Given these axioms, the GOLOG interpreter then tries to logically derive a linear sequence of primitive actions which are legally executable and which represent an execution trace of *catch_plane*. In our case, the only plausible candidates are *sense_gate·buy_paper·goto_gate·buy_coffee* and *sense_gate·buy_paper·buy_coffee·goto_gate*. The problem is that it will only be known at runtime and after the execution of *sense_gate* which of the two sequences is the actual one. Hence GOLOG, running off-line, is bound to fail since it cannot decide between the two.

If GOLOG allowed for branching in the plans it produces the problem could be overcome. In the example, we would need a plan that starts with *sense_gate* followed by *buy_paper* and then splits into two branches consisting of *goto_gate* followed by *buy_coffee* and the other way around, depending on whether *gate* $\geq$ 90 or not. This is in fact the main modification of GOLOG we propose in this paper. We call plans with branches *conditional action trees (CAT's)*, which are binary trees whose nodes can be thought of as situations with the root representing the initial situation. Every edge is labeled with a primitive action, which indicates how a situation is obtained from its predecessor. In addition, whenever branching occurs, the corresponding

---

[4] We will get back to [4] in Section 7.

node/situation is labeled by a formula, whose truth value at execution time determines which branch is taken. A CAT for the airport example could be drawn as follows:

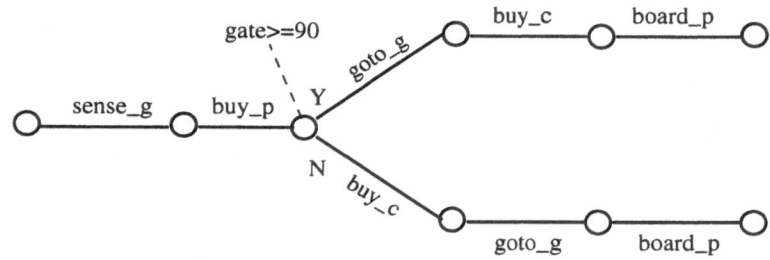

However, we will be writing it as a term using the following notation

sense_gate·buy_paper·[*gate* ≥ 90,
            goto_gate·buy_coffee·board_plane,
            buy_coffee·goto_gate·board_plane].

It turns out that the GOLOG interpreter which handles CAT's has a simple specification, which is very similar to the original one given in [12]. In our extension of GOLOG we allow sensing truth values as well as the referent of terms (as in the above example). Note also that branching need not occur immediately at the time of sensing. In contrast, [10], [4] only consider sensing truth values and branching happens immediately at the time of sensing.

The rest of the paper is organized as follows. In Sections 4 and 3, we give very brief introductions into the situation calculus and GOLOG. Section 4 introduces conditional action trees into the situation calculus. In Section 5, we define sGOLOG (= GOLOG + sensing) and in Section 6 we present a simple interpreter implemented in Prolog. In Section 7, we summarize our results and compare our work to [4].

2   The Situation Calculus

One increasingly popular language for representing and reasoning about the preconditions and effects of actions is the situation calculus [14]. We will only go over the language briefly here noting the following features: all terms in the language are one of three sorts, ordinary objects, actions or situations; there is a special constant S_0 used to denote the *initial situation*, namely that situation in which no actions have yet occurred; there is a distinguished binary function symbol *do* where $do(a, s)$ denotes the successor situation to s resulting from performing the action a; relations whose truth values vary from situation to situation, are called relational *fluents*, and are denoted by predicate symbols taking a situation term as their last argument; similarly, functions varying across situations are called functional fluents and are denoted analogously; finally, there is a special predicate $Poss(a, s)$ used to state that action a is executable in situation s. Throughout the paper we write action and situation variables using the letters a and s, respectively, possibly with sub- and superscripts. (The same convention applies to meta-variables for terms of the respective sorts.)

Within this language, we can formulate theories which describe how the world changes as the result of the available actions. One possibility is a *basic action theory* of the following form [17]:

- Axioms describing the initial situation, S_0.
- Action precondition axioms, one for each primitive action a, characterizing $Poss(a, s)$.
- Successor state axioms, one for each fluent F, stating under what conditions $F(x, do(a, s))$ holds as a function of what holds in situation s. These take the place of the so-called effect axioms, but also provide a solution to the frame problem [17].
- Domain closure and unique names axioms for the primitive actions.
- A collection of foundational, domain independent axioms.

In [13] the following foundational axioms are considered:[5]

1. $\forall s \forall a. S_0 \neq do(a, s)$.
2. $\forall a_1, a_2, s_1, s_2. do(a_1, s_1) = do(a_2, s_2) \supset \quad (a_1 = a_2 \wedge s_1 = s_2)$.
3. $\forall P. P(S_0) \wedge [\forall s \forall a.(P(s) \supset P(do(a, s)))] \supset \forall s P(s)$.
4. $\forall s. \neg(s < S_0)$.
5. $\forall s, s', a. (s < do(a, s') \equiv (Poss(a, s') \wedge s \leq s'))$,
   where $s \leq s'$ is an abbreviation for $s < s' \vee s = s'$.

The first three axioms serve to characterize the space of all situations, making it isomorphic to the set of ground terms of the form $do(a_n, \cdots, do(a_1, S_0) \cdots)$, which we also abbreviate as $do(a, S_0)$, where a stands for the sequence $a_1 \cdot a_2 \cdot \ldots \cdot a_n$. The third axiom ensures, by second-order induction, that there are no situations other than those accessible using do from S_0. The final two axioms serve to characterize a $<$ relation between situations.

Knowledge and Action So far the language allows us to talk only about how the actual world evolves starting in the initial situation S_0. With sensing, we also need an account of what the agent doing the actions knows about the world initially and in successor situations. Following [18], which in turn is based on [15], we introduce a binary fluent $K(s', s)$ which can be read as "in situation s, the agent thinks that s' is (epistemically) accessible."[6]

Given K, knowledge can then be defined in a way similar to possible-world semantics [8], [7] as truth in all accessible situations. We denote knowledge using

---

[5] In addition to the standard axioms of equality.

[6] This view requires, in general, that there are initial situations other than S_0, which also means that the above induction axiom no longer holds. See [9] for a way to handle many initial situations axiomatically.

the following macro, where α may contain the special situation symbol now. Let α_s^{now} refer to α with all occurrences of now replaced by s. Then

$$\mathbf{Knows}(\alpha, s) \doteq \forall s' K(s', s) \supset \alpha_{s'}^{now}$$

Given Knows we introduce further abbreviations which tell us whether the truth value of a formula or the value of a term is known:

$$\mathbf{Kwhether}(\alpha, s) \doteq \mathbf{Knows}(\alpha, s) \vee \mathbf{Knows}(\neg\alpha, s).$$
$$\mathbf{Kref}(\tau, s) \doteq \exists x \mathbf{Knows}(\tau = x, s).$$

To specify how actions and, in particular, sensing actions change what is known, we follow [10] and introduce a special function SF with two arguments, an action and a situation. As in the case of $Poss$ it is assumed that SF is user-defined, that is, the user writes down *sensed fluent axioms*, one for each action type. The idea is that $SF(a, s)$ gives the value sensed by action a in situation s. So we might have, for example,

$$SF(sense\_gate, s) = gate(s).$$

In case the action a has no sensing component (as in simple physical actions, like moving), the axiom should state that $SF(a, s)$ is some fixed value. If the action serves to sense whether or not some fluent $\phi(s)$ holds, two fixed values can be used such as 0 and 1.

In [18], Scherl and Levesque formulate a solution to the frame problem for knowledge by proposing a successor state axiom for K. Here we use the variant given in [10]:

Definition 1. $\forall a, s, s'. Poss(a, s) \supset K(s', do(a, s)) \equiv$
$$\exists s''. s' = do(a, s'') \wedge K(s'', s) \wedge Poss(a, s'')$$
$$\wedge [SF(a, s) = SF(a, s'')].$$

Roughly, after doing action a the agent thinks it could be in situation s' just in case s' results from doing a in some previously accessible situation s'', provided a is possible in s'' and both s and s'' agree on the value being sensed.

3   GOLOG

GOLOG [12] is a logic-programming language which, in addition to the primitive actions of the situation calculus, allows the definition of complex actions using programming constructs which are very much like those known from conventional programming languages. The procedure *catch_plane* introduced earlier is an example of such a complex action. Here is a list of the constructs available in GOLOG:

A	primitive action
$\phi?$	test a condition
$(\rho_1; \rho_2)$	sequence
$(\rho_1 \vert \rho_2)$	nondeterministic action choice
$(\pi x.\rho)$	nondeterministic argument choice
ρ^*	nondeterministic iteration
if ϕ **then** ρ_1 **else** ρ_2 **endif**	conditional
while ϕ **do** ρ **endwhile**	loop
proc $\rho(x)$ **endproc**	procedure

What is special about GOLOG is that the meaning of these constructs is completely defined by sentences in the situation calculus. For this purpose, a macro $Do(\rho, s, s')$ is introduced whose intuitive meaning is that executing the program ρ in situation s leads to situation s'. Here we provide some of the definitions needed for Do. See [12] for the complete list.

$Do(A, s, s') \doteq Poss(A, s) \wedge s' = do(A, s)$, where A is a primitive action.
$Do((\rho_1; \rho_2), s, s') \doteq \exists s'' Do(\rho_1, s, s'') \wedge Do(\rho_2, s'', s')$
$Do((\rho_1 \vert \rho_2), s, s') \doteq Do(\rho_1, s, s') \vee Do(\rho_2, s, s')$
$Do(\phi?, s, s') \doteq \phi(s) \wedge s = s'$
$Do(\textbf{if } \phi \textbf{ then } \rho_1 \textbf{ else } \rho_2 \textbf{ endif}, s, s') \doteq Do([(\phi?; \rho_1) \vert (\neg\phi?; \rho_2)], s, s')$

Here ϕ is a formula of the situation calculus with all situation arguments suppressed, which we also call a situation-free formula. $\phi(s)$ is then obtained from ϕ by reinserting s as the situation argument at the appropriate places. For example, if $\phi = (gate = A) \wedge am\_at(x, airport)$ with fluents $gate$ and am_at, then $\phi(s) = (gate(s) = A) \wedge am\_at(x, airport, s)$.

Given a situation calculus theory AX of the domain in question as sketched in the previous section, executing a program ρ means to first find a sequence of primitive actions a such that

$$AX \models Do(\rho, S_0, do(a, S_0))$$

and then handing the sequence a to an appropriate module that takes care of actually performing those actions in the real world.

4   Conditional action trees

In this section, we augment the situation calculus with conditional action trees. The idea is that, instead of having only linear action histories (i.e. situations), we have a tree of actions, where each path represents a situation.

We begin by introducing two new sorts, a sort formula for situation-free formulas and a sort CAT for conditional action trees. For each sort we add infinitely many variables to the language. We write ϕ and c with possible sub- or superscripts for variables of sort formula and CAT, respectively. Since there is a standard way of doing this, we gloss over the details of how to reify formulas as terms in the language. Given a term of sort formula, we even take the liberty to write $\phi(s)$ in place of a formula, where in fact we would need to say $Holds(\phi, s)$, where $Holds$ is appropriately axiomatized.

CAT terms are made up of a special constant ϵ, denoting the empty CAT, the primitive actions, and two constructors $a \cdot c$ and $[\phi, c_1, c_2]$, where a is an action, ϕ is a term of sort formula and c, c_1, and c_2 are themselves CAT's.[7] ϕ is also called a *branch-formula*. c_1 and c_2 are called the true- and false-branch (for ϕ), respectively. We saw an example CAT already in Section 1.

We can define CAT's within the situation calculus by adding the following foundational axioms, which are analogues of those introduced earlier in Section 4 for situations. (In the following, free variables in a formula are considered to be universally quantified.) The first five axioms make sure that CAT's are all distinct. The role of Axiom 6 is the same as the induction axiom for situations and minimizes the set of CAT's.

1. $a \cdot c \neq \epsilon$.
2. $[\phi, c_1, c_2] \neq \epsilon$.
3. $a \cdot c = a' \cdot c' \supset a = a' \wedge c = c'$
4. $[\phi, c_1, c_2] = [\phi', c_1', c_2'] \supset \phi = \phi' \wedge c_1 = c_1' \wedge c_2 = c_2'$.
5. $[\phi, c_1, c_2] \neq a \cdot c$.
6. $\forall P.P(\epsilon) \wedge \forall a P(a) \wedge [\forall a, c.P(c) \wedge c \neq \epsilon \supset P(a \cdot c)] \wedge$
   $[\forall \phi, c_1, c_2.P(c_1) \wedge P(c_2) \supset P([\phi, c_1, c_2])] \supset \forall c.P(c)$.

It is also convenient to define the following predicate ext, which will be needed later on to define how GOLOG, having already produced a CAT c in situation s, extends c by a CAT c^* to produce c'. Informally, $ext(c', c, c^*, s)$ holds if c' is a CAT which contains a path p from c extended by the CAT c^*. p is obtained by starting at situation s and then moving down the tree replacing s by successor situations according to the actions encountered along the path. p follows a particular branch in the tree depending on the truth value of the corresponding branch-formula relative to the current situation. Formally:

$$
\begin{aligned}
ext(c', c, c^*, s) \equiv\ & (c = \epsilon \supset c' = c^*) \wedge \\
& (c = a \supset c' = a \cdot c^*) \wedge \\
& (c = a \cdot c_1 \supset \exists c_1'.c' = a \cdot c_1' \wedge ext(c_1', c_1, c^*, do(a, s))) \wedge \\
& (c = [\phi, c_2, c_3] \supset \exists c_2', c_3'.c' = [\phi, c_2', c_3'] \wedge \\
& \qquad [\phi(s) \supset ext(c_2', c_2, c^*, s) \wedge \\
& \qquad \neg \phi(s) \supset ext(c_3', c_3, c^*, s)]
\end{aligned}
$$

For example, let $c = a_1 \cdot a_2 \cdot [p, a_3, \epsilon]$ and let p be false at $do(a_2, do(a_1, s))$. Then $ext(c', c, c^*, s)$ holds for $c^* = [q, \epsilon, a_4]$ and $c' = a_1 \cdot a_2 \cdot [p, a_5, [q, \epsilon, a_4]]$. Note that c' can differ arbitrarily in the branches which are not taken, in this case the true-branch for p.

---

[7] Logically, $\cdot$ and $[\_, \_, \_]$ are binary and ternary functions, respectively. We write them this way for better readability of CAT's.

Lastly, we introduce a two-place function *cdo*, which takes a CAT c and a situation s and returns a situation which is obtained from s using the actions along a particular path in c. The idea is that *cdo* follows a certain branch in the tree depending on the truth value of the respective branch-formula at the current situation.

$$cdo(\epsilon, s) = s.$$
$$cdo(a, s) = do(a, s).$$
$$cdo(a \cdot c, s) = cdo(c, do(a, s)).$$
$$cdo([\phi, c_1, c_2]), s) = \text{if } \phi(s) \text{ then } cdo(c_1, s) \text{ else } cdo(c_2, s).$$

5   sGOLOG

Programs in sGOLOG are those of GOLOG augmented by sensing actions for both formulas and terms. The main difference compared to the original GOLOG is that the interpreter now produces CAT's instead of situations. When constructing a CAT, a decision must be made as to when new branches, that is, constructs of the form $[\phi, c_1, c_2]$ are introduced. One possibility is to introduce them automatically whenever a sensing action occurs. This seems fine if we are sensing the truth value of a formula (like ϕ), but what should we branch on if we are sensing the value of a term, in particular if there are (infinitely) many potential values for the term? To overcome this problem, we leave the introduction of new branches under the control of the user by introducing a new special action $branch\_on(\phi)$, whose "effect" is to introduce a new CAT $[\phi, \epsilon, \epsilon]$. Since we want sGOLOG to produce CAT's which are ready for execution, we need to make sure that the truth value of the formula which decides on which branch to take is known. This is taken care of by attaching an appropriate Kwhether-term to the definition of $branch\_on$.

Technically, the sGOLOG interpreter is defined in a way very similar to the original GOLOG. We introduce a three place macro $Do(\rho, s, c)$ which expands into a formula of the situation calculus augmented by CAT's. It may be read as "executing the program ρ in situation s results in CAT c." Note the difference compared to the original Do, where the last argument was a situation rather than a CAT.

We begin with an auxiliary four-place macro $Do4$. Intuitively, $Do4(\rho, s, c, c')$ may be read as "starting in situation s, executing the CAT c and then the program ρ leads to c', which is an extension of c."

$$Do4(\rho, s, c, c') \doteq \exists c^* Do(\rho, cdo(c, s), c^*) \wedge ext(c', c, c^*, s).$$

$Do(\rho, s, c)$ is then defined as follows:

$$Do(a, s, c) \doteq Poss(a, s) \wedge c = a \text{ for every primitive action } a.$$
$$Do(branch\_on(\phi), s, c) \doteq \text{Kwhether}(\phi(now), s) \wedge c = [\phi, \epsilon, \epsilon].$$
$$Do(\phi?, s, c) \doteq \phi(s) \wedge c = \epsilon.$$
$$Do((\rho_1 | \rho_2), s, c) \doteq Do(\rho_1, s, c) \vee Do(\rho_2, s, c).$$
$$Do((\rho_1 ; \rho_2), s, c) \doteq \exists c' Do(\rho_1, s, c') \wedge Do4(\rho_2, s, c', c).$$
$$Do(\pi x \rho, s, c) \doteq \exists x Do(\rho, s, c).$$

$$Do(\rho^*, s, c) \doteq \forall P[P(\epsilon, \epsilon) \wedge \forall c_1, c_2, c_3.P(c_1, c_2) \wedge Do4(\rho, s, c_2, c_3)$$
$$\supset P(c_1, c_3)] \supset P(\epsilon, c).$$
$$Do(\textbf{proc } P \rho \textbf{ endproc}, s, c) \doteq Do(\rho, s, c)^8$$

Note that the definition of the various program constructs are not all that different from the original ones. In fact, if we confine ourselves to GOLOG programs without sensing and without occurrences of the special action $branch\_on$, it is not hard to show that the two interpreters coincide.

Theorem 1. *Let Do_{old} stand for the old definition of Do. Let ρ be a GOLOG program without sensing and $branch\_on$-actions and let $c = a_1 \cdot c_2 \ldots \cdot c_n$ for primitive actions a_i. Then $\models Do(\rho, s, c) \equiv Do_{old}(\rho, s, cdo(c, s))$.*

5.1   The airport example revisited

Let us now see how the airport example could be handled in sGOLOG. To keep the formalization brief, we make various simplifying assumptions. For example, we assume implicitly that the agent is at the airport and that buy_coffee, buy_paper, and $sense\_gate$ are always possible. $goto\_gate$ requires the referent of $gate$ to be known and $board\_plane$ requires being at the (right) gate.

$$Poss(buy\_coffee, s) \equiv \textbf{true}$$
$$Poss(buy\_paper, s) \equiv \textbf{true}$$
$$Poss(sense\_gate, s) \equiv \textbf{true}$$
$$Poss(goto\_gate, s) \equiv \texttt{Kref}(gate, s)$$
$$Poss(board\_plane, s) \equiv am\_at\_gate(s)$$

$sense\_gate$ is the only sensing action. Hence for the others SF always returns the same value:

$$SF(buy\_coffee, s) = 1$$
$$SF(buy\_paper, s) = 1$$
$$SF(goto\_gate, s) = 1$$
$$SF(sense\_gate, s) = gate(s)$$

$gate$ and am_at_gate are the only fluents, and $gate$ never changes its value:

$$Poss(a, s) \supset gate(do(a, s)) = y \equiv gate(s) = y.$$
$$Poss(a, s) \supset am\_at\_gate(do(a, s)) \equiv a = goto\_gate \vee am\_at\_gate(s)$$
(Assumes that boarding the plane leaves you at the gate.)

Finally, the sGOLOG program to catch the plane is like the one in the introduction except for the explicit $branch\_on$ action:

---

[8] For simplicity, we only consider simple nonrecursive procedures without parameters. See [12] for how to handle general procedures in GOLOG.

proc *catch_plane*
   sense_gate;
   buy_paper;
   branch_on(*gate* ≥ 90);
   if *gate* ≥ 90 **then** *goto_gate*;*buy_coffee*
                 else *buy_coffee*;*goto_gate*
   endif;
   board_plane
endproc

Assuming that AX consists of the foundational axioms of our extended situation calculus, the above airport axioms, and simple arithmetic to compare numbers, we obtain

AX $\models$ *Do*(*catch_plane*, S_0, *c*) with
   c = *sense_gate* · *buy_paper* · [*gate* ≥ 90,
                        goto_gate·*buy_coffee*·*board_plane*,
                        buy_coffee·*goto_gate*·*board_plane*].

Note that in the airport example we make use of the fluent K only in a very limited way, namely in the form of $\texttt{Kref}(gate, s)$ and $\texttt{Kwhether}(gate(now) \leq 90, s)$. (The latter results from interpreting *branch_on*(*gate* ≥ 90).) In particular, we do not have to deal with nested occurrences of Knows. For this reason, there is no need to stipulate any special properties of the K-relation such as reflexivity or transitivity. In principle, there is no problem adding such restrictions. Indeed Scherl and Levesque [18] have shown that it suffices to stipulate those for initial situations and that the successor state axiom for K (Definition 1) guarantees that these properties hold in all successor situations as well. However, as the work by Scherl and Levesque also shows, reasoning about K is not easy. Even if we restrict the use of K to the Knows-macro, we still need some form of modal reasoning to deal with it. When it comes to implementing sGOLOG, this seems like a high price to pay. Fortunately, as we will see in a moment, with reasonable restrictions on the use of sensing, there is a way to avoid this problem by not using the K-fluent at all.

6   A simple implementation

In this section we present a very simple implementation of sGOLOG in Eclipse-Prolog. Besides the usual constraints that come with the use of Prolog, like negation-as-failure and atomic facts only to describe the initial situation, we restrict sensing actions and their use as follows:

1. Only the truth value of atomic facts can be sensed. In particular, sensing actions have the form *sense*(*P*), where *P* is a fluent.
2. The truth value of a sensed fluent is never tested before the corresponding sensing action has been performed.
3. *branch_on*(*P*) actions are only allowed in case *P* is a sensed fluent.

4. Whenever a $branch\_on(P)$ action is reached, both truth values are conceivable for P.

Some remarks regarding each of these assumptions:

1. While restricting ourselves to atomic facts is essential, there is no problem in principle to allow for sensing the referent of a term as well. However, it would add some overhead to the implementation, and we have chosen to ignore this issue here for the sake of simplicity.
2. This is what De Giacomo and Levesque [4] call the *dynamic closed world assumption*. It lets us deal with incomplete information even in Prolog, which makes the closed world assumption. The idea is that whenever a fluent F is tested for the first time, complete information about F has been achieved.
3. Applying $branch\_on$ only to sensed fluents enables us to avoid having to work explicitly with a K-fluent in a very simple way. Testing whether a sensed fluent P is known can be reduced to testing whether the action $sense(P)$ was performed earlier. This test is easily implementable by introducing a new fluent $sensed(P, s)$ which becomes true when $sense(P)$ is executed and remains true from then on. In essence, under the above assumptions the truth value of $sensed(P, s)$ keeps track of whether P is known. Of course, the use of $sensed(P, s)$ to simulate $\texttt{Kwhether}(P, s)$ is not restricted to $branch\_on$. For example, it may also be part of a definition of $Poss(a, s)$.
4. If it is possible that P can take on either truth value, we can safely use hypothetical reasoning for both cases when evaluating $branch\_on(P)$.[9] The idea is that the true-branch $C1$ of the CAT $[P, C1, C2]$, which results from $branch\_on(P)$, is constructed by assuming that P is true in the current situation. Similarly, the false-branch $C2$ is developed by assuming that P is false in the current situation.

   We implement this by introducing new primitive actions $assm(P, 1)$ and $assm(P, 0)$ whose only effect is to turn P true and false, respectively. (See the definition of $do4$ below.) Note that that actions occurring after $assm$ are allowed to change the truth value of P. An example previously discussed in De Giacomo and Levesque [4] is an elevator controller, which first senses the value of a button and, if it is " on," resets it to " off" afterwards.

We now turn to the actual implementation. The reader familiar with the original paper on GOLOG [12] will notice the close similarity between the Prolog implementation of GOLOG and the one below for sGOLOG. Note, in particular, that the definitions of *do* in GOLOG and sGOLOG are practically identical except for sequence (:) and, of course, $branch\_on$, which does not exist in GOLOG.

```
:- dynamic(holds/2).
:- op(970,xfy, [:]).   /* Sequence.*/
:- op(950, xfy, [#]). /* Nondeterministic action choice.*/
```

---

[9] If it would follow that P is, say, true, then considering the case where P is false might lead to failure, even though that case never arises.

```
/* do4(P,S,C,C1) recursively descends the CAT C
                 (first two clauses). */
/* Once a leaf of the CAT is reached (third clause),
                 "do" is called, */
/* which then extends this branch according to P */
do4(E,S,[A|C],C1) :- primitive_action(A),C1=[A|C2],
                 do4(E,do(A,S),C,C2).
do4(E,S,[[P,C1,C2]],C) :- do4(E,do(assm(P,1),S),C1,C3),
                 do4(E,do(assm(P,0),S),C2,C4),
                 C = [[P,C3,C4]].
do4(E,S,[],C) :- do(E,S,C).


do(E1 : E2, S, C) :- do(E1,S,C1), do4(E2,S,C1,C).
                 /* sequence */
do(?(P),S,C) :- C=[],holds(P,S). /* test */
do(E1 # E2, S, C) :- do(E1,S,C) ; do (E2,S,C).
                 /* nond. act. choice */
do(if(P,E1,E2),S,C) :- do((?(P) : E1) # (?(neg(P)) : E2),S,C).
do(star(_),_,[]). /* nondet. iteration */
do(star(E),S,C) :- do(E : star(E),S,C).
do(while(P,E),S,S1):- do(star(?(P) : E) : ?(neg(P)),S,S1).
do(pi(V,E),S,S1) :- sub(V,_,E,E1), do(E1,S,S1).
                 /* nond. arg. choice */
do(E,S,C) :- proc(E,E1), do(E1,S,C). /* procedure */


/* the base cases: primitive actions and branch_on(P) */
do(E,S,[E]) :- primitive_action(E), poss(E,S).
do(branch_on(P),S,[[P,[],[]]]) :-  holds(sensed(P),S).


/* sub and sub_list are auxiliary predicates */
/* sub(Name,New,Term1,Term2): */
/*    Term2 is Term1 with Name replaced by New. */
sub(_,_,T1,T2) :- var(T1), T2 = T1.
sub(X1,X2,T1,T2) :- not var(T1), T1 = X1, T2 = X2.
sub(X1,X2,T1,T2) :- not T1 = X1,
                 T1 =..[F|L1], sub_list(X1,X2,L1,L2),
T2 =..[F|L2].
sub_list(_,_,[],[]).
sub_list(X1,X2,[T1|L1],[T2|L2]) :- sub(X1,X2,T1,T2),
                 sub_list(X1,X2,L1,L2).


/* Definition of holds for arbitrary nonatomic formulas */
holds(and(P1,P2),S) :- holds(P1,S), holds(P2,S).
holds(or(P1,P2),S) :- holds(P1,S); holds(P2,S).
holds(neg(P),S) :- not holds(P,S).  /* Negation by failure */
holds(some(V,P),S) :- sub(V,_,P,P1), holds(P1,S).
```

```
/* the successor state axiom for sensed */
holds(sensed(P),do(A,S))  :- A = sense(P) ;  holds(sensed(P),S).
```

Let us now consider an implementation of the airport example. Since we restrict ourselves to sensing the truth values of fluents, we need to adapt the example accordingly. For simplicity, let us assume that there are only two gates, *gate_A* and *gate_B*. We introduce a fluent *it_is_gate_A*, whose truth value can be sensed and which then tells us which gate to take. In contrast to the original example, we also use the slightly more general action *goto(x)* and fluent *am_at(x)* instead of *goto_gate* and *am_at_gate*, respectively. The general definition of sGOLOG requires the use of $SF(a, s)$ to define the result of a sensing action. However, SF is only needed for the definition of the epistemic K-fluent. Since the implementation does not use K, there is no need to use SF either.

```
/* declare the primitive actions */
primitive_action(goto(_)).
primitive_action(buy_paper).
primitive_action(buy_coffee).
primitive_action(board_plane).
primitive_action(sense(_)).

/* all actions are always possible except board_plane, */
/* which requires being at the right gate */
poss(goto(_),_).
poss(buy_paper,_).
poss(buy_coffee,_).
poss(sense(_),_).
/* boarding requires being at the right gate */
poss(board_plane,S) :- holds(it_is_gate_A,S) ->
                             holds(am_at(gate_A),S) ;
                             holds(am_at(gate_B),S).

/* successor state axioms */

/* I am at X if I just went there or */
/* if I was there and did not go anywhere else */
/* (assumes that boarding the plane leaves you at the gate)*/
holds(am_at(X),do(A,S)) :- A = goto(X) ;
                             (holds(am_at(X),S),
                              not (A = goto(_))).

holds(it_is_gate_A,do(A,S)) :- holds(it_is_gate_A,S) ;
                               A = assm(it_is_gate_A,1).

/* facts about the initial situation s0 */
/* none necessary here */
```

```
proc(catch_plane,
          (sense(it_is_gate_A):
           buy_paper:
           branch_on(it_is_gate_A):
           if(it_is_gate_A,
                  goto(gate_A):buy_coffee,
                  buy_coffee:goto(gate_B)):
           board_plane)
     ).


/* sample run, slightly reformatted for better readability */

[eclipse 3]: do(catch_plane,s0,C).

C = [sense(it_is_gate_A), buy_paper, [it_is_gate_A,
                  [goto(gate_A), buy_coffee, board_plane],
                  [buy_coffee, goto(gate_B), board_plane]]]
More? (;)

no (more) solution.
```

One might object that the implementation requires that the user is aware of the internals of the interpreter because the pseudo-action *assm* occurs in the user's code. In the example, it is part of the successor state axiom of holds (it_is_gate_A,do(A,S)). The objection can be dealt with by allowing the user to write successor state axioms without *assm* and then modifying the axioms automatically in the following way. Let F be a fluent whose truth value can be sensed. According to Reiter [17], the general form of the successor state axiom of F is (provided $Poss(a, s)$ holds):

$$Holds(F, do(a, s)) \equiv \Phi_F^+ \vee (Holds(F, s) \wedge \neg \Phi_F^-),$$

where Φ_F^+ and Φ_F^- describe the conditions which lead to F being true and false, respectively. We can compile the "effects" of the pseudo-action *assm* into the successor state axiom by replacing the above axiom by

$$Holds(F, do(a, s)) \equiv [\Phi_F^+ \vee a = assm(F, 1)] \vee$$
$$(Holds(F, s) \wedge \neg[\Phi_F^- \vee a = assm(F, 0)]).$$

Another issue that needs to be addressed is that of correctness. In other words, is the interpreter a faithful implementation of the specification of sGOLOG? Here the answer may not be that easy to come by. A reasonable intermediate step would be to first give a purely logical specification of our way to avoid the K-fluent using *sensed* and *assm* and then prove that the two formalizations coincide under the restrictions laid out at the beginning of this section. We leave this to future work.

7   Summary and discussion

In this paper we proposed sGOLOG, which extends GOLOG by adding sensing actions for sensing the truth values of formulas as well as the referents of terms.

We provided an off-line interpreter for sGOLOG, whose definition is simple and remarkably similar to the original one proposed for GOLOG. Instead of producing a linear sequence of primitive actions, the sGOLOG interpreter generates a tree of actions, if one exists, with the idea that branching is conditioned on the outcome of sensing actions.

In [4], De Giacomo and Levesque propose a different version of GOLOG with sensing. They advocate a combination of off-line and on-line interpretation. On-line interpretation means, roughly, that instead of verifying that the whole program is executable, the interpreter finds the next executable primitive action and commits to it by immediately executing it. The advantage is that, whenever a sensing action occurs, the outcome is immediately known and no branching is necessary. The authors argue that it is infeasible to verify very large programs off-line, in particular those that contain many nondeterministic actions and sensing actions. The authors certainly have a point here. In particular, programs with loops such as **while** ϕ **do** ... $sense(\phi)$ ... **endwhile** generally lead to infinite CAT's in our approach. Despite these shortcomings, we think there is a place for off-line interpretation of programs with sensing. For one, many programs with a moderate number of sensing actions can very well be handled by our approach. Also, as we have seen, sensing does not necessarily lead to branching. Furthermore, we believe that off-line interpreting is a valuable tool during program development, since we want to have some confidence that a program works before running it on an expensive robot. De Giacomo and Levesque seem to believe in off-line interpretation as well, at least partly. They allow a user to specify which parts of a program are to be handled off-line. Their version of off-line interpretation, however, is somewhat limited compared to ours. For one, they only verify that for all outcomes of sensing the program is executable without actually constructing a plan (like our CAT's), which could then be executed without further processing. Moreover, they do not handle sensing terms. It seems interesting to try and combine our ideas with theirs to have the best of both worlds.

References

1. Baral, C. and Son, T. C., Approximate reasoning about actions in presence of sensing and incomplete information., *Proc. of the International Logic Programming Symposium (ILPS'97)*, 1997.
2. Burgard, W., Cremers, A. B., Fox, D., Hähnel, D., Lakemeyer, G., Schulz, D., Steiner, W., Thrun, S., The Interactive Museum Tour-Guide Robot, *AAAI-98*.
3. De Giacomo, G. (ed.) *Proceedings of the AAAI Fall Symposium on Cognitive Robotics*, AAAI Technical Report FS-98-02, AAAI Press, 1998.
4. De Giacomo, G. and Levesque, H.J., An incremental interpreter for high-level programs with sensing. in: *Proceedings of the AAAI Fall Symposium on Cognitive Robotics*, AAAI Technical Report FS-98-02, AAAI Press, 1998, pp. 28–34.
5. Etzioni, O., Hanks, S., Weld, D., Draper, D., Lesh, N., and Williamsen, M., An approach to planning with incomplete information. *Proc. KR'92*, Morgan Kaufmann, 1992, pp. 115–125.
6. Golden, K. and Weld, D., Representing sensing actions: the middle ground revisited. *Proc. KR'96*, Morgan Kaufmann, 1996, pp. 174–185.

7. Hintikka, J., *Knowledge and Belief: An Introduction to the Logic of the Two Notions.* Cornell University Press, 1962.

8. Kripke, S. A., Semantical considerations on modal logic. *Acta Philosophica Fennica* **16**, 1963, pp. 83–94.

9. Lakemeyer, G. and Levesque, H. J., AOL: a logic of acting, sensing, knowing, and only knowing, *Proc. of the 6th International Conference on Principles of Knowledge Representation and Reasoning (KR'98)*, Morgan Kaufmann, 1998, pp. 316–327.

10. Levesque, H. J., What is Planning in the Presence of Sensing. AAAI-96, AAAI Press, 1996.

11. Levesque, H. J. and Reiter, R., High-level robotic control: beyond planning. Position Statement. *Working Notes of the AAAI Spring Symposium on Integrating Robotic Research: Taking the Next Leap*, AAAI Press, 1998.

12. H. J. Levesque, R. Reiter, Y. Lespérance, F. Lin, and R. B. Scherl. GOLOG: A logic programming language for dynamic domains. *Journal of Logic Programming*, **31**, 59–84, 1997.

13. Lin, F. and Reiter, R., State constraints revisited. *J. of Logic and Computation, special issue on actions and processes*, 4, 1994, pp. 665–678.

14. McCarthy, J., *Situations, Actions and Causal Laws.* Technical Report, Stanford University, 1963. Also in M. Minsky (ed.), *Semantic Information Processing*, MIT Press, Cambridge, MA, 1968, pp. 410–417.

15. Moore, R. C., A Formal Theory of Knowledge and Action. In J. R. Hobbs and R. C. Moore (eds.), *Formal Theories of the Commonsense World*, Ablex, Norwood, NJ, 1985, pp. 319–358.

16. Morgenstern, L., Knowledge preconditions for actions and plans. *Proc. IJCAI-87*, pp. 867–874.

17. Reiter, R., The Frame Problem in the Situation Calculus: A simple Solution (sometimes) and a Completeness Result for Goal Regression. In V. Lifschitz (ed.), *Artificial Intelligence and Mathematical Theory of Computation*, Academic Press, 1991, pp. 359–380.

18. Scherl, R. and Levesque, H. J., The Frame Problem and Knowledge Producing Actions. in *Proc. of the National Conference on Artificial Intelligence (AAAI-93)*, AAAI Press, 1993, 689–695.

Reactivity in a Logic-Based Robot Programming Framework (Extended Version)

Yves Lespérance, Kenneth Tam, and Michael Jenkin

Dept. of Computer Science, York University,
Toronto, ON Canada, M3J 1P3

Summary. A robot must often react to events in its environment and exceptional conditions by suspending or abandoning its current plan and selecting a new plan that is an appropriate response to the event. This paper describes how high-level controllers for robots that are reactive in this sense can conveniently be implemented in ConGolog, a new logic-based robot/agent programming language. Reactivity is achieved by exploiting ConGolog's prioritized concurrent processes and interrupts facilities. The language also provides nondeterministic constructs that support a form of planning. Program execution relies on a declarative domain theory to model the state of the robot and its environment. The approach is illustrated with a mail delivery application.[1]

In 1993, I [Yves Lespérance] was almost on my way to a new job in Germany when Hector Levesque and Ray Reiter asked me to come back to Toronto to work on a project called "cognitive robotics". When they explained their plans for this, I was immediately taken by the idea of developing tools based on formal knowledge representation (KR) techniques and using them to control a real embodied agent. KR was becoming more than a theoretical pursuit and getting concerned with interaction with the real world. I enjoyed my stint in the cognitive robotics group immensely. As I hope this paper shows, we have made a lot of progress in the last six years. Of course, a lot remains to be done. But I am still very enthused by Ray's vision. And I think I am not alone. KR has taken a turn towards the concrete in that last few years without sacrificing its dreams. And I think we are all better for it.

1 Introduction

Reactivity is usually understood as having mainly to do with strict constraints on reaction time. As such, much work on the design of reactive agents has involved non-deliberative approaches where behavior is hardwired [2] or produced from compiled universal plans [14], [13]. However, there is more to reacting to environmental

---

[1] The research described received financial support from Communications and Information Technology Ontario (and its earlier incarnation ITRC) and the Natural Science and Engineering Research Council of Canada. Hector Levesque came up with the idea of handling sensing through exogenous events; he also helped with the iterative deepening route planning procedure. We thank him as well as Ray Reiter, Jeff Lloyd, Mikhail Soutchanski, Giuseppe De Giacomo, and Daniele Nardi for helpful discussions related to this work. Many of our papers are available at http://www.cs.yorku.ca/~lesperan/.

events or exceptional conditions than reaction time. While some events/conditions can be handled at a low level, e.g., a robot going down a hallway can avoid collision with an oncoming person by slowing down and making local adjustments in its trajectory, others require changes in high-level plans. For example, an obstacle blocking the path of a robot attempting a delivery may mean that the delivery must be rescheduled. Here as in many other cases, the issue is not real-time response. What is required is *reconsideration of the robot's plans in relation to its goals and the changed environmental conditions*. Current plans may need to be suspended or terminated and new plans devised to deal with the exceptional event or condition.

To provide the range of responses required by environmental events and exceptional conditions, i.e. reactivity in the wide sense, the best framework seems to be a hierarchical architecture. Then, urgent conditions can be handled in real-time by a low-level control module, while conditions requiring replanning are handled by a high-level control module that models the environment and task, and manages the generation, selection, and scheduling of plans.

Synthesizing plans at run-time provides great flexibility, but it is often computationally infeasible in complex domains, especially when the agent does not have complete knowledge and there are exogenous events (i.e. actions by other agents or natural events). In [7], it was argued that *high-level program execution* was a more practical alternative. The idea, roughly, is that instead of searching for a sequence of actions that takes the robot from an initial state to some goal state, the task is to find a sequence of actions that constitutes a legal execution of some high-level program. By high-level program, we mean one whose primitive instructions are domain-dependent actions of the robot, whose tests involve domain-dependent predicates that are affected by the actions, and whose code may contain nondeterministic choice points where lookahead is necessary to make a choice that leads to successful termination. As in planning, to find a sequence that constitutes a legal execution of a high-level program, one must reason about the preconditions and effects of the actions within the program. However, if the program happens to be almost deterministic, very little searching is required; as more and more nondeterminism is included, the search task begins to resemble traditional planning. Thus, in formulating a high-level program, the user gets to control the search effort required.

In [7], *Golog* was proposed as a suitable language for expressing high-level programs for robots and autonomous agents. Golog was used to design a high-level robot control module for a mail delivery application [16]. This module was interfaced to systems providing path planning and low-level motion control, and successfully tested on several different robot platforms, including a Nomad 200, a RWI B21, and a RWI B12.

A limitation of Golog for this kind of applications is that it provides limited support for writing reactive programs. In [3], *GonGolog*, an extension of Golog that provides concurrent processes with possibly different priorities as well as interrupts was introduced. In this paper, we try to show that ConGolog is an effective tool for the design of high-level reactive control modules for robotics applications. We provide an example of such a module for a mail delivery application.

2 ConGolog

As mentioned, our high-level programs contain primitive actions and tests of predicates that are domain-dependent. Moreover, an interpreter for such programs must reason about the preconditions and effects of the actions in the program to find a legal terminating execution. We specify the required domain theories in the situation calculus [12], a language of predicate logic for representing dynamically changing worlds. In this language, a possible world history, which is simply a sequence of actions, is represented by a first order term called a *situation*. The constant S_0 is used to denote the initial situation – that in which no actions (of interest) have yet occurred. There is a distinguished binary function symbol *do* and the term $do(\alpha, s)$ denotes the situation resulting from action α being performed in situation s. Relations whose truth values vary from situation to situation, called *predicate fluents*, are denoted by predicate symbols taking a situation term as the last argument. For example, $Holding(o, s)$ might mean that the robot is holding object o in situation s. Similarly, functions whose value varies with the situation, *functional fluents*, are represented by function symbols that take a situation argument. The special predicate $Poss(\alpha, s)$ is used to represent the fact that primitive action α is executable in situation s. A domain of application will be specified by theory that includes the following types of axioms:

- Axioms describing the initial situation, S_0.
- Action precondition axioms, one for each primitive action α, which characterizes $Poss(\alpha, s)$.
- Successor state axioms, one for each fluent F, which characterize the conditions under which $F(x, do(a, s))$ holds in terms of what holds in situation s; these axioms may be compiled from effects axioms, but provide a solution to the frame problem [10].
- Unique names axioms for the primitive actions.
- Some foundational, domain independent axioms.

Thus, the declarative part of a ConGolog program implementing a high-level controller for a robot will be such a theory.

A ConGolog program also includes a procedural part which specifies the behavior of the robot. This is specified using the following constructs:

α,	primitive action
$\phi?$,	wait for a condition[2]
$(\sigma_1; \sigma_2)$,	sequence
$(\sigma_1 \mid \sigma_2)$,	nondeterministic choice between actions
$\pi\, x\, [\sigma]$,	nondeterministic choice of arguments
σ^*,	nondeterministic iteration

---

[2] Here, ϕ stands for a situation calculus formula with all situation arguments suppressed; $\phi(s)$ will denote the formula obtained by restoring situation variable s to all fluents appearing in ϕ.

if ϕ **then** σ_1 **else** σ_2 **endIf**,	conditional
while ϕ **do** σ **endWhile**,	loop
$(\sigma_1 \parallel \sigma_2)$,	concurrent execution
$(\sigma_1 \rangle\!\rangle \sigma_2)$,	concurrency with different priorities
$\sigma^{\parallel}$,	concurrent iteration
$< x : \phi \rightarrow \sigma >$,	interrupt
proc $\beta(x)$ σ **endProc**,	procedure definition
$\beta(t)$,	procedure call
noOp	do nothing

The nondeterministic constructs include $(\sigma_1 \mid \sigma_2)$, which nondeterministically choses between programs σ_1 and σ_2, $\pi x[\sigma]$, which nondeterministically picks a binding for the variables x and performs the program σ for this binding of x, and σ^*, which means performing σ zero or more times. Concurrent processes are modeled as interleavings of the primitive actions involved. A process may become blocked when it reaches a primitive action whose preconditions are false or a wait action ϕ? whose condition ϕ is false. Then, execution of the program may continue provided another process executes next. In $(\sigma_1 \rangle\!\rangle \sigma_2)$, σ_1 has higher priority than σ_2, and σ_2 may only execute when σ_1 is done or blocked. $\sigma^{\parallel}$ is like nondeterministic iteration σ^*, but the instances of σ are executed concurrently rather than in sequence. Finally, an interrupt $< x : \phi \rightarrow \sigma >$ has variables x, a trigger condition ϕ, and a body σ. If the interrupt gets control from higher priority processes and the condition ϕ is true for some binding of the variables, the interrupt triggers and the body is executed with the variables taking these values. Once the body completes execution, the interrupt may trigger again. With interrupts, it is easy to write programs that are reactive in that they will suspend whatever task they are doing to handle given conditions as they arise. A more detailed description of ConGolog and a formal semantics appear in [3]. We give an example ConGolog program in section 4.

A prototype ConGolog interpreter has been implemented in Prolog. This implementation requires that the axioms in the program's domain theory be expressible as Prolog clauses; note that this is a limitation of this particular implementation, not the framework.

In applications areas such as robotics, we want to use ConGolog to program an embedded system. The system must sense conditions in its environment and update its theory appropriately *as it is executing* the ConGolog control program.[3] This requires adapting the high-level program execution model presented earlier: the interpreter cannot simply search all the way to a final situation of the program. An adapted model involving incremental high-level program execution is developed in [4]. However in this paper, we sidestep these issues by making two simplifying assumptions:

1. that the interpreter immediately commits to and executes any primitive action it reaches when its preconditions are satisfied, and

---

[3] Here, the environment is anything outside the ConGolog control module about which information must be maintained; so the sensing might only involve reading messages from another module through a communication socket.

2. that there is a set of exogenous events detectable by the system's sensors (e.g. a mail pick up request is received or the robot has arrived at the current destination) and that the environment is continuously monitored for these; whenever such an exogenous event is detected to have occurred, it is immediately inserted in the execution.

We can get away with this because our application program performs very little search and the exogenous events involved are easy to detect.

3   Interfacing the High-Level Control Module

As mentioned earlier, we use a hierarchical architecture to provide both real-time response as well as high-level plan reconsideration when appropriate. At the lowest level, we have a reactive control system that performs time-critical tasks such as collision avoidance and straight line path execution. In a middle layer, we have a set of components that support navigation through path planning, map building and/or maintenance, keeping track of the robot's position, etc. and support path following by interacting with the low-level control module. On top of this, there is the ConGolog-based control module that supports high-level plan execution to accomplish the robot's tasks; this level treats navigation somewhat like a black box.

In this section, we describe how the ConGolog-based high-level control module is interfaced to rest of the architecture. The high-level control module needs to run asynchronously with the rest of the architecture so that other tasks can be attended to while the robot is navigating towards a destination. It also needs to interact with the navigation module to get tasks accomplished. To support this, we need to give the high-level control module a *model of the navigation module*. We have defined a simple version of such a model. With respect to navigation, the robot is viewed by the high-level control module as always being in one of the following set of states:

$$RS = \{Idle, Moving, Reached, Stuck, Frozen\}.$$

The current robot state is represented by the functional fluent $robotState(s)$. The robot's default state is $Idle$; when in this state, the robot is not moving towards a destination, but collision avoidance is turned on and the robot may move locally to avoid oncoming bodies. With the robot in $Idle$ state, the high-level control module may execute the primitive action $startGoTo(place)$; this changes the robot's state to $Moving$ and causes the navigation module to attempt to move the robot to $place$. If and when the robot reaches the destination, the navigation module generates the exogenous event $reachDest$, which changes the robot's state to $Reached$. If on the other hand the navigation module encounters obstacles it cannot get around and finds the destination unreachable, then it generates the exogenous event $getStuck$, which changes the robot's state to $Stuck$. In any state, the high-level control module may execute the primitive action $resetRobot$, which aborts any navigation that may be under way and returns the robot to $Idle$ state. Finally, there is the $Frozen$ state where collision avoidance is disabled and the robot will not move even if something approaches it; this is useful when the robot is picking up or dropping off things; humans may reach into the robot's carrying bins without it moving away. All other actions leave the robot's state unchanged. This is specified in the following successor state axiom for the *robotState* fluent:

$$robotState(do(a, s)) = i \equiv$$
$$\exists p \, a = startGoTo(p) \wedge i = Moving \vee$$
$$a = reachDest \wedge i = Reached \vee$$
$$a = getStuck \wedge i = Stuck \vee$$
$$a = resetRobot \wedge i = Idle \vee$$
$$a = frezeRobot \wedge i = Frozen \vee$$
$$i = robotState(s) \wedge \forall p \, a \neq startGoTo(p) \wedge$$
$$a \neq reachDest \wedge a \neq getStuck \wedge$$
$$a \neq resetRobot \wedge a \neq frezeRobot$$

We also have precondition axioms that specify when these primitive actions and exogenous events are possible. For example, the following says that the action of directing the robot to start moving toward a destination p is possible whenever the robot is in *Idle* state:

$$Poss(startGoTo(p), s) \equiv robotState(s) = Idle$$

The other precondition axioms are similar and appear in the appendix.

We also use two additional functional fluents: $robotDestination(s)$ refers to the last destination the robot was set in motion towards, and $robotPlace(s)$ refers to the current location of robot as determined from the model. Their successor state axioms are:

$$robotDestination(do(a, s)) = p \equiv$$
$$a = startGoTo(p) \vee$$
$$p = robotDestination(s) \wedge \forall p \, a \neq startGoTo(p)$$

$$robotPlace(do(a, s)) = p \equiv$$
$$\exists p' \, a = startGoTo(p') \wedge p = Unknown \vee$$
$$a = reachDest \wedge p = robotDestination(s) \vee$$
$$p = robotPlace(s) \wedge$$
$$\forall p \, a \neq startGoTo(p) \wedge a \neq reachDest$$

4 A Mail Delivery Example

To test our approach, we have implemented a simple mail delivery application. The high-level control module for the application must react to two kinds of exogenous events:

- new shipment orders, which are represented by the event $orderShipment(sender, recipient, priority)$, and

- signals from the navigation module, namely the *reachDest* event announcing that the destination has been reached and the *getStuck* event announcing that the robot has failed to reach its destination.

The first kind is typical of the communication interactions a robot may have with its environment, while the second kind is typical of the control interactions a task-level module may have with the rest of the robot's architecture. To require more reactivity from the robot, we assume that shipment orders come with different priority levels and that the system must interrupt service of a lower priority order when a higher priority one comes in. Also, we want the robot to make a certain number of attempts to get to a customer's mailbox as some of the obstacles it runs into may be temporary. This is handled by assigning a certain amount of credit to customers initially and reducing their credit when an attempt to go to their mailbox fails. When customers run out of credit, they are suspended and shipments sent to them are returned to the sender when possible.

In addition to the navigation primitive actions and exogenous events already described, the application uses the following primitive actions:

$ackOrder(n)$	acknowledge reception of servable order
$declineOrder(n)$	decline an unservable order
$pickUpShipment(n)$	pick up shipment n
$dropOffShipment(n)$	drop off shipment n
$cancelOrder(n)$	cancel an unservable order
$reduceCredit(c)$	reduce customer c's credit
$notifyStopServing(c)$	notify unreachable customer

Note that shipment orders are identified by a number n that is assigned from a counter when the *orderShipment* event occurs. We have precondition axioms for these primitive actions, for example:

$$Poss(pickUpShipment(n), s) \equiv$$
$$orderState(n, s) = ToPickUp \wedge$$
$$robotPlace(s) = mailbox(sender(n, s))$$

The primitive fluents for the application are:

$orderState(n, s) = i$	order n is in state i
$sender(n, s) = c$	sender of order n is c
$recipient(n, s) = c$	recipient of order n is c
$orderPrio(n, s) = p$	priority of order n is p
$orderCtr(s) = n$	counter for orders arriving
$credit(c, s) = k$	customer c has credit k
$Suspended(c, s)$	service to customer c is suspended

We have successor state axioms for these fluents. For example, the state of an order starts out as *NonExistent*, then changes to *JustIn* when the *orderShipment* event occurs, etc.; the following successor state axiom specifies this:

$$orderState(n, do(a, s)) = i \equiv$$
$$\exists c, r, p\; a = orderShipment(c, r, p) \wedge orderCtr = n$$
$$\wedge\, i = JustIn \vee$$
$$a = ackOrder(n) \wedge i = ToPickUp \vee$$
$$a = pickUpShipment(n) \wedge i = OnBoard \vee$$
$$a = dropOffShipment(n) \wedge$$
$$\quad robotPlace(s) = mailbox(recipient(n, s))$$
$$\quad \wedge\, i = Delivered \vee$$
$$a = dropOffShipment(n) \wedge$$
$$\quad robotPlace(s) = mailbox(sender(n, s))$$
$$\quad \wedge\, i = Returned \vee$$
$$a = dropOffShipment(n) \wedge$$
$$\quad robotPlace(s) = CentralOffice$$
$$\quad \wedge\, i = AtCentralOffice \vee$$
$$a = cancelOrder(n) \wedge i = Cancelled \vee$$
$$a = declineOrder(n) \wedge i = Declined \vee$$
$$i = orderState(n, s) \wedge$$
$$\quad \neg(\exists c, r, p\; a = orderShipment(c, r, p) \wedge orderCtr = n)$$
$$\quad \wedge\, a \neq ackOrder(n) \wedge a \neq pickUpShipment(n) \wedge$$
$$\quad a \neq dropOffShipment(n) \wedge a \neq cancelOrder(n)$$
$$\quad \wedge\, a \neq declineOrder(n)$$

The rest of the successor state and action precondition axioms appear in the appendix.

The initial state of the domain might be specified by the following axioms:

$$Customer(Yves) \quad Customer(Ken)$$
$$Customer(Hector) \quad Customer(Michael)$$
$$Customer(c) \supset credit(c, S_0) = 3$$
$$orderCtr(S_0) = 0$$
$$orderState(n, S_0) = NonExistent$$
$$robotState(S_0) = Idle$$
$$robotPlace(S_0) = CentralOffice$$

Let us now specify the behavior of our robot using a ConGolog program. Exogenous events are handled using prioritized interrupts. The main control procedure concurrently executes four interrupts at different priorities:

proc $mainControl$
$$< n : orderState(n) = JustIn$$
$$\quad \rightarrow handleNewOrder(n) >$$
$$\rangle\rangle$$
$$< n : (orderState(n) = ToPickUp \wedge Suspended(sender(n)))$$
$$\quad \rightarrow cancelOrder(n) >$$

$$\|$$
$$< \; n : \; (orderState(n) = ToPickUp$$
$$\qquad \lor \; orderState(n) = OnBoard$$
$$\qquad \lor \; robotPlace \neq CentralOffice)$$
$$\qquad \rightarrow robotMotionControl >$$
$$\rangle\rangle$$
$$< \; robotState = Moving \rightarrow \textbf{noOp} >$$
endProc

The top priority interrupt takes care of acknowledging or declining new shipment orders. This ensures that customers get fast feedback when they make an order. At the next level of priority, we have two other interrupts, one that takes care of cancelling orders whose senders have been suspended service, and another that controls the robot's motion. At the lowest priority level, we have an interrupt with an empty body that prevents the program from terminating when the robot is in motion and all other threads are blocked.

The top priority interrupt deals with a new shipment order n by executing the following procedure:

proc $handleNewOrder(n)$
  if $Suspended(sender(n)) \lor Suspended(recipient(n))$
    then $declineOrder(n)$
  else
    $ackOrder(n);$
    if $robotState = Moving \land$
      $orderPrio(n) > curOrderPrio$ **then**
     $resetRobot$ % abort current service
    endIf
  endIf
endProc

This sends a rejection notice to customers making an order whose sender or recipient is suspended; otherwise an acknowledgement is sent. In addition, when the new shipment order has higher priority than the order currently being served, the robot's motion is aborted, causing a reevaluation of which order to serve ($curOrderPrio$ is a defined fluent whose definition appears below).

The second interrupt in $mainControl$ handles the cancellation of orders when the sender gets suspended; its body executes the primitive action $cancelOrder(n)$.

The third interrupt in $mainControl$ handles the robot's navigation, pick ups, and deliveries. When the interrupt's condition is satisfied, the following procedure is called:

proc $robotMotionControl$
  if $\exists c \; CustToServe(c)$ **then** $tryServeCustomer$
  else $tryToWrapUp;$
  endIf
endProc

This tries to serve a customer when there is one to be served and tries to return to the central office and wrap up otherwise. $CustToServe(c, s)$ is a defined fluent:

$$CustToServe(c, s) \overset{\text{def}}{=} \exists n[$$
$$(orderState(n, s) = ToPickUp \wedge sender(n, s) = c$$
$$\wedge \neg Suspended(recipient(n, s), s)) \vee$$
$$(orderState(n, s) = OnBoard \wedge (recipient(n, s) = c$$
$$\vee sender(n, s) = c \wedge Suspended(recipient(n, s), s)))]$$
$$\wedge \neg Suspended(c, s)$$

To try to serve a customer, we execute the following:

proc $tryServeCustomer$
  $\pi c [BestCustToServe(c)?;$
    $startGoTo(mailbox(c));$
    $(robotState \neq Moving)?;$
    if $robotState = Reached$ **then**
      $freezeRobot;$
      $dropOffShipmentsTo(c);$
      $pickUpShipmentsFrom(c);$
      $resetRobot$
    else if $robotState = Stuck$ **then**
      $resetRobot;$ % abandon attempt
      $handleServiceFailure(c)$
    % else when service aborted nothing more to do
    endIf]
endProc

This first picks one of the best customers to serve, directs the robot to start navigating towards the customer's mailbox, and waits until the robot halts. If the robot reaches the customer's mailbox, then shipments for the customer are dropped off and shipments from him/her are picked up. If on the other hand, the robot halts because it got stuck, the $handleServiceFailure$ procedure is executed. Finally, if the robot halts because a higher priority order came in and the top priority interrupt executed a $resetRobot$, then there is nothing more to be done. $BestCustToServe(c, s)$ is a defined fluent that captures all of the robot's order scheduling criteria:

$$BestCustToServe(c, s) \overset{\text{def}}{=} CustToServe(c, s) \wedge$$
$$custPriority(c, s) = maxCustPriority(s) \wedge$$
$$credit(c, s) = maxCreditFor(maxCustPriority(s), s)$$

$$custPriority(c, s) = p \overset{\text{def}}{=}$$
$$\exists n \, OrderForCustAtPrio(n, c, p, s) \wedge$$
$$\forall n', p'(OrderForCustAtPrio(n', c, p', s) \supset p' \leq p)$$

$$OrderForCustAtPrio(n, c, p, s) \stackrel{\text{def}}{=} \neg Suspended(c, s) \wedge$$
$$[orderState(n, s) = ToPickUp \wedge sender(n, s) = c \wedge$$
$$orderPrio(n, s) = p \vee$$
$$orderState(n, s) = OnBoard \wedge orderPrio(n, s) = p \wedge$$
$$(recipient(n, s) = c \vee$$
$$sender(n, s) = c \wedge Suspended(recipient(n, s), s))]$$

$$maxCustPriority(s) = p \stackrel{\text{def}}{=} \exists c\, custPriority(c, s) = p$$
$$\wedge \forall c'\, custPriority(c', s) \leq p$$

$$maxCreditFor(p, s) = k \stackrel{\text{def}}{=}$$
$$\exists c[custPriority(c, s) = p \wedge credit(c, s) = k \wedge$$
$$\forall c'(custPriority(c', s) = p \supset credit(c', s) \leq k)]$$

This essentially says that the best customers to serve are those that have the highest credit among those having the highest priority orders. We can now also define the priority of the order currently being served as follows:

$$curOrderPrio(s) = p \stackrel{\text{def}}{=}$$
$$\forall c[robotState(s) = Moving \wedge$$
$$robotDestination(s) = mailbox(c)$$
$$\supset p = custPriority(c, s)] \wedge$$
$$[\neg(robotState(s) = Moving \wedge$$
$$\exists c\, robotDestination(s) = mailbox(c)) \supset p = -1]$$

The *handleServiceFailure* procedure goes as follows:

proc *handleServiceFailure(c)*
  reduceCredit(c);
  if *credit(c) = 0* **then**
    notifyStopServing(p);
  endIf;
endProc

When the robot gets stuck on the way to customer *c*'s mailbox, it first reduces *c*'s credit, and then checks whether it has reached zero and *c* has just become *Suspended*; if so, *c* is notified that he/she will no longer be served.

The *tryToWrapUp* procedure is similar to *tryServeCustomer*:

proc *tryToWrapUp*
  startGoTo(CentralOffice);
  (robotState ≠ Moving)?;
  if *robotState = Reached* **then**
    freezeRobot;
    dropOffToCentralOffice

$resetRobot$
else if $robotState = Stuck$ **then**
    $resetRobot$ % abandon attempt
% else when service aborted nothing more to do
endIf
endProc

It starts the robot on its way to the central office. The thread then waits until the robot halts. If the robot reaches the central office, then all undeliverable shipments on board are dropped off, and unless a new order comes in the program terminates. If the robot gets stuck, then the robot is reset and the procedure ends. If motion is aborted, the procedure ends immediately. In both cases control returns to the $mainControl$ procedure, which will serve a new order if one has come in or make a new attempt to return to the central office.

Procedure $dropOffShipmentsTo(c)$ delivers to customer c all shipments on board such that c is the shipment's recipient or c is the shipment's sender and the recipient has been suspended:

proc $dropOffShipmentsTo(c)$
  while $\exists n\ (orderState(n) = OnBoard\ \wedge$
      $(recipient(n) = c\ \vee$
      $sender(n) = c\ \wedge\ Suspended(recipient(n))))$ **do**
    $\pi\,n\ [(orderState(n) = OnBoard\ \wedge$
    $(recipient(n) = c\ \vee$
      $sender(n) = c\ \wedge\ Suspended(recipient(n))))?;$
      $dropOffShipment(n)]$
  endWhile
endProc

Procedure $pickUpShipmentsFrom(c)$ simply picks up all outgoing shipments from customer c's mailbox. Procedure $dropOffToCentralOffice$ drops off all shipments whose recipient and senders are both suspended to the central office. These are similar to $dropOffShipmentsTo(c)$ and are included in the appendix.

Note that by handling the cancellation of pick ups in a separate thread from that dealing with navigation and order serving, we allow the robot to be productive while it is in motion and waiting to reach its destination. This makes a better use of resources.

To run the system, we execute $mainControl$ after placing a few initial orders:

$$orderShipment(Yves, Ken, 0)\ \|$$
$$orderShipment(Ken, Hector, 1)\ \rangle\!\rangle$$
$$mainControl$$

5   Experimentation

The high-level control module for the mail delivery application has been ported to an RWI B12 mobile robot and tested in experiments (see figure 1). The other software

Fig. 1. Our robot facing an obstacle.

components for this were based on a system developed during an earlier project concerned with building an experimental vehicle to conduct survey/inspection tasks in an industrial environment [6]. The system supports point to point navigation in a previously mapped environment and can use pre-positioned visual landmarks to correct odometry errors. It relies on sonar sensors to sense unmodeled obstacles.

The system's architecture conforms to the general scheme described earlier. It provides two levels of control. An onboard low-level controller [12] performs all time-critical tasks such as collision avoidance and straight line path execution. The low-level controller assumes that the robot is always in motion and communicates with an offboard global path planner and user interface module known as the *Navigator*. The Navigator takes as inputs a metric/topological map of the environment in which the robot is located and the coordinates (as defined in the map) of the two end points, i.e., the source and the destination of a path. By making use of some predefined path-finding algorithms such as breadth-first search or A^* the Navigator identifies a feasible path between the source and the destination. To follow the path, the Navigator decomposes it into segments (a segment is a straight line between two adjacent way-points) and then forwards the segments to the low-level controller for execution. The Navigator supervises the low-level controller and identifies failures in the low-level controller's ability to execute a path segment.

The ConGolog-based high-level control module interacts with the rest of the architecture by communicating with the Navigator through a socket interface. The high-level controller, Navigator, and low-level controller all run asynchronously. The primitive actions in the ConGolog interface model are implemented using operations provided by the Navigator (currently, the mail pickup and drop off actions are only simulated). For example, the ConGolog primitive action $startGoTo(p)$ is implemented as [planPath(coordinatesOf(p)); followPath], where planPath and follow_path are operations supplied by the Navigator.

Our experiments confirmed the system's ability to deal with navigation failures and to interrupt the current task when an urgent shipment order is made. For more details on the implemented system, see [17].

6   Discussion

In this work, we have shown how ConGolog can be used to implement high-level robot controllers that can cope with dynamic and unpredictable environments – controllers that are reactive and support high-level plan reconsideration in response to exogenous events and exceptional conditions. Our work demonstrates that a logic-based approach can be used to build effective systems. In the ConGolog framework, application domains and their dynamics are specified declaratively and the axioms are used by the interpreter to automatically update its world model. This is less error-prone than having the user program his own ad-hoc world representation and update procedures. ConGolog controllers have clear formal specifications and are easier to extend and adapt to different environments or tasks.

The main limitation of the work accomplished so far is the lack of search/planning in the current high-level control program. ConGolog is designed to support such run-time planning through search over a nondeterministic program. This is the main reason for using a declarative representation of domain dynamics. So the next step in our work will be to extend the controller developed so far to perform run-time planning for dealing with failures (e.g., in navigation) and optimizing the scheduling of customer service calls.

It is not difficult to write search/planning code in ConGolog. For example, here's how one might implement an iterative deepening search to find the shortest route through customers needing service – one starts the search by invoking $planRoute(0)$:

proc $planRoute(n)$
  $serveAllCust(n) \mid planRoute(n + 1)$
endProc

proc $serveAllCust(n)$
  $\neg \exists c CustToServe(c)? \mid$
  $\pi c, d[(CustToServe(c) \wedge d = distanceTo(c) \wedge d \leq n)?;$
    $goServe(c);$
    $serveAllCust(n - d)]$
endProc

But adding such planning to a reactive control program raises a lot of complex issues. The planning/search must be interleaved with action execution and sensing the environment – as mentioned earlier, the standard ConGolog execution model does not support this. The generated plan must also be reevaluated when conditions change; it may no longer be executable, or achieve the goal, or be appropriate in the new conditions (e.g. when a more urgent order has just arrived). As well, when generating a plan, we may want to ignore the possibility that some actions may fail to achieve their objectives (e.g. navigating to a customer), and just deal with such failures when they occur.

Recently, De Giacomo and Levesque [4] have developed a new execution model for ConGolog that supports incremental high-level program execution in the presence of sensing. This should allow us to incorporate controlled search/planning in our programs while retaining a clean semantics. We are examining ways of dealing with the other issues mentioned. A Golog-based approach to execution monitoring and plan repair has been proposed in [5]. The use of Golog for planning is discussed in [11].

Another limitation of the work accomplished so far is that the system developed is rather small. We need to experiment with more complex systems to see whether the general approach and the use of prioritized interrupts to provide reactivity scales up. As well, interrupts support the suspension of the current plan but not its termination. The addition of a conventional exception throwing and catching mechanism that terminates the current plan is being investigated.

Another area under investigation is perceptual tasks. Such tasks often require sophisticated plan selection and involve information acquisition. We are working on an application where packages must be delivered to the recipient "in person" and where the robot must use sophisticated search strategies to locate the recipient, for example, asking whether a co-worker has seen the recipient [17].

The high-level program execution model of robot/agent control that underlies our approach is related to work on resource-bounded deliberative architectures [1], [9] and agent programming languages [15]. One difference is that in our approach, plan selection is coded in the program. This makes for a less declarative and in some cases more complex specification, but eliminates some overhead. On the other hand, the robot's world is modeled using a domain action theory and the world model is updated automatically using the successor state axioms; there is no need to perform asserts and retracts. Moreover, the evaluation of a test may involve arbitrary amounts of inference, although following logic programming philosophy, we take the programmer to be responsible for its efficiency/termination. Perhaps a more central difference is that our robots/agents can be understood as executing programs, albeit in a rather smart way – they have a simple operational semantics. Modeling the operation of an agent implemented using a resource-bounded deliberative architecture requires a much more complex account.

A   Additional Axioms and Procedure Definitions

The action precondition axioms not given in section 3 for the actions involved in the interface to the navigation module are:

$$Poss(reachDest, s) \equiv robotState(s) = Moving$$

$$Poss(getStuck, s) \equiv robotState(s) = Moving$$

$$Poss(resetRobot, s) \equiv True$$

$$Poss(freezeRobot, s) \equiv robotState(s) = Idle$$

For the mail delivery application domain described in section 4, we have the following additional axioms and procedure definitions:

1. Action precondition axioms:

$$Poss(orderShipment(sndr, rcpt, prio), s) \equiv$$
$$Customer(sndr) \wedge Customer(rcpt)$$
$$\wedge\, 0 \leq prio \leq 2 \wedge$$
$$orderState(orderCtr(s), s) = NonExistent$$

$$Poss(ackOrder(n), s) \equiv orderState(n, s) = JustIn$$

$$Poss(declineOrder(n), s) \equiv orderState(n, s) = JustIn$$

$$Poss(dropOffShipment(n), s) \equiv orderState(n, s) = OnBoard$$

$$Poss(cancelOrder(n), s) \equiv orderState(n, s) = ToPickUp$$

$$Poss(reduceCredit(c), s) \equiv credit(c) > 0$$

$$Poss(notifyStopServing(c), s) \equiv True$$

2. Successor state axioms:

$$recipient(n, do(a, s)) = r \equiv$$
$$\exists c, p\, a = orderShipment(c, r, p) \wedge orderCtr = n \vee$$
$$recipient(n, s) = r$$

$$orderPrio(n, do(a, s)) = p \equiv$$
$$\exists c, r\, a = orderShipment(c, r, p) \wedge orderCtr = n \vee$$
$$orderPrio(n, s) = p$$

$$orderCtr(do(a, s)) = n \equiv$$
$$\exists c, r, p\, a = orderShipment(c, r, p) \wedge orderCtr = n - 1 \vee$$
$$orderCtr(s) = n \wedge \forall c, r, p\, a \neq orderShipment(c, r, p)$$

$$credit(c, do(a, s)) = k \equiv$$
$$a = reduceCredit(c) \wedge credit(c, s) = k + 1 \vee$$
$$credit(c, s) = k \wedge a \neq reduceCredit(c)$$

$$Suspended(c, do(a, s)) \equiv$$
$$a = reduceCredit(c) \wedge credit(c, s) = 1 \vee$$
$$Suspended(c, s)$$

3. Procedure definitions:

proc $pickUpShipmentsFrom(c)$
  while $\exists n\,(orderState(n) = ToPickUp \wedge sender(n) = c)$ **do**
  $\pi\, n\,[(orderState(n) = ToPickUp \wedge sender(n) = c)?;$
     $pickUpShipment(n)]$
 endWhile
endProc

proc $DropOffToCentralOffice$
  while $\exists n\,(orderState(n) = OnBoard \wedge Suspended(sender(n))$
     $\wedge\, Suspended(recipient(n))$ **do**
  $\pi\, n\,[(orderState(n) = OnBoard \wedge Suspended(sender(n))$
     $\wedge\, Suspended(recipient(n)))?;$
     $DropOffShipment(n)]$
endProc

References

1. M.E. Bratman, D.J. Israel, and M.E. Pollack. Plans and ressource-bounded practical reasoning. *Computational Intelligence*, 4:349–355, 1988.
2. R.A. Brooks. A robust layered control system for a mobile robot. *IEEE Journal on Robotics and Automation*, 2(1):14–23, 1986.
3. Giuseppe De Giacomo, Yves Lespérance, and Hector J. Levesque. Reasoning about concurrent execution, prioritized interrupts, and exogenous actions in the situation calculus. In *Proceedings of the Fifteenth International Joint Conference on Artificial Intelligence*, pages 1221–1226, Nagoya, Japan, August 1997.
4. Giuseppe De Giacomo and Hector J. Levesque. An incremental interpreter for high-level programs with sensing. In *Cognitive Robotics – Papers from the 1998 AAAI Fall Symposium*, pages 28–34, Orlando, FL, October 1998. AAAI Press. AAAI Tech. Report FS-98-02.
5. Giuseppe De Giacomo, Raymond Reiter, and Mikhail E. Soutchanski. Execution monitoring of high-level robot programs. In *Principles of Knowledge Representation and Reasoning: Proceedings of the Sixth International Conference (KR'98)*, pages 453–464, Trento, Italy, June 1998.
6. M. Jenkin, N. Bains, J. Bruce, T. Campbell, B. Down, P. Jasiobedzki, A. Jepson, B. Majarais, E. Milios, B. Nickerson, J. Service, D. Terzopoulos, J. Tsotsos, and D. Wilkes. ARK: Autonomous mobile robot for an industrial environment. In *Proc. IEEE/RSJ IROS*, Munich,Germany, 1994.
7. Hector J. Levesque, Raymond Reiter, Yves Lespérance, Fangzhen Lin, and Richard B. Scherl. GOLOG: A logic programming language for dynamic domains. *Journal of Logic Programming*, 31(59–84), 1997.
8. John McCarthy and Patrick Hayes. Some philosophical problems from the standpoint of artificial intelligence. In B. Meltzer and D. Michie, editors, *Machine Intelligence*, volume 4, pages 463–502. Edinburgh University Press, Edinburgh, UK, 1979.
9. A.S. Rao and M.P. Georgeff. An abstract architecture for rational agents. In Bernhard Nebel, Charles Rich, and William Swartout, editors, *Principles of Knowledge Representation and Reasoning: Proceedings of the Third International Conference*, pages 439–449, Cambridge, MA, 1992. Morgan Kaufmann Publishing.

10. Raymond Reiter. The frame problem in the situation calculus: A simple solution (sometimes) and a completeness result for goal regression. In Vladimir Lifschitz, editor, *Artificial Intelligence and Mathematical Theory of Computation: Papers in Honor of John McCarthy*, pages 359–380. Academic Press, San Diego, CA, 1991.
11. Raymond Reiter. Knowledge in action: Logical foundations for describing and implementing dynamical systems. Draft Monograph, available at http://www.cs.toronto.edu/~cogrobo, 1998.
12. Matt Robinson and Michael Jenkin. Reactive low level control of the ARK. In *Proceedings, Vision Interface '94*, pages 41–47, Banff, AB, May 1994.
13. Stanley J. Rosenschein and Leslie P. Kaelbling. A situated view of representation and control. *Artificial Intelligence*, 73:149–173, 1995.
14. M. J. Schoppers. Universal plans for reactive robots in unpredictable environments. In *Proceedings of the Tenth International Joint Conference on Artificial Intelligence*, pages 1039–1046, 1987.
15. Yoav Shoham. Agent-oriented programming. *Artificial Intelligence*, 60(1):51–92, 1993.
16. K. Tam, J. LLoyd, Y. Lespérance, H. Levesque, F. Lin, D. Marcu, R. Reiter, and M. Jenkin. Controlling autonomous robots with GOLOG. In *Proceedings of the Tenth Australian Joint Conference on Artificial Intelligence (AI-97)*, pages 1–12, Perth, Australia, November 1997.
17. Kenneth Tam. Experiments in high-level robot control using ConGolog – reactivity, failure handling, and knowledge-based search. Master's thesis, Dept. of Computer Science, York University, 1998.

Success of Default Logic

Vladimir Lifschitz

Department of Computer Sciences, University of Texas, Austin, TX 78712, USA

Summary. Ray Reiter's *Logic for Default Reasoning* was published almost twenty years ago, but it is widely used today by researchers in knowledge representation, commonsense reasoning and logic programming. This note is a collection of random comments on aspects of this success story.

Someone defined success in science as writing a paper that is going to be referenced by colleagues for as long as ten years after its publication. Ray Reiter's *Logic for Default Reasoning* [15] was published almost twenty years ago, but it is widely used today by researchers in knowledge representation, commonsense reasoning and logic programming. This note is a collection of random comments on aspects of this success story.

1. Default logic, along with circumscription [9] and nonmonotonic logic [12], is one of the three original nonmonotonic formalisms described in the 1980 Special Issue of the journal *Artificial Intelligence*. These papers differed from all earlier research on formal logic in such a radical way that their authors could hardly be expected to define their systems "the right way" on the very first attempt.

Indeed, technical papers on circumscription today hardly ever refer to the definition given by John McCarthy in 1980: a few years later he came up with important enhancements and generalizations, such as prioritized circumscription [10], that turn his idea into a widely applicable tool. In place of the original definition of nonmonotonic logic due to Drew McDermott and Jon Doyle, researchers prefer to use its modification proposed in the mid-eighties by Bob Moore—autoepistemic logic [13].

Literature on default logic has its share of "revisionist" papers also. Many of them have to do with the use of bound variables in default theories (see [1] on problems caused by Skolemization). But even quantifier-free default logic is rich and useful. It is remarkable how much of the modern research related to default logic refers to Ray Reiter's original publication, and not to any of the ingenious attempts to improve on his definition.

2. In 1987, Nicole Bidoit and Christine Froidevaux observed that default logic can be used to model negation as failure in logic programs [2]. It follows that any Prolog system can be viewed as a partial implementation of default logic. For instance, given the program

```
p :- \+ q.
q :- \+ r.
```

Prolog responds no to queries p and r and yes to query q; this fact tells us that p and r do not belong to the extension for the default theory

$$\frac{: \neg q}{p} \quad \frac{: \neg r}{q}$$

and that q does. Since Prolog is older than default logic, we can say that default logic was partially implemented even before it was defined!

Other nonmonotonic formalisms mentioned above are closely related to logic programming too [4], [7], but in the case of default logic the match is particularly good. Extended logic programs in the sense of [5] are essentially the default theories that are built entirely from literals; an "answer set" (or "stable model") for a logic program is simply an extension in the sense of default logic intersected with the set of literals.

The idea that logic programming is a subset of default logic has had a major influence on declarative logic programming. Having proved a theorem about logic programs, a researcher can ask whether it can be generalized to default logic. Given a fact about default logic, we may inquire what it tells us about logic programs.

3. One of the goals of the theory of nonmonotonic reasoning is to solve the frame problem by formalizing the commonsense law of inertia. The method for doing that in default logic proposed in Reiter's 1980 paper uses the "frame default" which can be written, in the situation calculus notation, as follows:

$$\frac{P(s) \; : \; P(do(a, s))}{P(do(a, s))}. \tag{1}$$

It says, informally speaking, that if property P holds before executing action a and it is consistent to assume that P holds after executing a also then this assumption is indeed true. The definition of an extension shows how to make postulates like this precise.

Whether or not (1) provides a satisfactory solution to the frame problem may depend, of course, on how other properties of actions are formalized. There are several options here. For instance, we can express that an action A makes P true when a precondition Q is satisfied by axiom

$$Q(s) \supset P(do(A, s)) \tag{2}$$

or by an inference rule–a default without justifications:

$$\frac{Q(s)}{P(do(A, s))}. \tag{3}$$

Steve Hanks and Drew McDermott [6] investigated the use of formal nonmonotonic reasoning for solving the example of the frame problem known today as the

Yale Shooting story. Several formalizations discussed in their paper turned out to be inadequate, and they arrived at the grim conclusion that all nonmonotonic logics considered in their paper–including default logic–"are inherently incapable of representing this kind of default reasoning."

The flurry of publications on this subject in the late eighties described new nonmonotonic formalisms that might be more successful, as well as modifications of the existing formalisms that looked more promising and new ways of applying the existing formalisms to the frame problem. One interesting proposal, due to Paul Morris [14], was to use default logic in a different way: to express the commonsense law of inertia as a nonnormal default.

The amazing fact that transpired later is that there is no need to change anything in Reiter's original proposal. Normal default (1) provides a completely satisfactory solution to the frame problem provided that the rest of the default theory is set up correctly. This was demonstrated by Hudson Turner [16]; see Section 5.2 of that paper for a detailed comparison of his approach with the work done earlier by Hanks and McDermott and by Morris. Essentially, Turner makes three points:

- effects of actions should be described by inference rules like (3) rather than formulas like (2);
- we should include defaults

$$\frac{:\ P(S_0)}{P(S_0)} \quad \text{and} \quad \frac{:\ \neg P(S_0)}{\neg P(S_0)} \tag{4}$$

to guarantee that every extension contains a complete description of the initial situation;[1]
- any assumptions about situations other than S_0 should be stated as constraints:[2] instead of $P(do(A, S_0))$, we should postulate

$$\frac{\neg P(do(A, S_0))}{False}. \tag{5}$$

This way of using default logic is attractive for two reasons. First, it is applicable to actions with indirect effects. In other words, it solves the ramification problem. Second, the preconditions, justifications and conclusions in defaults (1) and (3)–(5) are literals. In other words, these defaults can be viewed as rules in an extended logic program and processed using computational methods of logic programming.

In a recent paper by Richard Watson [18] these ideas found their first practical application.

4. We need nonmonotonic reasoning for many reasons, and we learn of new reasons as time goes by.

In 1990, Hector Geffner [3] explained why causal reasoning is nonmonotonic. In a useful modification of Geffner's logic due to Norman McCain and Hudson

---

[1] This is not necessary for temporal projection problems with complete initial conditions.

[2] For an assumption about S_0, both ways of stating it are equivalent.

Turner [8], the existence of a causal dependency between propositional formulas F and G is represented by a "causal rule"

$$F \Rightarrow G. \tag{6}$$

The semantics of such rules defines when an interpretation of the underlying propositional language is "causally explained" by a set of causal rules. This definition uses a fixpoint construction somewhat reminiscent of the definition of an extension in default logic.

But this is more than an analogy: Turner's forthcoming paper [17] shows how to *embed* the causal logic from [8] into default logic. Replace every rule (6) in a causal theory by default

$$\frac{:\ F}{G}$$

and consider all complete consistent extensions for this set of defaults. These extensions turn out to be in a 1–1 correspondence with the causally explained interpretations.

Instead of restricting attention to complete extensions, we could have added defaults

$$\frac{:\ P, \neg P}{\textit{False}}$$

for all atoms P.

5. To sum up, default logic has brought us several pleasant surprises. In spite of all the controversy surrounding the Yale Shooting example, the original default logic approach to the frame problem turned out to be correct and quite general. Default logic provides a useful model of negation as failure in logic programming, and also of the relation between cause and effect, although it was not intended for any of these purposes. There is no doubt that other surprises like these expect us in the next century.

The author is grateful to Michael Gelfond and Norman McCain for comments on a draft of this note. This work was partially supported by National Science Foundation under grant IIS-9732744.

References

1. Franz Baader and Bernard Hollunder. Embedding defaults into terminological knowledge representation formalisms. *Journal of Automated Reasoning*, 14:149–180, 1995.
2. Nicole Bidoit and Christine Froidevaux. Minimalism subsumes default logic and circumscription. In *Proc. LICS-87*, pages 89–97, 1987.
3. Hector Geffner. Causal theories for nonmonotonic reasoning. In *Proc. AAAI-90*, pages 524–530, 1990.
4. Michael Gelfond. On stratified autoepistemic theories. In *Proc. AAAI-87*, pages 207–211, 1987.
5. Michael Gelfond and Vladimir Lifschitz. Classical negation in logic programs and disjunctive databases. *New Generation Computing*, 9:365–385, 1991.

6. Steve Hanks and Drew McDermott. Nonmonotonic logic and temporal projection. *Artificial Intelligence*, 33(3):379–412, 1987.
7. Vladimir Lifschitz. On the declarative semantics of logic programs with negation. In Jack Minker, editor, *Foundations of Deductive Databases and Logic Programming*, pages 177–192. Morgan Kaufmann, San Mateo, CA, 1988.
8. Norman McCain and Hudson Turner. Causal theories of action and change. In *Proc. AAAI-97*, pages 460–465, 1997.
9. John McCarthy. Circumscription–a form of non-monotonic reasoning. *Artificial Intelligence*, 13:27–39,171–172, 1980. Reproduced in [11].
10. John McCarthy. Applications of circumscription to formalizing common sense knowledge. *Artificial Intelligence*, 26(3):89–116, 1986. Reproduced in [11].
11. John McCarthy. *Formalizing Common Sense: Papers by John McCarthy*. Ablex, Norwood, NJ, 1990.
12. Drew McDermott and Jon Doyle. Nonmonotonic logic I. *Artificial Intelligence*, 13:41–72, 1980.
13. Robert Moore. Semantical considerations on nonmonotonic logic. *Artificial Intelligence*, 25(1):75–94, 1985.
14. Paul Morris. The anomalous extension problem in default reasoning. *Artificial Intelligence*, 35(3):383–399, 1988.
15. Raymond Reiter. A logic for default reasoning. *Artificial Intelligence*, 13:81–132, 1980.
16. Hudson Turner. Representing actions in logic programs and default theories: a situation calculus approach. *Journal of Logic Programming*, 31:245–298, 1997.
17. Hudson Turner. A logic of universal causation. *Artificial Intelligence*, 1999. To appear.
18. Richard Watson. An application of action theory to the space shuttle. In Gopal Gupta, editor, *Proceedings of the First International Workshop on Practical Aspects of Declarative Languages (Lecture Notes in Computer Science 1551)*, pages 290–304. Springer-Verlag, 1999.

Search Algorithms in the Situation Calculus

Fangzhen Lin

Department of Computer Science
Hong Kong University of Science and Technology
Clear Water Bay, Kowloon, Hong Kong
Email: flin@cs.ust.hk

Summary. In this paper we consider axiomatizing AI search algorithms in the situation calculus. Proper axiomatization of such algorithms is important for understanding control knowledge which is by definition search algorithm dependent: a control knowledge can be effective with respect to one search algorithm but ineffective with respect to another. The key idea here is to view search algorithms as strict linear orders on sets of situations in the situation calculus. There are several potential advantages of viewing search algorithms this way. One is that according to it, implementing a search strategy amounts to computing the "next" relation of the corresponding linear order. We found this perspective particularly helpful in understanding various memory-bounded implementations of breadth-first and best-first search as it separates the definition of a search strategy from its implementations. A more important advantage is that according to it, a particularly simple account of search pruning is to consider it as restriction of the original search algorithms, viewed as an ordering relation, to a smaller set of situations.

1 Dedication

This is a work on using the situation calculus to axiomatize control knowledge in problem solving. I can trace the basic idea to a time when I was working with Ray at the University of Toronto, and I have no doubt that it had been influenced by Ray's work and ideas. On a more technical note, this paper would not be possible without Ray's pioneering work on foundational axioms of the situation calculus.

2 Introduction

Meta-level knowledge is knowledge about knowledge. Control knowledge is the kind of meta-level knowledge that tells a problem solver how best to use domain knowledge to solve a problem. An example of control knowledge in planning in the blocks world is:

> If a block needs to be on the table in the goal state, and it is already on the table, then don't move it.

This advice indirectly tells the problem solver what to do by eliminating those actions that would change the location of a block that is already on the table.

Although the importance of control knowledge in problem solving has long been recognized (e.g. [3], [4], [1], [2]), there has not been much work on the subject.

In particular, there has not been a widely accepted formalism for representing and reasoning about such knowledge.

In the past few years, we have been advocating using the situation calculus as such a formalism: we have applied it to axiomatizing the cut operator in logic programming [9], subgoal ordering in planning [8], and temporally extended goals [10]

In this paper we consider axiomatizing AI search algorithms in the situation calculus. Proper axiomatization of such algorithms is important for understanding control knowledge which is by definition search algorithm dependent: it can be effective with respect to one search algorithm but ineffective with respect to another. For instance, the cut operator in Prolog is used to express control knowledge about when not to backtrack, thus is useful only for backtrack oriented search algorithms such as depth-first search, and useless for others such as the breadth-first search.

Our key idea is to view a search algorithm as a strict linear order on situations in the situation calculus: a situation s is ordered after another one s' if according to the search strategy s has to be examined first for s' to be examined. To illustrate, we consider depth-first, breadth-first, and best-first search strategies. As it turns out, the strict linear orders for these algorithms can be axiomatized succinctly.

We see several advantages of viewing search algorithms this way. One is that according to it, implementing a search strategy amounts to computing the "next" relation of the corresponding linear order. This perspective is particularly helpful for understanding memory-bounded implementations of breadth-first and best-first search. For instance, we show that iterative deepening is not really a new search strategy but an implementation of breadth-first search that trades time for space. A more important advantage is that according to it, a particularly simple account of search pruning is to consider it as the restriction of the original search algorithm, viewed as an ordering relation, to a smaller set of situations.

In the following, we first review the definition of search problems and the situation calculus, and then show how a search problem can be formulated as an action theory in the situation calculus.

3 Search Problems and Algorithms

A search problem consists of a set of operators, an initial state, a goal, a cost function, and for informed (heuristic) search problem, a state evaluation function.

A typical search problem is the 8-puzzle where we have a board with 3×3 cells that can be either blank or one of the numbers between 1 and 8:

- Operators: moving the blank up, down, left, and right.
- The initial state: any given configuration of the board.
- Goal test: check to see if the current configuration of the board is the same as that in the goal state in Figure 1.
- Cost function: number of actions in a solution, i.e. each action costs one unit.

Given a search problem, search algorithms start at the initial state, and keep applying operators until a solution is found. Examples of search problems and search

1	6	3
5		7
4	2	8

1	2	3
8		4
7	6	5

An initial state   The goal state

Fig. 1. The 8-puzzle

algorithms can be found in all introductory AI books (e.g. [13], [5], [17]). In the following we shall consider depth-first, breadth-first, and best-first search algorithms.

4   The Situation Calculus

The situation calculus (McCarthy and Hayes [12]) is a formalism for representing and reasoning about actions in dynamic domains. It is a many-sorted predicate calculus with some reserved predicate and function symbols. For example, in the blocks world, to say that block A is initially clear, we write:

$$Holds(clear(A), S_0),$$

where H is a reserved binary predicate and stands for "holds", and S_0 is a reserved constant symbol denoting the initial situation. As an another example, to say that after the action $stack(x, y)$ is performed, the proposition $on(x, y)$ will be true, we write:[1]

$$Poss(stack(x, y), s) \supset Holds(on(x, y), do(stack(x, y), s)),$$

where the reserved function $do(a, s)$ denotes the resulting situation of doing the action a in the situation s, and $Poss(a, s)$ is the precondition for a to be executable in s. This is an example of how the effects of an action can be represented in the situation calculus.

More precisely, we use a many-sorted second-order language with equality for the situation calculus. We assume the following sorts: *situation* for situations, *action* for actions, *fluent* for propositional fluents such as *clear* whose truth values depend on situations, and *object* for everything else. As mentioned above, we assume that S_0 is a reserved constant denoting the initial situation, H a reserved predicate for expressing properties about fluents in a situation, do a reserved binary function denoting the result of performing an action, and $Poss$ a reserved binary predicate for expressing action preconditions.

We assume that the space of situations is an infinite tree whose root is S_0 and whose nodes are situations obtained from S_0 by applying a sequence of actions.

---

[1] In this paper, free variables in a displayed formula are assumed to be universally quantified.

More precisely, we assume the following foundational axioms:

$$S_0 \neq do(a, s), \tag{1}$$

$$do(a_1, s_1) = do(a_2, s_2) \supset (a_1 = a_2 \wedge s_1 = s_2), \tag{2}$$

$$(\forall P)[P(S_0) \wedge (\forall a, s)(P(s) \supset P(do(a, s))) \supset (\forall s)P(s)], \tag{3}$$

The first two axioms are unique names assumptions for situations. The third axiom is second order induction which says that every situation has to be obtained from the initial one by repeatedly applying the function do. In other words, it says that a situation is either the initial situation S_0 or the result of performing a sequence of actions in the initial situation. These axioms imply that situations in the situation calculus are isomorphic to finite sequences of actions:[2] if s is a situation, then there is a unique finite sequence L of actions such that $s = do(L, S_0)$. This isomorphism is what makes the situation calculus a good formalism for representing control information because instead of talking about sequences of actions, which is often what control knowledge is about, we can then talk about situations which are objects in the situation calculus.

Frequently, instead of all situations, we only want to consider those that can be reached from the initial situation by a sequence of *executable* actions. To define these situations, we introduce a strict partial order $<$ into our language[3], and define it inductively as follows:

$$\neg s < S_0, \tag{4}$$

$$s < do(a, s') \equiv (Poss(a, s') \wedge s \leq s'), \tag{5}$$

where $s \leq s'$ is a shorthand for $s < s' \vee s = s'$.

Using this partial order, we can then define:

$$\mathsf{Exec}(s) \equiv S_0 \leq s.$$

It can be shown that $\mathsf{Exec}(s)$ iff there is a finite sequence of actions $[a_1, ..., a_n]$ such that $s = do([a_1, ..., a_n], S_0)$ and for each $1 \leq i \leq n$, a_i is executable: $Poss(a_i, do([a_1, ..., a_{i-1}], S_0))$. Thus the set $\{s \mid \mathsf{Exec}(s)\}$ is the set of situations reachable from the initial one by a sequence of executable actions. One can think of the set as the space of all situations that a robot can possibly end up in beginning in S_0. If, say, action A is not executable in S_0, then the situation $do(A, S_0)$, although exists as a logical object, is not realizable. In the following, we call a situation s such that $\mathsf{Exec}(s)$ an *executable situation*.

---

[2] Given a sequence of actions $[a_1, ..., a_n]$, we use $do([a_1, ..., a_n], s)$ to denote the resulting situation of performing the sequence of actions in s. Inductively, $do([], s) = s$ and $do([a|L], s) = do(L, do(a, s))$.

[3] The relation $<$ was first defined in the the situation calculus by Reiter [14]. Our presentation here follows (Lin and Reiter [11]) where some of its properties such as transitivity are shown. For a thorough discussion of foundational axioms of the situation calculus, see (Reiter [15]).

In the following, we shall denote by Σ the set of the above axioms (1) - (5). To make the problem non-trivial, we shall assume that there is at least one executable action: $(\exists a)Poss(a, S_0)$.

5   Search Problems in the Situation Calculus

Given a search problem, we shall represent its initial state by the initial situation S_0, and each of its operators by a unique action. Under this representation, information about the initial state of a search problem will be represented by a set of situation calculus formulas about S_0, and information about its operators will be axiomatized by a situation calculus action theory.

To express goals in search problems, we introduce *fluent formulas* which are situation calculus formulas with situation terms suppressed. Formally, we have:

1. Any formula that does not mention a situation term is a fluent formula.
2. If F is a n-ary fluent, and $t_1, ..., t_n$ are object terms, then $F(t_1, ..., t_n)$ is a fluent formula.
3. If ϕ and φ are fluent formulas, x an object variable, and p a fluent variable, then $\neg\varphi$, $\varphi \vee \phi$, $(\forall x).\varphi$ and $(\forall p).\phi$ are fluent formulas. (Other connectives such as $\wedge, \supset, \equiv$, and $\exists$ can be defined as usual.)

For example, $A1 \neq A2$, $on(block1, block2)$, $(\forall p)(p \supset p)$ are fluent formulas, where $A1$ and A are action constants and on a binary fluent.

Given a fluent formula φ and a situation term S, we define $H(\varphi, S)$ to be a shorthand for a situation calculus formula according to the following rules:

$$H(\varphi, S) \overset{\text{def}}{=} \varphi, \text{ provided } \varphi \text{ does not mention}$$
$$\text{any situation terms}$$
$$H(\neg\varphi, S) \overset{\text{def}}{=} \neg H(\varphi, S),$$
$$H(\varphi \vee \phi, S) \overset{\text{def}}{=} H(\varphi, S) \vee H(\phi, S),$$
$$H((\forall x)\varphi, S) \overset{\text{def}}{=} (\forall x)H(\varphi, S),$$
$$H((\forall p)\varphi, S) \overset{\text{def}}{=} (\forall p)H(\varphi, S).$$

For example,

$$H(a \neq b, s) \overset{\text{def}}{=} a \neq b,$$
$$H(on(a, b) \vee clear(b), s) \overset{\text{def}}{=} H(on(a, b), s) \vee H(clear(b), s).$$

Now given a search problem, suppose that T is a situation calculus theory that captures its initial state and operators, and that the goal of the problem corresponds to fluent formula G, then testing if the goal is achieved by a sequence of actions becomes logical entailment checking: a finite sequence L of actions is a solution iff $T \cup \Sigma \models H(G, do([a_1, ..., a_n], S_0))$.

To illustrate this conceptualization of search problems in the situation calculus, consider again the 8-puzzle. We have the following fluent:

- $at(x, i, j)$ - number x is in cell $i \times j$. We denote blank by number 0. So $at(0, 1, 2)$ means that blank is in row 1 and column 2.

and the following action:

- $move(i_1, j_1, x, i_2, j_2)$ - swap 0 and x, provided that 0 is in (i_1, j_1), x is in (i_2, j_2), and that they are next to each other.

The following precondition axiom ensures that at most four instantiations of the above action can be performed in any situation:

$$Poss(move(i_1, j_1, x, i_2, j_2), s) \equiv x \neq 0 \land H(at(0, i_1, j_1), s) \land H(at(x, i_2, j_2), s) \land$$
$$[i_1 = i_2 \land (j_1 = j_2 + 1 \lor j_2 = j_1 + 1)] \lor [j_1 = j_2 \land (i_1 = i_2 + 1 \lor i_2 = i_1 + 1)],$$

where we assume that i's and j's are numbers, and the operation "$+$" has its usual meaning. The effects of this action can be axiomatized by the following successor state axiom which says that the only action that has effect on at is $move$, and $move(i_1, j_1, x, i_2, j_2)$ swaps 0 and x and leaves the positions of all other numbers unchanged:

$$Poss(a, s) \supset H(at(x, i, j), do(a, s)) \equiv$$
$$x = 0 \land (\exists i', j', y)a = move(i', j', y, i, j) \lor$$
$$(\exists i', j', y)a = move(i, j, x, i', j') \lor$$
$$H(at(x, i, j), s) \land \neg (\exists i_1, i_2, j_1, j_2)a = move(i_1, j_1, x, i_2, j_2)$$

Now let

- T be the set of above action precondition and successor state axioms;
- D_{S_0} the following set of axioms that capture the initial state in Figure 1:

$\{H(at(1, 1, 1), S_0), H(at(6, 1, 2), S_0), H(at(3, 1, 3), S_0), H(at(5, 2, 1), S_0),$
$H(at(0, 2, 2), S_0), H(at(7, 2, 3), S_0), H(at(4, 3, 1), S_0), H(at(2, 3, 2), S_0),$
$H(at(8, 3, 3), S_0)\}$,

- D_{una} the set of following unique names axioms about actions:

$$move(i_1, j_1, x, i_2, j_2) = move(i_1', j_1', x', i_2', j_2') \supset$$
$$x = x' \land i_1 = i_1' \land j_1 = j_1' \land i_2 = i_2' \land j_2 = j_2',$$

- G the following fluent formula that captures the goal state in Figure 1:

$$at(1, 1, 1) \land at(2, 1, 2) \land at(3, 1, 3) \land at(8, 2, 1) \land at(0, 2, 2) \land$$
$$at(4, 2, 3) \land at(7, 3, 1) \land at(6, 3, 2) \land at(5, 3, 3).$$

then testing if a sequence of actions L is a solution becomes checking the validity of the following entailment:

$$T \cup \Sigma \cup D_{S_0} \cup D_{una} \models_{nat} \text{Exec}(do(L, S_0)) \land H(G, do(L, S_0)),$$

where $\models_{nat}$ stands for logical entailment with the arguments of at and $move$ given standard number interpretation. This entailment is an instance of temporal projection, and can be checked efficiently if the initial state is complete.

6   Search Algorithms in the Situation Calculus

Under this formulation of search problems in the situation calculus, the basic idea of this paper is to view a search algorithm as a strict linear order on a set of executable situations, and search pruning as a restriction on the domain of the linear order.

Definition 1. A binary relation R is a strict linear order if it is

1. irreflexive: $\neg R(x, x)$;
2. transitive: $R(x, y) \wedge R(y, z) \supset R(x, z)$; and
3. connected: $x = y \vee R(x, y) \vee R(y, x)$,

where x, y, z range over the domain of R, which is defined to be the following set:

$$D_R = \{x \mid (\exists y)(R(x, y) \vee R(y, x))\}$$

In the following, we write $x \in D_R$ as $D_R(x)$, and if the relation R is apparent from the context, write D_R as D.

Definition 2. A relation R with domain D is called a *search algorithm* if it is a strict linear order that is consistent with $<$, expands only executable situations, and includes S_0, that is, it satisfies the following axioms:

$$D(s) \supset \neg R(s, s), \tag{6}$$

$$D(s_1) \wedge D(s_2) \wedge D(s_3) \supset \{R(s_1, s_2) \wedge R(s_2, s_3) \supset R(s_1, s_3)\}, \tag{7}$$

$$D(s_1) \wedge D(s_2) \supset s_1 = s_2 \vee R(s_1, s_2) \vee R(s_2, s_1), \tag{8}$$

consistent with $<$:

$$D(s_2) \supset [s_1 \leq s_2 \supset R(s_1, s_2)] \tag{9}$$

expand only executable situations:

$$D(s) \supset \mathsf{Exec}(s), \tag{10}$$

includes S_0:

$$D(S_0). \tag{11}$$

Notice that axiom (9) implies that if $D(s)$ and $s' < s$, then $D(s')$. Notice also that axioms (9) – (11) imply that $D(s) \supset R(S_0, s)$. This means that for any search algorithm R, S_0 is the first situation that the algorithm will examine. The axioms also entail that if there is a next situation after S_0, it must be $do(A, S_0)$ for some action A, and the next after this one must be either $do(B, S_0)$ for some action $B \neq A$ or $do(A', do(A, S_0))$ for some action A', depending on, for example, whether the algorithm uses breadth-first or depth-first search strategy.

The following proposition is immediate:

Proposition 1. *Let D be the domain of R. If R is a search algorithm, then the restriction of R on a set $K \subseteq D$:*

$$R'(x, y) \equiv K(x) \wedge K(y) \wedge R(x, y)$$

is also a search algorithm. Notice that the domain of R' is a subset of K.

Given a search algorithm R, we can define a "next" relation Next_R such that $\mathsf{Next}_R(s, s')$ holds if s' is the situation immediately after s according to the linear order R:

$$\mathsf{Next}_R(s, s') \equiv R(s, s') \wedge \neg(\exists s'').R(s, s'') \wedge R(s'', s').$$

It's clear that for any s, there is at most one situation s' such that $\mathsf{Next}_R(s, s')$. Using Next, we can "execute" a search algorithm R as follows:

1. Let $S = S_0$;
2. If S is a goal situation, then exit with success;
3. If there is a situation S' such that $\mathsf{Next}_R(S, S')$, then let $S = S'$ and go back to step 2;
4. Exit with failure.

This procedure suggests that instead of a linear order, we can also capture a search algorithm by a "next" relation. Indeed these two approaches are essentially equivalent. We have seen how to define a "next" relation given a linear order. Conversely, we can also recover the essence of a linear order from a "next" relation by taking its transitive closure. However, we prefer to start with a linear order for two reasons. One is that for those search strategies that we have investigated, it is much easier and cleaner to capture their corresponding linear orders than their corresponding "next" relations. The other, more important reason is that linear orders provide a more general framework. This is because it is much easier to investigate various variants of a search strategy through a common linear order. For instance, as we shall see, depth first search with chronological backtracking and depth first search with backjumping share a common linear order. Similarly, it is much easier to explain search control and pruning by viewing a search algorithm as an ordering relation. Having said that, there are still some occasions when it is easier to characterize a search algorithm by its corresponding "next" relation, as we shall see when we consider depth-first search with backjumping.

It is clear that the key operation of the above procedure for executing a search algorithm is the computation of the partial function Next_R. So to implement a search algorithm, what we need to do is to find a way to compute this function. A generic way of doing this is as follows. Let S be an executable situation. To compute the next situation after S, do the following:

1. Compute the following set:

$$\mathcal{S} = \{s \mid R(S, s) \wedge \neg(\exists s')(s' < s \wedge R(S, s'))\}.$$

2. Sort S according to the linear order R.
3. If S is empty, then there is no next situation: $\neg(\exists s)\mathsf{Next}_R(S, s)$; otherwise, the first element S' in S is the next situation: $\mathsf{Next}_R(S, S')$.

The correctness of this procedure can be easily verified: If $\mathsf{Next}_R(S, S')$, then $R(S, S')$ and there are no situation in between S and S'. In particular, by (9), there does not exist an s such that $R(S, s)$ and $s < S'$. So $S' \in S$. Conversely, if S' is the first element of S, then $R(S, S')$. Furthermore, there cannot be any situation s such that $R(S, s)$ and $R(s, S')$. Otherwise, there is an s' such that $R(S, s')$, $s' \leq s$, and $\neg(\exists s^*)(s^* < s' \wedge R(S, s^*))$. Therefore $s' \in S$. By the transitivity of R, we also have $R(s', S')$, this contradicts with the assumption that S' is the first element of S.

Notice that theoretically, the condition $\neg(\exists s').s' < s \wedge R(S, s')$ in the definition of S is not necessary. It is included to keep the size of S down: without it, S would be an infinite set; with it, S will be finite for problems with finite number of operators.

Notice also that for this procedure to work, there must be a systematic way of generating and maintaining S. In particular, there must be an efficient way of checking whether $R(S, S')$ holds for any given two situations S and S'. For all search algorithms that we will consider in this paper, checking whether $R(S, S')$ can be done in time and space linear to the size of S and S'.

This procedure is in many ways similar to the implementation of search algorithms using an open and a closed list (see [13]) with S playing the role of the open list. A well-known limitation of this implementation is that it requires a lot of memory to store the open list. However, since what we really want is the first element of S, there is no real reason why we have to keep the entire set around. All we need is to keep the current closest situation C, and each time a new situation s in S is discovered, it is compared with C in the following way: if $R(C, s)$, then we do nothing; if $R(s, C)$, then we replace C by s. This seems to be the idea used by many memory-bounded implementation of best-first search algorithms ([16], [7]).

Another observation that we can make about the above procedure for executing a search algorithm is that not all executable situations will necessarily be visited by a search algorithm. Let us define $\mathsf{Reachable}_R$ to be the smallest relation satisfying the following two properties:

1. $\mathsf{Reachable}_R(S_0)$ and
2. if $\mathsf{Reachable}_R(s)$ and $\mathsf{Next}_R(s, s')$, then $\mathsf{Reachable}_R(s')$.

Formally,

$\mathsf{Reachable}_R(s) \equiv$
$$(\forall P).\{P(S_0) \wedge (\forall s_1, s_2)(P(s_1) \wedge \mathsf{Next}_R(s_1, s_2) \supset P(s_2))\} \supset P(s).$$

Then a situation s will be visited by an algorithm R iff $\mathsf{Reachable}_R(s)$. Using $\mathsf{Reachable}_R$, we can define when a search algorithm is complete, and when it is optimal:

Definition 3. Given a problem with goal G, a search algorithm R is complete for this problem if

$$(\exists s)[S_0 \leq s \wedge H(G, s)] \supset (\exists s)[\mathsf{Reachable}_R(s) \wedge H(G, s)]. \tag{12}$$

It is optimal if

$$\mathsf{First}_R(s) \supset \mathsf{Optimal}(s), \tag{13}$$

where $\mathsf{First}_R(s)$ if s is the first solution returned by R:

$\mathsf{First}_R(s) \equiv$
 $\quad \mathsf{Reachable}_R(s) \wedge H(G, s) \wedge \neg(\exists s')(\mathsf{Reachable}_R(s') \wedge R(s', s) \wedge H(G, s'))$,

and $\mathsf{Optimal}(s)$ if s is an optimal solution:

$\mathsf{Optimal}(s) \equiv S_0 \leq s \wedge H(G, s) \wedge \neg(\exists s')(S_0 \leq s' \wedge H(G, s') \wedge g(s') < g(s))$,

where g is the real valued cost function of the problem, and the relation "$<$" in $g(s') < g(s)$ has its usual meaning.

7   Depth-First Search

Depth-first search always chooses a node at the deepest level for expansion until it reaches a deadend, at that point it backtracks to a node at a shallower level. Like most search strategies, depth-first search stands for a family of search algorithms, depending on the order of actions used to expand a node and on the strategies used for backtracking.

   Let $\prec$ be a strict linear order on actions. One way of implementing depth-first search with chronological backtracking is as follows:

1. Construct a tree with S_0 as the only node, and let $S = S_0$.
2. If S is a goal situation, then exit with success.
3. Extend the tree by adding the situations in $\{do(a, S) \mid Poss(a, S)\}$ as the children of S, and order them from left to right according to $\prec$: if $a \prec b$, then $do(a, S)$ must be to the left of $do(b, S)$.
4. Let S be the leftmost leaf situation of the tree such that the set

$$\{do(a, S) \mid Poss(a, S)\}$$

   is not empty. If there is no such situation, then exit with failure.
5. Go back to step 2.

From this procedure, we can see that according to depth-first search, a situation s_1 is visited before s_2 if

1. s_1 and s_2 are on a same branch and $s_1 < s_2$; or

2. s_1 and s_2 share a common ancestor s and the branch that s_1 is on is to the left of the branch that s_2 is on.

This leads to the following axiom that captures the strict linear order for depth-first search:

$$\mathsf{DFS}(s_1, s_2) \equiv s_1 < s_2 \vee (\exists s, a, b)(do(a, s) \leq s_1 \wedge do(b, s) \leq s_2 \wedge a \prec b).$$
(14)

From DFS we get depth-first search with chronological backtracking DCB by restricting the domain of DFS to the set of executable situations:

$$\mathsf{DCB}(s_1, s_2) \equiv \mathsf{Exec}(s_1) \wedge \mathsf{Exec}(s_2) \wedge \mathsf{DFS}(s_1, s_2).$$
(15)

The following proposition shows that DCB so defined is indeed a search algorithm.

Proposition 2. *The relation* DCB *is a search algorithm.*

Proof. We need to show that properties (6)-(11) hold for DCB. First of all, it is easy to see that the domain of DCB is Exec:

$$(\exists s')(\mathsf{DCB}(s, s') \vee \mathsf{DCB}(s', s)) \equiv \mathsf{Exec}(s).$$

From this, properties (10) and (11) follow. Property (9) is also easy to see. In the following, all situations are assumed to be executable.
Property (6):

$$\neg s < s \wedge \neg (\exists s', a, b)(do(a, s') \leq s \wedge do(b, s') \leq s \wedge a \prec b).$$

The first conjunct holds because $<$ is a strict partial order (see [11] for a proof). The second conjunct holds because $\prec$ is a strict linear order on actions, and the fact that

$$do(a, s') \leq s \wedge do(b, s') \leq s \supset a = b.$$

Property (7): Suppose $\mathsf{DCB}(s_1, s_2)$ and $\mathsf{DCB}(s_2, s_3)$. We show that $\mathsf{DCB}(s_1, s_3)$. There are three cases:

1. $s_1 < s_2$ and $s_2 < s_3$: it follows that $s_1 < s_3$, thus $\mathsf{DCB}(s_1, s_3)$.
2. $s_1 < s_2$ and for some situation s and actions a and b,

$$do(a, s) \leq s_2 \wedge do(b, s) \leq s_3 \wedge a \prec b.$$

Either $s_1 < do(a, s)$ or $do(a, s) \leq s_1$. In the first case, we have $s_1 \leq s$, so $s_1 < s_3$, thus $\mathsf{DCB}(s_1, s_3)$. In the second case, we have

$$do(a, s) \leq s_1 \wedge do(b, s) \leq s_3 \wedge a \prec b.$$

Thus $\mathsf{DCB}(s_1, s_3)$.

3. $s_2 < s_3$ and for some situation s and actions a and b,

$$do(a, s) \leq s_1 \wedge do(b, s) \leq s_2 \wedge a \prec b.$$

We have

$$do(a, s) \leq s_1 \wedge do(b, s) \leq s_3 \wedge a \prec b.$$

Thus $\mathsf{DCB}(s_1, s_3)$.

Property (8): If either $s_1 \leq s_2$ or $s_2 \leq s_1$, then it is easy to see from (15) that (8). Suppose $\neg s_1 \leq s_2$ and $\neg s_2 \leq s_1$, then there must exist some situation s and actions $a \neq b$ such that

$$do(a, s) \leq s_1 \wedge do(b, s) \leq s_2.$$

Since $\prec$ is a strict linear order, either $a \prec b$ or $b \prec a$. In the former case, we have $\mathsf{DCB}(s_1, s_2)$, and the latter $\mathsf{DCB}(s_2, s_1)$.

The following proposition shows that DCB indeed corresponds to depth-first search with chronological backtracking:

Proposition 3. *Let S be an executable situation.*

1. If for some action A,

$$Poss(A, S) \wedge \neg(\exists a)(Poss(a, S) \wedge a \prec A), \tag{16}$$

then $\mathsf{Next}_{\mathsf{DCB}}(S, do(A, S))$. This is the case when S is not a deadend.
2. If for some actions A and B, and situation S',

$$\neg(\exists a)Poss(a, S), \tag{17}$$

$$do(B, S') \leq S \wedge Poss(A, S') \wedge B \prec A, \tag{18}$$

$$\neg(\exists a).Poss(a, S') \wedge B \prec a \prec A, \tag{19}$$

$$\neg(\exists s, a, b).do(B, S') < do(a, s) \leq S \wedge Poss(b, s) \wedge a \prec b \tag{20}$$

then $\mathsf{Next}_{\mathsf{DCB}}(S, do(A, S'))$. This is the case when search backtracks at S.
3. Otherwise, S has no next situation: $\neg(\exists s)\mathsf{Next}_{\mathsf{DCB}}(S, s)$.

Proof. **Case 1** Suppose (16). Clearly $\mathsf{DCB}(S, do(A, S))$. We show that

$$\neg(\exists s).\mathsf{DCB}(S, s) \wedge \mathsf{DCB}(s, do(A, S)).$$

Suppose otherwise, i.e. for some s, $\mathsf{DCB}(S, s) \wedge \mathsf{DCB}(s, do(A, S))$. There are three cases:

1. $S < s$ and $s < do(A, S)$: there cannot be such a situation s as $s < do(A, S)$ implies that $s \leq S$.

2. $S < s$ and $(\exists s', a, b).do(a, s') \leq s \wedge do(b, s') \leq do(A, S) \wedge a \prec b$. From $do(b, s') \leq do(A, S)$, there are two cases:
   (a) $s' = S$. In this case, $A = b$ and $a \prec A$. Since s is executable and $do(a, s') \leq s$, $Poss(a, S)$. This contradicts with our assumption (16).
   (b) $do(b, s') \leq S$. In this case, we have $\mathsf{DCB}(s, S)$, a contradiction with $S < s$.
3. $\mathsf{DCB}(s, do(A, S))$ and $(\exists s', a, b).do(a, s') \leq S \wedge do(b, s') \leq s \wedge a \prec b$. It follows that $do(a, s') \leq do(A, S)$. So $\mathsf{DCB}(do(A, S), s)$, a contradiction with $\mathsf{DCB}(s, do(A, S))$.

As we can see that none of these three cases are possible, so there cannot be a situation s such that $\mathsf{DCB}(S, s)$ and $\mathsf{DCB}(s, do(A, S))$. Thus $\mathsf{Next}_{\mathsf{DCB}}(S, do(A, S))$.

Case 2 Suppose (17) – (20) hold. It is clear that $\mathsf{DCB}(S, do(A, S'))$. We show that $\neg(\exists s)(\mathsf{DCB}(S, s) \wedge \mathsf{DCB}(s, do(A, S')))$. Suppose otherwise, i.e. for some s, both $\mathsf{DCB}(S, s)$ and $\mathsf{DCB}(s, do(A, S'))$. By (17), it is impossible that $S < s$. So for $\mathsf{DCB}(S, s)$ to hold, it must be the case that for some s' and a and b,

$$do(a, s') \leq S \wedge do(b, s') \leq s \wedge a \prec b.$$

This means that $s' < S$. Since we also have $S' < S$ from (18), so either $S' < s'$, $s' < S'$, or $s' = S'$:

1. $S' < s'$. This is impossible because it would imply $do(B, S') < do(a, s') \leq S$, a contradiction with (20). (Since $\mathsf{Exec}(s)$ and $do(b, s') \leq s$, we have $Poss(b, s')$ and $\mathsf{Exec}(s')$.)
2. $s' < S'$. In this case, we also have $do(a, s') \leq S'$, since both $do(a, s') \leq S$ and $S' < S$. Therefore $do(a, s') \leq do(A, S')$, thus $\mathsf{DCB}(do(A, S'), s)$, a contradiction with our assumption.
3. $s' = S'$. In this case, $a = B$. Since $\mathsf{DCB}(s, do(A, S'))$, we also have $\mathsf{DCB}(do(b, s'), do(A, S'))$. So $\mathsf{DCB}(do(b, S'), do(A, S'))$. This means that $b \prec A$. But we also have $B \prec b$, a contradiction with (19).

Case 3 Suppose neither Case 1 nor Case 2 hold, then $\neg(\exists s)\mathsf{Next}_{\mathsf{DCB}}(S, s)$. Suppose otherwise, i.e. for some s, $\mathsf{Next}_{\mathsf{DCB}}(S, s)$. Now if $(\exists a)Poss(a, S)$, then there must be some A such that $s = do(A, S)$, and the condition (16) holds. If $\neg(\exists a)Poss(a, S)$, then there must be an S' and two actions A and B such that

$$do(B, S') \leq S \wedge do(A, S') \leq s \wedge B \prec A.$$

In this case, $s = do(A, S')$, and the conditions (18)-(20) have to be true.

Notice here that DCB backtracks only when it hits a deadend, i.e. when none of the actions are executable. However, one can often tell that a situation has no hope of reaching a goal state before it hits a deadend. For instance, suppose our goal is to drive from one city to another, and we are in a situation where our car has run out of gas and there is no way for us to refuel it, then we can conclude that we can't achieve the goal although there are still some actions, such as kicking the car, that we can do.

Of course, this information is domain and goal dependent. In general, we can axiomatize it by a predicate $\text{Backtrack}(s)$, meaning that s cannot lead to a goal situation:

$$\text{Backtrack}(s) \supset \neg(\exists s').s \leq s' \wedge H(G, s'),$$

where G is the given goal. For example, in constraint satisfaction, if we have two constraints G_1 and G_2 to satisfy, then we can backtrack whenever one of them becomes false, although there may still be unassigned variables:

$$\neg G_1(s) \vee \neg G_2(s) \supset backtrack(s).$$

Once we have information about this predicate, then we can restrict search to the following set:

$$\text{Exec}(s) \wedge \neg(\exists s').s' < s \wedge \text{Backtrack}(s').$$

We have so far considered only depth-first search with chronological backtracking. We can also capture depth-first search with backjumping using the strict linear order DFS.

Although there is no precise definition of what backjumping is, the intuition is that if S is a deadend, then search should backtrack to the deepest level in the path from S_0 to S that may lead to a goal situation, given the information available from S. More precisely, given that

$$do([a_1, ..., a_k, ..., a_n], S_0)$$

is a deadend, if we know that for any $k < i \leq n$, there is no $s > do([a_1, ..., a_i], S_0)$ such that s is a goal situation, then there is no point backtracking to any of the situations $do([a_1, ..., a_i], S_0)$, for any $k < i \leq n$. Rather the search should backjump to $do([a_1, ..., a_k], S_0)$.

To axiomatize this intuition, we assume a predicate $\text{Prune}(s_1, s_2)$ whose intuitive meaning is that if one needs to backtrack at s_1, then don't backtrack to s_2. In order not to lose any solution, we require it satisfy the following axiom:[4]

$$\text{Prune}(s_1, s_2) \supset [\neg H(G, s_1) \supset \neg(\exists s)(s_2 < s \wedge H(G, s))], \qquad (21)$$

where G is the given goal of the problem. With this predicate, we can capture depth-first search with backjumping by axiomatizing its next relation:

$\text{Next}(s_1, s_2) \equiv$
    $\neg\text{Backtrack}(s_1) \supset \text{Next}_{\text{DCB}}(s_1, s_2) \wedge$
    $\text{Backtrack}(s_1) \supset \{\text{DCB}(s_1, s_2) \wedge \neg(\exists s)(s < s_2 \wedge \text{Prune}(s_1, s)) \wedge$
    $\neg(\exists s_2')[\text{DCB}(s_1, s_2') \wedge \text{DCB}(s_2', s_2) \wedge \neg(\exists s)(s < s_2' \wedge \text{Prune}(s_1, s))]\}.$

---

[4] One could just let $\text{Prune}(s_1, s_2)$ be $\neg(\exists s)(s_2 < s \wedge H(G, s))$. But then computing Prune will be the same as deciding if the goal is achievable or not, which is a very difficult task.

In other words, if s_1 is not a backtrack point, then backjumping is the same as chronological backtracking. If s_1 is a backtrack point, then the next situation should be the first situation s_2 (according to the order of DCB) which is not a descendent of a situation that should be pruned.

Notice that for any situations s and s', if Next(s, s'), then DCB(s, s'); furthermore, for any situation s'' such that DCB$(s, s'') \land$ DCB(s'', s'), s'' cannot be a goal situation. So backjumping will not miss any solution that can be returned by chronological backtracking. In this sense, backjumping is indeed a provably correct improvement on chronological backtracking. Of course, this does not mean that the former is always preferred over the latter as one has to take into account the cost of computing Prune.

Notice that if we use Next to define Reachable as in Section 6, and define

$$DBJ(s_1, s_2) \equiv \text{Reachable}(s_1) \land \text{Reachable}(s_2) \land DCB(s_1, s_2),$$

then the next relation defined using DBJ, Next$_{\text{DBJ}}$, is exactly the same one as the Next relation defined above. So DBJ can be considered to be DCB together with a pruning strategy that prunes some redundant branches of the search tree.

8 Breadth-First Search

In contrast to depth-first search, breadth-first search expands nodes at the same level first before expanding nodes at a deeper level.

The strict linear order BFS for breadth-first search is best defined inductively. In the base case, we know that S_0 needs to be the first situation:

$$S_0 < s \supset \text{BFS}(S_0, s), \tag{22}$$

$$\neg \text{BFS}(s, S_0), \tag{23}$$

Inductively, for two executable situations $do(a, s_1)$ and $do(b, s_2)$, the former is before the latter iff s_1 is before s_2 or $s_1 = s_2$ and $a \prec b$:

$$\text{BFS}(do(a, s_1), do(b, s_2)) \equiv \text{Exec}(do(a, s_1)) \land \text{Exec}(do(b, s_2)) \land$$
$$\text{BFS}(s_1, s_2) \lor [s_1 = s_2 \land a \prec b]. \tag{24}$$

Proposition 4. BFS *is a search algorithm.*

Proof. Properties (10) and (11) hold trivially because the domain of BFS is Exec. Again, in the following, we assume that all situations are executable.

Property (6): We prove by induction. Clearly, $\neg \text{BFS}(S_0, S_0)$. Inductively, suppose $\neg \text{BFS}(s, s)$. We show that $\neg \text{BFS}(do(a, s), do(a, s))$. This sentence is equivalently to

$$\neg(s = s \land a \prec a) \land \neg \text{BFS}(s, s)$$

which follows from inductive assumption and the assumption that $\prec$ is a strict linear order. This proves the inductive case, thus (6).

Property (7): we prove it by simultaneous induction on s_1, s_2, and s_3 using the following property:

$$s_1 = S_0 \vee s_2 = S_0 \vee s_3 = S_0 \vee$$
$$(\exists s_1', s_2', s_3', a, b, c).s_1 = do(a, s_1') \wedge s_2 = do(b, s_2') \wedge s_3 = do(c, s_3').$$

If

$$s_1 = S_0 \vee s_2 = S_0 \vee s_3 = S_0,$$

then (7) follows from (22) and (23). Inductively, assume that

$$\mathbf{BFS}(s_1, s_2) \wedge \mathbf{BFS}(s_2, s_3) \supset \mathbf{BFS}(s_1, s_3),$$

we show that

$$\mathbf{BFS}(do(a, s_1), do(b, s_2)) \wedge \mathbf{BFS}(do(b, s_2), do(c, s_3)) \supset \mathbf{BFS}(do(a, s_1), do(c, s_3)) \tag{25}$$

Suppose $\mathbf{BFS}(do(a, s_1), do(b, s_2))$ and $\mathbf{BFS}(do(b, s_2), do(c, s_3))$:

$$[s_1 = s_2 \wedge a \prec b] \vee \mathbf{BFS}(s_1, s_2),$$
$$[s_2 = s_3 \wedge b \prec c] \vee \mathbf{BFS}(s_2, s_3)$$

we have

$$[s_1 = s_3 \wedge a \prec c] \vee \mathbf{BFS}(s_1, s_3)$$

by the inductive assumption, and the fact that $\prec$ is a strict linear order. This proves the inductive case, thus (7).

Property (8): We prove by simultaneous induction on s_1 and s_2. If $s_1 = S_0 \vee s_2 = S_0$, then (8) follows from (23). Now inductively, suppose (8) holds for s_1 and s_2, we show that it holds for $do(a, s_1)$ and $do(a, s_2)$:

$$do(a, s_1) = do(b, s_2) \vee \mathbf{BFS}(do(a, s_1), do(b, s_2)) \vee \mathbf{BFS}(do(b, s_1), do(a, s_2))$$

is equivalent to

$$[a = b \wedge s_1 = s_2] \vee$$
$$[s_1 = s_2 \wedge a \prec b] \vee \mathbf{BFS}(s_1, s_2) \vee$$
$$[s_1 = s_2 \wedge b \prec a] \vee \mathbf{BFS}(s_2, s_1),$$

which follows from the inductive assumption and the fact that $\prec$ is a strict linear order.

Property (9): by induction on s_1 and s_2 using the following facts about $<$:

$$\neg s < S_0 \wedge do(a, s_1) < do(b, s_2) \supset s_1 < s_2.$$

Now consider the procedure for "executing" a search algorithm given in Section 6. When applied to BFS, it is as follows:

1. Let $S = S_0$;
2. If S is a goal situation, then exit with success;
3. If there is a situation S' such that $\mathsf{Next}_{\mathsf{BFS}}(S, S')$ then let $S = S'$ and go back to step 2;
4. Exit with failure.

As we mentioned, the key operation in this procedure is the computation of $\mathsf{Next}_{\mathsf{BFS}}$. The following proposition shows that we can actually use depth-first search to compute $\mathsf{Next}_{\mathsf{BFS}}$. In the following, for any situation S, we define the *depth* of S, written $\mathsf{Depth}(S)$, to be the number of actions in the sequence of actions L such that $S = do(L, S_0)$. Obviously, $\mathsf{Depth}(S)$ is a unique finite number.

Proposition 5. *Let S be an executable situation with depth n.*

1. *If S' is the first executable situation with depth n such that $\mathsf{DCB}(S, S')$:*

$$\mathsf{Depth}(S') = n \wedge \mathsf{DCB}(S, S'),$$

$$\neg(\exists s).\mathsf{Depth}(s) = n \wedge \mathsf{DCB}(S, s) \wedge \mathsf{DCB}(s, S'),$$

then $\mathsf{Next}_{\mathsf{BFS}}(S, S')$.
2. *If there is no such S', and S'' is the first executable situation of depth $n + 1$ after S according to DCB:*

$$\mathsf{Depth}(S'') = n + 1 \wedge \mathsf{DCB}(S, S''),$$

$$\neg(\exists s).\mathsf{Depth}(s) = n + 1 \wedge \mathsf{DCB}(S, s) \wedge \mathsf{DCB}(s, S''),$$

then $\mathsf{Next}_{\mathsf{BFS}}(S, S'')$.
3. *Otherwise, there is no next situation after S: $\neg(\exists s)\mathsf{Next}_{\mathsf{BFS}}(S, s)$.*

Proof. We do induction on n, the depth of S. If $n = 0$, this means that $S = S_0$. It is easy to see that $\mathsf{Next}_{\mathsf{BFS}}(S_0, S')$ iff for some action A,

$$S' = do(A, S_0) \wedge Poss(A, S_0) \wedge \neg(\exists a)(Poss(a, S_0) \wedge a \prec A)$$

iff $\mathsf{Next}_{\mathsf{DCB}}(S_0, S')$. From these, it is easy to see that the result holds for S_0.

For the inductive case, suppose that the results hold for n. We show that they hold for $n + 1$ as well. First of all, we have

$$(\exists a)(A \prec a \wedge Poss(a, S)) \supset \{\mathsf{Next}_{\mathsf{BFS}}(do(A, S), do(B, S')) \equiv$$
$$S = S' \wedge Poss(B, S) \wedge A \prec B \wedge \neg(\exists a)(A \prec a \prec B \wedge Poss(a, S))\}.$$

and

$$\neg(\exists a)(A \prec a \wedge Poss(a, S)) \supset \{\mathsf{Next}_{\mathsf{BFS}}(do(A, S), do(B, S')) \equiv$$
$$\mathsf{Next}_{\mathsf{BFS}}(S, S') \wedge Poss(B, S') \wedge \neg(\exists a)(Poss(a, S') \wedge a \prec B)\}.$$

So

$$\text{Next}_{\text{BFS}}(s, s') \supset \text{Depth}(s') = \text{Depth}(s) \vee \text{Depth}(s') = \text{Depth}(s) + 1.$$

From this, and the following two facts:

$$\text{BFS}(s, s') \supset \text{Depth}(s) \leq \text{Depth}(s'),$$
$$\text{Depth}(s) \leq \text{Depth}(s') \wedge \text{DCB}(s, s') \supset \text{BFS}(s, s'),$$

the results follow.

If we use this proposition to compute Next_{BFS}, then we'll get essentially iterative-deepening search. So under our interpretation of search algorithm, iterative-deepening is not really a new search strategy, but a way of implementing breadth-first search. It trades time for space.

9   Best-First Search

Given a search problem, let h be its heuristic function, g its path cost function, and f an evaluation function on situations composed of h and g. The idea behind best-first search is rather simple: at any point, the search always expands a situation with the smallest f value. Notice that sometimes there may be more than one situations all having the same smallest f value. In this case, a "tie-breaking" rule is needed. In the following, we assume that Pref is a strict linear order on situations used as a "tie-breaker".

Let's denote by Heur the linear order corresponding to best-first search. Since by definition, Heur needs to be transitive and consistent with $<$, so for any situations s_1 and s_2, if there is a situation s such that $s < s_2$ and $\text{HeurEq}(s_1, s)$, then $heur(s_1, s_2)$, where for any s' and s'', $\text{HeurEq}(s', s'')$ is a shorthand for $\text{Heur}(s', s'') \vee s' = s''$. If there is no such s, then $\text{Heur}(s_1, s_2)$ if either $f(s_1) < f(s_2)$ or $f(s_1) = f(s_2)$ and s_1 is before s_2 according to the tie-breaking rule:

$$\begin{aligned}
\text{Heur}(s_1, s_2) \equiv\ & \text{Exec}(s_1) \wedge \text{Exec}(s_2) \wedge \\
& (\exists s).s < s_2 \wedge \text{HeurEq}(s_1, s) \vee \\
& \neg(\exists s).s < s_1 \wedge \text{HeurEq}(s_2, s) \wedge \\
& [f(s_1) < f(s_2) \vee (f(s_1) = f(s_2) \wedge \text{Pref}(s_1, s_2))]
\end{aligned} \tag{26}$$

Proposition 6. Heur *is a search algorithm.*

Proof. The proof of this proposition is very tedious, and is omitted here for lack of space.

Recall that according to Definition 3, given a search problem with goal G and cost function g, a search algorithm R is optimal if the first solution returned by it is always an optimal solution:

$$\text{First}_R(s) \supset \text{Optimal}(s).$$

For best-first search, the following proposition shows that it is optimal if it always explores goal situations with less costs.

Proposition 7. *If*

$$\mathsf{Exec}(s_1) \wedge \mathsf{Exec}(s_2) \wedge H(G, s_1) \wedge H(G, s_2) \wedge g(s_1) < g(s_2) \supset \mathsf{Heur}(s_1, s_2)$$

then Heur *is optimal.*

Proof. Let s be any situation such that $\mathsf{First}_{\mathsf{Heur}}(s)$. We show that $\mathsf{Optimal}(s)$ holds as well. Suppose otherwise, then by the definition of Optimal in Definition 3, there must be a situation s' such that

$$H(G, s') \wedge S_0 \leq s' \wedge g(s') < g(s).$$

By the assumption in the proposition, $\mathsf{Heur}(s', s)$. Since $\mathsf{First}_{\mathsf{Heur}}$, we have $\mathsf{Reachable}(s)$. Thus $\mathsf{Reachable}(s')$, a contradiction with $\mathsf{First}_{\mathsf{Heur}}(s)$.

The following proposition says that if Heur is complete, i.e. always returns a solution when one exists, then the condition in the above proposition is also a necessary condition for Heur to be optimal for all search problems.

Proposition 8. *If* Heur *is complete:*

$$(\exists s)(S_0 \leq s \wedge H(G, s)) \supset (\exists s)\mathsf{First}_{\mathsf{Heur}}(s),$$

then Heur *is optimal for all search problems iff*

$$\mathsf{Exec}(s_1) \wedge \mathsf{Exec}(s_2) \wedge H(G, s_1) \wedge H(G, s_2) \wedge g(s_1) < g(s_2) \supset \mathsf{Heur}(s_1, s_2).$$

Proof. The sufficiency of this condition has been proved in last proposition. Now suppose the condition is not true, i.e. there are two situations s_1 and s_2 such that

$$S_0 \leq s_1 \wedge S_0 \leq s_2 \wedge H(G, s_1) \wedge H(G, s_2) \wedge g(s_1) < g(s_2) \wedge \neg\mathsf{Heur}(s_1, s_2).$$

By the connectedness of Heur, we have have $\mathsf{Heur}(s_2, s_1)$. Now define a new goal G' such that $H(G', s) \equiv s = s_1 \vee s = s_2$. By the completeness of Heur, for this new search problem, $\mathsf{First}(s_2)$. This means that Heur is not optimal for this new problem.

Using the first proposition, we can show that if the cost function is additive and the heuristic function h is admissible, then the resulting Heur with evaluation function $f(s) = g(s) + h(s)$, the traditional A^* algorithm, is optimal.

We can also show that among all complete and optimal Heur algorithms for admissible problems, A^* expands the least set of situations. We will not give precise proofs here as these results are not new and can be found in [6].

Finally, just like iterative deepening can be thought of as an implementation of breadth-first search, many memory bounded A^* variants such as iterative deepening A^* can be considered to be special strategies for implementing A^*. In fact, Korf [7] shows that if f is monotonically increasing along every path, then iterative deepening A^* expands nodes in best-first order. Korf also shows that his recursive procedure called RBFS always expands nodes in best-first order as well.

10 Conclusions

We have presented a situation calculus account of search algorithms by treating them as strict linear ordering relations. Although AI search is a more mature area of AI, we think it can still benefit from such a formal account. In particular, we believe this account will be good for proving properties of various search methods, for comparing them, for finding new ways of implementing them, and for studying various search control and pruning strategies such as backjumping and dependency-based backtracking.

Acknowledgements

This work was supported in part by the Research Grants Council of Hong Kong under Competitive Earmarked Research Grants HKUST6091/97E and HKUST6145/98E.

References

1. A. Bundy and L. Sterling. Meta-level inference: two applications. *Journal of Automatic Reasoning*, 4(1):15–27, 1988.
2. K. Currie and A. Tate. O-Plan: the open planning architecture. *Artificial Intelligence*, 51(1):49–86, 1991.
3. R. Davis. Meta-rules: Reasoning about control. *Artificial Intelligence*, 15(3):179–222, 1980.
4. R. Davis and B. G. Buchanan. Meta-level knowledge: overview and applications. In *Proceedings of the International Joint Conference on Artificial Intelligence (IJCAI-77)*, pages 920–927, 1977.
5. T. Dean, J. Allen, and Y. Aloimonos. *Artificial intelligence : theory and practice*. Menlo Park, Calif. : Addison-Wesley Pub. Co., 1995.
6. R. Dechter and J. Pearl. Generalized best-first search strategies and the optimality of A*. *Journal of ACM*, pages 505–536, 1985.
7. R. E. Korf. Linear-space best-first search. *Artificial Intelligence*, 62(1):41–78, 1993.
8. F. Lin. An ordering on subgoals for planning. *Annals of Mathematics and Artificial Intelligence*, 21:321–342, 1997.
9. F. Lin. Applications of the situation calculus to formalizing control and strategic information: the prolog cut operator. *Artificial Intelligence*, 103:273–294, 1998.
10. F. Lin. On measuring plan quality. In *Proceedings of the Sixth International Conference on Principles of Knowledge Representation and Reasoning (KR'98)*, pages 224–233, 1998.
11. F. Lin and R. Reiter. State constraints revisited. *Journal of Logic and Computation, Special Issue on Actions and Processes*, 4(5):655–678, 1994.
12. J. McCarthy and P. Hayes. Some philosophical problems from the standpoint of artificial intelligence. In B. Meltzer and D. Michie, editors, *Machine Intelligence 4*, pages 463–502. Edinburgh University Press, Edinburgh, 1969.
13. N. J. Nilsson. *Principles of Articifial Intelligence*. Morgan Kaufmann, Los Altos, CA., 1980.
14. R. Reiter. Proving properties of states in the situation calculus. *Artificial Intelligence*, 64:337–351, 1993.

15. R. Reiter. *Knowledge in Action: Logical Foundations for Describing and Implementing Dynamic Systems*. In preparation, 1997.
16. S. Russell. Efficient memory-bounded search methods. In *Proceedings of ECAI'92*, pages 1–5, 1992.
17. S. J. Russell and P. Norvig. *Artificial intelligence : a modern approach*. Englewood Cliffs, N.J. : Prentice Hall, 1995.

Logic and Databases: a 20 Year Retrospective – Updated in Honor of Ray Reiter

Jack Minker

Department of Computer Science, Institute for Advanced Computer Studies
University of Maryland, College Park, MD 20742, USA
minker@cs.umd.edu, http://www.cs.umd.edu/~minker

Summary. The field of deductive databases is considered to have started at a workshop in Toulouse, France. At that workshop, Gallaire, Minker and Nicolas stated that *logic and databases* was a field in its own right (see [174]). This was the first time that this designation was made. The impetus for this started approximately twenty three years ago in 1976 when I visited Gallaire and Nicolas in Toulouse, France, which culminated in the Toulouse workshop in 1977. Ray Reiter was an attendee at the workshop and contributed two seminal articles to the book that resulted from the workshop. In this article I provide an assessment as to what has been achieved since the field started as a distinct discipline. I review developments in the field, assess contributions, consider the status of implementations of deductive databases and discuss future work needed in deductive databases.

As noted in [298], the use of logic and deduction in databases started in the late 1960s. Prominent among developments was work by Levien and Maron and Kuhns Green and Raphael [199] were the first to realize the importance of the Robinson Resolution Principle [369] for databases. Early uses of logic in databases are reported upon in [299] and are not covered here. Detailed descriptions of many of the accomplishments made in the 1960s can be found in [311].

---

[1]This article is a combination of a tutorial paper [303], "Logic and Databases: Past, Present, and Future," ©1997, American Association for Artificial Intelligence, which was a condensation with some additions of an invited keynote address presented at the *Workshop on Logic in Databases*, San Miniato, Italy, 1996. The longer version of the article appears in [302]. A number of my colleagues contributed to what they considered to be the significant developments in the field for the original articles, including: Robert Demolombe, Hervé Gallaire, Georg Gottlob, John Grant, Larry Henschen, Bob Kowalski, Jean-Marie Nicolas, Raghu Ramakrishnan, Kotagiri Ramamohanarao, Ray Reiter and Carlo Zaniolo. Many of my former and current students and Ph.D. students also contributed thoughts, including Sergio Alvarez, Chitta Baral, José Alberto Fernández, Terry Gaasterland, Parke Godfrey, Jarek Gryz, Jorge Lobo, Sean Luke, and Carolina Ruiz. I would like to thank Juergen Dix, Thomas Eiter, Georg Gottlob, Nicola Leone, Witek Marek, Mirek Truszczyński, and V.S. Subrahmanian for their assistance in new material contained in this updated paper. Although many of the views reflected in the article may be shared by those who made suggestions, I take full responsibility for them. This version of the paper is dedicated to my friend and colleague Ray Reiter. I have known and admired Ray's work since 1970. He has greatly influenced my own work.

1   Introduction

As noted in [298], the use of logic and deduction in databases started in the late 1960s. Prominent among developments was work by Levien and Maron [256], [259], [257] and Kuhns [238,239]. Green and Raphael [199] were the first to realize the importance of the Robinson Resolution Principle [369] for databases. Early uses of logic in databases are reported upon in [299] and are not covered here. Detailed descriptions of many of the accomplishments made in the 1960s can be found in [311].

A major influence of the use of logic in databases was the development of the field of logic programming through the work of Kowalski [234], who promulgated the concept of logic as a programming language, and by Colmerauer and his students who developed the first *Prolog* interpreter [107]. I refer to logic programs that are function-free as *deductive databases (DDBs)*, or as *Datalog*. I do so since databases are finite structures. Most of the results discussed can be extended to include logic programming.

The impetus for the use of logic in databases came about through meetings in 1976 in Toulouse, France, when I visited Gallaire and Nicolas while on sabbatical leave. The idea of a workshop on "Logic and Data Bases" was conceived at that time. It is clear that a number of individuals had the idea of using logic as a mechanism to handle databases and deduction at around that time and they were invited to participate in the workshop. The appearance of the book, edited by Gallaire and Minker, *Logic and Data Bases*[2] [174] was highly influential in the development of the field, as were the books [175], [176] that were a result of two subsequent workshops held in Toulouse. Another influential development was the article by Gallaire, Minker and Nicolas [177], which surveyed work in the field up to 1984.

The use of logic in databases was received by the database community with much skepticism. It is reported that a key researcher stated, "...logic and databases had little to do either with the theory or the practice of databases." It might also be well to recall some of the comments made in the review of the book *Logic and Data Bases*, [203]. The reviewer took issue with the statement in the Foreword to the book, where Gallaire and I called *logic and databases a field*. The reviewer believed that a *field* had to satisfy criteria that he set forth and that these criteria were not met by the book. He stated,

> More significant are the conscious efforts of the editors and most of the authors to promote the work in the volume as representing a field – a well-defined area of research, complete with important past achievements and stimulating areas for future work.

At the time the review came out, and in retrospect, I believe that Gallaire and I were correct in our assessment that *logic and databases* is, indeed, a *field*. In this chapter I describe 'important past achievements and stimulating areas for future work.'

---

[2] Nicolas was not one of the editors of the book by his choice. He played a key role in organizing the workshop.

The present volume is a collection of papers on a topic, some more interesting than others but none really outstanding...

While each individual may have different criteria for what is outstanding, that several papers in the book have become *classics*: the article by Reiter on the *Closed World Assumption (CWA)* [362] and by Clark on *Negation-as-Failure (NAF)* [103]. Indeed, several other important papers appeared in the book, such as those by Nicolas and Gallaire [323] on theory vs. interpretation, by Kowalski [235] on logic for data description, by Nicolas and Yazdenian [325] on integrity constraints, and several early implementations [87], [224], [295] of *DDBs*.

As confirmed by Papadimitriou, it is clear that logic has everything to do with the theory of databases, and many of those who were then critical of the field have changed their position.

Datalog, and its two main issues of query optimization and negation, took the field by storm (possibly because they had been brewing in other communities for some time).

In the remainder of this paper I shall describe what I believe to be the major intellectual developments in the field, the status of commercial implementations and future trends. As will be seen, the field of logic and databases has been very prolific. I apologize in advance to any author whose work I have inadvertently not covered or referenced.

2   Intellectual Contributions of Deductive Databases

In describing the contributions of logic to databases, it is necessary to note work in relational databases by Codd [106] who formalized databases in terms of the relational calculus and the relational algebra. He provided a logic language, the relational calculus, and described how to compute answers to questions in the relational algebra and the relational calculus. Both the relational calculus and the relational algebra provide declarative formalisms to specify queries. This was a significant advance over network [104] and hierarchic systems [417], [418] which only provided procedural languages for databases. The relational algebra and the relational calculus permitted individuals who were not computer specialists to write declarative queries, and to have the computer answer the queries. The development of syntactic optimization techniques (see [89], [417], [418] for references) permitted relational database systems to retrieve answers to queries efficiently and to compete with network and hierarchic implementations. Relational systems have been enhanced to include "views". A view as has been used in relational databases is, essentially, a non-recursive procedure. There are, today, numerous commercial implementations of relational database systems for large database manipulation and for personal computers. Relational databases are a forerunner of logic in databases.

Although relational databases used the language of logic in the relational calculus, it was not formalized in terms of logic. The formalization of relational databases in terms of logic and the extensions that have been developed are the focus of this paper. Indeed, *formalizing databases through logic has played a significant role in our understanding of what constitutes a database, what is meant by a query, what is meant by an answer to a query, and how databases may be generalized for knowledge bases.* It has also provided tools and answers to problems that would have been extremely difficult without the use of logic.

In the remainder of the paper I focus on some of the more significant aspects that have been contributed by logic in databases. I discuss:

1. A formalization of a database, a query, and an answer to a query.
2. A realization that logic programming extends relational databases.
3. A clear understanding of the semantics of large classes of databases that include alternative forms of negation, and disjunction.
4. An understanding of relationships between model theory, fixpoint theory and proof procedures.
5. An understanding of the properties that alternative semantics may have and their complexity.
6. An understanding of integrity constraints their use for updates, semantic query optimization, and cooperative answering.
7. A formalization and solutions to the update and view update problems.
8. An understanding of bounded recursion and recursion, and how they may be implemented in a practical manner.
9. An understanding of the relationship between logic based systems and knowledge base systems.
10. A formalization of incomplete information in knowledge bases.
11. A correspondence that relates alternative formalisms of nonmonotonic reasoning to databases and knowledge bases.

I address implementations of *DDBs* in section 3, where commercial developments have not progressed as rapidly as the intellectual developments. I then discuss trends and future directions in section 4. In Table 1 acronyms used in the chapter are listed.

2.1 Formalizing Database Theory

The first formalization of databases in terms of logic was due to Reiter [364] who noted that underlying relational databases there were a number of assumptions that were not made explicit. With respect to negation, an assumption was being made that facts not known to be *true* are assumed *false*. This assumption is Reiter's well-known *Closed World Assumption (CWA)*, expounded earlier in 1978 [362]. A second assumption, the *unique name assumption*, states that any item in a database has a unique name, and that individuals with different names are not the same. The last assumption, the *domain closure assumption*, states that there are no other individuals than those in the database.

Reiter then formalized relational databases as follows. He stated that a relational database is a set of ground assertions over a language $\mathcal{L}$ together with a set

CWA	Closed World Assumption
DB	Database
DDB	Deductive Database
DDDB	Disjunctive Deductive Database
D-WFS	Disjunctive Well-Founded Semantics
EDB	Extensional Database
EDDDB	Extended Disjunctive Deductive Database
FD	Functional Dependency
GCWA	Generalized Closed World Assumption
G_p	Dependency Graph
HCF	Head Cycle Free
KB	Knowledge Base
SQO	Semantic Query Optimization
WFS	Well-Founded Semantics
IC	Integrity Constraint
IDB	Itensional Database
UC	User Constraint

Table 1. List of Acronyms used in this Article

of axioms. The language $\mathcal{L}$ does not contain function symbols. These assertions and axioms are of the following form:

- **Assertions:** $R(a_1, \cdots, a_n)$, where R is an n-ary relational symbol in $\mathcal{L}$, and $a_1, \cdots, a_n$ are constant symbols in $\mathcal{L}$.

- **Unique Names Axiom:** If $a_1, \cdots, a_p$ are all the constant symbols of $\mathcal{L}$, then

$$(a_1 \neq a_2), \cdots, (a_1 \neq a_p), (a_2 \neq a_3), \cdots, (a_{p-1} \neq a_p)$$

- **Domain Closure Axiom:** If $a_1, \cdots, a_p$ are all the constant symbols of $\mathcal{L}$, then

$$\forall X((X = a_1) \vee \cdots \vee (X = a_p))$$

- **Completion Axioms:** For each relational symbol R, if $R(a_1^1, \cdots a_n^1), \cdots, R(a_1^m, \cdots, a_n^m)$ denote all facts under R, the completion axiom for R

$$\forall X_1 \cdots \forall X_n (R(X_1, \cdots, X_n) \rightarrow$$
$$(X_1 = a_1^1 \wedge \cdots \wedge X_n = a_n^1) \vee \cdots \vee (X_1 = a_1^m \wedge \cdots \wedge X_n = a_n^m))$$

- **Equality Axioms:**
$$\forall X \ (X = X)$$
$$\forall X \forall Y \ ((X = Y) \rightarrow (Y = X))$$
$$\forall X \forall Y \forall Z ((X = Y) \wedge (Y = Z) \rightarrow (X = Z))$$
$$\forall X_1 \cdots \forall X_n (P(X_1, \cdots, X_n) \wedge$$
$$(X_1 = Y_1) \wedge \cdots \wedge (X_n = Y_n) \rightarrow P(Y_1, \cdots Y_n))$$

Example 1, translates a small database to logic. Handling such databases through conventional techniques clearly leads to more efficient implementations. However, it serves to formalize previously unformalized databases.

Example 1. Consider the family database to consist of the *FATHER* relation with schema *FATHER(father,child)* and the *MOTHER* relation with schema *MOTHER(mother,child).* Let the database be:

FATHER	father	child
	j	m
	j	s

MOTHER	mother	child
	r	m
	r	s

The database translated to logic is given as follows, where we do not include the *Equality Axioms,* as they are obvious.

Assertions: $\{Father(j,m), Father(j,s), Mother(r,m), Mother(r,s)\}$, where $Father$ and $Mother$ are predicates and j, m, s, and r are constants.

Unique Name Axiom: $((j \neq m), (j \neq s), (j \neq r), (r \neq m), (r \neq s), (m \neq s))$

Domain Closure Axiom): $(\forall X)((X = j) \vee (X = m) \vee (X = s) \vee (X = r))$

Completion Axioms:
$(\forall X_1 \forall X_2)(Father(X_1, X_2) \leftarrow ((X_1 = j) \wedge (X_2 = m)) \vee ((X_1 = j) \wedge (X_2 = s)))$
$(\forall X_1 \forall X_2)(Mother(X_1, X_2) \leftarrow ((X_1 = r) \wedge (X_2 = m)) \vee ((X_1 = r) \wedge (X_2 = s)))$

The *completion axiom* was proposed by Clark [103] as the basis for his Negation-as-Failure rule. It states that the only tuples that a relation can have are those that are specified in the relational table. This statement is implicit in every relational database. The *completion axiom* makes this explicit. Another contribution of logic programs and databases is that:

Formalizing relational databases in terms of logic permits the definition of a query and an answer to a query to be defined precisely. A query is a statement in the first-order logic language $\mathcal{L}$. Q(a) is an answer to a query, Q(X), over a database DB if Q(a) is a logical consequence of DB.

2.2   Deductive Databases (DDBs)

Relational databases have been shown to be a special case of deductive databases. A deductive database may be considered as a theory, *DDB*, in which the database consists of a set of ground assertions, referred to as the *extensional database (EDB)*, and a set of axioms, referred to as the *intensional database (IDB)*, of the form:

$$P \leftarrow Q_1, \ldots, Q_n, \tag{1}$$

where $P, Q_1, \dots, Q_n$ are atomic formulae in the language $\mathcal{L}$. Databases of this form are termed *Datalog* databases [417], [418]. A *Datalog* database is a particular instance of a more general Horn logic program that permits function symbols in clauses given by Formula (1). The recognition that logic programs are significant for databases was understood by a number of individuals in 1976. Many of them have articles in the book [174]. The generalization permits views to be defined that are recursive.

The recognition that logic programming and databases are fundamentally related has led to more expressive and powerful databases than is possible with relational databases defined in terms of the relational algebra.

That logic programming and *DDBs* are fundamentally related is a consequence of the fact that databases are function-free logic programs. As shown in many papers and in particular, in Gottlob [190], the expressive power of logic programming extends that of relational databases.

ICs play an important role in database systems. In addition to defining a database in terms of an *EDB* and an *IDB*, it is necessary to formalize what is meant by an *IC*. Kowalski [382], [235] suggests that an integrity constraint is a formula that is *consistent* with the *DDB*, while for Reiter, and Lloyd and Topor [364], [271], an *IC* is a theorem of the *DDB*. Reiter proposed other definitions of *ICs* in [367], [368]. He states that *ICs* should be statements about the content of a database. *ICs* can be written in a modal logic to use a belief operator to express beliefs that the database must satisfy. He explores Levesque's KFOPCE [255], an epistemic modal logic, as a suitable framework. Demolombe and Jones [122], view *ICs* as statements *true* about the world, whereas a *DB* is a collection of *beliefs* about the world. *ICs* then can be used to qualify certain information in the *DB* as *valid* or *complete*. They define these properties formally in the framework of *doxatic logic*.

In *DDBs*, the semantics of the *DB* design are captured by *ICs*. Information about *functional dependencies (FDs)*–that a relation's key functionally determines the rest of the relation's attributes–can be written via *ICs*. For example, assume the predicate *flight* for an airline database, and that the attributes *AIRLINE* and *NO.* are a composite key for the relation. The *FD*, that departure time is functionally determined by airline and flight number, is given by:

$$DTIME[1] = DTIME[2] \Leftarrow flight(AIRLINE, NO., DTIME[1], -, \dots, -),$$
$$flight(AIRLINE, NO., DTIME[2], -, \dots, -),$$

where the symbol $\Leftarrow$ is used to distinguish a rule from an integrity constraint.

Likewise, *inclusion dependencies*, semantic information about a *DB's* design, are easily represented. Let the predicate *airport* record information about airports known to the *DB*. We want to ensure that any airport which serves as a departure or an arrival of any flight known to the database is also in the *airport* relation. The first of these–that the departure airport is known–could be represented as follows.

$$airport(-, \dots, -, FIELDCODE) \Leftarrow flight(-, \dots, -, FIELDCODE)$$

The major use of *ICs* in the DB community has been in updating to assure database consistency. Nicolas [322] has shown how, using techniques from *DDBs*, improvements can be made to the speed of update Reiter [362] showed that one can query a Horn *DB* with or without *ICs* and the answer to the query is the same. However, this does not preclude the use of *ICs* in the query process. While *ICs* do not affect the result of a query, they may affect computational efficiency. *ICs* provide *semantic* information about the data in the database. If a query requests a join for which there will never be an answer because of system constraints, this can be used to advantage by not performing the query and returning an empty answer set. This avoids an unnecessary join on two potentially large relational tables in a relational database system, or a long deduction in a *DDB* system. The process of using *ICs* to constrain a search is called *semantic query optimization (SQO)* [81,82]. McSkimin and Minker [289] were the first to use *ICs* for *SQO* in *DDBs*. Hammer and Zdonik [202] and King [229] were the first to apply *SQO* to relational databases. Chakravarthy, Grant and Minker [81,82] formalized *SQO* and developed the *partial subsumption algorithm* and method of *residues*. These provide a general technique applicable to any relational or *DDB* that is able to perform *SQO*. In [188] is is shown how to apply *SQO* bottom-up. Gaasterland and Lobo [168] extended the work to include *DBs* with negation in the body of clauses, and Levy and Sagiv [263] showed how to handle recursive *IDB* rules in *SQO*. [263] show that *SQO* can be done in recursive rules provided that order constraints or negated EDB subgoals appear only in the recursive rules, but not in the *ICs*. If either order constraints or negated EDB subgoals are introduced in *ICs*, then the problem of *SQO* becomes undecidable. Their result also applies to the containment problem of a *Datalog* program in a union of conjunctive queries. For *SQO* to be an effective tool, it will need to be integrated with conventional syntactic query optimization techniques ([89]) used for *SQL* queries in relational databases.

Early work on *SQO* by [289] relates to constraint logic programming by Jaffar and Lassez [215]. See [214] for a survey of constraint logic programming.

A topic related to *SQO* is that of *cooperative answering systems*. A cooperative answering system provides information to a user as to why a query succeeds or fails. When a query fails, the user, in general, cannot tell why the failure occurred. There may be several reasons: the database currently does not contain information to respond to the user, or there will never be an answer to the query. The distinction could be important to the user. Another aspect related to *ICs* is that of *user constraints (UCs)*. A user constraint is a formula, that models a user's preferences. It may constrain providing answers to queries in which the user may have no interest (e.g., stating that in developing a route of travel, the user does not want to pass through a particular city), or provide other constraints that may restrict the search. As shown by [173,172,169,171], *UCs* which are identical in form to integrity constraints, can be used for this purpose. While *ICs* provide the semantics of the entire database, *UCs* provide the semantics of the user. *UCs* may be inconsistent with the database and hence, a separation of these two semantics is essential. To maintain the consistency of the database, only *ICs* are relevant. A query may be thought of as the conjunction of the query and the *UCs*.Hence, a query can be semantically optimized based both on *ICs* and *UCs*.

As noted above, *ICs* are more versatile than just to represent dependencies. General semantic information can be captured as well. Assume that at the national airport in Washington, D.C., (DCA) that no flights are allowed (departures or arrivals) after ten at night or before eight in the morning as the airport is downtown so night flights would disturb city residents. This can be captured as an integrity constraint.

Such knowledge, captured and recorded via *ICs*, can be used to answer queries more intelligently and more informatively. If someone asks for flights out of DCA to, say, LAX leaving between 10:30 at night and midnight, a *DB* usually returns the empty answer set. (There will be no such flights if the *DB* is consistent with its constraints.) Better, however, would be for the *DB* system to inform the querier that there *can be no* such flights, because of Washington, D.C. flight regulations.

With *UCs* and *ICs* it is possible to develop a system that provides responses to users that inform them as to the reasons why queries succeed or fail [189]. Other features may be built into a system, such as the ability to relax a query given that it fails, so that an answer to a related request may be found. This has been termed *query relaxation* [171]. A survey of work in cooperative answering systems may be found in [170].

SQO, user constraints and cooperative answering systems are important contributions both for relational and DDB systems. I believe that they will eventually be incorporated into commercial relational and DDB systems.

Indeed, I cannot imagine a *DDB* developed for commercial systems to be successful if it does not contain both *SQO* and cooperative answering capabilities. How can users understand why deductions succeed or fail if such information is not provided? How can queries doomed to fail because they violate integrity or user constraints be allowed to take up significant search time if the query cannot possibly succeed? I also believe that these techniques must be incorporated into relational technology. As discussed in section 3, this is beginning to happen. Practical considerations of performing *SQO* bottom-up have been addressed by [188].

2.3   Extended Deductive Database Semantics

The first generalization of relational databases was to permit recursive Horn rules. A *Horn rule* is one in which the head of a rule is an atom and the body of a rule is a conjunction of atoms. These databases are referred to as *DDBs* or *Datalog DBs*. Subsequently, other *DDBs* that may contain negated atoms in the body of rules were permitted. In the following sections I discuss extensions made to *DDBs* and describe their significance.

Horn Semantics and Datalog   One of the early developments was due to van Emden and Kowalski [421] who wrote a seminal paper that discussed the semantics of Horn theories. I believe that a significant contribution to logic programming and *DDBs* was *the recognition by van Emden and Kowalski that the semantics of Horn* *databases can be characterized in three distinct ways by model, fixpoint or proof theory. These three characterizations lead to the same semantics.*

Model theory deals with the definition of a collection of models of a *DB* that captures the intended meaning of the *DB*. Fixpoint theory deals with the definition of a fixpoint operator that constructs the collection of all atoms that can be inferred to be *true* from the *DB*. Proof theory deals with the definition of a procedure that finds answers to queries with respect to the *DB*. van Emden and Kowalski [421] also showed that if one considers all Herbrand models of a Horn *DDB*, the intersection is a unique minimal model. The unique minimal model is the same as all of the atoms in the fixpoint, and are the only atoms provable from the theory.

To find if the negation of a ground atom is *true*, subtract, from the Herbrand base (the set of all atoms that can be constructed from the constants and the predicates in the *DB*), the minimal Herbrand model. An atom in this set is assumed to be *false* and its negation is *true*. Alternatively, answering queries that consist of negated atoms that are ground may be achieved using negation-as-finite failure as described by [362,103].

The first approaches to answering queries in *DDBs* did not handle recursion and were primarily top-down (or backward reasoning) or top-down using sets [174]. The approach to answering queries in relational database systems was a bottom-up (or forward reasoning) approach, since all answers are usually required and it is more efficient to do so in a bottom-up approach. Pioneering work in recursion in *DDBs* was done by Chang [87,88], McKay and Shapiro [394] and Henschen and Naqvi [204]. Both Chang, and McKay and Shapiro used a bottom-up approach to handle recursion, while Henschen and Naqvi used an iterated top-down approach. This work was followed by a large number of papers on computing recursion including the *naive* and *semi-naive evaluation methods* [20]. These methods are based on primitive deduction techniques and are generally inefficient. They perform bottom-up reasoning proceeding from the database. Hence, they do not use constants that may appear in a query to guide the search. Top down reasoning takes into account constants that appear in a query. The *QSQ* method of Vieille [426,428] and the extension tables method of Dietrich and Warren [123] are key representatives of this approach. On the other hand, the renaming of the Alexander [370] and *magic set* [21,32] methods make use of the constants for bottom-up reasoning. Bry [66] reconciles the bottom-up and top-down methods to compute recursive queries. He shows that the Alexander and magic set methods based on rewriting and the methods based on resolution implement the same top-down evaluation of the original database rules by means of auxiliary rules processed bottom-up. Based on the work by Bry, Brass [52] developed a rewriting method for *Datalog-programs* which simulates *SLD-resolution* more closely than the usual magic set method. It improves upon the method of Ross [374], and can save joins. Minker and Nicolas [305] were the first to show that there are forms of rules that lead to *bounded recursion*. That is, the deduction process using these rules must terminate in a finite number of steps. This work has been extended by Naughton and Sagiv [320]. Example 2 illustrates a rule that terminates finitely regardless of the state of the database.

Example 2. Bounded Recursion. If a rule satisfies the condition that it is *singular*, then it must terminate in a finite number of steps independent of the database state. A recursive rule is *singular* if it is of the form

$$R \leftarrow F \wedge R_1 \wedge \wedge \ldots \wedge R_n,$$

where F is a conjunction of possibly empty base (i.e. *EDB*) relations and $R, R_1, R_2, \ldots, R_n$ are atoms that have the same relation name iff:

1. each variable that occurs in an atom R_i and does not occur in R only occurs in R_i;
2. each variable in R occurs in the same argument position in any atom R_i where it appears, except perhaps in at most one atom R_1 that contains all of the variables of R.

Thus, the rule

$$R(X, Y, Z) \leftarrow R(X, Y', Z), R(X, Y, Z')$$

is singular since (a) Y' and Z' appear respectively in the first and second atoms in the head of the rule (condition 1), and (b) the variables X, Y, Z always appear in the same argument position (condition 2).

The efficient handling of recursion and the recognition that some recursive cases may inherently be bounded contributes to the practical implementation of deductive databases. An understanding of the relationship between resolution-based (top-down) and fixpoint-based (bottom-up) techniques and how the search space of the latter can be made identical to top-down resolution with program transformation is another contribution of DDBs.

Extended Deductive Databases and Knowledge Bases The ability to develop a semantics for theories in which there are rules with a literal (i.e., an atomic formula or the negation of an atomic formula) in the head and literals with possibly negated-by-default literals in the body of a clause, has significantly expanded the ability to write and understand the semantics of complex applications. Such clauses, referred to as *extended clauses*, are given by:

$$L \leftarrow M_1, \cdots, M_n, not \ M_{n+1}, \cdots not \ M_{n+k}, \tag{2}$$

where L and the $M_j, j = 1, \cdots, (n + k)$ are literals. *DBs* with such clauses combine both *classical negation* and *default negation* (represented by *not* immediately preceding a literal), and are called *extended deductive databases*. This extension provides users with greater expressive power.

Logic programs where default negation may appear in the body of a clause first appeared in a workshop held in 1986 [297]. Selected papers from the workshop appeared in 1988 [298]. The concept of stratification was discussed first by Chandra and Harel [85] and introduced to logic programs by Apt, Blair and Walker [13], and by Van Gelder [422] who considered stratified theories in which L and the M_j in Formula (2) are atomic formulas and there is no recursion through negation. For such theories they show there is a unique preferred minimal model, computed from strata to strata. Przymusinski [338] terms this minimal model the *perfect model* and extends the concept to *locally stratified* theories. Given a stratified theory, one can place clauses in different strata, where predicates in the head of a rule are in a higher stratum than predicates negated in the body of the clause. Thus, one can compute the positive predicates in a lower stratum and the negated predicate's complement is *true* in the body of the clause if the positive atom has not been computed in the lower stratum.

Example 3. Stratified Program. The rules,
$r_1 : p \leftarrow q, not\ r$
$r_2 : q \leftarrow p$
$r_3 : q \leftarrow s$
$r_4 : s$
$r_5 : r \leftarrow t$

comprise a stratified theory. Rule r_5 is in the lowest stratum, the other rules are in a higher stratum. The predicate p is in a higher stratum than $r's$ stratum as it depends negatively on r. q is in $p's$ stratum as as it depends upon p. s is also in the same stratum as q. The meaning of the program is: $\{s, q, p\}$ are *true*, while $\{t, r\}$ are *false*. t is *false* since there is no defining rule for t. Since t is *false*, r is *false*. s is given as *true*, and hence, q is *true*. Since q is *true*, and r is *false*, from *rule* r_1, p is *true*.

The theory of stratified *DDBs* was followed by permitting recursion through negation in Formula (2) where the L and M_j are atomic formulae. Such *DDBs* are called *normal deductive databases*. There have been a large number of papers devoted to defining the semantics of these databases. Some semantics, drawn from [314], are summarized in the second and third columns of Table 2. The most prominent of these for the Horn case, are the *well-founded semantics (WFS)* of Van Gelder, Ross and Schlipf [180], and the *stable semantics* of Gelfond and Lifschitz [182]. The *WFS* leads to a unique three-valued model. Stable semantics may lead to a collection of minimal models. For some *DDBs* this collection may be empty. Fitting [164] defined a three-valued model to capture the semantics of *normal DBs*.

Example 4. Non-Stratifiable Database. Consider the database given by:
$r_1 : p(X) \leftarrow not\ q(X)$
$r_2 : q(X) \leftarrow not\ p(X)$
$r_3 : r(a) \leftarrow p(a)$
$r_4 : r(a) \leftarrow q(a)$

Clauses r_1 and r_2 are recursive through negation. Hence, the database is not stratifiable. According to the WFS, $\{p(a), q(a), r(a)\}$ are assigned *unknown*. However, for the *stable model semantics*, there are two minimal models: $\{\{p(a), r(a)\}, \{q(a), r(a)\}\}$. Hence one can conclude that $r(a)$ is *true*, while the disjunct, $p(a) \lor q(a)$ is *true* in the stable model semantics.

A number of normal *DDB* semantics have been shown to be equivalent. You and Yuan [436], show that a number of extensions to the stable model semantics coincide: regular model semantics [434], partial stable model [381], preferential semantics [137], and a stronger version of the stable class semantics [30]. R-stable models have been proposed in [217]. See [146] for a discussion of how to introduce subprograms and modules into logic programming with stable models. [194] investigate the complexity of approximating the stable model semantics and show that unless P=NP, no approximation exists that uniformly bounds the intersection (union) of stable models.

There have been several implementations of the *WFS*. Chen and Warren [92] developed a top-down approach to answer queries in this semantics, while Leone and Rullo [249] developed a bottom-up method for *Datalog* databases. Sagonas, Swift and Warren [383] show that a fixed order of computation does not suffice for answering queries in the *WFS*. They introduce a variant of *SLG* resolution [412], SLG_{strat} which uses a fixed computation rule to evaluate ground left-to-right dynamically stratified programs. Warren, Swift and their associates [354] developed an efficient deductive logic programming system, *XSB*, that computes the well-founded semantics. *XSB* is supported to the extent of answering questions and fixing bugs within the time schedule of the developers. The system extends the full-functionality of *Prolog* to the *WFS*. *XSB* forms the core technology of a start-up company, XSB, Inc. whose current focus is application work in data cleaning and mining. In [413] it is shown how nonmonotonic reasoning may be done within *XSB* and describes mature applications in medical diagnosis, model checking and parsing.

Several methods have been developed to compute answers to queries in stable model semantics. Fernández et al. [162] develop a bottom-up approach based on *model trees*. Every branch of a model tree is a model of the database, where a node in a tree is an atom shared by each branch below that node. Nerode et al. [33,34] developed a method based on linear programming. [408] relate *WFS* and branch-and-bound to stable models. Also, Inoue et al., [210] developed a method to compute stable model semantics. Ruiz and Minker, [376], devised a procedure to construct the collection of partial (3–valued) stable models of a *DDB* and, with Seipel, developed a characterization of the partial stable models of a disjunctive deductive database. Bagai and Sunderraman [18] developed a bottom-up method to compute the Fitting model. For a discussion of the use of *WFS* for default logic, see [63]. See [192,152] for an approach to realize stable and *WF* semantics for the non-ground case, and [219] for computing minimal models without grounding. Leone, Rossi [248] extend traditional logic programming to *Ordered Logic (OL)* programming, to support classical negation and object-oriented constructs. See [68], [250] for a se-

Semantics	Horn		Disjunctive	
	Theory	Reference	Theory	Reference
Positive Consequences				
Fixpoint	T_P	[421]	T_P^I	[308]
Model	Least	[421]	Min. Models	[296]
	Model		Model-State	[348]
Procedure	SLD	[206]	SLI/SLO	[312,348]
			Case Based	[360]
Negation				
Theory	CWA	[362]	GCWA	[296,308]
			WGCWA	[375,349]
Rule	NAF	[103]	SN-rule	[306]
			NAFFD-rule	[349]
Procedure	SLDNF	[103]	SLONF	[348]
Stratified Programs				
Fixpoint	T_P	[13]	T_P^C	[308]
			T_P^I	[375,307]
Model	Standard	[13]	Stable State	[307]
	Perfect	[342]	Perfect	[338]
Procedure	SLS	[339]	SLP	[161]
Normal Programs				
	Well-Founded		Strong/Weak Well-F/Stationary	
Fixpoint	I^∞	[181]		
Model	$M_{WF}(P)$	[181]	$M_{WF}^{S/W}(P)$	[371]
			M_P	[337]
Procedure	SLS	[372,343]		
	General Well-Founded		General Disj. Well-Founded	
Fixpoint	I^E	[27]	S^{ED}	[28]
Model	M_P^E	[29]	MS_P^{ED}	[29]
Procedure	SLIS	[28]	SLIS	[337]
	Stable Models		Stable Models	
Fixpoint	T_P^M	[162,210]	T_P^M	[162,210]
Model	Stable	[182]	Stable	[340,341]
Procedure	SLP	[161]	SLP	[161]
Extended Programs				
	Stable Models		Stable Models	
Fixpoint				
Model	Stable	[183,184]	Stable	[340,341]
Procedure				
	Stationary		Stationary	
Fixpoint	S^{SE}	[314]	S^{SE}	[314]
Model	Stationary	[9]		
Procedure				
	Arbitrary Semantics SEM		Arbitrary Semantics SEM	
Fixpoint	T_P^E	[314]	T_P^E	[314]
Model	$\mathcal{M}_P^{SEM}$	[314]	$\mathcal{M}_P^{SEM}$	[314]
Procedure	EPP_{SEM}	[314]	EPP_{SEM}	[314]

Table 2. (Taken from [314]) Semantics of Horn and Disjunctive Logic Programs

mantics and algorithm to compute stable models and *WFS* to obtain a nonmonotonic *DDB* system.

Three important implementations of stable model semantics have been implemented by Marek and Truszczyński [95],[96],[97],[285], by Niemelä and Simons [327],[328], [326], and by Eiter and Leone [147], [148]. Marek and Truszczyński developed a program, *Default Reasoning System (DeReS)*, that implements Reiter's default logic. It computes extensions of default theories. As logic programming with stable model semantics is a special case of Reiter's default logic, *DeReS* also computes stable models of logic programs. To test *DeReS*, a system, called *Theory-Base*, was built to generate families of large default theories and logic programs that describe graph problems such as existence of colorings, kernels and hamiltonian cycles.

Niemelä and Simons developed a system, *smodels*, to compute stable models of programs in *Datalog* with negation. At present, smodels is considered the most efficient implementation of stable model computation. However, comparative studies are needed. The *smodels* are based on two important ideas: intelligent grounding of the program, limiting the size of the grounding, and use of the *WFS* computation as a pruning technique.

Eiter and Leone developed a system, *dlv (DataLog with Or)*, that computes answer sets (in the sense of Gelfond and Lifschitz) for disjunctive logic programs in the syntax generalizing *Datalog* with negation. As *smodels*, *dlv* also uses a very powerful grounding engine and some variants of *WFS* computation as a pruning mechanism. The method used to compute *disjunctive stable models* is described in [247].

A further extension of normal *DDBs*, proposed by Gelfond and Lifschitz [183] and by Pearce and Wagner [330], permits clauses in Formula (2) where L and M_j are literals and therefore combines classical and default negation in one database. Blair and Subrahmanian [40] use the same idea as in [183], [330] to handle literals in paraconsistent logic programs. The semantics for normal deductive databases has been described by [309], [314].

These notions of default negation have been used as separate ways to interpret and to deduce default information. That is, each application has chosen one of these notions of negation and has applied it to every piece of data in the domain of the application. In [310], [377], and Ruiz define a new class of *DDBs* that allow for the combination of several forms of default negation in the same database. In this way different pieces of information in the domain may be treated appropriately. They introduce the *well–founded stable* semantics that characterizes the meaning of DDBs that combine the well–founded and the stable semantics.

Work on the implementation of semantics relating to extended DDBs has been very impressive. These systems can also be used for nonmonotonic reasoning. Brewka and Niemelä [64] state,

> *The participants in the plenary panel identified the following major trends in the field: First, serious systems for nonmonotonic reasoning are now available (XSB, SMODELS, DLV). Second, people outside the community*

are starting to use these systems with encouraging success (for example, in planning). Third, nonmonotonic techniques for reasoning about action are used in highly ambitious long-term projects (for example, the WITAS Project, www.ida.liu.se/ext/witas/eng.html). Fourth, causality is still an important issue; some formal models of causality have surprisingly close connections to standard nonmonotonic techniques. Fifth, the nonmonotonic logics being used most widely are the classical ones: default logic, circumscription, and autoepistemic logic.

Schlipf [389] has written a comprehensive survey that summarizes complexity results for deductive databases. Some results, taken from [389], are listed in Table 3. A user may wish to determine which semantics to be used based upon its complexity.

The development of the semantics and complexity results of extended DDBs that permit a combination of classical negation and multiple default negations in the same DDB are important contributions to database theory. They permit wider classes of applications to be developed.

Knowledge bases are important for A.I. and expert system developments. A general way to represent knowledge bases is through logic. All work developed for extended *DDBs* concerning semantics and complexity apply directly to knowledge bases. Baral and Gelfond [22] describe how extended *DDBs* may be used to represent knowledge bases. For an example of a knowledge base, see Example 5. Extended *DDBs*, together with *ICs* permit a wide range of *knowledge bases* to be implemented. Many papers devoted to knowledge bases consider them to consist of facts and rules. Certainly, this is one aspect of a knowledge base, as is the ability to extract proofs. However, *ICs* supply another aspect of knowledge and differentiate knowledge bases which may have the same rules, but different *ICs*. I believe that one should *define a knowledge base to consist of an extended deductive database plus integrity constraints.*

Since alternative extended deductive semantics have been implemented, the knowledge base expert should focus on specifying rules and *ICs* that characterize the database, selecting the particular semantics that meets the needs of the problem, and employing a *DDB* system that uses the required semantics.

The field of deductive databases has contributed to providing an understanding of knowledge bases and their implementation.

2.4   Extended Disjunctive Deductive Database Semantics

In databases discussed above, information is definite. However, there are many situations where our knowledge of the world is incomplete. For example, when a *null value* appears as an argument of an attribute of a relation, the value of the attribute is unknown. Uncertainty in databases may be represented by probabilistic information. See [241], [409], [321,117], [116] for additional work on probabilistic logic programming. Another area of incompleteness arises when it is unknown as to which among several facts are *true*, but it is known that one or more are *true*. It is therefore necessary to be able to represent and understand the semantics of theories that

include incomplete data. A natural way to extend *DBs* is to permit disjunctive statements as part of the language. This leads to deductive *DBs* which permit clauses
with disjunctions in their heads, represented as,

$$L_1 \vee L_2 \vee \dots \vee L_m \leftarrow M_1, \dots, M_n, not\ M_{n+1}, \dots not\ M_{n+k}, \qquad (3)$$

and referred to as *extended disjunctive clauses*. A database that consists of such
clauses is referred to as an *extended disjunctive deductive database (EDDDB)*. The
book by Lobo, Minker and Rajasekar [273] describes the theory of disjunctive logic
programs and includes several chapters devoted to *disjunctive deductive databases
(DDDBs)*, which consist of clauses of the form 3, where there is no default negation operator in the right hand side of the clause and all literals are atoms, are also
referred to as *Disjunctive Datalog*. Eiter, Gottlob and Mannila [145] study three different semantics for *Disjunctive Datalog*: the minimal model, the perfect model and
the stable model semantics. They show that disjunctive Datalog is more expressive
than normal logic programming with negation. The following example illustrates
the use of such a theory of databases

Example 5. Knowledge Base [22]. Consider the following database, where the predicate $p(X, Y)$ denotes that *X is a professor in department Y*, $a(X, Y)$ denotes that
individual X has an account on machine Y, $ab(W, Z)$ denotes that *it is abnormal in
rule W to be individual Z.*

We wish to represent the following information where *mike* and *john* are *professors* in the *computer science department.*:

(i) As a rule, professors in the computer science department have Vax accounts. This rule is not applicable to Mike. He may or may not have an
account on that machine.
(ii) Every computer science professor has one of the Vax or IBM accounts,
but not both.

These rules can be captured in the following disjunctive database.

1. $p(mike, cs) \leftarrow$
2. $p(john, cs) \leftarrow$
3. $\neg p(X, Y) \leftarrow not\ p(X, Y)$
4. $a(X, vax) \leftarrow p(X, cs), not\ ab(r4, X), not\ \neg a(X, vax)$
5. $ab(r4, mike) \leftarrow$
6. $a(X, vax) \vee a(X, ibm) \leftarrow p(X, cs), ab(r4, X)$
7. $\neg\, a(X, ibm) \leftarrow p(X, cs), a(X, vax)$
8. $\neg\, a(X, vax) \leftarrow p(X, cs), a(X, ibm)$
9. $a(X, ibm) \leftarrow \neg a(X, vax), p(X, cs)$

Rule 3. states that if by default negation predicate $p(X, Y)$ fails, then $p(X, Y)$ is
logically *false*. The other rules encode the satements listed above.

From this formalization one can deduce that *john* has a *vax* account, while *mike*
has either a *vax* or an *ibm* account, but not both.

Below I discuss the semantics of *DDDBs*, where clauses are given by Formula (3), where the literals are restricted to atoms and there is no default negation in the body of a clause. I then discuss the semantics of *EDDDBs*, where there are no restrictions on clauses in Formula (3).

Disjunctive Deductive Databases (DDDBs)   Work in disjunctive theories was pursued seriously after a workshop I organized in 1986 [297]. The field of *disjunctive deductive databases (DDDBs)* started approximately in 1982 with the the paper by Minker [296], who described how one can answer both positive and negated queries in such databases. For a historical perspective of disjunctive logic programming and *DDDBs*, see [299]. There is a major difference between the semantics of *DDDBs* and those for *DDDBs*. Whereas *DDBs* usually have a unique minimal model that describes the meaning of the database, *DDDBs* generally have multiple minimal models.

As shown in [296] it is sufficient to answer positive queries over *DDDBs* by showing that the query is satisfied in every minimal model. Thus, in the *DDDB*, $\{a \vee b\}$, there are two minimal models, $\{\{a\}, \{b\}\}$. The query, $a?$, is not satisfied in model $\{b\}$, and hence, a cannot be *true*. However, the query, $(a \vee b)$ is satisfied in both minimal models and hence the answer to the query $\{a \vee b\}$ is *yes*. To answer negated queries, it is not sufficient to use Reiter's *CWA* [362] since, as he noted, from the theory $DB = \{a \vee b\}$, it is not possible to prove a, and it is not possible to prove b. Hence, by the *CWA*, *not a* and *not b* follow. But, $\{a \vee b, not\ a, not\ b\}$ is not consistent. The *Generalized Closed World Assumption (GCWA)*, [296] resolves this problem by specifying that a negated atom be considered *true* if the atom does not appear in any minimal model of the database. This provides a model theoretic definition of negation. An equivalent proof theoretic definition, also in [296], is that an atom a may be considered to be *false* if, whenever $a \vee C$ may be proven from the database, then C may be proven from the database, where C is an arbitrary positive clause.

For related work on negation in disjunctive theories see [431], [186], [83], [384], [411], [47], [375], [349]. For surveys on negation see [395], [14], [131], [301].

In *DDDBs*, it is natural for the fixpoint operator to map atoms to atoms. However, for *DDDBs*, it is natural to map positive disjunctions to positive disjunctions. A set of positive disjunctions is referred to as a *state*. A *model state* is a state all of whose minimal models satisfy the *DDDB*. The concept of a state was defined by Minker and Rajasekar [308] as the domain of a fixpoint operator T_P whose least fixpoint characterizes the semantics of a disjunctive logic program P. The operator is shown to be monotonic and continuous, and hence converges in ω iterations. The fixpoint computation operates bottom-up and yields a minimal model state logically equivalent to the set of minimal models of the program. The Minker/Rajasekar fixpoint operator is an extension of the van Emden/Kowalski fixpoint operator. If one considers all model states of a *DDDB* and intersects them, the resultant is a model state, and among all model states it is minimal. Hence, one obtains a unique minimal model in a Horn database, while one obtains a unique model state in a *DDDB*.

Decker [115] develops an alternative fixpoint operator for *DDDBs* which reduces to the Minker/Rajasekar fixpoint operator [308]. At each iteration of his operator, he finds partial models of the database. In the limit, he obtains the set of minimal models of the database. If one takes an atom from each minimal model and forms a disjunction, the resulting set of all such disjunctions is equivalent to the minimal model state of the *DDDBs*.

Answering queries in *DDDBs* has been studied by a number of individuals. Grant and Minker [304] were among the first to address the problem of computing answers to queries in *DDDBs*. They investigated the case where the database consists exclusively of ground positive disjuncts. Yahya and Henschen [431] developed a deductive method to determine whether or not a conjunction of ground atoms can be assumed *false* in a *DDDB* under the *Extended Generalized Closed World Assumption (EGCWA)*. The *EGCWA* is an extension of the *GCWA*. Bossu and Siegel [47] developed a deductive method to answer a query by subimplication (a generalization of the *GCWA* that handles databases that have no minimal models). Henschen and Park [205] answer *yes/no* questions in a database that consists of an *EDB*, an *IDB* and *ICs* that are all function-free. In addition, they allow negated unit clauses to be part of the database. The axioms in the *IDB* may be recursive. Yahya [432] discusses how to answer queries defined as sets of clauses in implication form in a *DDDB*. Liu and Sunderraman [269] generalize the relational model to represent disjunctive data. They develop a data-structure, called *M-table* to represent the data. Their generalized relational algebra operates on *M-tables*, however, it is sound, but not complete. Yuan and Chiang [437] developed a generalized relational algebra that is a sound and complete query evaluation algorithm for *DDDBs* that do not contain recursive *IDB* rules.

Fernández and Minker [163] developed the concept of a *model tree*. They incrementally compute sound and complete answers to queries in *hierarchical DDDBs*. An example of a *model tree* is shown in Figure 7. A *DDDBs* is hierarchical if it contains no recursion. In [162] they develop a fixpoint operator over trees to capture the meaning of a *DDDB* that includes recursion. The tree representation of the fixpoint is equivalent to the Minker/Rajasekar fixpoint [308]. They compute the model tree of the extensional *DDDB* once for all queries. To answer queries *intensional database* rules may be invoked. Their approach to compute answers generalizes both to stratified and normal *DDDBs*.

Example 6 (Model Tree).
    Consider the following example given by the database: $\{a(1); a(2) \vee b(2); b(1) \vee b(2)\}$. There are two minimal models: $\{\{a(1), a(2), b(1)\}, \{\{a(1), b(2)\}\}$, written in tree form in Figure 7:

The above approaches to answering queries in *DDDBs* have the following limitations. [304] can only compute answers to queries that contain a disjunctive extensional database. [431], [205], [93] can only answer *yes/no* questions. [268], [269] provide sound, but not complete answers to queries. [437] essentially compute the fixpoint of the entire *DDB* to answer each query. [163] compute the model tree of

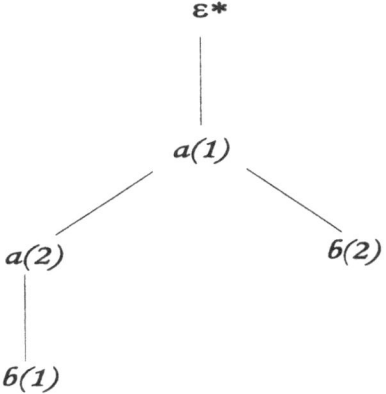

Fig. 1. Model Tree

the extensional *DDDB* once. To answer queries *IDB* rules may be invoked. However, the models of the extensional disjunctive part of the database do not have to be generated for each query.

Loveland et al. [272], [277], [399], [359], [360] developed a top-down case-based reasoner that uses *Prolog* when the database is *near Horn*. Loveland et al. [276] introduced a relevancy detection algorithm to be used with *SATCHMO*, developed by Manthey and Bry [282], for automated theorem proving. Their system, *SATCHMORE (SATCHMO with RElevancy)*, improves *SATCHMO* by limiting the uncontrolled use of forward chaining. Another approach is by Stickel using a theorem prover *PTTP (Prolog Technology Theorem Prover)* [405]. Seipel [391] developed a system *DisLog* that incorporates different disjunctive theories and strategies.

Imielinski and Vadaparty [209], Vardi [424] and Imielinski [207] have investigated the complexity of answering queries in disjunctive logic programs. Chomicki and Subrahmanian [99] discuss the complexity of the *GCWA*. For disjunctive theories that are tractable, see [46], [35], [110], [274]. For complexity results for disjunctive propositional logic programs see Eiter and Gottlob [141] and [112]. A summary of complexity results, drawn from [141], is given in Table 4.

The development of model theoretic, fixpoint and proof procedures has placed the semantics of DDDBs on a firm foundation. Methods to handle DDDBs have started and should lead to enhanced implementations. The GCWA and alternative theories of negation have enhanced our understanding of default negation in DDDBs. Complexity results provide an understanding of the difficulties to find answers to queries in such systems.

Extended Disjunctive Deductive Databases Fernández and Minker [159] present a new fixpoint characterization of the minimal models of *DDDBs* and stratified *DDDBs*. They prove that by applying the operator iteratively, in the limit, it constructs the perfect models semantics (Przymusinski [338]) of stratified *DDDBs*.

Given the equivalence between the *perfect models semantics of stratified programs* and *prioritized circumscription* [338] their fixpoint characterization captures the meaning of the corresponding circumscribed theory. They present a bottom-up evaluation algorithm for stratified *DDDBs* using the *model-tree* data structure to represent and to compute answers to queries. In [157], [158], they develop the theory of *DDDBs* using model trees. Work on updates in *DDDBs* is described in [198],[196], [160].

Four alternative semantics were developed for non-stratifiable normal *DDDBs* at approximately the same time: Ross [371], Baral et al. [29], [28], and two semantics by Przymusinski [337], [340]. Ross termed his semantics the *strong well founded semantics*, Baral et al. defined their semantics the *Generalized Disjunctive Well-Founded Semantics (GDWFS)*. They defined a fixpoint operator, and gave model and proof theoretic semantics for such *DDDBs*. Przymusinski [340] extends *stable model semantics* for normal *DDDBs*. He also defined in [337] the *stationary semantics*. As in the case of normal *DDBs* it will be necessary to develop effective bottom-up computing techniques to answer queries in these theories.

In addition, other important semantics have been developed. Przymusinski [344] describes a new *semantic framework* for disjunctive logic programs and introduces the *static expansions* of disjunctive programs. The class of static expansions extends both the classes of stable, well-founded and stationary models of normal programs and the class of minimal models of disjunctive programs. Any static expansion of a program P provides the corresponding semantics for P consisting of the set of all sentences logically implied by the expansion. The stable model semantics has also been extended to disjunctive programs [184], [341]. Leone et al., [252], develop an efficient algorithm for solving the (co-NP-hard decision) problem of checking if a model is stable. It runs in polynomial space and single exponential time (in the worst case) on the class of *head-cycle free programs* (discussed below), and in the case of general disjunctive logic programs limits the inefficient part of the computation only to components of the program which are not head-cycle free. Leone et al. [251], [247] extend the notion of unfounded sets from normal to disjunctive logic programs and provide a declarative characterization of disjunctive stable models in terms of unfounded sets. They define an algorithm to compute stable model semantics of disjunctive logic programs. [69], [70] extend *DisjunctiveDatalog* to include *ICs* and so-called weak constraints that are satisfied if possible.

The D-WFS semantics of Brass and Dix [53], [58] is of interest as it permits a general approach to bottom-up computation in disjunctive programs. In addition, their transformation approach leads to several confluent calculi ([57]) which leads both to a better understanding ([56]) and an efficient computation of such semantics ([132]). In [133] this approach was extended to first-order programs and coupled with constraint logic programming techniques. A restriction of the transformation approach to normal programs yields an implementation of the *WFS* which is provably better than the alternating fixpoint procedure and is linear for almost all programs occurring in practice ([59]).

Extensions of disjunctive programs to more general belief programs are considered in [61], [60]. Buccafurri et al. extend Ordered Logic programming (OL) (see section 2.3) to disjunctive theories. They relate the work to knowledge base systems, show the language *DOL (Disjunctive Ordered Logic)* to be useful for diagnostic processes based on stepwise refinements and study the expressive power and complexity of DOL.

Systems that implement disjunctive semantics are:

1. *near Horn*, headed by Loveland and implemented in *Prolog* [272], [277], [399], [359], [360].
2. *DisLoG*, headed by Seipel [390], which aims at implementing several *SLI resolution* based semantics introduced in [273].
3. *DisLoP*, headed by Dix and Furbach [15], [16], which aims at extending the *restart model elimination* and *hyper tableau calculi*, for disjunctive logic programming under the D-WFS and stable semantics.
4. *dlv*, headed by Eiter, Gottlob and Leone [147], [148], which implements the disjunctive stable semantics and is based on subtle complexity-theoretic considerations.

Inoue and Sakama develop comparisons between abductive and disjunctive programs [211]. They give translations between abductive theories and disjunctive programs.

The last two columns of Table 2 summarize work in the semantics of *DDDBs*, and is drawn from [314]. See Eiter et al. [149], [150] for complexity results for partial models of *DDDBs*. They also summarize results for partial stable models [341]; maximum stable models (M-stable) which are partial stable models under set inclusion [149], [378]; regular models [435] which are similar in spirit to M-stable models, but based on a weaker concept; and least undefined stable models [149], [378] which are the partial stable models with the minimal degree of undefinedness. See [151] for complexity results on partial stable models.

As noted previously, there are semantics both for extended *DDBs* and extended *DDDBs*. A user of such a system has the problem of selecting the appropriate semantics for his needs. Which semantics should be used, and under what circumstances? There have been no guidelines developed. However, many complexity results have been obtained for these semantics. Schlipf [389] and Eiter and Gottlob [141] summarize complexity results known for alternative semantics. Some of these results, taken from [139], [141], are listed in Table 3. A user may wish to determine the semantics to be used based upon the complexity expected to find answers to queries.

Ben-Eliyahu and Dechter [35] investigate tractable cases of disjunctive theories. They introduced the concept of a *head-cycle free (HCF)* clause as follows. Let a clause consist of a disjunction of literals. A *dependency graph G_P* is associated with each program P as follows:

- each clause of the form, Formula (2) and each predicate in P is a node.

- there is a positive (negative) arc from a predicate node p to a rule node δ iff p appears positive (negative) in the body of δ, and an arc from δ to p (resp., and also an arc from p to δ) if p appears in the head of δ.

The *positive dependency graph* of P is a subgraph of G_P containing only positive arcs. A directed cycle in G_P is called *negative* if it contains at least one negative arc. A *DDDB P* is *head-cycle free (HCF)* if for every two predicate names p and q, if p and q are on a positive directed cycle in the dependency graph G_P then there is no rule in P in which both p and q appear in the head. They show in [37] that answers to queries expressed in this language can be computed in polynomial time. The language is sufficiently powerful to express all polynomial time queries. It is shown in [36] that there is an algorithm that performs, in polynomial time, minimal model finding and minimal model checking if the theory is *HCF*. An efficient algorithm to solve the (co-NP-hard) problem of checking if a model is stable in function-free disjunctive logic programs is developed in [252]. The algorithm runs in polynomial time on *HCF* programs and in the case of general disjunctive logic programs, it limits the inefficient part of the computation only to the components of the program which are not *HCF*.

Dix et al. [125] describe *causal* programs, where disjunction is simulated by negation-as-failure. Disjunctive programs are reduced to *stratified nondisjunctive* programs by a series of *shift-operations*. They show *causal* semantics belongs to the first level of the polynomial hierarchy unlike minimal model semantics (GCWA), which is Π_2^P-complete for positive disjunctive programs. *Causal semantics* are also *cumulative* and *rational*. The class of *positive causal* programs extends the class of *positive HCF* programs [35].

Consideration has been given to approximate reasoning. In such reasoning, one may give up soundness or completeness of answers. Efforts have been developed both for deductive and disjunctive deductive databases by Selman and Kautz [392], [223], [393], who developed lower and upper bounds for Horn (*Datalog*) databases and compilation methods, by Cadoli [74], who developed computational and semantical approximations, and by del Val [118], [119], who developed techniques for approximating and compiling databases. See Cadoli [75] for references on compilation, approximation and tractability of knowledge bases.

A second way to determine the semantics to be used for and application is through their properties. Dix in [126], [127] proposed criteria useful in determining the appropriate semantics to be used. He developed semantics both for normal *DDBs* [129], [130] and normal *DDDBs* [128], [54] that satisfy some of the properties that he describes. While some properties are adaptations and extensions to those developed by Kraus et al. [237] to compare nonmonotonic theories, *relevance*, *partial evaluation* and *modularity* were newly developed.

A property an arbitrary semantics, *SEM*, might have is that its semantics should not be changed if a tautology is eliminated from its database. Table 5 summarizes other useful properties of semantics of *DDDBs* and specifies for alternative semantics the properties that they satisfy. This table is adapted from tables in [124], [131], [55], [56]. Although complexity results and properties that a semantics satisfy are

Semantics	Propositional		First Order over Herbrand models		First Order no function symb. over Herbrand models	
	Complexity	Ref.	Complexity	Ref.	Data Complexity	Ref.
Positive Consequences						
Minimal Model	$\mathcal{O}(|P|)$	[135], [213]	r.e.–complete	[400], [401,11]	polynomial in $|E|$	[84]
Negation						
CWA	$\mathcal{O}(|P|)$	[135], [213]	co–r.e.–complete	[400], [401], [11]	co–r.e.–complete	[84]
Stratified Programs						
Perfect	$\mathcal{O}(|P|)$		complete arithmetic	[12]	polynomial in $|E|$	[85]
Locally Stratified Programs						
Perfect	$\mathcal{O}(|P|)$		Δ_1^1–complete over ω	[39]	N/A	
Normal Programs						
2–valued completion	co–NP–complete	[230]	Π_1^1–complete over ω	[230]	co–NP–complete	[230]
3–valued completion	$\mathcal{O}(|P|)$	folklore	Π_1^1–complete over ω	[164]	polynomial in $|E|$	[164]
Stable	co–NP–complete	[284]	Π_1^1–complete over ω	[283], [32]	co–NP–complete	[284]
Well–Founded	$\mathcal{O}(|A||P|)$	folklore	Π_1^1–complete over ω	[179], [32]	polynomial in $|E|$	[181], [179]
Extended Programs						
Stable	co–NP–complete	[314]	Π_1^1–complete over ω	[314]	co–NP–complete	[314]
Well–Founded	$\mathcal{O}(|A||P|)$	[314]	Π_1^1–complete over ω	[314]	polynomial in $|E|$	[314]

Table 3. (Adapted from [389]) Complexity of Horn Logic Programs

Notation: The complexity results in the above table refer to worst case analysis for skeptical reasoning, i.e. to determining if a given literal is *true* in every canonical model (with respect to a particular semantics) of the program. For logic programs with no function symbols, the data complexity over an *EDB* E is presented. The notation used is the following: $|P|$ denotes the length of the program P; $|A|$ denotes the number of propositional letters in P; $|E|$ denotes the total number of symbols that occur in the *EDB* E.

Semantics	Propositional		First Order over Herbrand models		First Order no function symb. over Herbrand models	
	Complexity	Ref.	Complexity	Ref.	Data Complexity	Ref.
Positive Consequences						
Minimal Models	Π_2^P–complete	[139]			Π_2^P–complete	[144]
Negation						
GCWA	Π_2^P–complete	[139]	Π_2^0–complete	[99]	Π_2^P–complete	[144]
WGCWA	co–NP–complete	[83]				
Stratified Programs						
Perfect	Π_2^P–complete	[139]			Π_2^P–complete	[144]
Locally Stratified Programs						
Perfect	Π_2^P–complete	[139]			Π_2^P–complete	[144]
Normal Programs						
Stable	Π_2^P–complete	[139]			Π_2^P–complete	[144]
Partial Stable	Π_2^P–complete	[139]			Π_2^P–complete	[144]
Extended Programs						
Stable	Π_2^P–complete	[140]			Π_2^P–complete	[144]
Partial Stable	Π_2^P–complete	[314]			Π_2^P–complete	[144]

Table 4. (Taken from [139], [141]) Complexity of Disjunctive Logic Programs (with Integrity Constraints)

extremely useful, no generally accepted criteria exist as to why one semantics should be used over another. A semantics may have all the properties one may desire, be computationally tractable and yet not provide answers that a user expected. If in Example 4, the user expected an answer $r(a)$ in response to a query $r(X)$, and the semantics were the *WFS*, the user would receive the answer, $r(a)$ is *unknown*. However, if the *stable model semantics* had been used, the answer returned would be $r(a)$. Perhaps the best that can be expected is to provide users with complexity results and criteria so they may decide which semantics meets the needs of their problems.

Understanding the semantics of disjunctive theories is related to nonmonotonic reasoning. The field of nonmonotonic reasoning has resulted in several alternative approaches to perform default reasoning [287], [363], [288], [315], [316]. The articles [301], [141], [78] cite results where alternative theories of nonmonotonic reasoning can be mapped into extended disjunctive logic programs and databases.

Property	Condition on a Semantics SEM to satisfy the Property							
	Clark's Compl.	GCWA	WGCWA	Perfect	Stable	WFS	D-WFS	Static
Elimination of Tautologies	If a rule $\mathcal{A} \leftarrow \mathcal{B}, \mathbf{not}\ \mathcal{C}$ with $\mathcal{A} \cap \mathcal{B} \neq \emptyset$ is eliminated from a program P, then the resulting program is SEM–equivalent to P.							
	No	Yes	No	Yes	Yes	Yes	Yes	Yes
Generalized Principle of Partial Evaluation (GPPE)	If a rule $\mathcal{A} \leftarrow \mathcal{B}, \mathbf{not}\ \mathcal{C}$, where $\mathcal{B}$ contains an atom B, is replaced in a program P by the n rules $\mathcal{A} \cup (\mathcal{A}^i - \{B\}) \leftarrow ((\mathcal{B} - \{B\}) \cup \mathcal{B}^i), \mathbf{not}\ (\mathcal{C} \cup \mathcal{C}^i)$ where $\mathcal{A}^i \leftarrow \mathcal{B}^i, \mathbf{not}\ \mathcal{C}^i$ $(i = 1, \dots, n)$ are all rules for which $B \in \mathcal{A}^i$, then the resulting program is SEM–equivalent to P.							
	Yes	Yes	Yes	Yes	Yes	Yes	Yes	Yes
Positive/ Negative Reduction	If (1) a rule $\mathcal{A} \leftarrow \mathcal{B}, \mathbf{not}\ \mathcal{C}$ is replaced in a program P by $\mathcal{A} \leftarrow \mathcal{B}, \mathbf{not}\ (\mathcal{C} - C)$ where C appears in no rule head, and (2) a rule $\mathcal{A} \leftarrow \mathcal{B}, \mathbf{not}\ \mathcal{C}$ is deleted from P if there is a fact $\mathcal{A}' \leftarrow$ in P such that $\mathcal{A}' \subseteq \mathcal{C}$, then the resulting program is SEM–equivalent to P.							
	Yes	N/A	N/A	Yes	Yes	Yes	Yes	Yes
Elimination of Non–Minimal Rules	If a rule $\mathcal{A} \leftarrow \mathcal{B}, \mathbf{not}\ \mathcal{C}$ is deleted from a program P if there is another rule $\mathcal{A}' \leftarrow \mathcal{B}', \mathbf{not}\ \mathcal{C}'$ such that $\mathcal{A}' \subseteq \mathcal{A}$, $\mathcal{B}' \subseteq \mathcal{B}$, and $\mathcal{C}' \subseteq \mathcal{C}$, where at least one $\subseteq$ is proper, then the resulting program is SEM–equivalent to P.							
	Yes	Yes	No	Yes	Yes	Yes	Yes	Yes
Consistency	If $\mathrm{SEM}(P) \neq \emptyset$ for all disjunctive deductive database P.							
	No	Yes	Yes	Yes	No	Yes	Yes	Yes
Independence	If for every literal l, l is **true** in every $M \in \mathrm{SEM}(P)$ iff l is **true** in every $M \in \mathrm{SEM}(P \cup P')$ provided that the language of P and P' are disjoint and l belongs to the language of P.							
	No	Yes	Yes	Yes	No	Yes	Yes	Yes

Table 5. (Adapted from [124]) Properties of the semantics of disjunctive deductive databases.

Hence, *DDDBs* may be used to compute answers to queries in such theories. In [62] priority information on extended logic programs and principles that an approach to handling priorities should satisfy are discussed. The expressive power of a query language over a disjunctive ground database is studied in [42]. They show there exist simple queries that cannot be expressed by any preferential semantics (including minimal model semantics and various forms of circumscription), while they can be expressed in default and autoepistemic logic. Default logic, autoepistemic logic and some of their fragments are shown to express the same class of Boolean queries, which turns out to be a strict subclass of the Σ_2^p-recognizable Boolean queries. They prove that under the assumption that the database consists of clauses whose length is bounded by some constant, default logic and autoepistemic logic express *all* of the Σ_2^p-recognizable Boolean queries, while preference-based logics cannot. Eiter and Gottlob [142] show that over the standard infinite Herbrand universe, disjunctive logic programming and normal logic programming under the (cautious) stable model semantics coincide. See Cadoli and Lenzerini for complexity results concerning circumscription and closed world reasoning [73], [77]. See Yuan and You [438] for relationships between autoepistemic circumscription and logic programming. They use two different belief constraints to define two semantics, the *stable circumscriptive semantics* and *the well-founded circumscriptive semantics*, for autoepistemic theories. The work in [438] and on *static semantics* developed by Przymusinski [344] appear to be related. As shown in [61], [60], these approaches, though differently defined, are also related to the D-WFS approach [53],[58]

DDDBs have also contributed to the null value problem. If an attribute of a relation may have a null value, where this value is part of a known set, then one can represent this as a disjunction of relations, where, in each disjunction a different value is given to the argument. For papers on the null value problem both in relational and *DDBs*, see [38], [105], [195], [197], [208], [267],[270], [365], [425,439].

There are several significant contributions of *DDDBs*:

1. *Greater expressive power is provided to the user to develop knowledge base systems.*
2. *Alternative concepts of negation have been developed as evidenced by the different semantics for logic programs (e.g., well-founded semantics and stable semantics for extended logic programs and alternative semantics for disjunctive logic programs).*
3. *Complexity results have been found for alternative semantics of disjunctive databases including alternative theories of negation.*
4. *Methods have been developed to permit prototype systems to be implemented.*
5. *Disjunctive databases can be used as the computational vehicle for a wide class of nonmonotonic reasoning theories.*

In Section 2, we showed how relational databases are formalized in logic. This permitted databases to be extended beyond what is possible with relational databases. Various extensions were discussed, such as *Datalog* extended *Datalog*, Disjunctive *Datalog* and extended disjunctive databases. Alternative theories of negation were discussed, and the semantics of the alternative databases, including negation, were described. *These extensions were shown to be useful for developing complex knowledge bases. The role of* ICs *and other constraints for such systems was described.*

3   Implementation Status of Deductive Databases

The field of deductive databases has made significant intellectual contributions since 1976. However, these have not been matched by implementations available in the commercial market. In the early 1970's, when Codd introduced the relational model [106], there were numerous debates as to the efficacy of such systems relative to network and hierarchic systems [417], [418]. These debates ended when an effective relational system was implemented and shown to be comparable to these systems. Now, some individuals prominent in relational databases claim that *DDBs* are not effective and are not needed. Although I believe otherwise, these comments can be addressed better either when a full commercial implementation of a *DDB* is available, or when many techniques introduced in *DDBs* find their way into relational databases. Both of these are beginning to happen.

In the following subsection I discuss stages through which implementations of *DDBs* have progressed. I describe contributions made in each stage. I then discuss reasons why I believe no current systems are commercially marketed and speculate on how this might change.

3.1   Deductive Database Systems

There have been three stages of implementations of *DDBs*: pre-1970, 1970-1980, and 1980-present. Each stage contributed towards understanding the problems inherent in developing a *DDB* system.

First Stage: Pre 1970s. Two efforts stands out during this period: work by Levien, Maron and Kuhns [256], [259], [257], [238], [239] who developed a prototype system that demonstrated the feasibility of performing deduction in databases, and by Green and Raphael [199], [200] who recognized that the resolution method of Robinson [369] was a uniform procedure based on a single rule of inference that could be used for *DDBs*. This was the first general approach to deductive databases. The work at the Rand Corporation, *Relational Data File (RDF)*, started in 1963. A procedural language, *INFEREX*, was used to execute inference routines. Plausible, formal, and temporal inferencing were performed in *RDF*. The system was implemented on a file consisting of some 55,000 statements. The work by Green and Raphael resulted in a system termed *Question Answering System (QA-3.5)*. It was an outgrowth of Raphael's thesis, *Semantic Information Retrieval (SIR)* [355], that performed deduction. *QA-3.5* generalized *SIR* and included a natural language component. Another system, *Relational Store Structure (RSS)*, started in 1966, performed deduction, and was developed by Marrill [108], [109]. *RSS* had twelve deductive rules in the program. It was possible to incorporate other deductive rules into the system. The *Association Store Processor (ASP)*, developed by Savitt et al.,

[387], performed deduction over binary relational data. The inference rules, specified as relational statements, were handled by breadth-first, followed by depth-first search. These efforts, and those cited in [311], were important pre-cursors to *DDBs*. In Table 6, adapted from [311], I list some capabilities of systems developed in the first stage.

Second Stage: 1970-1980. Whereas the first stage used ad hoc techniques for deduction (except for the work by Green and Raphael), the second stage systems were based on the Robinson resolution principle. The *SYNTEX* system built by Nicolas and Syre [324] used logic as the basis for deduction. It was when I heard Nicolas lecture at the IFIP Congress in 1974 that I decided to visit him in Toulouse on my sabbatical leave the following year. It was there that Nicolas, Gallaire and I conceived of holding our "Workshop on Logic and Data Bases." Work by Chang [87] on *DEDUCE 2*, by Kellogg et al. [224] on the *Deductively Augmented Data Management (DADM)* system, and by Minker [295] on the *Maryland Refutation Proof Procedure System (MRPPS 3.0)*, represent work during the second stage of development of deductive databases. Table 7 provides a brief summary of some of the features of these systems.

DADM precomputed unifications among premises so as to not have to recompute them during deduction. Variables were typed. Inference plans and database access strategies were created from a premise file without requiring access to database values.

MRPPS performed top-down searches for large databases. It perpermitted function symbols as arguments of predicates and had a knowledge base index to access data. The system performed deduction and used a typed unification algorithm and a semantic network. The *SQO* method described in [289], answer extraction, natural language, and voice output were part of the system.

The *DEDUCE 2* system performed deduction over databases. Non-recursive Horn rules were compiled in terms of base relations. *ICs* were used to perform *SQO* on queries. Chang discussed problems that arise with respect to recursive rules and termination [88].

Third Stage: 1980-Present. Many prototype *DDBs* were developed during this period and most are described in [351]. I briefly discuss several major efforts during this period: work at ECRC led by J.-M. Nicolas, work at MCC led by S. Tsur and C. Zaniolo, work at the University of Wisconsin led by R. Ramakrishnan, and work at Stanford led by J. Ullman. They were attempts to develop operational and possibly commercial *DDBs*. They contributed significantly to both the theory and implementation of *DDBs*. See [351] for descriptions of contributions made by these and other systems. I briefly describe some highlights here. In Table 8, taken from [351], I list some capabilities of systems developed in the third stage.

Implementation efforts at ECRC on *DDBs* started in 1984 led to the study of algorithms and prototypes: deductive query evaluation methods (QSQ/SLD and others) [426], [427], [428], [243]; integrity checking (Soundcheck) [114]; the deductive database system EKS(-V1) by Vieille et al. [429]; hypothetical reasoning and *ICs* checking [430]; and aggregation through recursion [244]. The EKS system used top-

Name	QA–3.5 [199], [200] Question–Answering System	ASP [387] Association Storing Processor	RDF [256] Relational Data File	RSS [108] Relational Structures System
Organization	Stanford Research Inst.	Hughes Aircraft Corp.	RAND Corp.	Computer Corp. of America
Designers	Raphael, Green & Coles	Savitt, Love & Troop	Levien & Maron	Marrill
Computer	PDP 10	IBM 360/65	IBM 7044 & 360/65	IBM 360/75
Pgmg. Lang.	LISP 1.5	Assembly Language	Assembly Language	Assembly Language
Input Language Model	"Near–natural" language model based on simple transformations and context–free grammar	Stylized input form & a procedural language	Stylized input forms analyzed by FOREMAN Language	"Near–natural" language model based on matching sentence templates
Syntactic Analysis Technique	Earley algorithm for context–free grammar	N/A	N/A	Match of sentence against stored templates
Semantics Analysis Technique	Semantics stack built during syntax analysis phase	N/A	N/A	"Pattern ⇒ action" operation invoked as a result of template match
Intermediate Language	First–order predicate calculus	Binary relations	Binary relations	n–ary relations ($n \leq 7$ as implemented)
Data Structures	LISP chained list structures	Relational statement elements randomized (coded) and replicated statements stored under each element	Files quadruplicated and ordered by statement number and three elements	Statement elements are hash–coded and "open" statements linked to corresponding "closed" statements
Inference Procedures	Formal theorem proving by Robinson Resolution Procedure	Inference rules specified as relational statements are handled by "breadth–first–followed–by–depth"	"Plausible" inference rules are specified in a procedural language called INFEREX	Twelve general rules of deductive logic are used
Output Language	"Near–natural" language generated in a synthesis phase	Relational statements	Relational statements	"Near–natural" language generated from n–ary relational statements

Table 6. First Stage DDB Implementations (Adapted from [311])

Name	MRPPS 3.0 [295] Maryland Refutation Proof Procedure System	DADM [224] Deductively Augmented Data Management	DEDUCE 2 [87]
Organization	University of Maryland	System Development Corp.	IBM San Jose
Designers	Minker, McSkimin, Wilson & Aronson	Kellogg, Klahr, & Travis	Chang
Computer	UNIVAC 1108		
Pgmg. Lang.	SIMPL		
Input Language Model	Multi–sorted Well–Formed formulae	Primitive Conditional Statements and Natural Language	DEDUCE [86] (based on Symbolic Logic)
Intermediate Language	Clausal Form	Primitive Conditional Statements	DEDUCE
Data Structures	Semantic Networks, Knowledge Base Index	Predicate Array, Premise Array, Semantic Network, Predicate Connection Graph [232], [397]	Connection Graph
Inference Procedures	SL–resolution [233] and LUSH–resolution [206]	Connection Graph	Connection Graph
Output Language	Natural Language Voice and English [334]	Primitive Condition Statements	DEDUCE
Features	Semantic Query Optimization, Multi–Sorted Variables, No Recursion, Non–Horn Clauses, Clauses need not be Function–Free, Relations not in First Normal Form	Semantic Query Optimization, Multi–Sorted Variables, No Recursion	Semantic Query Optimization

Table 7. Second Stage DDB Implementations

down evaluation and was released to ECRC shareholder companies in 1990. ECRC was also engaged in research on a deductive object-oriented database.

Implementation efforts at MCC on a *DDB* started in 1984 and emphasized bottom-up evaluation [416]; query evaluation using semi-naive evaluation, magic sets and counting methods [21], [32], [379], [380]; semantics for stratified negation and set-grouping [31]; investigation of safety; the finiteness of answer sets; and join order optimization. A system, *LDL*, was implemented in 1988, and released with refinements during 1989-1991. It was among the first operational *DDBs* that was widely available. It was distributed to universities and shareholder companies of MCC.

Implementation efforts at the University of Wisconsin on the *Coral DDBs* started in the late 1980s. Bottom-up and magic set methods were implemented. The system, written in *C* and *C++*, is extensible and provides aggregation modularly stratified databases. *Coral* supports a declarative language, and an interface to C++ which allows for declarative and imperative programming. The declarative query language supports general Horn clauses augmented with complex terms, set-grouping, aggregation, negation, and relations with tuples that contain universally quantified variables. *Coral* supports a wide range of evaluation strategies, and automatically chooses an efficient evaluation strategy. Users can guide query optimization by selecting from among alternative control choices. *Coral* provides imperative constructs such as update, insert and delete rules. Disk-resident data is supported using the *EXODUS* storage manager, which also provides transaction management in a client-server environment.

Implementation at Stanford University on a *DDBs* started in 1985 on *NAIL! (Not Another Implementation of Logic!)*. The work led to the first paper on recursion using the magic sets method [21]. Other contributions were aggregation in logical rules, and theoretical contributions to negation: stratified negation by Van Gelder [422], well-founded negation [181] and modularly stratified negation [373]. A language, *GLUE* [331], [317], was developed for logical rules that has the power of *SQL* statements, together with a conventional language that permits the construction of loops, procedures and modules.

Implementations of DDBs in the first, second and third stages of their development have demonstrated the feasibility and practicality of the technology. Tools and techniques have been developed to produce efficient DDBs.

3.2   Prospects for Commercial Implementation of DDBs

One might address why, after 23 years of research in *DDBs*, no commercial systems exist. To place this statement in perspective, it is well to recall that it required approximately twelve years for relational systems to become a reality. Additionally, as Ullman has stated, *DDB* theory is more subtle than relational database theory. Nevertheless, there have been many prototypes, starting from the 1960s until the present (see section 3.1). However, none of the systems in [351] are likely to become commercial products with possibly one exception, *VALIDITY* [166], [266]. In [303] I speculated that there might be a second possible commercial deductive database, *Aditi* [353], and stated,

Name	Developed	Recurs.	Negation	Aggregat.	Upd	ICs	Optimizat	Storage	Interfaces
Aditi [420]	U. Melbourne	General	Stratified	Stratified	No	No	Magic Sets, Seminaive	EDB, IDB	Prolog
COL [2]	INRIA	?	Stratified	Stratified	No	No	None	Main memory	ML
Concept-Base [218]	U. Aachen	General	Locally Stratified	No	Yes	No	Magic sets, Seminaive	EDB only	C, Prolog
CORAL [352]	U. Wisconsin	General	Modular Stratified	Modular Stratified	No	No	Magic sets, Seminaive, Context factoring, Projection pushing	EDB, IDB	C,C++, Extensible
EKS–V1 [429]	ECRC	General	Stratified	General	Yes	Yes	Query–subquery, left/right linear	EDB, IDB	Persistent Prolog
DECLARE	MAD Intelligent Systems	General	Locally Stratified	General	No	No	Magic sets, Seminaive, Projection pushing	EDB only	C, Common Lisp
LDL, LDL++, Salad [94]	MCC	General	Stratified	Stratified	Yes	No	Magic sets, Seminaive, left/right linear, Projection pushing	EDB only	C, C++, SQL
LOGRES [72]	Polytech. of Milan	Linear	Inflation. Semant.	Stratified	Yes	Yes	Seminaive Algebraic Xforms	EDB, IDB	INFORMIX
NAIL/ GLUE	Stanford U.	General	Well–Founded	Glue only	Glue only	No	Magic sets, Seminaive, Right–linear	EDB only	None
Starburst [319]	IBM Almaden	General	Stratified	Stratified	No	No	Magic sets, Seminaive (variant)	EDB, IDB	Extensible

Table 8. Existing Implementations of DDBs (Adapted from [351])

According to a personal communication from Ramamohanarao, the leader of the *Aditi* effort, *Aditi* is perhaps one year from being a commercial product. Whether or not it will become a product remains to be seen. I believe that it will depend upon moving the system from a university setting to industry. Implementors and applications specialists will be required, rather than university researchers.

Based on a communication from one of the developers of *Aditi*[3], it is doubtful that *Aditi* will ever become a product. The state of *Aditi* is that the "final" release of the "first" version of *Aditi* was circa 1993 and is still the only one publically available version. It may be in use in at least one educational institution for teaching purposes. It is neither being supported nor being sold, and the developers are not aware of it being used for real applications. A "second" version, is still under development. Progress has been made towards a release, with simple queries running. Holdups have been mostly software engineering related rather than research. There are currently no plans to sell this version.

At the Bull Corporation, J.-M. Nicolas headed an effort with L. Vieille, to develop a deductive object-oriented database system, *VALIDITY*, an outgrowth of work at *ECRC*. *VALIDITY* is available for use on applications, but is not sold as a product [4]. The effort has been moved from Bull to a new company, Next Century Media, Inc. (NCM), in which Bull has some equity interests. NCM is now responsible for its maintenance, marketing and improvements. The current focus is "targeted television advertising". A *VALIDITY* based product suite named *Opti*Mark (Optimized Marketing)* has been developed and is in use by cable operators, cable TV networks, system integrators, advertising agencies, and advertisers in the USA and in France. The *VALIDITY* team has written a summary state of the art report on deductive and deductive object-oriented databases [414].

There are several reasons why we may not have had commercial systems earlier. Most prototype systems developed have been at universities. Unless there is commercial backing for the venture, universities are not in a position either to develop or to support maintenance required for large system developments. Those systems developed in research organizations controlled by consortia (ECRC and MCC) have not had full backing of consortia members. Second, the implementation effort to develop a commercial product was vastly underestimated by some of the organizations. A large investment must be made to develop a *DDB* that both competes with and extends relational technology. According to industry standards, an investment in the order of $30 to $50 million dollars is required to develop and place in the market a database system, whatever technology it relies upon. Furthermore, researchers tend to change their interests rather than to consolidate their work and invest into technology transfer to favor development of commercial software. Third, until recently, there had not been a convincing demonstration of a large commercial problem that requires a *DDB*. This may have been a reason why the *LDL (Logical Data*

---

[3] Received January 13, 1999
[4] Personal communication from Jean-Marie Nicolas dated January 12, 1999

Language) system at *MCC*, and the *ECRC* developments were terminated. However, today, there are many applications that could take advantage of this technology as evidenced by the book, " Applications of Logic Databases," [350]. In addition, Levy et al. [261] study the problem of computing answers to queries using materialized views and note that this work is related to applications such as Global Information Systems, Mobile Computing, view adaptation and maintaining physical data independence. In [262] it is describe how *DDBs* can be used to provide uniform access to a heterogeneous collection of more than 100 information sources on the World Wide Web. Work on *STARBASE*, [345], an automatic method that reorders literals in the body of a rule so that the next literal to be processed is guaranteed to be the most instantiated, might be useful in conjunction with this work. Fourth, none of the university researchers tried to obtain venture capital to permit them to build a product outside the university. Efforts by some from MCC to obtain venture capital did not succeed. The *VALIDITY* effort to develop a system in an industrial setting is the only successful effort, but it is not generally available.

Does lack of a commercial system at this date forebode the end of logic and databases? I believe that such a view would be naive. First, *VALIDITY* exists and may become commercially marketed. Second, when one considers the fact that it took over twelve years before relational database technology entered the market place from the late 1960s, when research into relational systems started (see [106]), there is no need to be alarmed. Third, as the following developments portend, relational databases are incorporating techniques stemming from research in *DDBs*.

Indeed, many techniques introduced within *DDBs* are finding their way into relational technology. The new SQL standards for relational databases are adopting many powerful features of *DDBs*. Although there are always delays and discrepancies between specifications of SQL standards and their effective support by commercial systems, this is a good sign. In the SQL-2 standards (also known as SQL-92) [292], a general class of *ICs* called *asserts* allow for arbitrary relationships between tables and views to be declared. These constraints exist as separate statements in the database, and are not attached to a particular table or view. This extension is powerful enough to express the types of *ICs* generally associated with *DDBs*. However, only the full SQL-2 standard includes assert specifications. The intermediate SQL-2 standard, the basis for most current commercial implementations, does not include asserts. The next generation *SQL* relational language, *SQL3*, will provide an operation called *recursive union* that directly supports recursive processing of tables [291]. As noted in [291],

> The use of the recursive union operator allows both linear (single-parent, or tree) recursion and non-linear (multi-parent, or general directed graph) recursion. This solution will allow easy solutions to bill-of-material and similar applications.

Linear recursion is currently a part of the client server of *IBM*'s *DB2 (DataBase 2)* system. It appears that they are using the *magic sets* method to perform linear recursion. There are indications that the *ORACLE* database system will also support some form of recursion.

A further development is that some form of *SQO* is being incorporated into relational databases. In some instances, such as *DB2*, cases are recognized when only one answer is to be found and search is terminated. In other systems, equalities and other arithmetic constraints are being added to optimize search. It will not be too long, in my view, before join elimination will be considered and introduced to relational technology. One can now estimate when it will be useful to eliminate a join [188]. The tools and techniques already exist and it is merely a matter of time before the users and the system implementers have them as part of their database systems. In addition, *SQO* techniques are being added to *DB2*, as noted to me in a private communication from Jarek Gryz.

Another technology available for commercial use is cooperative databases. The tools and techniques exist, as evidenced by the *COBASE* [101], [102] and *CARMIN* [170], [189] work. With the introduction of recursion and *SQO* techniques into relational database technology, users will require cooperative responses to understand why certain queries fail or succeed. It will also permit queries to be relaxed when the original query fails. This will permit reasonable, if not logically correct answers to be provided to users. Since user constraints may be handled in the same way *ICs* we will see relational systems incorporate the needs of individual users into a query, as represented by their constraints.

Although the relational database community may ignore that features they are adding to their systems are outgrowths of developments in *DDBs*, we should not be concerned. We should be pleased that our work in developing logic-based theories has come to fruition to improve the efficiency and power of relational systems.

Two significant developments have taken place in the implementation of commercial DDBs. The first is the incorporation of techniques developed in DDDBs into relational technology. Recursive views that use the magic set technique for implementation are being permitted and methods developed for SQO are being applied. The second is the development of a deductive object–oriented DBs, VALIDITY, that is in commercial use, but only for applications developed for its owner company. It remains to be seen how long one can make patches to relational technology to simulate the capabilities of DDBs.

4   Emerging Areas and Trends

In the previous sections I described theories for negation both in extended *DDBs* and in *DDDBs*. Many different semantics have been proposed based on these theories. We understand a great deal about negation, except for how and when one should use a given theory. This will be a major area of confusion when users become cognizant of what is available for them to use. Much more work has to be done if the areas of implementation and application are to match intellectual developments that have taken place over the past 23 years. We have saturated the market for alternative theories of semantics and should move to more fertile topics. Unless we do so, funding for logic and databases will start to wane, as I believe it has in the United States. That does not mean that we should abandon all research in theories of negation and alternative semantics, but we have to take stock of what we have accomplished and make it accessible for users.

The role of logic will be of increasing importance due to the need to handle highly complex data (partly due to the advances in networking technology and the reduction of cost in both processing time and primary, secondary and tertiary memory). This data will require more complex models of data access and representation. Advances will require formal models of logic rather than ad-hoc solutions. Below, I discuss fertile areas for exploration.[5] This is not intended to be an exhaustive listing of important areas to investigate.[6] Useful books containing additional information on logic and databases are given in [79]], [3], [417], [418], [442], [100].

Temporal databases, which deal with time, are important for historical databases, real-time databases and other aspects of databases. Work in this area is reported on in [404], [402], [403], [98], [398]. For a paper on applying transition rules to such databases for *IC* checking, see [286].

Transactions and updates have not had sufficient attention. There exist semantic models of updates [155], [196], [160] that assure that views and data are updated correctly. Leone et al. [246], [245] propose an update language, *ULL*, for knowledge systems. They discuss the expressive power of the language, its implementation, and provide an interpreter, proven correct with respect to the semantics. More work is required on transactions which require sequences of updates. In emerging applications of *DB* systems, transactions are viewed as sequences of nested, and interactive subtransactions that may sparsely occur over long time periods. In this scenario new

---

[5] Revisions made to this section subsequent to the Workshop on Logic in Databases, held in San Miniato, Italy, July 1-2, 1996, incorporate comments made in the panel session, *Deductive Databases: Challenges, Opportunities and Future Directions*, by A. Siebes, S. Tsur, J. Ullman, L. Vieille and C. Zaniolo, and in a personal communication by J.-M. Nicolas. I am responsible for any views expressed here.

[6] In a yet to be published report that can be found on the internet at http://xxx.lanl.gov/html/cs.DB/9811013, entitled *The Asilomar Report on Database Research*, written in September, 1998 by P. Bernstein, M. Brodie, S. Ceri, D. DeWitt, M. Franklin, H. Garcia-Molina, J. Gray, J. Held, J. Hellerstein, H.V. Jagadish, M. Lesk, D. Maier, J. Naughton, H. Pirahesh, M. Stonebraker, and J. Ullman, the executive summary specifies new issues that need to be addressed in the area of databases.

The database research community is rightly proud of success in basic research, and its remarkable record of technology transfer. Now the field needs to radically broaden its research focus to attack the issues of capturing, storing, analyzing, and presenting the vast array of online data. The database research community should embrace a broader research agenda – broadening the definition of database management to embrace all the content of the Web and other online data stores, and rethinking our fundamental assumptions in light of technology shifts. To accelerate this transition, we recommend changing the way research results are evaluated and presented. In particular, we advocate encouraging more speculative and long-range work, moving conferences to a poster format, and publishing all research literature on the Web.

complex transaction systems must be designed. Logic-based transaction systems will be essential to assure appropriate and correct transactions. Work in this area is described in [43], [44], [265], [264], [231], [10], [156], [178], [279]. A language for specifying, analyzing, and scheduling of workflows using *Concurrent Transaction Logic (CTR)* has been proposed. An algorithm and runtime complexity results has been developed for scheduling workflows in the presence of temporal constraints.

Active databases consist of data that protects its own integrity and describes the database semantics. Active rules are represented by the formalism, *Event-Condition-Action (ECA)* [415] and denote that whenever an event E occurs, if condition C holds, then trigger action A. It has a behavior of its own beyond passively accepting statements from users or applications. Upon recognition of certain events, it invokes commands and monitors and manages itself. It may invoke external actions that interact with systems outside the *DB* and may activate a potentially infinite set of triggers. The *termination* problem for recursive active database rules is investigated in [19]. Although declarative constraints are provided for such systems, the *ECA* formalism is essentially procedural in nature. Zaniolo noted the need for declarative semantics of triggers [440], [441]. He developed a unified semantics for active and deductive databases and showed how active database rules relate to transaction-conscious stable model semantics. Work in [26] proposes a first step towards characterizing active databases. A formal framework for studying the semantics and expressiveness of active databases is provided in [333]. The power of various abstract trigger languages is characterized and related to several major active database prototypes. In [216] a declarative mechanism based on meta-rules to control the interaction and execution of multiple rules is proposed. One can determine, in polynomial time, if a rule will never execute, whether two rules can ever be executed together, and whether a rule system is guaranteed to have a unique execution set for all possible rules that become firable. An abstract logical framework for rule-based maintenance of referential integrity is proposed in [280]. They show that the declarative semantics of the resulting logic program captures the intended logic program. The well-founded model yields a unique set of updates, which is a safe, sceptical approximation of the set of all maximal admissible updates; the truth value *undefined* is assigned to all controversial updates. Paton [329], has edited a book devoted to active databases. In [193], properties a semantics for processing active rules should have are discussed and a new semantics is defined that yields a fixpoint semantics. The approach can be combined with standard *Datalog*. Work is required to develop a clear semantics, sound implementations and a better understanding of complexity issues in active databases. Work in the situation calculus and *Datalog* extensions apply here.

Data mining and inductive inference deal with finding generalizations that may be extracted from a database. Such generalizations may be *ICs* that must be *true* with respect to the database, or generalizations that may be *true* of the current state, but may change if there are updates. A database administrator must determine which of the generalizations are *ICs* and which are applicable only to the current state and must be checked upon update. *SQO* can handle either case, and inform the user which constraint applied to the query. As demonstrated in [318], logic programming may be used to find inductive inferences. The book [332] covers work on knowledge data mining. [242] discuss how to couple *DDBs* and inductive logic programming to learn query rules for optimizing databases with update rules. See [191] for results concerning complexity of inductive inference.

Integrating production systems with DDBs is future work that is needed. A formal approach to performing such integrations and to developing the semantics of rule-based systems will result. As rule-based systems have played a role in active systems, this work will benefit active databases. Related work in this area is described in [201], [281], [357], [358], [356]

Logical foundations of object-oriented deductive databases is needed. In many ways *object-oriented databases (OODBs)* relate to hierarchic and network systems. It is essential that *OODBs* have a formal theory and a semantics. This is difficult as there does not appear to be a formal definition of an *OODB*. Kifer et al. [227] developed a formal foundation for *OODBs*. Paris C. Kanellakis, who died tragically in an airplane crash in December 1995, did fundamental work with Abiteboul in the development of a formal foundation of *OODBs* [4]. They extend the techniques of database theory to understand the concepts of "object-identity, types and type inheritance" in *OODBs*. They introduce a language *IQL* based on logic programming for this purpose. Work is needed to develop *SQO* techniques, a formal theory of updating and all of the tools and techniques developed for *DDBs*.

Description logics are concerned with restricting knowledge representation so that deduction over a knowledge base is tractable, but is still powerful enough to represent the knowledge. *KRYPTON*, an early hybrid knowledge representation system which combined a functional knowledge representation with a specialized connection graph theorem prover [50], [51]. *KL-ONE* is perhaps the best known knowledge representation system based on a description logic [48]. *CLASSIC* is a modern successor of KL-ONE which allows for the definition of complex objects and moves towards database support [45], [49]. In *DDBs* representational power is also limited to allow for more tractable deduction (via specific proof procedures). Some of these limits are a restriction to Horn clauses, no logical terms, and no existential quantification. For a discussion of description logics and the materialized view problem, see [69].

Heterogeneous databases integrate into one system multiple databases that do not necessarily share the same data models. There is need for a common logic-based language for mediation and a formal semantics of such databases. Work on the *Hermes* system [406], the *Tsimmis* system [91], and the work in [294] illustrate work on this subject. See [278,8] for additional work on the use of logic for information mediation. Kero and Tsur [225] describe the $\mathcal{IQ}$ system that uses the deductive database $\mathcal{LDL}++$ to reason about textual information. Language extensions for the semantic integration of *DDBs* is proposed by [17]. The language allows *mediators* to be constructed using a set of operators for composing programs and message passing features.

Multi-media databases [407] have special problems such as manipulating geographic databases; picture retrieval where a concept orthogonal to time may appear in the database, and video databases, where space and time are combined. Temporal and spatial reasoning will be needed. Logic will play a role in the development of query languages for these new data models, and a formal semantics will provide a firm theoretical basis for them.

Combining databases relates to both heterogeneous and multi-media systems. One is trying to combine databases that share the same *ICs* and schema. Such databases arise in distributed systems work and in combining knowledge bases. In addition to handling problems that arise because the combined databases may be inconsistent, one has to handle priorities that may exist among facts. For a formal treatment of this subject, see [24], [25], [336], [335].

Using Materialized Views to Answer Queries. With databases distributed widely over a network, it has become useful to materialize views to optimize queries. The problem is to determine if a given query can be answered in terms of these views. An algorithm to rewriting conjunctive queries using materialized views over non-recursive databases was given in [346]. Work on conjunctive query optimization was done in information integration by [419], [260], [262], [138], and query optimization by [90], [433]. Levy et al. [261] showed that the question of determining whether a conjunctive query can be rewritten to an equivalent conjunctive query that only uses views is NP-complete and was extended in [347] to include binding patterns in view definitions. [138] extend the work to recursive Datalog programs. The complexity of answering queries in materialized views for conjunctive queries with inequality, positive queries, *Datalog* and first-order logic is addressed in [1]. Work is required to develop a uniform framework to handle the materialized view problem and to determine conditions under which materialized views can be used to provide sound, but not necessarily complete answers to queries.

Integrity constraints, semantic query optimization and constraint logic programming are related topics. *SQO* uses *ICs* to prune the search space. These *ICs* introduce equalities, inequalities, and relations into a query to optimize the search [82]. The use of *ICs* to perform *SQO* is being incorporated into current relational technology, as noted in section 3. Constraint logic programming introduces domain constraints. These constraints may be equalities, inequalities, and may even be relations [215], [214], [222]. Constraint databases and constraint-intensive queries are required in many advanced applications. Constraints capture spatial and temporal behavior which is not possible in existing *DBs*. Relationships between these areas need to be explored further and applied to *DDBs*. Spatial databases defined in terms of polynomial inequalities are investigated by [240], who consider termination properties of *Datalog* programs. See [187] for a comprehensive discussion of the semantics and applications of *ICs* in *DDBs*.

Abductive reasoning is the process of finding explanations for observations in a given theory. Selman and Levesque [396] developed an approach using logic, described in Kakas et al. [220]. Given a set of sentences T (a theory), and a sentence

G (an observation), the abductive task is to find a set of sentences Δ (abductive explanation for G) such that

(1) $T \cup \Delta \models G$,

(2) $T \cup \Delta$ is consistent.

(3) Δ is minimal with respect to set inclusion [396], [220], [78].

A survey of the extension of logic programming to abductive reasoning (abductive logic programming) is given in [220]. They outline the framework of abduction and its applications to default reasoning; and introduce an augmentation theoretic approach to the use of abduction as an interpretation for negation-as-failure. They show abduction has strong links to extended disjunctive logic programming and generalizes negation-as-failure to include not only negative but also positive hypotheses, and to include general *ICs* and also is related to the justification based truth maintenance system of Doyle [136]. In [221], *Abductive Constraint Logic Programming (ACLP)* is described, which integrates abduction and constraint solving in Logic Programming. Inoue and Sakama [212] develop a fixpoint semantics for abductive logic programs in which the belief models are characterized as the fixpoint of a disjunctive program obtained by a suitable program transformation. For complexity results on abductive and nonmonotonic reasoning see [78], [143] for additional results.

High-level robotics use knowledge bases to solve problems in cognition required to plan actions for robots and to deal with multiple agents in complicated environments. *DDBs* and *DDDBs* relate to this problem. In some instances a robot may have several options that can be represented by disjunctions. Additional information derived from sensors may serve to disambiguate the possibilities. Several are groups engaged in this research: the University of Toronto [253], [254], the University of Texas at El Paso [23], the University of Texas at Austin [185] and the University of Linkoping [385], [386]. Kowalski and Sadri [236], discuss a unified agent architecture that combines rationality with reactivity, and relates to this topic. Eiter and Subrahmanian [154], [153] develop a logic-based semantics, algorithm and complexity results for heterogeneous active agents.

Applications of DDB techniques will become more prevalent. *DDBs* have been shown to be important both for relational and deductive systems on such topics as *SQO* [82], cooperative answering [170], global information systems and mobile computing [261], [113]. Reiter developed a theory of diagnosis from first principles [366]. [165] shows that Reiter's approach to consistency-based diagnosis can be easily and profitably implemented in *Prolog*. For additional work on diagnosis see [111]. Loke and Davison [275] developed a model of the world wide web, where web pages are rephrased as logic programs, and hypertext links are relationships between these programs. A logic language based on *LogicWeb* has been developed and applied to: web search, and the structuring of web information using *DDBs*. For additional work on logic and its application to the web, see [293], [6], [7]. A query language, called *Sequence Datalog*, extends *Datalog* with interpreted function symbols for manipulating sequences, such as genome and text databases [290]. An application to spatial databases is provided in [423]. Bonatti et al. [41] apply logic to develop secure databases. Logic has also been applied to planning in AI systems [410], [65].

Commercial Implementations of Deductive (DDB) and Object-Oriented Deductive Databases (DOOD). It is important to develop a commercial *DDB* or *DOOD* system. The deductive model of databases is beginning to take hold as evidenced by the textbook [5]. The merger of object-oriented and deductive formalisms is taking place as illustrated by the proceedings of the DOOD conference series [228], [120], [80], [266]. That the *VALIDITY* [167], [166], system is in use by customers at the present time is an indication that the object-oriented and deductive formalisms are available commercially. The strategy used by the company that now owns *VALIDITY* is not to make it available for general use, but for applications they will develop for customers. Additional features will be required for commercial systems such as cooperative answering [170] and arithmetic and aggregate operators, as described by Dobbie and Topor [134].

It is clear from the above that logic and databases can contribute significantly to many exciting new topics. Hence, the field of logic and databases will continue to be a productive area of research and implementation.

5   Summary

In section 1, I discussed the significant pre-history of the field of *logic and databases*, [238], [239], by Green and and referred to [298], where I discussed the pre-history of the field. In sections 2 and 3, I discussed the major accomplishments that have taken place in this field since 1976. Among these accomplishments are the extension of relational databases, the development of the semantics and complexity of these alternative databases, permit knowledge base systems to be represented and developed, and nonmonotonic reasoning systems to be implemented. In section 4, I discussed many exciting new areas that will be important in the near and long-term future. *It is clear that the field of logic and databases has had a significant pre-history before 1970, and a well-defined area of research, complete with past achievements and continued future areas of work.*

Since 1976 we have seen logic and databases progress from a fledgling field, to a fully developed, mature field. The new areas that I cited (4) that need further investigation show that we have not exhausted work in this field. *Logic and databases has contributed to the field of databases being a scientific endeavor rather than an ad-hoc collection of techniques. We understand what constitutes a database, a query, an answer to a query, and where knowledge has its place.* I look forward to the next twenty years of this field. To remain vibrant, we will have to take on some of the new challenges, rather than to be mired in the semantics of more exotic databases. We will have to address implementation issues and we will have to be able to provide guidance to practitioners who will need to use the significant developments in logic and databases.

In Tribute to Ray Reiter

It is with great pleasure that I am able to contribute a chapter in this collection to honor Ray Reiter. The field of *DDBs* is deeply indebted to Ray's pioneering work. As I note in this Chapter, Ray made many contributions to this field and to others as well. He provided the first formal definition of a relational database in terms of logic, elucidated the problem of negation in *DDBs* and contributed the closed world and open world assumptions with respect to negation. He specified several possible definitions as to what is meant by an *IC* in a *DDB* and developed a sound and sometimes complete method to query null values. These and other contributions have laid the foundations of *DDB*1 theory. His early work on *DDBs* was an unpublished technical report [361], from which his paper on the closed world assumption was drawn. The technical report is a classic and, unfortunately, is not widely known. After having read it in 1977, I told Ray that he should publish it as a monograph. However, this was not to be.

On a personal level, Ray's work on negation led me to investigate negation in *DDDBs*. Indeed, I have been influenced in all of my work in *DDBs* and *DDDBs* by Ray's work. We met in 1973, when I visited him while I was on vacation in Vancouver. I had read some of his papers and wanted to meet him. We have been friends from the time we met.

In 1992 I e-mailed Ray and told him that I was nominating him for the *Research Excellence Award* given by the International Joint Conference on Artificial Intelligence (IJCAI), which he was awarded in 1993. He e-mailed me back and told me not to waste my time. I responded that I was not asking for his permission, but was merely informing him that I was proceeding with the nomination. He was upset with me for wasting my time, but if I did, and by some strange miracle he won the award, he said he would treat me to dinner at the best restaurant in France. Thus, it was not only my good fortune to nominate him for the award, which he of course did win, but to enjoy a great meal and wonderful company at the three star restaurant *Paul Bocuse* outside Lyon, France.

References

1. S. Abiteboul and O.M. Duschka. Complexity of answering queries using materialized views. In *Proceedings Seventeenth ACM SIGACT-SIGMOD-SIGART Symposium on Principles of Database Systems (PODS 98)*, pages 254–263, Seattle, Washington, June 1-3 1998.
2. S. Abiteboul and S. Grumback. A rule-based language with functions and sets. *ACM Transactions on Database Systems*, 16(1):1–30, 1991.
3. S. Abiteboul, R. Hull, and V. Vianu. *Foundations of Databases*. Addison-Wesley Publishing Comp, 1995.
4. S. Abiteboul and P.C. Kanellakis. Object identity as a query language primitive. *Journal of the ACM*, 45(5):798–842, 1998.
5. S. Abiteboul, Y. Sagiv, and V. Vianu. *Foundations of Databases*. Addison-Wesley, 1995.

6. S. Abiteboul and V. Vianu. Regular path queries with constraints. In *Proc. 16th ACM SIGACT-SIGMOD-SIGART Symp.on Principles of Database Systems (PODS 97)*, pages 122–133, May 1997.
7. S. Abiteboul, V. Vianu, B. Fordham, and Y. Yesha. Relational transducers for electronic commerce. In *Proceedings Seventeenth ACM SIGACT-SIGMOD-SIGART Symposium on Principles of Database Systems (PODS 98)*, pages 179–187, Seattle, WA, June 1-3 1998.
8. S. Adali and V.S. Subrahmanian. Amalgamating knowledge bases, iii: Algorithms, data structures and query processing. *Journal of Logic Programming*, 28(1):57–100, July 1996.
9. J.J. Alferes and L.M. Pereira. On logic program semantics with two kinds of negation. In K. Apt, editor, *Proceedings of the Joint International Conference and Symposium on Logic Programming*, pages 574–588, Washington, D.C. USA, Nov 1992. The MIT Press.
10. P. Ammann, S. Jajodia, and I. Ray. Using formal methods to reason about semantics-based decompositions of transactions. In *Proc. of 21st VLDB Conference*, pages 218–227, 1995.
11. H. Andreka and I. Nemeti. The generalized completeness of Horn predicate logic as a programming language. *Acta Cybernetica*, 4(1):3–10, 1978.
12. K.R. Apt and H.A. Blair. Arithmetic classification of perfect models of stratified programs. *Fundamenta Informaticae*, XIII:1–18, 1990. With addendum in vol. XIV: 339-343. 1991.
13. K.R. Apt, H.A. Blair, and A. Walker. Towards a theory of declarative knowledge. In J. Minker, editor, *Foundations of Deductive Databases and Logic Programming*, pages 89–148. Morgan Kaufmann, 1988.
14. K.R. Apt and R.N. Bol. Logic programming and negation: a survey. *Journal of Logic Programming*, 19/20:9–71, May/June 1994.
15. A. Chandrabose, , J. Dix, and I. Niemelä. Dislop: A research project on disjunctive logic programming. *AI Communications*, 10(3/4):151–165, 1997.
16. A. Chandrabose, J. Dix, and I. Niemelä. DisLoP: Towards a Disjunctive Logic Programming System. In J. Dix, U. Furbach, and A. Nerode, eds., *Logic Programming and Nonmonotonic Reasoning, Proc. of the 4th Int. Conf.*, LNAI 1265, pages 342–353, 1997, Springer.
17. P. Asirelli, C. Renso, and F. Turini. Language extensions for semantic integration of deductive databases. In D. Pedreschi and C. Zaniolo, editors, *Logic in Databases (LID'96)*, pages 425–444, July 1-2 1996.
18. R. Bagai and R. Sunderraman. Bottom-up computation of the fitting model for general deductive databases. *Intelligent Information Systems*, 6(1):59–75, Jan. 1996.
19. J. Bailey, G. Dong, and K. Ramamohanarao. Decidability and undecidability results for the termination problem of active database rules. In *Proc. 17th ACM SIGACT-SIGMOD-SIGART Symp. on Principles of DB Systems (PODS 98)*, pages 264–273, Seattle, WA, June 1998.
20. F. Bancilhon. Naive evaluation of recursively defined relations. In M. Brodie and J. Mylopoulos, editors, *On Knowledge Base Management Systems–Integrating Database and AI Systems*, pages 165–178. Springer-Verlag, 1986.
21. F. Bancilhon, D. Maier, Y. Sagiv, and J. Ullman. Magic sets and other strange ways to implement logic programs. In *Proc. ACM Symp. on Principles of Database Systems*, March 1986.
22. C. Baral and M. Gelfond. Logic programming and knowledge representation. *Journal of Logic Programming*, 19/20:73–148, July 1994.

23. C. Baral, M. Gelfond, and A. Provetti. Representing Actions: Laws, Observations and Hypothesis. *Journal of Logic Programming*, 1996.

24. C. Baral, S. Kraus, and J. Minker. Combining multiple knowledge bases. *IEEE Transactions on Knowledge and Data Engineering*, 3(2):208–220, July 1991.

25. C. Baral, S. Kraus, J. Minker, and V.S. Subrahmanian. Combining default logic databases. *Intl. Journal of Intelligent and Cooperative Info. Systems*, 3(3):319–348, 1994.

26. C. Baral and J. Lobo. Formal characterization of active databases. In *Logic in Databases (LID'96)*, San Miniato, Italy, July 1996. Springer.

27. C. Baral, J. Lobo, and J. Minker. Generalized well-founded semantics for logic programs. In M. E. Stickel, editor, *Proc. of Tenth Internatinal Conference on Automated Deduction*, pages 102–116, Kaiserslautern, Germany, July 1989. Springer-Verlag.

28. C. Baral, J. Lobo, and J. Minker. Generalized disjunctive well-founded semantics for logic programs: Procedural semantics. In *Proceedings of the Fifth International Symposium on Methodologies for Intelligent Systems*, pages 456–464, Knoxville TN, USA, 1990.

29. C. Baral, J. Lobo, and J. Minker. Generalized disjunctive well-founded semantics for logic programs: Declarative semantics. In *Proceedings of the Fifth International Symposium on Methodologies for Intelligent Systems*, pages 465–473, Knoxville TN, USA, 1990.

30. C. Baral and V.S. Subrahmanian. Stable and extension class theory for logic programs and default theory. *Journal of Automated Reasoning*, pages 345–366, 1992.

31. C. Beeri, S. Naqvi, O. Shmueli, and S. Tsur. Set constructors in a logic database language. *Jnl. of Logic Programming*, 10 (3&4), 1991.

32. C. Beeri and R. Ramakrishnan. On the power of magic. *Journal of Logic Programming*, 10(3/4):255–300, 1991.

33. C. Bell, A. Nerode, R. Ng, and V.S. Subrahmanian. Implementing stable model semantics by linear programming. In *Proc., 1993 Int. Workshop on Logic Programming and Nonmonotonic Reasoning*, June 1993.

34. C. Bell, A. Nerode, R. Ng, and V.S. Subrahmanian. Implementing deductive databases by mixed integer programming. *ACM Transactions on Database Systems*, 21(2):238–269, 1996.

35. R. Ben-Eliyahu and R. Dichter. Propositional semantics for disjunctive logic programs. *Annals of Mathematics and AI*, 12:53–87, 1994.

36. R. Ben-Eliyahu and L. Palopoli. Reasoning with minimal models: efficient algorithms and applications. In *Proc. of the Fourth Int. Conf. on Principles of Knowledge Representation and Reasoning*, pages 39–50, 1994. Full paper submitted for journal publication.

37. R. Ben-Eliyahu, L. Palopoli, and V. Zemlyanker. The expressive power of tractable disjunction. In W. Wahlster, editor, *ECAI96. 12th European Conference on Artificial Intelligence*, pages 345–349, 1996.

38. J. Biskup. A foundation of Codd's relational maybe-operations. University Park, 1981. Pennsylvania State Univ.

39. H. Blair, W. Marek, and J. Schlipf. The expressiveness of locally stratified programs. Technical report, Mathematical Sciences Institute, Cornell University, 1992. Available as technical report 92-8.

40. H. Blair and V.S. Subrahmanian. Paraconsistent logic programming. *Theoretical Computer Science*, 68, 135–154, Feb. 1989.

41. P.A. Bonatti, S. Kraus, and V.S. Subrahmanian. Foundations of secure deductive databases. *IEEE Transactions on Knowledge and Data Engineering*, 7(3), June 1995.

42. P.A. Bonatti and T. Eiter. Querying disjunctive databases through nonmonotonic logics. *Theoretical Comp. Sci.*, 160:321–363, 1996.
43. A. Bonner and M. Kifer. Transaction logic programming. In D. S. Warren, editor, *Logic Programming: Proc. of the 10th International Conf.*, pages 257–279, 1993.
44. A.J. Bonner and M. Kifer. Concurrency and communication in transaction logic. In D. Pedreschi and C. Zaniolo, editors, *Logic in Databases (LID'96)*, pages 153–172, July 1-2 1996.
45. A. Borgida, R. J. Brachman, D. L. McGuiness, and L. A. Resnick. CLASSIC: A structural data model for objects. *ACM SIGMOD Record*, 18(2):58, June 1989. Also in: 19 ACM SIGMOD Conf. on the Management of Data, (Portland OR), May–Jun 1989.
46. A. Borgida and D.W. Etherington. Hierarchical knowledge bases and efficient disjunction. In *Proceedings of the First International Conference on Principle of Knowledge Representation and Reasoning (KR-89)*, pages 33–43, Toronto, Ontario, CANADA, 1989.
47. G. Bossu and P. Siegel. Saturation, nonmonotonic reasoning and the closed-world assumption. *Artificial Intelligence*, 25(1):13–63, Jan 1985.
48. R. J. Brachman and J. G. Schmolze. An overview of the KL-ONE knowledge representation system. *Cognitive Science*, pages 171–216, August 1985.
49. R.J. Brachman, A. Borgida, D. McGuinness, P. Patel-Schneider, and L. Resnick. The Classic knowledge representation system of KL-ONE: The next generation. In *International Conference on Fifth Generation Computer Systems*, pages 1036–1043, ICOT, Japan, 1992.
50. R.J. Brachman, R.E. Fikes, and H.J. Levesque. KRYPTON: A functional approach to knowledge representation. *IEEE Computer*, 16(10):67–73, October 1983.
51. R.J. Brachman, V. Pigman Gilbert, and H.J. Levesque. An essential hybrid reasoning system: Knowledge and symbol level accounts of KRYPTON. In *International Joint Conference on Artificial Intelligence*, pages 532–539, August 1985.
52. S. Brass. Sldmagic — an improved magic set technique. In B. Novikov and J.W. Schmidt, editors, *Advances in Database and Information Systems - ADBIS'96*, Sept 1996.
53. S. Brass and J. Dix. A general approach to bottom–up computation of disjunctive semantics. In J. Dix, L.M. Pereira, and T.C. Przymusinski, editors, *Nonmonotonic Extensions of Logic Programming*, pages 127–155. Lecture Notes in CS 927. Springer-Verlag, 1995.
54. S. Brass and J. Dix. Disjunctive Semantics based upon Partial and Bottom-Up Evaluation. In L. Sterling, editor, *Proc. of the 12th Int. Conf. on Logic Programming*, pages 199–213. MIT Press, June 1995.
55. S. Brass and J. Dix. Characterizing D-WFS: Confluence and Iterated GCWA. In L.M. Pereira and E. Orlowska, editors, *JELIA '96*, LNCS 1111, Berlin, 1996. Springer.
56. S. Brass and J. Dix. Characterizations of the Disjunctive Stable Semantics by Partial Evaluation. *Journal of Logic Programming*, 32(3):207–228, 1997.
57. S. Brass and J. Dix. Characterizations of the Disjunctive Well-founded Semantics: Confluent Calculi and Iterated GCWA. *Journal of Automated Reasoning*, 20(1):143–165, 1998.
58. S. Brass and J. Dix. Semantics of Disjunctive Logic Programs Based on Partial Evaluation. *Journal of Logic Programming*, 38(3):167–213, 1999.
59. S. Brass, J. Dix, B. Freitag, and U. Zukowski. Transformation-based bottom-up computation of the well-founded model. *Journal of Logic Programming*, to appear, 1999.

60. S. Brass, J. Dix, I. Niemelä, and T.. C. Przymusinski. A Comparison of the Static and the Disjunctive Well-founded Semantics and its Implementation. In A. G. Cohn, L. K. Schubert, and S. C. Shapiro, editors, *Principles of Knowledge Representation and Reasoning: Proc. of the Sixth Int. Conf. (KR '98)*, pages 74–85, May 1998.

61. S. Brass, J. Dix, and T.C. Przymusinski. Super Logic Programs. In L. C. Aiello, J. Doyle, and S. C. Shapiro, editors, *Principles of Knowledge Representation and Reasoning: Proc. of the Fifth Int. Conf. (KR '96)*, pages 529–541. San Francisco, CA, Morgan Kaufmann, 1996.

62. G. Brewka and T. Eiter. Preferred answer sets for extended logic programs. In A.G. Cohn, L. Schubert, and S.C. Shapiro, editors, *Proc. Sixth Int. Conf. on Principles of Knowledge Representation and Reasoning (KR-98)*, pages 86–97, June 2–4 1998.

63. G. Brewka and G. Gottlob. Well-founded semantics for default logic. *Fundamenta Informaticae*, 31(3/4):221–236, 1997.

64. G. Brewka and I. Niemelä. Report on the Seventh International Workshop on Nonmonotonic Reasoning. *AI Mag.*, 19(4):139–139, 1998.

65. A. Brogi, V.S. Subrahmanian, and C. Zaniolo. The logic of total and partial order plans: A deductive database approach. *Annals of Math and Artificial Intelligence*, 19(1,2):27–58, March 1997.

66. F. Bry. Query evaluation in recursive databases: bottom-up and top-down reconciled. *Data and Knowledge Engineering*, 5:289–312, 1990.

67. F. Buccafurri, N. Leone, and P. Rullo. Disjunctive ordered logic: Semantics and expressibility. In *Proc. of the Int. Conf. on Principles of Knowledge Representation and Reasoning - KR'98*, 1998.

68. F. Buccafurri, N. Leone, and P. Rullo. Stable models and their computation for logic programming with inheritance and true negation. *Journal of Logic Programming*, 27(1):5–43, April 1996.

69. F. Buccafurri, N. Leone, and P. Rullo. Strong and weak constraints in disjunctive datalog. In *Proc. of the 4th Int. Conf. on Logic Programming and Nonmonotonic Reasoning (LPNMR-97)*, LNCS 1265, pages 2–17. Springer, 1997.

70. F. Buccafurri, N. Leone, and P. Rullo. Enhancing disjunctive datalog by constraints. *IEEE Transactions on Knowledge and Data Engineering*, 1999. To appear.

71. F. Buccafurri, N. Leone, and P. Rullo. Semantics and expressive power of disjunctive ordered logic. *Annals of Mathematics and Artificial Intelligence*, 1999. To appear.

72. F. Cacace, S. Ceri, S. Crespi-Reghizzi, and R. Zicari. Integrating object-oriented data modeling with a rule-based programming paradigm. In *Proc. of ACM SIGMOD Conference on Management of Data*, May 1990.

73. M. Cadoli. The complexity for model checking for circumscriptive formulae. *Information Processing Letters*, 44:113–118, Oct 1992.

74. M. Cadoli. Semantical and computational aspects of Horn approximations. In *Proc. of IJCAI-93*, pages 39–44, 1993.

75. M. Cadoli. Panel on "Knowledge compilation and approximation": terminology, questions, references. In *Proc. of the 4th Int. Symp. on A.I. and Mathematics, AI/Math-96*, pages 183–186, Jan 1996.

76. M. Cadoli, F. Donini, P. Liberatore, and M. Schaerf. The size of a revised knowledge base. In *Proc. 14th ACM SIGACT-SIGMOD-SIGART Symp. on Principles of DB Systems (PODS 95)*, pages 151–161, May 1995.

77. M. Cadoli and M. Lenzerini. The complexity of closed world reasoning and circumscription. *J. Comp. and Syst. Sci.*, 43:165–211, 1994.

78. M. Cadoli and M. Schaerf. A survey of complexity results for non-monotonic logics. *Journal of Logic Programming*, 13:127–160, 1993.

79. S. Ceri, G. Gottlob, and L. Tanca. Logic programming and databases. 1990.
80. S. Ceri, K. Tanaka, and S. Tsur, editors. *Proc. of the 3rd Int. Conf. on Deductive and Object-Oriented Databases - DOOD'93*, December 1993. In LNCS 760, Springer-Verlag, Heidelberg, Germany.
81. U. S. Chakravarthy, J. Grant, and J. Minker. Foundations of semantic query optimization for deductive databases. In Jack Minker, editor, *Proc. of the Workshop on Foundations of Deductive Databases and Logic Programming*, pages 67–101, Washington, D.C., Aug 1986.
82. U. S. Chakravarthy, J. Grant, and J. Minker. Logic based approach to semantic query optimization. *ACM Transactions on Database Systems*, 15(2):162–207, June 1990.
83. E. Chan. A possible world semantics for disjunctive databases. *IEEE Trans. Data and Knowledge Eng.*, 5(2):282–292, 1993.
84. A. Chandra and D. Harel. Structure and complexity of relational queries. *Journal of Computer System Sciences*, 25:99–128, 1982.
85. A. Chandra and D. Harel. Horn clause queries and generalizations. *Journal of Logic Programming*, 2(1):1–15, April 1985.
86. C.L. Chang. DEDUCE—a deductive query language for relational databases. In C.H. Chen, editor, *Pattern Recognition and Artificial Intelligence*, pages 108–134. Academic Press, New York, 1976.
87. C.L. Chang. Deduce 2: Further investigations of deduction in relational databases. In H. Gallaire J. Minker, editor, *Logic and Databases*, pages 201–236. Plenum, New York, 1978.
88. C.L. Chang. On evaluation of queries containing derived relations. In H. Gallaire J. Minker J-M. Nicolas, editor, *Advances in Database Theory, Volume 1*, pages 235–260. Plenum Press, New York, 1981.
89. S. Chaudhuri. An overview of query optimization in relational systems. In *Proc. 17th ACM SIGACT-SIGMOD-SIGART Symp. on Principles of Database Systems (PODS 98)*, pages 34–43, Seattle, WA, June 1998.
90. S. Chaudhuri, R. Krishnamurthy, S. Potamianos, and K. Shim. Optimizing queries with materialized views. In *Proceedings of the 11th ICDE*, pages 190–200, 1995.
91. S. Chawathe, H. Garcia-Molina, J. Hammer, K. Ireland, Y. Papakonstantinou, J. Ullman, and J. Widom. The TSIMMIS project: Integration of heterogeneous information sources. In *Proc., IPSJ Conf.*, 1994.
92. W. Chen and D.S. Warren. A goal-oriented approach to computing well founded semantics. In K.R. Apt, editor, *Proc., Joint Int. Conf. and Symp. on Logic Programming*, Washington, D.C., Nov. 1992.
93. S. Chi and L. Henschen. Recursive query answering with non-Horn clauses. In E. Lusk and R. Overbeek, editors, *Proc. 9th Int. Conf. on Automated Deduction*, pages 294–312, Argonne, IL, May 1988.
94. D. Chimenti, R. Gamboa, R. Krishnamurthy, S. Naqvi, S. Tsur, and C. Zaniolo. The LDL system prototype. *IEEE Transactions on Knowledge and Data Engineering*, 2(1):76–90, 1990.
95. P. Cholewiński, W. Marek, A. Mikitiuk, and M. Truszczyński. Experimenting with nonmonotonic reasoning. In *Proc. of the 12th Int. Conf. on Logic Programming*, pages 267–281, Cambridge, MA, 1995.
96. P. Cholewiński, W. Marek, A. Mikitiuk, and M. Truszczyński. Programming with default logic. Submitted for publication, 1998.
97. P. Cholewiński, W. Marek, and M. Truszczyński. Default reasoning system deres. In *Proceedings of KR-96*, pages 518–528, 1996.

98. J. Chomicki. Efficient checking of temporal integrity constraints using bounded history encoding. *ACM Transactions on Database Systems*, 20(2):111–148, May 1995.

99. J. Chomicki and V.S. Subrahmanian. Generalized closed world assumption is π_2^0−complete. *Inf. Proc. Letters*, 34:289–291, May 1990.

100. J. Chomicki and G. Saake, editors. *Logics for Databases and Information Systems*. Kluwer International Series in Engineering and Computer Science, 436. Kluwer Publishers, Boston, March 1998.

101. W. W. Chu, Q. Chen, and A. Y. Hwang. Query answering via cooperative data inference. *Journal of Intelligent Information Systems (JIIS)*, 3(1):57–87, Feb 1994.

102. W. W. Chu, Q. Chen, and M. A. Merzbacher. CoBase: A cooperative database system. In Demolombe and Imielinski [121], chapter 2, pages 41–73.

103. K. L. Clark. Negation as Failure. In H. Gallaire and J. Minker, editors, *Logic and Data Bases*, pages 293–322. Plenum, NY, 1978.

104. CODASYL. *CODASYL Data Base Task Group April 71 Report*. ACM, New York, 1971.

105. E. F. Codd. Extending the database relational model to capture more meaning. *ACM Trans. on Database Syst.*, 4(4):397–434, 1979.

106. E.F. Codd. A relational model of data for large shared data banks. *Comm. ACM*, 13(6):377–387, June 1970.

107. A. Colmerauer, H. Kanoui, R. Pasero, and P. Roussel. Un systeme de communication homme-machine en francais. TR, Groupe de Intelligence Artificielle Universitae de Aix-Marseille II, 1973.

108. Computer Corporation of America. Relational structures research. Technical report, Computer Corporation of America, August 5 1967. For Contract Period April 6, 1966-July 5, 1967.

109. Computer Corporation of America. Relational structures applications research. Technical report, Computer Corporation of America, July 11 1969. For Contract Period May 5, 1967-March 31, 1969.

110. M. Dalal. Some tractable classes of disjunctive logic programs. Technical report, Rutgers University, 1992.

111. C.V. Damasio, L.M. Pereira, and M. Schroeder. REVISE: Logic programming and diagnosis. In J. Dix, U. Furbach, and A. Nerode, editors, *Proc. of the 4th Int. Conf. on Logic Programming and Nonmonotonic Reasoning (LPNMR '97)*, LNAI 1265, Berlin, 1997. Springer.

112. E. Dantsin, T. Eiter, G. Gottlob, and A. Voronkov. Complexity and expressive power of logic programming. In *Proc. of the 12th IEEE Int. Conf. on Comp. Complexity (CCC '97)*, pages 82–101, 1997.

113. S. Dar, H.V. Jagadish, A.Y. Levy, and D. Srivastava. Answering queries with aggregation using views. In *Proc. of the 22nd Int. Conf. on Very Large Databases, VLDB-96*, Sept 1996.

114. H. Decker. Integrity enforcement on deductive databases. In *Proc. 1st Int. Conf. on Expert Database Systems*, April 1986.

115. H. Decker. On the declarative, operational and procedural semantics of disjunctive computational theories. In *Proc. of the 2nd Int. Workshop on the Deductive Approach to Information Systems and Databases*, Aiguablava, Spain, Sept 1991. Invited paper.

116. A. Dekhtyar, M. Dekhtyar, and V.S. Subrahmanian. Hybrid probabilistic programs: Algorithms and complexity. *Journal of Theoretical Computer Science*. Submitted.

117. A. Dekhtyar and V.S. Subrahmanian. Hybrid probabilistic programs. *Journal of Logic Programming*, Jan 1999 (submitted).

118. A. del Val. Tractable databases: how to make propositional unit resolution complete through compilation. In *Proc. of KR-94*, pages 551–561, 1994.

119. A. del Val. An analysis of approximate knowledge compilation. In *Proc. of IJCAI-95*, 1995.

120. C. Delobel, M. Kifer, and Y. Masunaga, editors. *Proc. 2nd Int. Conf. on Deductive and Object-Oriented Databases - DOOD'91*. Springer-Verlag, Heidelberg, Germany, December 1991.

121. R. Demolombe and T. Imielinski, editors. *Nonstandard Queries and Nonstandard Answers*. Studies in Logic and Computation 3. Clarendon Press, Oxford, 1994.

122. R. Demolombe and A.J.I. Jones. Integrity constraints revisited. *Journal of the IGPL (Interest Group in Pure and Applied Logics): An Electronic Journal on Pure and Applied Logic*, 4(3):369–383, 1996.

123. S.W. Dietrich and D. S. Warren. Extension tables: memo relations in logic programming. In *Proc. Symp. on Logic Programming*, pages 264–273, San Francisco, Ca, 1987.

124. J. Dix. Disjunctive deductive databases: theoretical foundations and operational semantics. PhD thesis, Institut für informationssysteme abteilung wissensbasierte systeme, Technische Universität Wien, Sept 1995. Habilitation Thesis.

125. J. Dix, G. Gottlob, and V. Marek. Reducing disjunctive to nondisjunctive semantics by shift operations. *Fundamenta Informaticae*, 28(1,2):87–100, 1996.

126. J. Dix. Classifying Semantics of Logic Programs. In A. Nerode, W. Marek, and V. S. Subrahmanian, editors, *Logic Programming and Non-Monotonic Reasoning, Proc. of the 1st Int. Workshop*, pages 166–180, Cambridge, Mass., July 1991. Washington D.C, MIT Press.

127. J. Dix. A Framework for Representing and Characterizing Semantics of Logic Programs. In B. Nebel, C. Rich, and W. Swartout, editors, *Principles of Knowledge Representation and Reasoning: Proc. of the 3rd Int. Conf. (KR '92)*, pages 591–602. San Mateo, CA, 1992.

128. J. Dix. Classifying Semantics of Disjunctive Logic Programs. In K. R. Apt, editor, *Proc., 1992 Joint Int. Conf. and Symp. on Logic Programming*, pages 798–812, Cambridge, Mass., Nov. 1992. MIT Press.

129. J. Dix. A Classification-Theory of Semantics of Normal Logic Programs: I. Strong Properties. *Fundamenta Informaticae*, XXII(3):227–255, 1995.

130. J. Dix. A Classification-Theory of Semantics of Normal Logic Programs: II. Weak Properties. *Fundamenta Informaticae*, XXII(3):257–288, 1995.

131. J. Dix. Semantics of Logic Programs: Their Intuitions and Formal Properties. An Overview. In A. Fuhrmann and H. Rott, editors, *Logic, Action and Information – Essays on Logic in Philosophy and Artificial Intelligence*, pages 241–327. DeGruyter, 1995.

132. J. Dix, U. Furbach, and I. Niemelä. Nonmonotonic Reasoning: Towards Efficient Calculi and Implementations. In A. Voronkov and J.A. Robinson, editors, *Handbook of Automated Reasoning*. Elsevier-Science-Press, to appear 1999.

133. J. Dix and F. Stolzenburg. A Framework to incorporate Nonmonotonic Reasoning into Constraint Logic Programming. *Journal of Logic Programming*, 37(1,2,3):47—76, 1998.

134. G. Dobbie and R. Topor. Arithmetic and aggregate operators in deductive object-oriented databases. In D. Pedreschi and C. Zaniolo, editors, *Logic in Databases (LID'96)*, pages 399–407, July 1-2 1996.

135. W.F. Dowling and J. H. Gallier. Linear time algorithms for testing the satisfiability of propositional Horn formulae. *Journal of Logic Programming*, 1:267–284, 1984.

136. J. Doyle. Truth Maintenance System. *Artificial Intelligence*, 13, 1980.

137. P. Dung. Negations as hypothesis: an abductive foundation for logic programming. In *Proc., 8th Int. Conf. on Logic Programming*, 1991.
138. O.M. Duschka and M.R. Genesereth. Answering recursive queries using views. In *Proc. 16th ACM SIGACT-SIGMOD-SIGART Symp. on Principles of Database Systems (PODS 97)*, pages 109–116, 1997.
139. T. Eiter and G. Gottlob. Complexity aspects of various semantics for disjunctive databases. In *Proc. of the 12th ACM SIGART–SIGMOD–SIGART Symp. on Principles of Database Systems (PODS-93)*, pages 158–167. May 1993.
140. T. Eiter and G. Gottlob. Complexity results for disjunctive logic programming and application to nonmonotonic logics. In D. Miller, editor, *Proc. of the Int. Logic Programming Symp. ILPS'93*, pages 266–278, 1993.
141. T. Eiter and G. Gottlob. On the computation cost of disjunctive logic programming: Propositional case. *Annals of Mathematics and Artificial Intelligence*, 15(3-4):289–323, Dec. 1995.
142. T. Eiter and G. Gottlob. Expressiveness of stable model semantics for disjunctive logic programs with functions. *Journal of Logic Programming*, 33(2):167–178, 1997.
143. T. Eiter, G. Gottlob, and N. Leone. Abduction from logic programs: Semantics and complexity. *Theoretical Computer Science*, 189(1-2):129–177, December 1997.
144. T. Eiter, G. Gottlob, and H. Mannila. Adding Disjunction to Datalog. In *Proc. of the 13th ACM SIGACT SIGMOD-SIGART Symp. on Principles of DB Systems (PODS '94)*, pages 267–278, May 1994.
145. T. Eiter, G. Gottlob, and H. Mannila. Disjunctive datalog. *ACM Transactions on Database Systems*, 22(3):364–418, September 1997.
146. T. Eiter, G. Gottlob, and H. Veith. Modular logic programming and generalized quantifiers. In J. Dix, U. Furbach, and A. Nerode, editors, *Proc. of the 4th Int. Conf. on Logic Programming and Nonmonotonic Reasoning (LPNMR-97)*, LNCS 1265, pages 290–309. Springer, 1997.
147. T. Eiter, N. Leone, C. Mateis, G. Pfeifer, and F. Scarcello. A deductive system for nonmonotonic reasoning. In J. Dix, U. Furbach, and A. Nerode, editors, *Proc. of the 4th Int. Conf. on Logic Programming and Nonmonotonic Reasoning (LPNMR97)*, pages 364–375, San Francisco, California, 1997. Springer LNCS 1265.
148. T. Eiter, N. Leone, C. Mateis, G. Pfeifer, and F. Scarcello. The kr system dlv: Progress report, comparisons, and benchmarks. In A.G. Cohn, L. Schubert, and S.C. Shapiro, editors, *Proc. 6th Int. Conf. on Principles of Knowledge Representation and Reasoning (KR-98)*, pages 406–417, San Francisco, CA, June 1998. Springer LNCS 1265.
149. T. Eiter, N. Leone, and D. Saccà. The expressive power of partial models for disjunctive databases. In D. Pedreschi and C. Zaniolo, editors, *Logic in Databases (LID'96)*, pages 261–280, July 1-2 1996.
150. T. Eiter, N. Leone, and D. Saccà. On the partial semantics for disjunctive deductive databases. *Annals of Mathematics and Artificial Intelligence*, 17(1/2):59–96, 1997.
151. T. Eiter, N. Leone, and D. Saccà. Expressive power and complexity of partial models for disjunctive deductive databases. *Theoretical Computer Science*, 206(1-2):181–218, 1998.
152. T. Eiter, J. Lu, and V.S. Subrahmanian. Computing non-ground representations of stable models. In J. Dix, U. Furbach, and A. Nerode, editors, *Proc. of the 4th Int. Conf. on Logic Programming and Nonmonotonic Reasoning (LPNMR-97)*, LNCS 1265, pages 198–217. Springer, 1997.
153. T. Eiter and V.S. Subrahmanian. Heterogeneous active agents, ii: Algorithms and complexity. *Artificial Intelligence Journal*. To appear.

154. T. Eiter, V.S. Subrahmanian, and G. Pick. Heterogeneous active agents, i: Semantics. *Artificial Intelligence Journal.* To appear.

155. R. Fagin, J.D. Ullman, and M.Y. Vardi. On the semantics of updates in databases. In *Proc. Senth ACM SIGACT/SIGMOD Symposium on Principles of Database Systems,* pages 352–365, 1983.

156. A. Farrag and M. Ozsu. Using semantic knowledge of transactions to increase concurrency. *ACM, TODS,* 14(4):503–525, 1989.

157. J.A. Fernández and J. Minker. Semantics of disjunctive deductive databases. In *Proceedings of the International Conference on Database Theory,* pages 332–356, 1992. (Invited Paper).

158. J.A. Fernández and J. Minker. Theory and algorithms for disjunctive deductive databases. *Programmirovanie,* N 3:5–39, 1993. (also appears as University of Maryland Technical Report,CS-TR-3223, UMIACS-TR-94-17,1994. Invited Paper in Russian).

159. J.A. Fernández and J. Minker. Bottom-up computation of perfect models for disjunctive theories. *Journal of Logic Programming,* 25(1):33–51, October, 1995.

160. J.A. Fernández, J. Grant, and J. Minker. Model theoretic approach to view updates in deductive databases. *Journal of Automated Reasoning,* 17(2):171–197, 1996.

161. J.A. Fernández and J. Lobo. A proof procedure for stable theories. Technical Report, University of Illinois, Chicago Circle, 1993.

162. J.A. Fernández, J. Lobo, J. Minker, and V.S. Subrahmanian. Disjunctive LP + integrity constraints = stable model semantics. *Annals of Mathematics and Artificial Intelligence,* 8(3–4):449–474, 1993.

163. J.A. Fernández and J. Minker. Bottom-up evaluation of Hierarchical Disjunctive Deductive Databases. In K. Furukawa, editor, *Logic Programming Proc. of the Eighth Int. Conf.,* pages 660–675, 1991.

164. M. Fitting. A Kripke-Kleene semantics for logic programs. *Journal of Logic Programming,* 2:295–312, 1985.

165. G. Friedrich, G. Gottlob, and W. Nejdl. Generating efficient diagnostic procedures from model based knowledge using logic programming techniques. *Computers and Mathematics with Applications,* 20(9/10):57–72, 1990.

166. O. Friesen, G. Gauthier-Villars, A. Lefebvre, and L. Vieille. Applications of deductive object-oriented databases using del. In R. Ramakrishnan, editor, *Applications of Logic Databases.* Kluwer, 1995.

167. O. Friesen, A. Lefebvre, and L. Vieille. VALIDITY: Applications of a dood system. In *Proc. 5th Int. Conf. on Extending DB Technology - EDBT'96 (LNCS 1057),* Avignon, March 1996, Springer.

168. T. Gaasterland and J. Lobo. Processing negation and disjunction in logic programs through integrity constraints. *Journal of Intelligent Information Systems,* 2(3), 1993.

169. T. Gaasterland. *Cooperative Answers for Database Queries.* PhD thesis, Univ. of MD, Dept. of Computer Science, College Park, 1992.

170. T. Gaasterland, P. Godfrey, and J. Minker. An overview of cooperative answering. *Journal of Intelligent Information Systems,* 1(2):123–157, 1992. Invited paper.

171. T. Gaasterland, P. Godfrey, and J. Minker. Relaxation as a platform for cooperative answering. *Journal of Intelligent Information Systems,* 1:293–321, 1992.

172. T. Gaasterland and J. Minker. User needs and language generation issues in a cooperative answering system. In P. Saint-Dizier, editor, *ICLP'91 Workshop: Adv. Logic Programming Tools and Formalisms for Language Processing,* pages 1–14, 1991.

173. T. Gaasterland, J. Minker, and A. Rajasekar. Deductive database systems and knowledge base systems. In *Proceedings of VIA 90,* Barcelona, Spain, October 1990.

174. H. Gallaire and J. Minker, editors. *Logic and Databases*. Plenum Press, New York, April 1978.

175. H. Gallaire, J. Minker, and J-M. Nicolas, editors. *Advances in Database Theory*, volume 1. Plenum Press, 1981.

176. H. Gallaire, J. Minker, and J-M Nicolas, editors. *Advances in Database Theory*, volume 2. Plenum Press, 1984.

177. H. Gallaire, J. Minker, and J-M. Nicolas. Logic and databases: A deductive approach. *ACM Comp. Surveys*, 16(2):153–185, 1984.

178. H. Garcia-Molina. Using semantic knowledge for transaction processing in a distributed database. *ACM, TODS*, 8(2):186–213, 1983.

179. A. Van Gelder. The alternating fixpoint of logic programs with negation. In *8th ACM Symp. on Prin. of Database Syst.*, pages 1–10, 1989.

180. A. Van Gelder, K. Ross, and J.S. Schlipf. Unfounded Sets and Well-founded Semantics for General Logic Programs. In *Proc. 7th Symposium on Principles of Database Systems*, pages 221–230, 1988.

181. A. Van Gelder, K.A. Ross, and J.S. Schlipf. The well-founded semantics for general logic programs. *Journal of the Association for Computing Machinery*, 38(3):620–650, July 1991.

182. M. Gelfond and V. Lifschitz. The stable model semantics for logic programming. In R. Kowalski and K. Bowen, editors, *Proc., 5th Int. Conf. and Symp. on Logic Programming*, pages 1070–1080, 1988.

183. M. Gelfond and V. Lifschitz. Logic programs with classical negation. In D.H.D. Warren and P. Szeredi, editors, *Proc., 7th Int. Conf. on Logic Programming*, pages 579–597, Jerusalem, 1990.

184. M. Gelfond and V. Lifschitz. Classical negation in logic programs and disjunctive databases. *New Generation Computing*, 9:365–385, 1991.

185. M. Gelfond and V. Lifschitz. Representing actions and change by logic programs. *Jnl. of Logic Programming*, 17(2,3,4):301–323, 1993.

186. M. Gelfond, H. Przymusinska, and T.C. Przymusinski. The extended closed world assumption and its relation to parallel circumscription. *Proc. Fifth ACM SIGACT-SIGMOD Symposium on Principles of Database Systems*, pages 133–139, 1986.

187. P. Godfrey, J. Grant, J. Gryz, and J. Minker. Integrity constraints: Semantics and applications. In J. Chomicki and G. Saake [100], chapter 9, pages 265–306.

188. P. Godfrey, J. Gryz, and J. Minker. Semantic query evaluation for bottom-up evaluation. In *Proc. ISMIS96*, June 1996.

189. P. Godfrey, J. Minker, and L. Novik. An architecture for a cooperative database system. In W. Litwin and T. Risch, editors, *Proc. of the First Int. Conf. on Applications of Databases*, LNCS 819, pages 3–24. Springer Verlag, Vadstena, Sweden, June 1994.

190. G. Gottlob. Complexity and expressive power of disjunctive logic programming (research overview). In M. Bruynooghe, editor, *International Logic Programming Symposium ILPS'94*, pages 23–42, 1994.

191. G. Gottlob, N. Leone, and F. Scarcello.

192. G. Gottlob, Sh. Marcus, A. Nerode, G. Salzer, and V.S. Subrahmanian. Non-ground realization of the stable and well-founded semantics. *Theoretical Computer Science*, 166(1&2):221–262. 1996.

193. G. Gottlob, G. Moerkotte, and V.S. Subrahmanian. The PARK semantics for active rules. In *Proc. International Conference on Extending Database Technology, EDBT'96*, LNCS, Springer, 35–95, 1996.

194. G. Gottlob and M. Truszczyński. Approximating stable models is hard. *Fundamenta Informaticae*, 28(1,2):123–128, 1996.

195. J. Grant. Incomplete information in a relational database. In *Proc. Fund Inf III*, pages 363–378, 1980.
196. J. Grant, J. Horty, J. Lobo, and J. Minker. View updates in stratified disjunctive databases. *Journal Automated Reasoning*, 11:249–267, March 1993.
197. J. Grant and J. Minker. Answering queries in indefinite databases and the null value problem. In P. Kanellakis, editor, *Advances in Computing Research: The Theory of Databases*, pages 247–267. 1986.
198. J. Grant, J. Gryz, and J. Minker. Updating disjunctive databases via model trees. CS-TR-3407, UMIACS-TR-95-11, Department of Computer Science, University of Maryland, Feb 1995.
199. C.C. Green and B. Raphael. Research in intelligent question answering systems. *Proc. ACM 23rd National Conf.*, pages 169–181, 1968.
200. C.C. Green and B. Raphael. The use of theorem-proving techniques in question-answering systems. *Proc. 23rd National Conf. ACM*, 1968.
201. H. Groiss. A formal semantics for a rule-based language. In *IJCAI Workshop on Production Systems and their Innovative Applications*, 1993.
202. M.T. Hammer and S.B. Zdonik. Knowledge-based query processing. *Proc. 6th Int. Conf. on Very Large Data Bases*, pages 137–147, 1980.
203. D. Harel. Review no.36,671 of Logic and Data Bases by H. Gallaire and J. Minker. *Computing Reviews*, 21(8):367–369, Aug 1980.
204. L.J. Henschen and S.A. Naqvi. On compiling queries in recursive first-order databases. *J.ACM*, 31(1):47–85, January 1984.
205. L.J. Henschen and H. Park. Compiling the GCWA in Indefinite Databases. In J. Minker, editor, *Foundations of Deductive Databases and Logic Programming*, pages 395–438, 1988.
206. R. Hill. Lush resolution and its completeness. TRDCL Memo 78, Department of Artificial Intelligence, Univ. of Edinburgh, Aug 1974.
207. T. Imielinski. Incomplete deductive databases. *Annals of Mathematics and Artificial Intelligence*, 3:259–293, 1991.
208. T. Imielinski and W. Lipski. Incomplete information in relational databases. *J. ACM*, 31(4):761–791, 1984.
209. T. Imielinski and K. Vadaparty. Complexity of query processing in databases with OR-objects. In *Proc. 7th ACM SIGACT/SIGMOD Symp. on Principles of Database Systems*, pages 51–65, 1989.
210. K. Inoue, M. Koshimura, and R. Hasegawa. Embedding negation as failure into a model generation theorem prover. In D. Kapur, editor, *Proc., 11th Int. Conf. on Automated Deduction*, pages 400–415, 1992.
211. K. Inoue and C. Sakama. Transforming abductive logic programs to disjunctive programs. In *Proceedings of the 10th International Conference on Logic Programming*, pages 335–353, 1993.
212. K. Inoue and C. Sakama. A fixpoint characterization of abductive logic programs. *Jnl. of Logic Programming*, 27(2):107–136, May 1996.
213. A. Itai and J. A. Makowsky. On the complexity of Herbrand's theorem. TR, Dept. of Computer Science, Israel Inst. of Tech., Haifa, 1982.
214. J. Jaffar and M. Maher. Constraint logic programming:a survey. *Journal of Logic Programming*, 19-20:503–581, May-July 1994.
215. J. Jafffar and J-L. Lassez. Constraint logic programming. In *Proc. of the 14th ACM Symposium on Principles of Programming Languages*, pages 111–119, München, Germany, Jan 1987.

216. H.V. Jagadish, A.O. Mendelzon, and I.S. Mumick. Managing conflicts between rules. In *Proc. 15th ACM SIGACT-SIGMOD-SIGART Symp. on Principles of Database Systems (PODS 96)*, pages 192–201, 1996.

217. H. Jakobovits and D. Vermier. R-stable models for logic programs. In D. Pedreschi and C. Zaniolo, editors, *Logic in Databases (LID'96)*, pages 251–259, July 1-2 1996.

218. M. Jeusfeld and M. Staudt. Query optimization in deductive object bases. In G. Vossen, J. C. Feytag, and D. Maier, editors, *Query Processing for Advanced Database Applications*, 1993.

219. V. Kagan, A. Nerode, and V.S. Subrahmanian. Computing minimal models by partial instantiation. *Theoretical Computer Science*, 155:157–177, 1996.

220. A. C. Kakas, R. A. Kowalski, and F. Toni. Abductive logic programming. *Journal of Logic and Computation*, 6(2):719–770, 1993.

221. A.C. Kakas and C. Mourlas. ACLP: Flexible solutions to complex problems. In J. Dix, U. Furbach, and A. Nerode, editors, *Proc. of the 4th Int. Conf. on Logic Programming and Nonmonotonic Reasoning (LPNMR '97)*, LNAI 1265, Berlin, 1997. Springer.

222. P. Kanellakis. Constraint programming and database languages. In *Proc. 14th ACM SIGACT-SIGMOD-SIGART Symp. on Principles of Database Systems (PODS 95)*, pages 46–57, 1995.

223. H.A. Kautz and B. Selman. Forming concepts for fast inference. In *Proc. of AAAI-92*, pages 786–793, 1992.

224. C. Kellogg, P. Klahr, and L. Travis. Deductive planning and pathfinding for relational data bases. In H. Gallaire and J. Minker, editors, *Logic and Data Bases*, pages 179–200. Plenum Press, New York, 1978.

225. B. Kero and S. Tsur. The $\mathcal{IQ}$ system: a deductive database information lens for reasoning about textual information. In D. Pedreschi and C. Zaniolo, eds., *Logic in Databases (LID'96)*, pages 377–395, 1996.

226. W. Kiebling and H. Schmidt. DECLARE and SDS: Early efforts to commercialize deductive database technology. 1993.

227. M. Kifer, G. Lausen, and J. Wu. Logical Foundations of Object-Oriented and Frame-Based Languages. *Journal of ACM*, 1993.

228. W. Kim, J-M. Nicolas, and S. Nishio, editors. *Proc. 1st Int. Conf. on Deductive and Object-Oriented Databases - DOOD'89*. North-Holland Publishing Co., Amsterdam, The Netherlands, Dec 1990.

229. J.J. King. Quist: A system for semantic query optimization in relational databases. *Proc. 7th International Conference on Very Large Data Bases*, pages 510–517, September 1981.

230. P. Kolaitis and C. Papadimitriou. Why not negation by fixpoint? *JCSS*, 43:125, 1991.

231. H. Korth and G. Speegle. Formal aspects of concurrency control in long duration transaction systems using the NT/PV model. *ACM, TODS*, 19(3):492–535, 1994.

232. R. A. Kowalski. A proof procedure using connection graphs. *Journal of the ACM*, 22(4):572–595, Oct. 1975.

233. R. A. Kowalski and D. Kuehner. Linear Resolution with Selection Function. *Artificial Intelligence*, 2:227–260, 1971.

234. R.A. Kowalski. Predicate logic as a programming language. *Proc. IFIP 4*, pages 569–574, 1974.

235. R.A. Kowalski. Logic for data description. In H. Gallaire J. Minker, editor, *Logic and Data Bases*, pages 77–102. Plenum Press, NY, 1978.

236. R.A. Kowalski and F. Sadri. Towards a unified agent architecture that combines rationality with reactivity. In D. Pedreschi and C. Zaniolo, editors, *Logic in Databases (LID'96)*, pages 131–150, July 1-2 1996.

237. S. Kraus, D. Lehmann, and M. Magidor. Nonmonotonic Reasoning, Preferential Models and Cumulative Logics. *Artificial Intelligence*, 44(1):167–207, 1990.

238. J.L. Kuhns. Logical aspects of question answering by computer. *Third Int. Symp. on Computer and Information Sciences*, Dec 1969.

239. J.L. Kuhns. Interrogating a relational data file: Remarks on the admissibility of input queries. TR, The Rand Corporation, Nov 1970.

240. B. Kuipers, J. Paredaens, M. Smits, and J. Van den Bussche. Termination properties of spatial Datalog programs. In D. Pedreschi and C. Zaniolo, eds., *Logic in Databases (LID'96)*, pages 95–109, 1996.

241. L.V.S. Laksmanan and N. Shiri. A parametric approach to deductive databases with uncertainty. In D. Pedreschi and C. Zaniolo, editors, *Logic in Databases (LID'96)*, pages 55–73, July 1-2 1996.

242. D. Laurent and Ch. Vrain. Learning query rules for optimizing databases with update rules. In D. Pedreschi and C. Zaniolo, editors, *Logic in Databases (LID'96)*, pages 173–192, July 1-2 1996.

243. A. Lefebvre and L. Vieille. On deductive query evaluation in the DedGin* system. In W. Kim, J-M. Nicolas, and S. Nishio, editors, *1st Int. Conf. on Deductive and Object-Oriented Databases*, Dec 1989.

244. L. Lefebvre. Towards and efficient evaluation of recursive aggregates in deductive databases. In *Proc. 4th Int. Conf. on Fifth Generation Computer Systems (FGCS)*, June 1992. extended version in. New Generation Computing 12, Ohmsha Ltd. & Springer-Verlag, 1994.

245. N. Leone, L. Palopoli, and M. Romeo. A language for updating logic programs. *Journal of Logic Programming*, 23(1):1–61, April 1995.

246. N. Leone, L. Palopoli, and M. Romeo. Modifying intensional logic knowledge. *Fundamenta Informaticae*, 21(3):183–203, 1994.

247. N. Leone, P. Rullo, and F. Scarcello. Disjunctive stable models: Unfounded sets, fixpoint semantics and computation. *Information and Computation*, 135:69–112, 1997.

248. N. Leone and G. Rossi. Well-founded semantics and stratification for ordered logic programs. *New Generation Computing*, 12(1):91–121, Nov 1993.

249. N. Leone and P. Rullo. The safe computation of the well-founded semantics for logic programming. *Inf. Systs.*, 17(1):17–31, Jan. 1992.

250. N. Leone, P. Rullo, A. Mecchia, and G. Ross. A deductive environment for dealing with objects and non-monotonic reasoning. *IEEE Trans. on Knowledge and Data Engineering*, 9(4), July/Aug 1997.

251. N. Leone, P. Rullo, and F. Scarcello. Declarative and fixpoint characterizations of disjunctive stable models. In *Proc. of Int. Logic Programming Symposium–ILPS'95*, pages 399–413. MIT Press, Dec 1995.

252. N. Leone, P. Rullo, and F. Scarcello. Stable model checking for disjunctive programs. In D. Pedreschi and C. Zaniolo, editors, *Logic in Databases (LID'96)*, pages 281–294, July 1-2 1996.

253. Y. Lesperance, H. Levesque, F. Lin, D. Marcu, R. Reiter, and R. Scherl. A logical approach to high level robot programming – a progress report. In *Working notes, 1994 AAAI Fall Symp. on Control of the Physical World by Intelligent Systems*, Nov 1994.

254. H.J. Levesque, R. Reiter, Y. Lesperance, F. Lin, and R. Scherl. Golog: A logic programming language for dynamic domains. *Journal of Logic Programming*, 1996.

255. H.J. Levesque. Foundations of a functional approach to knowledge representation. *Artificial Intelligence*, 23:155–212, March 1984.

256. R. Levien and M.E. Maron. *Relational Data File: A Tool for Mechanized Inference Execution and Data Retrieval*. The Rand Corporation, December 1965.

257. R.E. Levien. Relational data file ii: Implementation. *Proc. Third Annual National Colloquium on Information Retrieval*, pages 225–241, May 1967.

258. R.E. Levien. Relational data file: Experience with a system for propositional data storage and inference execution. Technical report, The Rand Corporation, April 1969.

259. R.E. Levien and M.E. Maron. A computer system for inference execution and data retrieval. 10(11):715–721, Nov. 1967.

260. A. Y. Levy, A. Rajaraman, and J. J. Ordille. Query-answering algorithms for information agents. In *Proceedings of the Thirteenth National Conference on Artificial Intelligence (AAAI)*, 1996.

261. A.Y. Levy, A.O. Mendelzon, Y. Sagiv, and D. Srivastava. Answering queries using views. In *Proc., 14th ACM SIGACT-SIGMOD-SIGART Symp. on Principles of Database Systems, PODS-95*, 1995.

262. A.Y. Levy, A. Rajaraman, and J.J. Ordille. Querying heterogeneous information sources using source-descriptions. In *Proc. of the 22nd Int. Conf. on Very Large Databases*, Bombay, India, 1996.

263. A.Y. Levy and Y. Sagiv. Semantic query optimization in datalog programs. In *Principles of Database Systems 1995 (PODS95)*, pages 163–173, 1995.

264. F. Lin and R. Reiter. How to progress a database II: The STRIPS connection. TR, Dept of Comp. Sci., Univ. of Toronto, 1993.

265. F. Lin and R. Reiter. How to progress a database (and Why) I: Logical foundations. In *KR94*, pages 425–436, 1994.

266. T-W. Ling, A. Mendelzon, and L. Vieille, editors. *Proc. 4th Int. Conf. on Deductive and Object-Oriented Databases - DOOD'95*. Springer-Verlang, December 1995. LNCS 1013, Heidelberg, Germany.

267. W. Lipski. On databases with incomplete information. volume 28, pages 41–70. ACM, New York, 1981.

268. K.-C. Liu and R. Sunderraman. Indefinite and maybe information in relational databases. *ACM Trans. on DB Systems*, 15(1):1–39, 1990.

269. K.-C. Liu and R. Sunderraman. On representing indefinite and maybe information in relational databases: A generalization. In *Proc. of IEEE Data Engineering*, pages 495–502, Los Angeles, Feb 1990.

270. Y. Liu. Null values in definite programs. In S. Debray and M. Hermenegildo, editors, *Proc. of North American Conf. on Logic Programming*, pages 273–288, Austin, Texas, October 1990. MIT Press.

271. J. W. Lloyd and R. W. Topor. A basis for deductive database systems. *Journal of Logic Programming*, 2(2):93–109, July 1985.

272. J.W. Lloyd. *Foundations of Logic Programming*. Springer–Verlag, second edition, 1987.

273. J. Lobo, J. Minker, and A. Rajasekar. *Foundations of Disjunctive Logic Programming*. MIT Press, 1992.

274. J. Lobo, C. Yu, and G. Wang. Computing the transitive closure in disjunctive databases. Technical report, Department of Electrical Engineering and Computer Science, Univ. of Illinois at Chicago, 1992.

275. S.W. Loke and A. Davison. Logicweb: Enhancing the Web with logic programming. *The Journal of Logic Programming*, 36(3):195–240, September 1998.

276. D. Loveland, D. Reed, and D. Wilson. Satchmore: Satchmo with relevancy. Technical report, Duke University, Durham, North Carolina, April 1993.

277. D.W. Loveland. Near-Horn prolog. In J.L. Lassez, editor, *Proc. 4th Int. Conf. on Logic Programming*, pages 456–459, 1987.

278. J. Lu, A. Nerode, and V.S. Subrahmanian. Hybrid knowledge bases. *IEEE Transactions on Knowledge and Data Engineering*, 8(5):773–785, Oct. 1996.

279. B. Ludascher, W. May, and G. Lausen. Nested transactions in a logical language for active rules. In D. Pedreschi and C. Zaniolo, editors, *Logic in Databases (LID'96)*, pages 217–242, July 1-2 1996.

280. B. Ludäscher, W. May, and G. Lausen. Referential actions in logic rules. In *Proceedings Sixteenth ACM SIGACT-SIGMOD-SIGART Symposium on Principles of Database Systems (PODS 97)*, pages 217–227, Tucson, Arizona, May 12-14 1997.

281. C. Maindreville and E. Simon. Modeling non-deterministic queries and updates in deductive databases. In *Proc. of VLDB*, 1988.

282. R. Manthey and F. Bry. Satchmo: A theorem prover implemented in prolog. In *Proc. 9th Int. Conf. on Automated Deduction (CADE)*, May 1988.

283. V. W. Marek, A. Nerode, and J.B. Remmel. The stable models of a predicate logic program. In K. Apt, editor, *Proc. of the Joint Int. Conf. and Symposium on Logic Programming*, pages 446–460, Washington D.C., USA, Nov 1992. The MIT Press.

284. V.W. Marek and M. Truszczyński. Autoepistemic logic. *Journal of the ACM*, 38(3):588–619, 1991.

285. W. Marek and M. Truszczyński. Stable logic programming and an alternative logic programming paradigm. In K.R. Apt, V.W. Marek, M. Truszczyński, and D.S. Warren, editors, *The Logic Programming Paradigm: a 25-Year Perspective*. Springer, Berlin, 1999, to appear.

286. C. Martin and J. Sistac. Applying transition rules to bitemporal deductive databases for integrity constraint checking. In D. Pedreschi and C. Zaniolo, editors, *Logic in Databases (LID'96)*, pages 111–128, July 1-2 1996.

287. J. McCarthy. Circumscription - a form of non-monotonic reasoning. *Artificial Intelligence Journal*, 13:27–39, 1980.

288. D. McDermott and J. Doyle. Non-monotonic logic i. *Artificial Intelligence Journal*, 13:41–72, 1980.

289. J.R. McSkimin and J. Minker. The use of a semantic network in deductive question-answering systems. *Proc. IJCAI 5*, pages 50–58, 1977.

290. G. Mecca and A.J. Bonner. Sequences, datalog and transducers. In *Proc. 14th ACM SIGACT-SIGMOD-SIGART Symp. on Principles of Database Systems (PODS 95)*, pages 23–35, May 1995.

291. J. Melton. An SQL3 snapshot. In *Twelfth International Conference on Data Engineering*, pages 666–672, 1996.

292. J. Melton and A. R. Simon. *Understanding the New SQL: A Complete Guide*. Morgan Kaufmann, San Mateo, CA, 1993.

293. A.O. Mendelzon and T. Milo. Formal models of web queries. In *Proc. 16th ACM SIGACT-SIGMOD-SIGART Symp. on Principles of Database Systems (PODS 97)*, pages 134–143, May 1997.

294. R. Miller, Y. Ioannidis, and R. Ramakrishnan. Translation and integration of heterogeneous schemas: Bridging the gap between theory and practice. *Information Systems*, 19(1):3–31, Jan. 1994.

295. J. Minker. Search strategy and selection function for an inferential relational system. *Trans. on Data Base Systems*, 3(1):1–31, 1978.

296. J. Minker. On indefinite databases and the closed world assumption. In *Proceedings of the Sixth Conference on Automated Deduction*, pages 292–308, 1982. Also in: *LNCS* 138, pages 292-308, Springer, 1982.

297. J. Minker, editor. *Proceedings of Workshop on Foundations of Deductive Databases and Logic Programming*, August 1986.

298. J. Minker, editor. *Foundations of Deductive Databases and Logic Programming.* Morgan-Kaufmann, 1988.

299. J. Minker. Perspectives in deductive databases. *Journal of Logic Programming*, 5:33–60, 1988.

300. J. Minker. Toward a foundation of disjunctive logic programming. In *Proceedings of the North American Conference on Logic Programming*, pages 121–125. MIT Press, 1989. Invited Banquet Address.

301. J. Minker. An overview of nonmonotonic reasoning and logic programming. *Journal of Logic Programming*, 17(2, 3 and 4):95–126, November 1993.

302. J. Minker. Logic and databases:a 20 year retrospective. In D. Pedreschi and C. Zaniolo, editors, *Logic in Databases*, pages 3–57. Springer, July 1996. Proc. Int. Workshop LID'96, San Miniato, Italy.

303. J. Minker. Logic and databases: Past, present, and future. *AI Magazine*, 16(3):21–47, Fall 1997.

304. J. Minker and J. Grant. Answering queries in indefinite databases and the null value problem. In P. Kanellakis, editor, *Advances in Computing Research*, pages 247–267. JAI Press, 1986.

305. J. Minker and J-M. Nicolas. On recursive axioms in deductive databases. *Information Systems*, 7(4):1–15, 1982.

306. J. Minker and A. Rajasekar. Procedural interpretation of non-Horn logic programs. In E.L. Lusk and R.A. Overbeek, eds., *Proc. of the 9th Int. Conf. on Automated Deduction*, pages 278–293, 1988.

307. J. Minker and A. Rajasekar. Disjunctive logic programming. In *Proceedings of the International Symposium on Methodologies for Intelligent Systems*, pages 381–394, 1989. (Invited Lecture).

308. J. Minker and A. Rajasekar. A fixpoint semantics for disjunctive logic programs. *Journal of Logic Programming*, 9(1):45–74, July 1990.

309. J. Minker and C. Ruiz. On extended disjunctive logic programs. In J. Komorowski and Z.W. Raś, editors, *Proc. of the Seventh International Symposium on Methodologies for Intelligent Systems*, pages 1–18. LNAI, Springer-Verlag, June 1993. (Invited Paper).

310. J. Minker and C. Ruiz. Mixing a default rule with stable negation. In *Proceedings of the Fourth International Symposium on Artificial Intelligence and Mathematics*, pages 122–125, Jan. 1996.

311. J. Minker and J.D. Sable. Relational data system study. RADC-TR-70-180, Rome Air Development Center, Griffiss AF Base, NY, Sept 1970. Auerbach Corp. Report AD 720-263.

312. J. Minker and G. Zanon. An Extension to Linear Resolution with Selection Function. *Inf. Proc. Letters*, 14(3):191–194, June 1982.

313. J. Minker, editor. *Foundations of Deductive Databases and Logic Programming.* Morgan Kaufmann Publishers, 1988.

314. J. Minker and C. Ruiz. Semantics for disjunctive logic programs with explicit and default negation. *Fundamenta Informaticae*, 20(3/4):145–192, 1994. Anniversary Issue edited by H. Rasiowa.

315. R.C. Moore. Possible-world semantics for autoepistemic logic. In *Proc. AAAI Workshop on Nonmonotonic Reasoning*, pages 396–401, 1984.

316. R.C. Moore. Semantical considerations on nonmonotonic logic. *Artificial Intelligence* 25, pages 75–94, 1985.

317. S. Morishita, M. Derr, and G. Phipps. Design and implementation of the Glue-Nail database system. In *Proc. ACM-SIGMOD'93 Conf.*, pages 147–167, May 1993.

318. S. Muggleton and L. De Raedt. Inductive logic programming: theory and methods. *Journal of Logic Programming*, 19/20:629–679, May/July 1994.

319. I.S. Mumick, S. Finkelstein, H. Pirahesh, and R. Ramakrishnan. Magic is relevant. In *Proc. of the ACM SIGMOD Intl. Conf. on Management of Data*, May 1990.

320. J.F. Naughton and Y. Sagiv. A decidable class of bounded recursions. *Proc. of the Sixth ACM SIGACT-SIGMOD-SIGART Symposium on Principles of Database Systems*, pages 227–236, March 1987.

321. R. Ng and V.S. Subrahmanian. Probabilistic logic programming. *Information and Computation*, 101(2):150–201, 1993.

322. J.-M. Nicolas. Logic for improving integrity checking in relational databases. *Acta Informatica*, 18(3):227–253, Dec. 1979.

323. J.-M. Nicolas and H. Gallaire. Data base: Theory vs. interpretation. In H. Gallaire and J. Minker, editors, *Logic and Data Bases*, pages 33–54. Plenum Press, New York, 1978.

324. J.-M. Nicolas and J.-C. Syre. Natural question - answering and automatic deduction in system syntex. *Proc. IFIP Congress 1974*, pages 595–599, 1974.

325. J.-M. Nicolas and K. Yazdanian. Integrity checking in deductive databases. In H. Gallaire and J. Minker, editors, *Logic and Data Bases*, pages 325–599. Plenum, New York, 1978.

326. I. Niemelä. Logic programs with stable model semantics as a constraint programming paradigm. In I. Niemelä and T. Schaub, editors, *Proceedings of the Workshop on Computational Aspects of Nonmonotonic Reasoning*, pages 72–79, June 2–4 1998.

327. I. Niemelä and P. Simons. Efficient implementation of the well-founded and stable model semantics. In I. Niemelä and T. Schaub, editors, *Proceedings of JICSLP-96*, Cambridge, MA, 1996. MIT Press.

328. I. Niemela and P. Simons. Smodels - an implementation of the stable model and well-founded semantics for normal logic programs. In *Logic Programming and Nonmonotonic Reasoning - 4th International Conference, LPNMR '97*, pages 420–429. Springer, 1997.

329. N.W. Paton, Jr. (editor). *Active Rules in Database Systems*. Springer-Verlag, New York, 1999. ISBN 0-387-98529-8.

330. P. Pearce and G. Wagner. Logic programming with strong negation. In P. Schroeder-Heister, editor, *Proc. of the Int. Workshop on Extensions of Logic Programming*, pages 311–326, Tübingen, FRG, Dec 1989. Lecture Notes in AI, Springer -Verlag.

331. G. Phipps, M. Derr, and K.A. Ross. Glue-Nail: A deductive database system. In *Proc. ACM-SIGMOD'91 Conf.*, May 1991.

332. G. Piatetsky-Shapiro and W.J. Frawley, editors. *Knowledge Discovery in Databases*. AAAI Press and MIT Press, Menlo Park, CA., 1991.

333. P. Picouet and V. Vianu. Semantics and expressiveness issues in active databases. In *Proc. 14th ACM SIGACT-SIGMOD-SIGART Symp. on Principles of Database Systems (PODS 95)*, pages 126–138, San Jose, CA, May 1995.

334. P. Powell. Answer-Reason extraction in a parallel relational data base system. MS thesis, Dept. of Comp. Sci., Univ. of Maryland, 1977.

335. S. Pradhan. Combining datalog databases using priorities. In *Advances in Data Management '94*, pages 355–375, 1995.

336. S. Pradhan, J. Minker, and V.S. Subrahmanian. Combining databases with prioritized information. *Journal of Intelligent Information Systems*, 4(3):231–260, May 1995.

337. T.C. Przymusinski. Stationary semantics for disjunctive logic programs and deductive databases. In S. Debray and M. Hermenegildo, editors, *Proc., No. Amer. Conf. on Logic Programming*, pages 40–62, 1990.

338. T. C. Przymusinski. On the declarative semantics of deductive databases and logic programming. In J. Minker, editor, *Foundations of Deductive Databases and Logic Programming*, chapter 5, pages 193–216. Morgan Kaufmann Pub., Washington, D.C., 1988.

339. T. C. Przymusinski. On the declarative and procedural semantics of logic programs. *Journal of Automated Reasoning*, 5:167–205, 1989.

340. T. C. Przymusinski. Extended stable semantics for normal and disjunctive programs. In D.H.D. Warren and P. Szeredi, editors, *Proceedings of the 7th International Logic Programming Conference*, pages 459–477, Jerusalem, 1990. MIT Press. Extended Abstract.

341. T. C. Przymusinski. Stable semantics for disjunctive programs. *New Generation Computing*, 9:401–424, 1991.

342. T.C. Przymusinski. Perfect model semantics. In R. Kowalski and K. Bowen, editors, *Proc. of the 5th Int. Conf. and Symposium on Logic Programming*, pages 1081–1096, Aug 1988.

343. T.C. Przymusinski. Every logic program has a natural stratification and an iterated fixed point model. In *Proc., 8th ACM SIGACT-SIGMOD-SIGART Symp. on Principle of DB Systems*, pages 11–21, 1989.

344. T.C. Przymusinski. Static semantics for normal and disjunctive logic programs. *Annals of Mathematics and Artificial Intelligence*, 14:323–357, 1995. Festschrift in honor of Jack Minker.

345. E. Pudilo. Database query evaluation with the STARBASE method. In D. Pedreschi and C. Zaniolo, editors, *Logic in Databases (LID'96)*, pages 335–354, July 1-2 1996.

346. X. Qian. Query folding. In *Proceedings of the 12th International Conference on Data Engineering*, pages 48–55, 1996.

347. A. Rajaraman, Y. Sagiv, and J.D. Ullman. Answering queries using templates with binding patterns. In *Proc. 8th ACM SIGACT-SIGMOD-SIGART Symp. on Principles of DB Systems (PODS 95)*. 1995.

348. A. Rajasekar, J. Lobo, and J. Minker. Skeptical reasoning and disjunctive programs. In *Proc. of 1st Int. Conf. on Knowledge Representation and Reasoning*, pages 349–357. Morgan-Kaufmann, 1989.

349. A. Rajasekar, J. Lobo, and J. Minker. Weak generalized closed world assumption. *Journal of Automated Reasoning*, pages 293–307, 1989.

350. R. Ramakrishnan. *Applications of Logic Databases*. 1995.

351. R. Ramakrishnan and J.D. Ullman. A survey of research on deductive database systems. *Journal of Logic Programming*, 23(2):125–149, May 1995.

352. R. Ramakrishnan, D. Srivastava, and S. Sudarshan. CORAL—control, relations and logic. In L.-Y. Yuan, editor, *Proc. of the 18th Int. Conf. on Very Large Databases*, pages 238–250, 1992.

353. K. Ramamohanarao. An implementation overview of the Aditi deductive database system. In *LNCS 760, Third Int. Conf., DOOD'93*, pages 184–203, Phoenix, AZ, December 1993. Springer-Verlag.

354. P. Rao, K. Sagonas, T. Swift, D.S. Warren, and J. Friere. XSB: A system for efficiently computing well-founded semantics. In J. Dix, U. Ferbach, and A. Nerode, editors, *Logic and Nonmonotonic Reasoning - 4th Int. Conf., LPNMR'97*, pages 430–440, July 1997.

355. B. Raphael. A computer program for semantic information retrieval. In M. Minsky, editor, *Semantic Information Processing*, pages 33–134. MIT Press, 1968.

356. L. Raschid. A semantics for a class of stratified production system programs. *Journal of Logic Programming*, 21(1):31–57, 1994.

357. L. Raschid and J. Lobo. A semantics for a class of non-deterministic and causal production system programs. *Journal of Automated Reasoning*, 12:305–349, 1994.

358. L. Raschid and J. Lobo. Semantics for update rule programs and implementation in a relational database management system. *ACM Transactions on Database Systems*, 1996.

359. D.W. Reed and D.W. Loveland. A comparison of three prolog extensions. TR CS-1989-8, Department of Conputer Science, Duke University, March 1990. Appears in Journal of Logic Programming.

360. D.W. Reed, D.W. Loveland, and B.T. Smith. An alternative characterization of disjunctive logic programs. In *Proc. of the Int. Logic Programming Symposium*, Cambridge, MA, 1991. MIT Press.

361. R. Reiter. An approach to deductive question-answering. Technical report, Bolt, Beranek and Newman, Inc., Cambridge, 1977.

362. R. Reiter. On closed world data bases. In H. Gallaire and J. Minker, editors, *Logic and Data Bases*, pages 55–76. Plenum, New York, 1978.

363. R. Reiter. A logic for default reasoning. *Artificial Intelligence Journal*, 13:81–132, 1980.

364. R. Reiter. Towards A Logical Reconstruction of Relational Database Theory. In M.L. Brodie, J.L. Mylopoulos, and J.W. Schmit, editors, *On Conceptual Modelling*, pages 163–189. Springer, NY, 1984.

365. R. Reiter. A sound and sometimes complete query evaluation algorithm for relational databases with null values. *J.ACM*, 33(2):349–370, April 1986.

366. R. Reiter. A thoery of diagnosis from first principles. *Artificial Intelligence*, 32:57–95, 1980.

367. R. Reiter. On integrity constraints. In M.Y. Vardi, editor, *Proc. of the 2nd Conf. on the Theoretical Aspects of Reasoning about Knowledge*, pages 97–111, San Francisco, CA, March 1988.

368. R. Reiter. On asking what a database knows. In J.W. Lloyd, editor, *Computational Logic*, Basic research Series. Springer-Verlag Publishers, 1990. DG XIII Commission of the European Communities.

369. J.A. Robinson. A machine-oriented logic based on the resolution principle. *J.ACM*, 12(1), January 1965.

370. J. Rohmer, R. Lescoeur, and J-M. Kerisit. The Alexander method: a technique for the processing of recursive axioms in deductive databases. *New Generation Computing*, 4(3), 1986.

371. K.A. Ross. Well-founded semantics for disjunctive logic programs. In *Proc. of the first International Conference on Deductive and Object Oriented Databases*, pages 352–369, Kyoto, Japan, December 1989.

372. K.A. Ross. A procedural semantics for well-founded negation in logic programs. In *Proc. of the 8th ACM SIGACT-SIGMOD-SIGART Symp. on Principle of Database Systems*, Philadelphia, PA., 1989.

373. K.A. Ross. Modular stratification and magic sets for datalog programs with negation. In *Proc. ACM Symp. on Principles of Database Systems*, April 1990.

374. K.A. Ross. Modular acylicity and tail recursion in logic programs. In *Proc. of the Tenth ACM SIGACT-SIGMOD-SIGART Symp. on Principles of Database Systems (PODS'91)*, pages 92–101, 1991.

375. K.A. Ross and R.W. Topor. Inferring negative information from disjunctive databases. *Journal of Automated Reasoning*, 4(2):397–424, December 1988.

376. C. Ruiz and J. Minker. Computing stable and partial stable models of extended disjunctive logic programs. In J. Dix, L.M. Pereira, and T.C. Przymusinski, editors, *Nonmonotonic Extensions of Logic Programming*, pages 205–229. LNCS 927. Springer-Verlag, 1995.

377. C. Ruiz and J. Minker. Logic knowledge bases with two default rules. *Journal of Mathematics & AI*, 22(3,4):363–361, July 1998.

378. D. Saccà. The expressive power of stable models for bound and unbound DATALOG queries. *The Journal of Computer and System Sciences*, 1996.

379. D. Saccà and C. Zaniolo. On the implementation of a simple class of logic queries. In *Proc. ACM Symp. on Principles of Database Systems*, March 1986.

380. D. Saccà and C. Zaniolo. The generalized counting method for recursive logic queries. *Theoretical Computer Science*, 62, 1988.

381. D. Saccà and C. Zaniolo. Stable models and non-determinism to logic with negation. In *Proc. Workshop on Logic Programming and Nonmonotonic Reasoning*, pages 87–101, 1991.

382. F. Sadri and R.A. Kowalski. Database integrity. In Minker [313], chapter 9, pages 313–362.

383. K. Sagonas, T. Swift, and D.S. Warren. The limits of fixed-order computation. In D. Pedreschi and C. Zaniolo, editors, *Logic in Databases (LID'96)*, pages 355–374, July 1-2 1996.

384. C. Sakama. Possible model semantics for disjunctive databases. In *Proc. First International Conference on Deductive and Object Oriented Databases*, pages 337–351, 1989.

385. E. Sandewall. Features and fluents: A systemetic approach to the representation of knowledge about dynamical systems. Technical report, Institutionen for datavetenskap, Universitet och Tekniska hogskolan i Linkoping, Sweden, 1992.

386. E. Sandewall. The range of applicability of some non-monotonic logics for strict inertia. *Journal of Logic and Computation*, 4(5):581–616, Oct. 1994.

387. D.A. Savitt, H.H. Love, and R.E. Troop. ASP: A new concept in language and machine organization. In *1967 Spring Joint Computer Conference*, pages 87–102, 1967.

388. J.S. Schlipf. The expressive powers of the logic programming semantics. *JCSS*, 1990. A preliminary version appeared in Ninth ACM Symposium on Principles on Database Systems, pages 196-204, 1990.

389. J.S. Schlipf. Complexity and undecideability results for logic programming. *Annals of Mathematics and AI*, 15(3-4):257–288, Dec 1995.

390. D. Seipel. An efficient computation of the extended generalized closed world assumption by support-for-negation sets. In *Proc. Int. Conf. on Logic Programming and Automated Reasoning (LPAR'94)*, number 822 in LNAI, pages 245–259, Berlin, 1994. Springer.

391. D. Seipel. *Efficient reasoning in disjunctive Deductive Databases*. PhD thesis, University of Tübingen, 1995.

392. B. Selman and H.A. Kautz. Knowledge compilation using Horn approximations. In *Proc. of AAAI-91*, pages 904–909, 1991.

393. B. Selman and H.A. Kautz. Knowledge compilation and theory approximation. *Journal of the ACM*, 1996.

394. S.E. Shapiro and D.P. McKay. Inferences with recursion. In *Proc. of the 1st Annual National Conference on Artificial Intelligence*, 1980.

395. J.C. Shepherdson. Negation in Logic Programming. In J. Minker, editor, *Foundations of Deductive Databases and Logic Programming*, pages 19–88. Morgan Kaufman Pub., 1988.

396. A. Sheth and J. Larson. Federated database systems for managing distributed, heterogeneous, and autonomous databases. *ACM Computing Surveys*, 22(3):183–236, September 1990.

397. S. Sickel. A search technique for clause interconnectivity graphs. *IEEE Transactions on Computers*, C-25(8):823–835, Aug. 1976.

398. P. Sistla and O. Wolfson. Temporal conditions and integrity constraint checking in active database systems. In *Proc. of the 1995 ACM SIGMOD Int. Conf. on Management of Data*, 1995.

399. B.T. Smith and D. Loveland. A simple near-Horn Prolog interpreter. In R.A. Kowalski and K.A. Bowen, editors, *Proc. 5th Int. Conf. and Symp. on Logic Programming*, pages 794–809, Seattle, WA, Aug 1988.

400. R. M. Smullyan. Elementary formal system (abstract). *Bull, AMS62*, page 600, 1956.

401. R.M. Smullyan. On definability by recursion (abstract). *Bull, AMS62*, page 601, 1956.

402. R. Snodgrass. The temporal query language TQuel. *ACM Transactions on Database Systems*, 12(2):247–298, June 1987.

403. R. Snodgrass, editor. *Data Engineering*. IEEE Computer Society, December 1988. Special issue on temporal databases.

404. R. Snodgrass and E. McKenzie. Research concerning time in databases. *SIGMOD Record*, 15(4):19–52, December 1986.

405. M. Stickel. A PROLOG technology theorem prover: Implementation by an extended PROLOG compiler. *Journal of Automated Reasoning*, 4(4):353–380, 1988.

406. V.S. Subrahmanian, S. Adali, A. Brink, R. Emery, J. Lu, A. Rajput, T.J. Rogers, and R. Ross. Hermes: A heterogeneous reasoning and mediator system, 1994.

407. V.S. Subrahmanian and S. Jajodia. *Multimedia Database Systems*. Springer Verlag, 1995.

408. V.S. Subrahmanian, D.S. Nau, and C. Vago. Wfs + branch and bound = stable models. *IEEE Transactions on Knowledge and Data Engineering*, 7(3):362–377, June 1995.

409. V.S. Subrahmanian and C. Ward. A deductive database approach to planning in uncertain environments. In D. Pedreschi and C. Zaniolo, editors, *Logic in Databases (LID '96)*, pages 77–92, July 1-2 1996.

410. V.S. Subrahmanian and C. Zaniolo. Relating stable models and ai planning domains. In *Proc. 1995 Intl. Conf. on Logic Programming*, 1995.

411. M. A. Suchenek. Minimal models for closed world databases. In Z.W. Ras, editor, *Proc. of ISMIS 4*, pages 515–522, 1989.

412. T. Swift and D.S. Warren. An abstract machine for SLG resolution: definite programs. In *Proc. of the 1994 ILPS*, pages 633–652. The MIT Press, 1994.

413. T. Swift. Tabling for non-monotonic programming. Technical report, SUNY Stony Brook, 1999. Submitted for publication.

414. The Validity Team. Summary state of the art on deductive and deductive object-oriented databases - dood. Technical report, Bull Corporation, April 1996. Groupe Bull report (limited distribution).

415. Xerox Advanced Information Technologies. HIPAC: a research project in active, time-constrained databases. Technical Report 187, Xerox Advanced Information Technologies, 1989.

416. S. Tsur and C. Zaniolo. LDL: A logic-based data-language. In *Proceedings of the 12th VLDB Conf*, August 1986.

417. J.D. Ullman. *Principles of Database and Knowledge-Base Systems, Volume I*. Principles of Computer Science Series. Computer Science Press, Incorporated, Rockville, MD, 1988.

418. J.D. Ullman. *Principles of Database and Knowledge-Base Systems, Volume II: The New Technologies.* Principles of Computer Science Series. Computer Science Press, Incorporated, Rockville, MD, 1989.

419. J.D. Ullman. Information integration using logical views. In *Proceedings of the Sixth International Conference on Database Theory (ICDT'97)*, Delphi, Greece, Jan 1997.

420. J. Vaghani, K. Ramamohanarao, D. B. Kemp, and P. J. Stuckey. Design overview of the Aditi deductive database system. In *Proc. of the 7th Intl. Conf. on Data Engineering*, pages 240–247, April 1991.

421. M.H. van Emden and R.A. Kowalski. The Semantics of Predicate Logic as a Programming Language. *J.ACM*, 23(4):733–742, 1976.

422. A. Van Gelder. Negation as failure using tight derivations for general logic programs. In J. Minker, editor, *Foundations of Deductive Databases and Logic Programming*, pages 149–176, 1988.

423. L. Vandeurzen, M. Gyssens, and D. Van Gucht. An expressive language for linear spatial databases. In *Proc. 17th ACM SIGACT-SIGMOD-SIGART Symp. on Principles of Database Systems (PODS 98)*, pages 109–118, Seattle, WA, June 1998.

424. M.Y. Vardi. The complexity of relational query languages. In *Proc., 14th ACM Symposium on Theory of Computing*, pages 137–146, May 1982.

425. Y. Vassiliou. Null values in data base management: A denotational semantics approach. In *Proc. of the ACM SIGMOD Int. Symp. on Management of Data*, pages 162–169, Boston, MA, 1979.

426. L. Vieille. Recursive axioms in deductive databases: the Query/SubQuery approach. In *Proc. 1st. Int. Conf. on Expert Database Systems*, April 1986.

427. L. Vieille. Database-complete proof procedures based on SLD-resolutions. In *Proc. 4th Int. Conf. on Logic Programming*, May 1987.

428. L. Vieille. Recursive query processing: The power of logic. *Theoretical Computer Science*, 69, 1989.

429. L. Vieille, P. Bayer, V. Kuechenhoff, and A. Lefebvre. EKS-V1, a short overview. *AAAI'90 Workshop on Knowledge Base Management Systems*, July 1990.

430. L. Vielle, P. Bayer, and V. Kuechenhoff. Integrity checking and materialized view handling by update propagation in the EKS-V1 system. Technical report, ECRC, 1991. Appears in: "Materialized Views", A. Gupta and I. Mumick (eds), MIT Press, Cambridge, MA, 1996.

431. A. Yahya and L.J. Henschen. Deduction in Non-Horn Databases. *J. Automated Reasoning*, 1(2):141–160, 1985.

432. A.H. Yahya. Generalized query answering in disjunctive deductive databases: Procedural and nonmonotonic aspects. In J. Dix, U. Furbach, and A. Nerode, editors, *Proc., 4th Int. Conf. on Logic Programming and Nonmonotonic Reasoning (LPNMR '97)*, LNAI 1265, 1997.

433. H. Z. Yang and P.-Å. Larson. Query transformation for PSJ-queries. In *Proceedings of the Thirteenth International Conference on Very Large Data Bases*, pages 245–254, 1987.

434. J.-H. You and L.Y. Yuan. Three-valued formalization of logic programming. In *Proc. of the 9th ACM PODS*, pages 172–182, 1990.

435. J.-H. You and L.Y. Yuan. A three-valued semantics for deductive databases and logic programs. *Journal of Computer and System Sciences*, 49:334–361, 1994.

436. J.-H. You and L.Y. Yuan. On the equivalence of semantics for normal logic programs. *Journal of Logic Programming*, pages 211–222, 1995.

437. L. Y. Yuan and D.-A. Chiang. A sound and complete query evaluation algorithm for relational databases with disjunctive information. In *Proceedings of the Eighth Symposium on Principles of Database Systems*, pages 66–74. ACM Press, March 1989.
438. L.Y. Yuan and J.-H. You. Autoepistemic circumscription and logic programming. *Journal of Automated Reasoning*, 10:143–160, 1993.
439. C. Zaniolo. Database relations with null values. *JCSS*, 28:142–166, 1984.
440. C. Zaniolo. A unified semantics for active and deductive databases. In *Proceedings of 1st international workshop on rules in database systems*, pages 271–287. Springer-Verlag, 1993.
441. C. Zaniolo. Active database rules with transaction-conscious stable models semantics. In *Proceedings of DOOD 1996*, pages 55–72, 1996.
442. C. Zaniolo, S. Ceri, C. Faloutsos, R.T. Snodgrass, V.S. Subrahmanian, and R. Zicari. *Advanced Database Systems*, 1997.

Action Inventory for a Knowledge-Based Colloquium Agent. Preliminary Version

Erik Sandewall

Department of Computer and Information Science, Linköping University, Linköping, Sweden

*Dedicated to Ray Reiter
on his 60th birthday*

Summary. The article describes the resources and the requirements for an intelligent autonomous agent that assists the moderator-editor of an electronic colloquium, that is, an online facility for communication between researchers. Such an agent should receive information about actual and planned events in the "world" of researchers in a specific area, and it should infer what operations need to be performed in the hierarchical structure containing the colloquium's information base. The article proposes that this would be a useful test domain for the situation calculus and GOLOG, as well as for other approaches to reasoning about actions and change.

Research on reasoning about actions and change has now arrived to a stage where it is important to have a number of well understood domains of non-trivial size and complexity, where different approaches can be tested, compared, and developed. This being so is to no small extent due to Ray Reiter's contributions in the development of the situation calculus and the Golog language to the point where such application domains can be addressed.

If one and the same domain is going to be studied by different researchers and possibly using different formal approaches, it becomes important to have an initial domain description that is accessible to all, and that is kept distinct from the details of how it is realized in formalisms. Toy domains can typically be described in one or two pages, but as the complexity of domains grow we face the need for separate articles that document promising proposed domains in their pre-formalized or semi-formalized stages.

As one contribution towards this end, the present article describes the reportoire of actions that arises in the domain of *electronic colloquium maintenance*. In this Internet-based domain, the state of the world is constituted by the contents of directories and files in a computer network, and the actions are those that change the state of the world in attempts to achieve specified goals. In comparison with traditional cognitive robotics domains where the robot operates in the physical world, the present domain has considerable differences, but also surprisingly many similarities. The major difference seems to be that we do not encounter any continuous or hybrid properties, continuous time being the only non-discrete type that is involved.

The present domain description is based on actual experience with working with the domain without intelligent agent support. It is preliminary in the sense that the catalogue of actions is not proposed to be complete for the application. The article is therefore intended as a source of examples for situation calculus and alternative formalisms, but not as a comprehensive specification towards an implemented system.

1   The Electronic Colloquium

An electronic colloquium is an Internet-based activity and the information structure on the net that it maintains; the colloquium serves as a communication medium and communication resource for researchers in a specific research area. Our particular experience stems from working with the [1]*[Electronic Colloquium on Reasoning about Actions and Change]* which has been organized in conjunction with the [2]*[Electronic Transactions on Artificial Intelligence]* (where Ray is also an editor and contributor). The most salient activity of the colloquium is to organize discussions about research articles, research results, and current topics of general interest for the colloquium members. Other activities and services may include a calendar service, bibliographic services, and the like.

A conventional academic colloquium meets in order to listen to seminars, discuss their contents, and exchange recent news. The electronic colloquium is a forum for exchange of the same kind of information, but over the net and with minimal constraints with respect to spatial location and time of interaction.

The Electronic Transactions on Artificial Intelligence is an electronic journal based on open reviewing (that is, without anonymity) in a number of participating colloquia. Each colloquium has its own membership, its own area editor who also organizes and moderates the colloquium, and its own webpage structure.

From a workflow point of view, the colloquium can be seen as an information conveyor that receives information in the form of *messages*, organizes and reformats this information, and then distributes it in two distinct modes. Colloquium information is *sent out* to the members e.g. as email messages called newsletters. It is also *made available* to the membership through a webpage structure that is set up and maintained for this purpose.

In an extended design, the colloquium does not only receive information through messages that are sent to it; it may also *acquire* information by actively going out and retrieving relevant information from the net. For example, given that the colloquium information structure keeps track of the URL:s for its members' home pages, it may actively download those pages and retrieve the links to the full text of articles of each particular member (after having obtained the owner's approval, of course).

---

[1] Ref: http:/www.etaij.org/rac/
[2] Ref: http:/www.etaij.org/

The parsing of the webpages that would be required for this task is not at all trivial, however.

The task of moderating a colloquium and editing its information structure is very interesting because of its complexity. It is primarily a social task: the liveliness of the colloquium depends a lot on the efforts of the moderator and editor. Few people send contributions spontaneously to a colloquium. The role of moderator is therefore the primary one, but it is intrinsically connected with the role as editor. He or she has to decide where the arriving information is to be located in the colloquium information structure, if at all. He also has to deal with a number of technical aspects of the documents, such as the existence of multiple document editing systems (e.g. Word, Framemaker) and multiple markup schemes (e.g. Latex, HTML), as well as the richness of modes of expression in documents (text, formulae, tables, figures, etc).

For the purposes of ETAI and of our colloquium within the ETAI, we have developed a fairly extensive software system that provides support for both the reception and the presentation side of colloquium information. This software is used to generate the web pages as well as the latex-produced ETAI journal and ENRAC news journal (the latter containing monthly discussion protocols). Especially the news journal production is somewhat complex since each issue contains not only a few articles, but also the summaries of discussions each of which consists of several contributions.

In our experience, the presentation side of this information conveying process can be quite well automated, if one proceeds from the assumption that the colloquium information is stored in uniform, structured, and systematic form in the colloquium information base. The major problem, of course, is how to manage the reception process (and when applicable, the acquisition process) so that the received information can be transformed to the uniform structure in a cost-effective and work-effective way. This is where intelligent agent technology may come in.

2   Additional application aspects

Before we proceed, we need to comment on whether it is necessary to allow incoming information in a wide variety of formats. Wouldn't it be easier to *require* the information providers to use particular markup schemes, style files, and the like, so that information is standardized already at the source? This, after all, is what commercial publishers of scientific journals tend to do.

There are at least two reasons for allowing a multiversity of input document formats. Doing so broadens the "author base" of possible contributors, thereby providing a distinct competitive advantage for the colloquium that has this capability. Furthermore, when the colloquium software is developed so that it makes active acquisition of information from the net, it becomes imperative of course that it can deal with the information in the form that it actually appears at the source.

It is also interesting to view automatic information acquisition in a colloquium as a kind of data mining, but using a method that *focusses on developing a knowledge base:* rather than searching the web for the answer to each particular question, it attempts to build its knowledge base in a systematic way using information from the web; it then uses the knowledge base for presenting what it has found, or for answering questions. The hypothesis is that the web search can be more effective and efficient if it is supported by such a knowledge base.

3 Some examples of tasks

Here are some examples of typical situations that arise in the course of the moderator-editor's work.

1. An information provider sends a contribution to the discussion about one partic-ular article that has been submitted to the colloquium's open-reviewing journal. Article discussion contributions are organized as a set of *threads*, each consist-ing of an initial question, the answer from the authors, and optionally a list of follow-up comments. The contribution is to be assigned to the correct article and the correct thread; it is to be marked with the name of who sent it in; this name has to be annotated with the URL to the person's webpage (taken from the database); and the contribution has to be included in the Newsletter of the day and the News Journal of the month.

2. The same information provider writes back and says that unfortunately there was a mistake in the contribution that he sent the day before, so could the editor please replace the second-last paragraph as follows...

3. The call for papers arrives for a conference that is relevant to the present collo-quium. It arrives as an E-mail message containing also the URL's of the confer-ence information webpage and of the cfp itself. The conference calendarium of the colloquium is to be updated with the name of the conference, its venue and important dates, links to its webpage(s), etc.

4. The discussion about a particular article has diverged (or ascended) into a dis-cussion about a quite general issue. It shall therefore be shifted over to a separate structure for that general discussion, with appropriate cross-links between the article discussion and the topic discussion. Some of the later contributions to the article discussion are to be moved over and considered as part of the topic discussion. This possibility was not foreseen when the webpage structure and the database were set up.

5. Given the lists of contents for each of the newsletters during the past month, rearrange and combine them into a news journal for the month where all discus-sion contributions on the same topic appear together. Generate the news journal issue, complete with table of contents, subscriber information, and the like.

6. A discussion contribution by a colloquium member refers to a number of earlier articles, but only in general terms like "my paper at IJCAI 93" or "Anderson's AI Journal paper on nonmonotonic spatial logic". Replace these phrases by cor-rect bibliographic references, including also direct clickable links to the articles in question and/or to their authors's web pages.

The list can be continued, but this ought to give the general flavor of what is involved. Notice, however, that these descriptions of the examples only capture the structural or " database" side of the problem. In addition, there are problems having to do with the encodings and markups used by the various contributions.

4   Agent support for electronic colloquia

Returning to the question of software support for electronic colloquia, it is clear that the task of receiving colloquium contributions in the forms they tend to take with little or no regulation, is not likely to be automated successfully with conventional software techniques. At the same time, the domain is reasonably well contained, and in particular all the relevant information objects are available in principle, as long as one can assume that the information travels over Internet channels. It is likely, therefore, that knowledge-based techniques can be a useful contribution towards a solution.

This does not mean that the colloquium moderator and editor can be replaced altogether. The social aspect of the moderator activity is very important, as we have emphasized above, so we need a technique whereby the moderator-editor can obtain efficient help with the chore side of her or his duty. In other words, a colloquium agent ought to be a software system that oversees the reception of colloquium contributions, assigns preliminary locations to them in the colloquium information base, and prepares them so that they can easily be read by the editor and so that they are as ready as possible for the distribution phase. The software agent should facilitate work for the moderator-editor, but it should also be able to ask her for advise when the need arises. In short, this has all the typical characteristics of an intelligent software agent application. Not surprisingly, it also makes sense to think of a group of agents with specialized skills that are able to cooperate for dealing with the situations that arise.

The examples in the previous section can be taken as a suggestion that two distinct types of agents are appropriate: an agent that specializes in markup and format conversions of single documents, and an agent that understands how to incorporate documents of various kinds into the colloquium information structure, and what indirect effects ought to follow from those updates. Such a distinction is indeed appropriate up to a point, but in any case those two agents will have to cooperate at times. Structured information about a document or related to it may be helpful for the processes of decoding, but the document may also contain information that properly belongs to the database and to the information *about* the document, rather than *in* it. For example, the customary Latex commands for declaring the author and the title of an article do in fact refer to bibliographic metadata.

5   Present organization of the colloquium software

The design of a software agent must be based on the architecture of an existing, more conventional software system. There is a fairly large set of services that *can* be handled by conventional techniques, after all, and the knowledge-based system should cooperate with those facilities rather than compete with them.

The most important requirement on the base software for the colloquium is that it must handle a combination of structured data and document processing – editing, document assembly, and formatting operations – and it must of course be capable of handling fairly large document and data volumes. For the ETAI and the RAC Colloquium, we are using the AIMS system (Academic Information Management System) which has the following characteristic properties:

- The system uses a hierarchical structure of *information objects* on different levels, where the individual document (for example an article) is often at the second-lowest level, and document fragments are at the lowest level. A diagram, a table, or a contribution to a discussion are examples of fragments.
- Each information object has properties and relations, as usual, but it can also have a number of document files associated with it.
- Information objects can always be accessed by starting from the root and navigating the tree so as to get to the desired node. The arcs from a node to its descendant nodes are labelled by symbols (short strings, "identifiers") and these labels are used for the navigation.
- In most cases, each information object is represented as its own directory in the operating system. In particular, a document is uniformly represented as a directory, and not as a named file within a directory as is usually the case. Exception is made for document fragments, which do not obtain their own directories.
- The structured data belonging to an object is represented by one or a few files with a simple textual format belonging to the object directory.
- There is no separate database management system. All processing is done by programs that operate on the object directory structure, that is, directly on the operating system (UNIX) level. It is therefore straightforward to mix uses of different programming languages in the program library.
- Distributed processing is obtained by using object directories that are WWW accessible, so that files can be transferred using the HTTP protocol. This has previously been described as the World Wide Data Base approach.

Tasks such as those described in section 3 are honored by simple procedures. During the past phase of system development and early use, the actual software has provided a command-oriented dialogue through which one has been able to navigate the object hierarchy and make local changes at the position presently being visited. These local changes involve assigning or reassigning properties of current nodes, operating on the document files associated with those nodes, and checking the results.

Standardized procedures or plans for common tasks have evolved gradually, and were first documented by notes that gave advise to the operator. The next step has been to formalize these procedures into actual programs in a simple and quite restricted language for these database operations. At the same time we have noticed that it is not always possible to leave the job to the formal procedure; human supervision and sometimes intervention has been necessary.

This is exactly the point where current methods for autonomous agents can fill a role. It is natural to view the task procedures as "plans" also in the KR sense, and to consider techniques for constructing plans and for modifying and repairing plans that have been obtained from the plan library. In fact, task procedures share many of the properties of the GOLOG language pioneered by Reiter. One of the first presentations of the principles of GOLOG was made by Ray in his invited lecture at the 1993 IJCAI conference in Chambéry, France, and we certainly belong to those who were inspired by that lecture as well as by much of the continued work on GOLOG.

Notice, however, that these procedures and plans evolved in a quite pragmatic fashion and through bottom-up development of the system. They were not the result of *imposing* KR techniques on the application.

6   States and Actions in the Colloquium Domain

Although it is strikingly natural to view the AIMS task procedures in terms of plans in the logics and languages for actions and change, such a view still begs the question of how we ought to *specify the goals* that the plans attempt to achieve. Are we able to specify the structure of the state space and the target set of states for each of the task types that occur in this application?

Analyzing the application domain, it appears that another perspective is much more fruitful, namely to use a temporal representation of events in the colloquium community as the reference. The application is then understood in terms of *domain events*, such as in the following examples:

- Conference XYZ will take place in (venue) during (days), (year)
- The call for papers (d) for conference (name) in (year) was issued on (date)
- Conference (name)(year) accepted paper (d) submitted by author (name)
- Colloquium member (name) made discussion statement (d) in the open discussion about article (d')
- After the open colloquium discussion about article (d) finished, its author (name) decided to revise the paper and intends to come back with a revised version.
- Author (name) has now submitted the revised version (d') of paper (d) for which the open discussion has been concluded.

Each of these kinds of tasks is associated with a procedure for updating the colloquium information base accordingly. Note that the information base changes in an accumulative way, most of the time. For example, the state of the information base at a given point in time will contain the threads of review-discussion comments that have been made until that time in the discussion about a particular article. Some time earlier, it was in a different and less extensive state. Past states are not being preserved, but the current state contains traces of past history in the domain world, as well as some predictions of the future e.g. for the occurrence of conferences.

The purpose of the task procedures is to update the state of the information base in adequate ways, but it should not be necessary to ask for direct specifications of the postconditions of thosee tasks. Instead, it is appropriate to start with the events in the domain world and along the real-world timeline: *communication events* such as the arrival of a call-for-papers through the electronic mail coming to the moderator-editor of the colloquium.

Every such communication event *implies* that certain data objects, properties, and relationships are to be present in the information base. This then is how we would prefer to specify tasks: a set of logical statements allow us to infer, for every incoming communication event, a number of consequences that must hold in the information base; this constitutes the goal statements for the task procedure. As usual it may be handled either by retrieving an appropriate plan from the plan library and executing it as is, or by adapting plans or constructing them from scratch.

One more observation can be made on the basis of the current experience from the bottom-up implementation. In principle, the task procedures operate on the current state of the hierarchical information structure, but in practice there is a bit more to it. Each task (for example, each incoming communication event) is itself characterized as an object, consisting of a number of properties as well as one or a few associated documents, as shown in the examples above. The *current state* from the point of view of the plans has three components: the current state of the hierarchy of objects, the current position of the agent in that hierarchy, and the current task object. Graphically speaking, the agent "carries the task object with it" as it moves around the hierarchy. The most important actions are as follows:

- go up to the nearest ancestor node
- go down to a descendant node along a path with a given (constant) label
- go down to a descendant node along a path whose label is determined by the properties of the present task object and/or the node in the present position
- assign a property to the current node
- assign a property to the task object
- perform a document operation on one or more of the documents associated with the present node or task object

The property assignment operations may use properties of the current node and task object, and the document operation may use those properties as parameters.

The design that emerges from these observations, therefore, is one where the agent has to construct and execute a plan, composed of operations such as those just described, whereby the logical consequences of the incoming communication events are satisfied in the information base.

7 Discussion

We have proposed a way of specifying the logical requiremennts on an intelligent agent system for assisting the moderator-editor of an electronic colloquium. The specification is easily understandable in logicist terms, but we observe that none of the well-known contemporary approaches to reasoning about actions and change is being used in this way. We hope, therefore, that the present analysis of this particular application area can help fill the need for interesting, nontrivial application domains for logics of actions and change, including both GOLOG and other approaches with which it is in friendly and constructive competition.

The Electronic Colloquium on Reasoning about Actions and Change contains descriptions of a number of proposed applications for action logics, and it is intended that the application that has been outlined here will be added to that structure as soon as the present birthday souvenir has been awarded. The ETAI and ENRAC support software will also be made available to interested parties to the extent that its documentation permits, and this will add further detail to the present scenario example. All such additional information about the present documentation can be obtained from the colloquium web page, http://www.etaij.org/rac/.

A GOLOG Specification of a Hypertext System

Richard B. Scherl

Department of Computer and Information Sciences
New Jersey Institute of Technology
Newark, NJ 07102-1982 U.S.A.
email: scherl@cis.njit.edu

Dedicated to Ray Reiter
on his 60th birthday

Summary. GOLOG, a logic programming language for dynamic domains is utilized to give a specification of a hypertext system. The situation calculus is used to specify the basic actions such as traversing a link, while the overall control structure is modelled with the complex actions of GOLOG. The result serves as both a logical specification of a software system and as an executable model of the very same system.[1]

1 Introduction

From early on [17], work in A.I. on representing and reasoning about actions and their effects has taken the central problem to be the construction of a robotic agent that can both carry out actions and reason about the effects of those actions. Work on the situation calculus[21] and GOLOG [14] has been no exception. Ray Reiter has always [23], [25] taken a broader perspective towards these tools and has encouraged others to look at a wide range of "non-traditional" applications. In using GOLOG to model a hypertext system, this paper is presented in the same spirit.

Hypertext and related systems are becoming ubiquitous. A wide variety of systems have been built. Each of these systems [11]:

> provides its users with the ability to create, manipulate, and/or examine a network of information-containing nodes interconnected by relational links.

There is both a database of interconnected pieces of information and facilities for navigating and modifying this database.

Bieber and Kimbrough [2] have developed an initial logic model of a generic hypertext system and then built upon that model a notion of generalized hypertext.

---

[1] This paper includes portions of "A Situation Calculus model of Hypertext," written by Richard Scherl, Michael Bieber and Fabio Vitali. It was published in the proceedings of the Thirty-First Hawaii International Conference on System Sciences (HICSS-31), January 1998. I thank IEEE Press and also both Michael and Fabio for permission to prepare this version.

Generalized hypertext uses logical inferencing rather than manual (human) specification to create links, nodes, and impose views. In addition to serving as the foundation for generalized hypertext, the logic model also proved useful for the specification and coding of the generalized hypertext system Max [2]. Logic models of hypertext systems can in general play a useful role in the comparison of various hypertext systems, as well as in the specification and coding of specific systems, much in the same way as a specification in a formal language such as Z. Once inferencing is added, as in generalized hypertext, the logic model becomes essential.

But the logic model developed by Bieber and Kimbrough [2] does not have a means of representing the dynamic aspects that are essential to a hypertext systems. It really is only a logic model for certain elements of the hypertext system. The changes that occur as links are traversed, is left outside of the model. The use of a logic of action such as the situation calculus is necessary to capture the dynamic aspects

The situation calculus provides a formalism for reasoning about actions and their effects on the world. Axioms are used to specify the prerequisites of actions as well as their effects, that is, the fluents that they change. In general, it is also necessary to provide frame axioms to specify which fluents remain unchanged by the actions. In the worst case this might require an axiom for every combination of action and fluent. Recently, Reiter [21] (generalizing the work of Haas [9], Schubert [27] and Pednault [20]) has given a set of conditions under which the explicit specification of frame axioms can be avoided. Under these circumstances a relatively simple solution suffices. We utilize the formulation of Reiter in this paper.

Building upon Reiter's work, the Cognitive Robotics group at the Univesity of Toronto [14], [13], [12] has further developed the situation calculus to both model a robotic agent in a dynamic world and to develop a high-level programming language called GOLOG for declaratively defining complex actions (such as iteration, conditionals, and loops) that are built upon the basic primitive actions of the situation calculus. Thus specification and implementation are accomplished in a unified framework.

In this paper, we utilize both the situation calculus and GOLOG to represent the dynamics of a hypertext system. Even though the situation calculus was initially developed and studied with A.I. problems in mind, it and the approach to the frame problem has been applied to the formalization of software systems. For example it has been used to formalize the evolution of a database under the effect of an arbitrary sequence of update transactions [23]. Additionally, the approach has been used as the foundations for a language for software specification [3]

We utilize the situation calculus to develop a logical model of a core sort of hypertext system, very close to the basic hypertext of [2]. It does not, for example, have all of the features of the Dexter model [11]. These can all be readily added. For example, the only links considered in this paper are binary and we do not consider composite nodes.

In [11], a hypertext system is divided into three layers. These are the *run-time* layer, the *storage* layer, and the *within-component* layer. The run-time layer consists of the mechanisms that enable the user to interact with the hypertext system. The storage layer is the network of nodes and the links between them. The internal structure of the nodes is captured by the within-component layer. In this paper nothing more will be said about the within-component layer as we will not be considering composite nodes.

In the next section, the situation calculus background is given. Then, in Section 3, a model of the core hypertext system is developed. This discussion includes the representation of the storage layer, and the basic actions needed by the run-time layer. But it does not include a representation of the overall operation of the run-time layer, because that requires a discussion of the complex actions that form the basis of GOLOG.

In Section 4 complex actions and the GOLOG programming language are discussed and in Section 5 these features are used to specify the run-time component. At this point, the model can be used to reason about what is true after the execution of a particular sequence of actions and to represent the hypertext system as an ongoing process, a read-evaluate-print loop. Finally, in Section 6, the paper is summarized and future work is discussed.

2   The Situation Calculus

The situation calculus (following the presentation in [21]) is a first-order language for representing dynamically changing worlds in which all of the changes are the result of named *actions* performed by some agent. For example DROP(x) represents the action of dropping some object x. Terms are used to represent states of the world–i.e. *situations*. If α is an action and s a situation, the result of performing α in s is represented by $do\,(\alpha, s)$. The constant S_0 is used to denote the initial situation. Relations whose truth values vary from situation to situation, called *fluents*, are denoted by predicate symbols taking a situation term as the last argument. For example, BROKEN (x, s) means that object x is broken in situation s. Functions whose denotations vary from situation to situation are called *functional fluents*. They are denoted by function symbols with an extra argument taking a situation term, as in POSITION($robot, s$), i.e., the position of the *robot* in s.

It is assumed that the axiomatizer has provided for each action $\alpha(\boldsymbol{x})$, an *action precondition axiom* of the form given in 1, where $\pi_\alpha(\boldsymbol{x}, s)$ is a formula specifying the preconditions for action $\alpha(\boldsymbol{x})$.

$$\text{POSS}(\alpha(\boldsymbol{x}), s) \equiv \pi_\alpha(\boldsymbol{x}, s) \tag{1}$$

An action precondition axiom for the action DROP is given below.

$$\text{POSS}(\text{DROP}(x), s) \equiv \text{HOLDING}(x, s) \tag{2}$$

The axiom states that the drop action is possible if and only if the agent is holding an object.

The core of the method of axiomatization is the construction of *successor state axioms*. Their general form is given below:

$$\text{POSS}\,(a, s) \;\rightarrow\; [F(\boldsymbol{x}, do(a, s)) \;\equiv\; \\ \gamma_F^+(\boldsymbol{x}, a, s) \;\vee\; (F(\boldsymbol{x}, s) \;\wedge\; \neg\gamma_F^-(\boldsymbol{x}, a, s))] \tag{3}$$

Similar successor state axioms may be written for functional fluents. A successor state axiom is needed for each fluent F, and an action precondition axiom is needed for each action a.

Reiter[21] shows how to derive a set of *successor state axioms* of the form given in 3 from the usual positive and negative effect axioms, a completeness assumption[2], and the restriction that there are no ramifications, i.e., indirect effects of actions[3]. Often it is possible to code axioms directly in the form of successor state axioms. That is what we will do in the next section of this paper.

Given such an axiomatization, one can give the axiomatization of the initial situation $\mathcal{F}_0$, the axiomatization of the hypertext system $\mathcal{F}_{ss}$ and then ask whether or not the axiomatization entails that a particular sentence G will be true after the execution of a particular sequence of actions (contained in s_{gr}).

$$\mathcal{F}_0 \bigcup \mathcal{F}_{ss} \models G(s_{gr})$$

Methods for efficiently automating such queries are discussed in [21].

3 Hypertext

Following [2], a *hypertext* consists of an arbitrary number of interrelated *nodes*, *links*, and *buttons*. *Nodes* are objects that are declared in a data base and, when displayed, are represented as text on the screen. *Links*, which describe relationships between pairs of nodes[4] (called the source and sink), are also declared in a data base.

The essential fluents (adapted from [2]) and their intended interpretations are as follows:

1. $\text{NODE}(x, y, z, s)$ x is a node with content expression y and semantic type z in situation s.
2. $\text{LINK}(u, v, w, x, y, z, s)$ u is a link from source node v to sink node w with semantic type x, operation type y, and either display mode or procedure identifier z in situation s.

---

[2] Reiter[21] also discusses the need for unique name axioms for actions and situations.

[3] This last condition can be ensured by prohibiting state constraints, i.e., sentences that specify an interaction between fluents. An example of such a sentence is $\forall s P(s) \equiv Q(s)$. If an action a made $P(do(a, s))$ true, then the action would have the indirect effect of making $Q(do(a, s))$ true as well since the constraint specifies that Q must have the same truth value as P in every situation.

[4] The links are directional — pointing from the source node to the sink node.

3. BUTTON(x, y, z, s) x is a button representing link y with context expression z in situation s.
4. WINDOW(s) denotes that which is displayed in the main window in situation s (a functional fluent).
5. CURRENT_NODE(x, s) x is the current node in situation s.
6. POP_UP_WINDOW(s) denotes the display in the pop_up_window in situation s (a functional fluent).

Consider the following simple hyperdocument[5] designed to be of use to someone interested in both travel and butterflies. All of these declarations are facts declared to be true at the initial situation S_0.

NODE(1,[' Pieridae:',' This is a large family',' of more than 1,000', 'species of butterfly.', ' Most of the species', ' are predominantly', ' white, yellow, or orange', ' in color and are often', ' referred to collectively', ' as whites, yellows, or sulphurs.', ' Pigments that are derived', ' from the body's waste', ' products explain the', ' distinct coloring, which', ' is a feature', ' peculiar to this family', ' of butterflies. Species:' button(1), button(2), button(3)], DESCRIPTION,S_0)

NODE(2,[' Orange Albatross', ' This striking butterly', ' is probably the only species', ' in the world that', ' is entirely orange in color.', ' Females look similar', ' to males but have a black', ' border around the wings', ' and a black band on the hindwing.', ' Males are often seen', ' drinking from the moist sand of riverbanks.', ' Females are much less bold', ' and tend to keep high in the tree canopy.', ' They are known to feed from', ' the flowers of a variety of trees.', ' Region: Indo-Australian'], DESCRIPTION,S_0)

NODE(3,['Bali is a small island,', ' midway along the string', ' of islands which makes', ' up the Indonesian archipelago.', ' The island offers', ' sandy beaches as well', ' as dramatically mountainous terrain.', ' Bali is also has rich culture ', ' with distinctive traditions', ' in the arts ',button(4), button(5), ' and religion', button(6)' .'],DESCRIPTION,S_0)

NODE(4,[' Balinese music is based', ' around an ensemble known as a gamelan,', 'which can comprise from four to', 'as many as 50 or more instruments.', 'It is derived from the Javanese', ' gamelan, though', ' the playing style is quite different.'],DESCRIPTION,S_0)

LINK(1,1,2,MEMBERS_OF,DISPLAY, FULL_WINDOW,S_0)

LINK(2,3,4,MORE_INFORMATION,DISPLAY,FULL_WINDOW,S_0)

LINK(3,3,2,MORE_INFORMATION,DISPLAY,FULL_WINDOW,S_0)

---

[5] The information contained in this hyperdocument is taken from [5] and [16].

BUTTON(1,1,'Appias nero',S_0)

BUTTON(2,2,'Mylothris chloris',S_0)

BUTTON(3,3,'Neophasia menapia',S_0)

BUTTON(4,4,'Music',S_0)

BUTTON(5,5,'Painting',S_0)

BUTTON(6,6,'Hindu Influence',S_0)

BUTTON(7,1,'Family',S_0)

BUTTON(0,0,'Exit',S_0)

CURRENT_NODE(3,S_0)

This database declares nodes 1, 2, 3, and 4. Link 1 has node 1 as a source node and node 2 as a sink node. Link 2 has node 3 as a source node and node 4 as a sink node. Link 3 has node 3 as a source node and node 2 as a sink node. Links 4, 5, and 6 are not given here. Button 1 is a button for link 1 and is displayed with node 1. Similarly for buttons 2, 3, 4, 5, and 6. Button 7 is displayed with node 1. Node 1 is set to be initially the current node. Finally, button 0 is always displayed so that the user can signal to exit and terminate the session.

The available commands (adapted from [2]) are:

1. TRAVERSE_SINK(x, y) moves the current node from x to y.
2. TRAVERSE_SOURCE(x, y) moves the current node from y to x.
3. DISPLAY_NODE_ATTRIBUTE(x) displays in a pop-up window the attribute of the node x.
4. DISPLAY_LINK_ATTRIBUTE(x) displays in a pop-up window the attribute of the link x.
5. MAKE_DISPLAY_TEXT(x) displays in a window the text encoded in content expression x.
6. CREATE_NODE(x, y, z) adds to the database node x with content expression y and semantic type z.
7. DELETE_NODE(x) deletes from the database node x.
8. CREATE_LINK(u, v, w, x, y, z) adds to the database link u from source node v to sink node w with semantic type x, operation type y, and either display mode or procedure identifier z.
9. DELETE_LINK(u) deletes from the database link u.

The action MAKE_DISPLAY_TEXT creates the display in the window. Here we do not give a full specification of what appears in the window. But it is required that the action properly displays all buttons in the content of the node and also button 0 is always displayed. Also, we assume that there can only be one main window and one pop_up_window displayed at a time.

The following is the successor state axiom for the predicate CURRENT_NODE.

$$
\text{POSS}(a, s) \rightarrow [\text{CURRENT\_NODE}(x, do(a, s)) \equiv \\
a = \text{TRAVERSE\_SINK}(y, x) \ \lor \ a = \text{TRAVERSE\_SOURCE}(x, y) \ \lor \\
(\neg(a = \text{TRAVERSE\_SINK}(y, x) \ \lor \ a = \text{TRAVERSE\_SOURCE}(x, y)) \\
\land \ \text{CURRENT\_NODE}(x, s))]
\tag{4}
$$

Sentence 4 specifies that the current node in a particular situation that results from performing an action is either the sink node on a link if the action was a TRA-VERSE_SINK action, or the source node on a link if the action was a TRAVERSE_SOURCE action or otherwise the current node in the situation prior to the action.

The axioms[6] capturing the possibility conditions for the two actions TRAVERSE_SINK and TRAVERSE_SOURCE are as follows:

$$
\text{POSS}(\text{TRAVERSE\_SINK}(y, x), s) \equiv \\
\text{CURRENT\_NODE}(y, s) \ \land \ \exists u \ \text{LINK}(u, y, x, \_, \_, \_, s)
\tag{5}
$$

$$
\text{POSS}(\text{TRAVERSE\_SOURCE}(x, y), s) \equiv \\
\text{CURRENT\_NODE}(y, s) \ \land \ \exists u \ \text{LINK}(u, x, y, \_, \_, \_, s)
\tag{6}
$$

Sentence 5 specifies that the TRAVERSE_SINK action from y to x is possible in a particular situation s if and only if y is the current node and there exists a link from y to x. Sentence 6 specifies that the TRAVERSE_SOURCE action from y to x is possible in a particular situation s if and only if y is the current node and there exists a link from x to y.

The following is the successor state axiom for WINDOW.

$$
\text{POSS}(a, s) \rightarrow [\text{WINDOW}(do(a, s)) = x \equiv \\
a = \text{MAKE\_DISPLAY\_TEXT}(x) \ \lor \ (a \neq \text{MAKE\_DISPLAY\_TEXT}(y) \\
\land \ \text{WINDOW}(s) = x)]
\tag{7}
$$

Some object x is displayed in the main window if either the previous action was a MAKE_DISPLAY_TEXT(x) action or x was already in the window and the action was something other than one that displayed text in the window.

The following is the specification of the possibility condition for the action MAKE_DISPLAY_TEXT.

---

[6] We note that for the sake of notational perspicuity we are using the underscore as in Prolog's anonymous variable. No logical generality is lost by this, for an equivalent formula can always be had by replacing each anonymous variable with a unique variable and universally quantifying over the entire formula.

$$\text{POSS}(\text{MAKE\_DISPLAY\_TEXT}(x), s) \equiv$$
$$\exists z \, \text{CURRENT\_NODE}(z, s) \ \wedge \ \text{NODE}(z, x, \_, s) \tag{8}$$

The displayed material must be the content expression of the current node.

Sentence 9 is the successor-state axiom for what is displayed in the pop_up_window.

$$\text{POSS}(a, s) \rightarrow [\text{POP\_UP\_WINDOW}(do(a, s)) = x \equiv$$
$$(\exists y \, a = \text{DISPLAY\_LINK\_ATTR}(y) \ \wedge \ \text{LINK}(y, \_, \_, x, \_, \_, s)) \vee$$
$$(\exists y a = \text{DISPLAY\_NODE\_ATTR}(y) \ \wedge \ \text{NODE}(y, \_, x, s)) \vee \tag{9}$$
$$(\neg(a = \text{DISPLAY\_LINK\_ATTR}(y) \ \vee \ a = \text{DISPLAY\_NODE\_ATTR}(y))$$
$$\wedge \ \text{POP\_UP\_WINDOW}(s) = x))]$$

The following are the two axioms defining the possibility conditions for the actions DISPLAY_LINK_ATTR and DISPLAY_NODE_ATTR.

$$\text{POSS}(\text{DISPLAY\_NODE\_ATTR}(y), s) \equiv$$
$$\text{CURRENT\_NODE}(y, s) \ \wedge \ \exists x \, \text{BUTTON}(x, y, \_, s) \tag{10}$$

The command DISPLAY_NODE_ATTR(y) is possible if and only if y is the current node and is represented by a button.

$$\text{POSS}(\text{DISPLAY\_LINK\_ATTR}(y), s) \equiv$$
$$\exists z \, \text{CURRENT\_NODE}(z, s) \ \wedge$$
$$(\text{LINK}(y, z, \_, \_, \_, \_, s) \vee \text{LINK}(y, \_, z, \_, \_, \_, s)) \tag{11}$$
$$\wedge \ \exists x \, \text{BUTTON}(x, y, \_, s)$$

The command DISPLAY_LINK_ATTR(y) is possible if and only if z is the current node, y is a link with z as either the source or a the sink node, and there is a button representing y.

We also need successor state axioms for the fluents LINK and NODE.

$$\text{POSS}(a, s) \rightarrow [\text{NODE}(x, y, z, do(a, s)) \equiv$$
$$a = \text{CREATE\_NODE}(x, y, z) \vee \tag{12}$$
$$(a \neq \text{DELETE\_NODE}(x) \ \wedge \ \text{NODE}(x, y, z, s))]$$

$$\text{POSS}(a, s) \rightarrow [\text{LINK}(u, v, w, x, y, z, do(a, s)) \equiv$$
$$a = \text{CREATE\_LINK}(u, v, w, x, y, z) \vee (a \neq \text{DELETE\_LINK}(u) \wedge \tag{13}$$
$$\text{LINK}(u, v, w, x, y, z, s))]$$

The following[7] are the four axioms defining the possibility conditions for the actions CREATE_NODE, DELETE_NODE, CREATE_LINK, and DELETE_LINK.

---

[7] We are assuming some method for specifying an arbitrary link or node as in the Dexter reference model [11].

$$\text{POSS}(\text{CREATE\_NODE}(x, y, z), s) \equiv \tag{14}$$
$$\neg(\exists\, u, v, w\ \text{NODE}(u, v, w, s)\ \land u = x)$$

$$\text{POSS}(\text{DELETE\_NODE}(x), s) \equiv$$
$$\exists u, y, z\ \text{NODE}(u, y, z, s)\ \land\ \neg\exists v, w\ \text{LINK}(v, w, u, \_, \_, \_, s)\ \land\ u = x \tag{15}$$

$$\text{POSS}(\text{CREATE\_LINK}(x, v, w, x, y, z), s) \equiv \tag{16}$$
$$\neg(\exists\, u, n, o, p, q, r\ \text{LINK}\ (u, n, o, p, q, r, s)\ \land u = x)$$

$$\text{POSS}(\text{DELETE\_LINK}(x), s) \equiv$$
$$\exists u, n, o, p, q, r\ \text{LINK}\ (u, n, o, p, q, r, s)\ \land\ u = x \tag{17}$$

One can not create a link or node that already exists. Furthermore one can not delete a node if there is a link that points to it. This is requirement is designed to prevent dangling links. Additionally, one can only delete a link if it exists.

We have not axiomatized a session log – a history of the commands executed since the current session was initiated. This is necessary to model backtracking, a common feature of hypertext systems. Note that the name of each situation contains within it a history of all of the commands executed since S_0. Thus backtracking can easily be added utilizing the situation itself as a session log.

Using the foundational axioms for the situation calculus [15], [22], one can prove inductive consequences of a situation calculus axiomatization of a domain. An example in the area of hypertext is Sentence 18 which expresses the property that there are no dangling links.

$$\forall\, s, u, v, w, x, y, z\ \text{LINK}(u, v, w, x, y, z, s)\ \rightarrow \tag{18}$$
$$\exists x, m, n\ \text{NODE}(x, m, n, s)\ \land\ x = w$$

It can be shown that Sentence 18 is an inductive consequence of the axiomatization presented in this paper. See [26] for a fuller discussion.

4   GOLOG: Complex Actions

The actions discussed in the previous section are the basic primitive actions needed to specify the hypertext system. We need complex actions to construct a program that performs a series of primitive actions, tests predicates for their truth values, and then performs other actions depending on the results of the test. An example of such a program is the run-time environment. This set of complex action expressions is available in the logic programming language that is called GOLOG (alGOl in LOGic) [14].

Only a sketch of the constructs of GOLOG is given here. See [14], a full discussion of the language and also a Prolog interpreter. Complex actions are abbreviations for situation calculus expressions. The macros are defined through the predicate predicate Do as in $Do(\delta, s, s')$ where δ is a complex action expression. $Do(\delta, s, s')$ is intended to mean that the agent's doing action δ in situation s leads to a (not necessarily unique) situation s'. The inductive definition of Do includes the following cases:

- $Do(a, s, s') \overset{\text{def}}{=} \text{POSS}(a, s) \wedge (s' = do(a, s))$ — simple actions
- $Do(\phi?, s, s') \overset{\text{def}}{=} \phi[s] \wedge (s = s')$ — tests
- $Do([\delta_1; \delta_2], s, s') \overset{\text{def}}{=} \exists s''(Do(\delta_1, s, s'') \wedge Do(\delta_2, s'', s'))$ — sequences
- $Do([\delta_1 | \delta_2], s, s') \overset{\text{def}}{=} Do(\delta_1, s, s') \vee Do(\delta_2, s, s')$ — nondeterministic choice of actions
- $Do((\Pi x)\delta, s, s') \overset{\text{def}}{=} \exists x \, Do(\delta, s, s')$ — nondeterministic choice of parameters
- $Do(\textbf{if } \phi \textbf{ then } \delta_1 \textbf{ else } \delta_2, s, s') \overset{\text{def}}{=}$

$$(\phi[s] \rightarrow Do(\delta_1, s, s')) \wedge (\neg\phi[s] \rightarrow Do(\delta_2, s, s'))$$

- $Do(\delta^*, s, s') \overset{\text{def}}{=}$ — nondeterministic iteration
  $\forall P[(\forall s_1 \, P(s_1, s_1) \rightarrow$
  $\quad \forall s_1, s_2, s_3[P(s_1, s_2) \wedge Do(\delta, s_2, s_3) \rightarrow P(s_1, s_3)]$
  $\quad \rightarrow P(s, s')]$
- $Do(\textbf{while } \phi \textbf{ do } \delta, s, s') \overset{\text{def}}{=}$
  $\forall P($
  $\quad (\forall s_1 \, \neg\phi[s_1] \rightarrow P(s_1, s_1)) \wedge$
  $\quad (\forall s_1, s_2, s_3 \, (\phi[s_1] \wedge Do(A, s_1, s_2) \wedge P(s_2, s_3)) \rightarrow P(s_1, s_3))$
  $\quad) \rightarrow P(s, s')$

The notation $\phi[s]$ means that a situation argument is added to all fluents in ϕ, if one is missing. The definition of while loops could be simplified by utilizing the definition of nondeterministic iteration.

The following is the definition of the procedure construct:

- (Recursive) procedure P with formal parameters $x_1, \ldots, x_n$ and actual parameters $t_1, \ldots, t_n$ and body α is defined as follows:
  $Do([\textbf{proc}$
  $\quad P(x_1, \ldots, x_n) \, \alpha \textbf{ end}](t_1, \ldots, t_n), s_1, s_2) \overset{\text{def}}{=}$
  $\quad (\forall P)\{(\forall x_1, \ldots, x_n, s_1', s_2')$
  $\quad [P(x_1, \ldots, x_n, s_1', s_2') \equiv Do(\alpha, s_1', s_2')]$
  $\quad \rightarrow P(t_1, \ldots, t_n, s_1, s_2)\}.$

Executing procedure P on actual parameters $t_1, \ldots, t_n$ takes you from s_1 to s_2 iff $(t_1, \ldots, t_n, s_1, s_2)$ is in the smallest set of tuples $(x_1, \ldots, x_n, s_1', s_2')$ such that executing α on $x_1, \ldots, x_n$ takes you from s_1' to s_2'.

5 Run-time Layer

The complex action macros of GOLOG can now be used to define the run-time layer. But we first need some additional actions. These are:

- RIGHTCLICK(x)
- MIDDLECLICK(x)
- LEFTCLICK(x)

The intended meaning of RIGHTCLICK(x) is that the user has clicked the button x with the right side of the mouse indicating that the request is for a traverse action (either traverse to source or traverse to sink). The intended meaning of LEFTCLICK(x) is that the user has clicked the button x with the left side of the mouse indicating that the request is to display the link attributes. A MIDDLECLICK(x) action is an indication that the user wants to have the node (current node) attributes displayed. These actions are really exogenous actions in that they are not performed by the program (agent), but rather by the outside world (user of the program). The choice of which one occurs is external and not part of the logic model.

Additionally, we need the following action:

- DEACTIVATE(x)

After executing the action requested by the user, the system will need to deactivate the button with this action.

Also needed are three fluents to indicate the status of the button.

- ACTIVEBUTTONRIGHT(x)
- ACTIVEBUTTONMIDDLE(x)
- ACTIVEBUTTONLEFT(x)

The successor state axioms for these fluents are relatively simple:

$$\text{POSS}(a, s) \rightarrow [\text{ACTIVEBUTTONRIGHT}(x, do(a, s)) \equiv$$
$$a = \text{RIGHTCLICK}(x) \vee (a \neq \text{DEACTIVATE}(x) \wedge \quad (19)$$
$$\text{ACTIVEBUTTONRIGHT}(x, s))]$$

$$\text{POSS}(a, s) \rightarrow [\text{ACTIVEBUTTONLEFT}(x, do(a, s)) \equiv$$
$$a = \text{RIGHTCLICK}(x) \vee (a \neq \text{DEACTIVATE}(x) \wedge \quad \cdot \quad (20)$$
$$\text{ACTIVEBUTTONLEFT}(x, s))]$$

$$\text{POSS}(a, s) \rightarrow [\text{ACTIVEBUTTONMIDDLE}(x, do(a, s)) \equiv$$
$$a = \text{RIGHTCLICK}(x) \vee (a \neq \text{DEACTIVATE}(x) \wedge \quad (21)$$
$$\text{ACTIVEBUTTONMIDDLE}(x, s))]$$

These axioms specify that the button is active (in the appropriate fashion — right, left, or middle) if the previous action was a click (of the appropriate type — right, left, or middle), or if the previous action was not a deactivate action and the button was on (in the appropriate fashion) in the previous situation.

The following complex actions define the run-time environment. The procedure FIND_BUTTON determines what button is pressed and then calls the procedure

PROCESS_COMMAND with the button as an argument and finally deactivates the button.

$$\textbf{proc } \text{FIND\_BUTTON}(n)$$
$$(\pi\, n)[(\text{ACTIVEBUTTONRIGHT}(n) \lor \text{ACTIVEBUTTONLEFT}(n) \lor$$
$$\text{ACTIVEBUTTONMIDDLE}(n))?;$$ $$(22)$$
$$\text{PROCESS\_COMMAND}(n);\; \text{DE\_ACTIVATE}(n)]$$
$$\textbf{end.}$$

The procedure PROCESS_COMMAND takes a button as an argument. If the middle mouse key has been clicked, then the attributes of the current node are displayed. Otherwise, if the left mouse button has been clicked, the attributes of the link associated with the button are displayed. Otherwise, the procedure TRAVERSE is called with the link as an argument.

$$\textbf{proc } \text{PROCESS\_COMMAND}(n)$$
$$\textbf{if } \text{ACTIVEBUTTONMIDDLE}(n)$$
$$\textbf{then } (\pi\, u)[\text{CURRENT\_NODE}(u)?;$$
$$\text{DISPLAY\_NODE\_ATTR}(u)$$
$$\textbf{else } [\textbf{if } \text{ACTIVEBUTTONLEFT}(n)$$
$$\textbf{then } (\pi\, y)\text{BUTTON}(n, y, z)?;$$ $$(23)$$
$$\text{DISPLAY\_LINK\_ATTR}(y)$$
$$\textbf{else}(\pi\, y)\text{BUTTON}(n, y, z)?;$$
$$\text{TRAVERSE}(y)]]$$
$$\textbf{end.}$$

The procedure TRAVERSE takes as an argument the link that is associated with the button that was activated. If the current node is the source node of that link, then a traverse to sink is performed. Otherwise a traverse to source is performed.

$$\textbf{proc } \text{TRAVERSE}(x)\; (\pi\, y, z)[\text{LINK}(x, y, z, \_, \_, \_)?;$$
$$\textbf{if } \text{CURRENTNODE}(z)$$
$$\textbf{then } \text{TRAVERSE\_SOURCE}(y, z)$$ $$(24)$$
$$\textbf{else } \text{TRAVERSE\_SINK}(y, z)]$$
$$\textbf{end.}$$

The procedure CONTROL is the top level routine. It first calls OPEN_SESSION and then request. Note that request is not really definable within the logic model because it is basically a request for an exogenous action. But it is easily implemented as a prompt to the user to enter an action and then the execution of that action. Then the procedure loops as long as button 0 is not activated. This button is taken to represent the command to exit the program. Each time through the loop, first there is a call to FIND_BUTTON and then another request.

$$
\begin{array}{ll}
\textbf{proc } \text{CONTROL} \\
\quad \text{OPEN\_SESSION;} \\
\quad \text{request;} \\
\quad [\textbf{while } \neg \text{ACTIVE}(0) \textbf{ do} \\
\qquad \text{FIND\_BUTTON;} & (25) \\
\qquad \text{request }]; \\
\quad \textit{CLOSE\_SESSION} \\
\textbf{end.}
\end{array}
$$

The procedure CLOSE_SESSION is quite simple. It calls the commands (undefined in our model) clearscreen and terminate.

$$
\begin{array}{ll}
\textbf{proc } \text{CLOSE\_SESSION} \\
\quad \text{clearscreen;} \\
\quad \text{terminate} & (26) \\
\textbf{end.}
\end{array}
$$

The procedure OPEN_SESSION, first performs clearscreen and then displays the current node.

$$
\begin{array}{ll}
\textbf{proc } \text{OPEN\_SESSION} \\
\quad \text{clearscreen;} \\
\quad (\pi\, n)[\text{CURRENT\_NODE}(n)?; & (27) \\
\quad \text{MAKE\_DISPLAY\_TEXT}(n)] \\
\textbf{end.}
\end{array}
$$

This specification has been implemented [1] using the GOLOG interpreter presented in [14]. The interpreter accepts input (representing mouse clicks on buttons) from the user and returns a final situation containing the node traversal route followed. Currently a window based interface is being developed. The result will be a hypertext system driven by GOLOG that responds to actual mouse clicks on the displayed material.

6   Conclusions and Future Work

A situation calculus model for hypertext systems has been presented. This model captures the dynamic aspects of hypertext – the changes that occur in moving from one situation to another when commands are executed. It has the additional advantage over other logical specifications of hypertext systems [2] of being executable by an existing Prolog interpreter [14] for the language (GOLOG) in which the model is expressed.

The model is only of the simplest or basic hypertext system. Many more details can be incorporated. In particular, the example of hypertext can make interesting use of a number of features that have more recently been added to GOLOG. These include concurrency (support for multiple users who may be concurrently reading and editing the hypertext) [8], [24], explicit representations for time (e.g., support for a request to revisit the node that the user was reading 5 minutes ago) [24], and also planning (e.g., finding a sequence of nodes to display that satisfy certain requirements) [25].

Furthermore, the logical specification and implementation in Prolog facilitates the addition of features that have been more recently proposed in various hypertext models. Many of these involve inferencing in one form or another[8]. One is performing *structure search*; a search for subnetworks that match a given pattern of nodes (including aspects of their content) and links (including their types) [10].

Also, some of these features center around the notion of context. One may want to model applications that determine which links to display based on the dynamic situation of the user interacting with the database. A number of hypertext systems incorporate such features. For example, in Trellis [7] it is possible to control the navigational possibilities available to the user, and to let him/her access and activate links based on the previous steps that the user has taken. Context is also utilized in [28], [4].

An additional issue of interest is the ability to infer links rather than require that the author of the hyperdocument explicitly construct all links[9]. Consider again the example hyperdocument presented earlier. Ideally, authors would only write separate hyperdocuments for travel information on different areas of the world and also separate hyperdocuments for the various other topics of potential interest such as butterflies. For a reader of the travel guide who has an interest in butterflies, the system should automatically infer the appropriate links between pages on a particular region of the world and the pages on those butterflies that can be found in that region. A system that performs such inferences needs to have the appropriate deductive mechanisms integrated with the hypertext machinery. The GOLOG specification presented in this paper is particularly suitable as the foundation for such a project.

References

1. Dennis Batich. A hypertext model in GOLOG. unpublished undergraduate project, Department of Computer Science, New Jersey Insitute of Technology, 1998.
2. Michael P. Bieber and Steven O. Kimbrough. On the logic of generalized hypertext. *Decision Support Systems*, 11:241–257, 1994.
3. Alex Borgida, John Mylopoulos, and Raymond Reiter. On the frame problem in procedure specificiations, 1993. Presented at the Fifteenth International Conference on Software Engineering, Baltimore, Maryland.
4. M.C. Buchanan and P.T. Zellweger. Specifying temporal behavior in hypermedia documents. In *ECHT '92 Proceedings*, pages 262–271, Milano, November 1992. ACM Press.
5. David Carter. *Butterflies and Moths*. DK Publishing, New York, N.Y., 1992.

---

[8] For a survey of the issues discussed in the literature on hypertext systems, see [6], [18], [10], [19].

[9] The bridge laws of [2] is one approach to automatically constructing links.

6. Jeff Conklin. Hypertext: An introduction and survey. *IEEE Computer*, 20(9):17–41, 1987.
7. R. Furuta and P.D. Stotts. Programmable browsing semantics in Trellis. In *Hypertext '89 Proceedings*, pages 27–42, Pittsburgh, November 1989. ACM Press.
8. Giuseppe De Giacomo, Yves Lespérance, and Hector J. Levesque. Reasoning about concurrent execution, prioritized interrupts, and exogeneous actions in the situation calculus. In *Proceedings of the Fifteenth International Joint Conference on Artificial Intelligence (IJCAI-97)*, pages 1221–1226, Nagoya, Japan, 1997.
9. A. R. Haas. The case for domain-specific frame axioms. In F. M. Brown, editor, *The Frame Problem in Artificial Intelligence. Proceedings of the 1987 Workshop*, pages 343–348. Morgan Kaufmann Publishers, Inc., San Mateo, California, 1987.
10. Frank Halasz. Reflections on notecards: Seven issues for the next generation of hypermedia systems. *Communications of the ACM*, 31(7):836–852, 1988.
11. Frank Halasz and Mayer Schwartz. The dexter hypertext reference model. *Communications of the ACM*, pages 30–39, 1994.
12. Yves Lespérance, Hector Levesque, Fangzhen Lin, Daniel Marcu, Ray Reiter, and Richard Scherl. A logical approach to high-level robot programming — a progress report. Appears in *Control of the Physical World by Intelligent Systems*, Working Notes of the 1994 AAAI Fall Symposium, New Orleans, LA, November 1994.
13. Yves Lespérance, Hector J. Levesque, F. Lin, Daniel Marcu, Raymond Reiter, and Richard B. Scherl. Foundations of a logical approach to agent programming. In *Proceedings of the IJCAI-95 Workshop on Agent Theories, Architectures, and Languages*, August 1995.
14. Hector Levesque, Raymond Reiter, Yves Lespérance, Fangzhen Lin, and Richard B. Scherl. Golog: A logic programming language for dynamic domains. *Journal of Logic Programming*, 1997.
15. Fangzhen Lin and Raymond Reiter. State constraints revisited. *Journal of Logic and Computation*, 4(5):655–678, 1994.
16. James Lyon and Tony Wheeler. *Bali & Lombok: a Lonely Planet Travel Survival Kit*. Lonely Planet Publications, Hawthorn, Victoria, Australia, 1997.
17. J. McCarthy and P. Hayes. Some philosophical problems from the standpoint of artificial intelligence. In B. Meltzer and D. Michie, editors, *Machine Intelligence 4*, pages 463–502. Edinburgh University Press, Edinburgh, UK, 1969.
18. Jakob Nielsen. The art of navigating through hypertext. *Communications of the ACM*, 33(3):296–310, 1990.
19. Jakob Nielsen. *Multimedia and Hypertext: The Internet and Beyond*. Academic Press, Boston, Massachusetts, 1995.
20. E.P.D. Pednault. ADL: exploring the middle ground between STRIPS and the situation calculus. In R.J. Brachman, H. Levesque, and R. Reiter, editors, *Proceedings of the First International Conference on Principles of Knowledge Representation and Reasoning*, pages 324–332. Morgan Kaufmann Publishers, Inc., San Mateo, California, 1989.
21. Raymond Reiter. The frame problem in the situation calculus: A simple solution (sometimes) and a completeness result for goal regression. In Vladimir Lifschitz, editor, *Artificial Intelligence and Mathematical Theory of Computation: Papers in Honor of John McCarthy*, pages 359–380. Academic Press, San Diego, CA, 1991.
22. Raymond Reiter. Proving properties of states in the situation calculus. *Artificial Intelligence*, pages 337–351, December 1993.
23. Raymond Reiter. On specifying database updates. *The Journal of Logic Programming*, pages 53–91, 1995.

24. Raymond Reiter. Natural actions, concurrency and continuous time in the situation calculus. In L.C. Aiello, J. Doyle, and S.C. Shapiro, editors, *Principles of Knowledge Representation and Reasoning: Proceedings of the Fifth International Conference on Principles of Knowledge Representation and Reasoning (KR'96)*, pages 2–13, San Francisco, CA, 1996. Morgan Kaufmann Publishing.
25. Raymond Reiter. Knowledge in action: Logical foundations for describing and implementing dynamical systems. Unpublished book draft, 1999.
26. Richard Scherl, Michael Bieber, and Fabio Vitali. A situation calculus model of hypertext. In *Proceedings of the Thirty-First Hawaii International Conference on System Sciences (HICSS-31)*, 1998.
27. L.K. Schubert. Monotonic solution of the frame problem in the situation calculus: an efficient method for worlds with fully specified actions. In H. E. Kyberg, R.P. Loui, and G.N. Carlson, editors, *Knowledge Representation and Defeasible Reasoning*, pages 23–67. Kluwer Academic Press, Boston, Massachusetts, 1990.
28. F.W. Tompa, G.E. Blake, and D.R. Raymond. Hypertext by link-resolving components. In *Hypertext '93 Proceedings*, pages 118–130, Seattle, November 1993. ACM Press.

Explanation Closure, Action Closure, and the Sandewall Test Suite for Reasoning about Change

Lenhart K. Schubert

Department of Computer Science, University of Rochester
Rochester, NY 14627-0226

Summary. *Explanation closure* (EC) axioms were previously introduced as a means of solving the frame problem. This paper provides a thorough demonstration of the power of EC combined with *action closure* (AC) for reasoning about dynamic worlds, by way of Sandewall's test suite of 12-or-so problems [29], [30], [31]. Sandewall's problems range from the "Yale turkey shoot" (and variants) to the "stuffy room" problem, and were intended as a test and challenge for nonmonotonic logics of action. The EC/AC-based solutions for the most part do not resort to nonmonotonic reasoning at all, yet yield the intuitively warranted inferences in a direct, transparent fashion. While there are good reasons for ultimately employing nonmonotonic or probabilistic logics – e.g., pervasive uncertainty and the qualification problem – this does show that the scope of monotonic methods has been underestimated. Subsidiary purposes of the paper are to clarify the intuitive status of EC axioms in relation to action effect axioms; and to show how EC, previously formulated within the situation calculus, can be applied within the framework of a temporal logic similar to Sandewall's "discrete fluent logic", with some gains in clarity.

1   Introduction

Explanation closure (EC) axioms are complementary to effect axioms. For instance, just as we can introduce effect axioms stating that painting or wallpapering a wall (with appropriate preconditions) changes its color, we can also introduce an EC axiom stating that a change in wall color implies that it was painted or wallpapered. The "closure" terminology signifies that the alternatives given are exhaustive.

This complementarity extends to their use: effect axioms allow the inference of change, and EC axioms the inference of non-change (persistence). For instance, if I know that no-one has painted or wallpapered the wall, then I can conclude that its color has remained unaltered. As first noted by Haas [15], EC-based persistence reasoning provides a very good handle on the *frame problem*.[1] In [32] (henceforth Sch90) I extended Haas' work, showing that EC-based techniques generalize to worlds with continuous and agentless change and concurrent actions, and support extremely efficient STRIPS-like methods for tracking effects of successive actions. Moreover, these methods are entirely monotonic as long as we are only concerned

---

[1] A number of other writers have made closely related proposals, e.g., Lansky [20], Georgeff [9], Morgenstern & Stein [27].

with inference of those changes and those explanations for change that are a matter of "practical certainty" (given our theory of the domain).

In view of their potency, it is surprising that EC-based approaches did not surface much sooner in the history of the frame problem. A commonly expressed qualm about EC axioms is that any enrichment of the (micro)world under consideration is likely to necessitate their revision. For instance, while in a simple world a change in wall color may be attributable to painting or wallpapering, in a more complex world the change may also be due to spraying, tiling, or panelling (or even decay, etc.). True enough – but it is equally true that enrichment of a microworld complicates the effect axioms. For instance, having paint and a brush may be sufficient for successful wall-painting in a simple, benign world, but in a more realistic one, the painter may be thwarted by dried-out paint, an undersize or oversize brush, injury, interference by other agents, etc. (i.e., the *qualification problem* crops up). Yet the fallibility of simple effect axioms has deterred few – not even nonmonotonic theorists – from relying on them! For instance, most formalizations of the Yale Turkey Shoot include axioms asserting that loading a gun makes it loaded, and firing the loaded gun at Fred kills him. This is generally done without comment or apology (except perhaps for a perfunctory gesture toward the qualification problem, which is thereafter ignored).[2] Yet the idea of turning this around and applying the same strategy to inference of explanations, given a change, seems to occur to almost no-one, and if raised, is met with skepticism.

I am led to believe that there are deep-seated prejudices against the idea of *reasoning deductively against the causal arrow*, perhaps stemming in part from the philosophical tradition on explanation. This tradition holds that physical theories enable us to deduce *resultant* states and events from given ones; while going from results to their causes is not a matter of deduction, but a matter of generating assumptions *from* which we can deduce the results. But while reasoning against the arrow of time and causation (retrodiction, explanation) is apt to generate more alternatives than reasoning with it (prediction), there is no *a priori* physical or logical reason for confining deduction to the forward direction.[3]

But the best argument against these prejudices lies in the practical efficacy of EC reasoning, of which there is growing awareness (as evidenced not only in [15], Sch90 and herein, but also in [5], [7], and [16]). Here we should also take note of an elegant and useful extension of the EC-based approach to the frame problem developed by Reiter [28]. Rather like Morgenstern and Stein [27], he focuses on cases

---

[2] A notable exception is [21], which explicitly addresses the qualification problem through circumscriptive minimization of preconditions. Also [13] addresses the qualification problem via a "possible worlds approach" (see the "stuffy room" scenario below).

[3] It is interesting that people versed in formal logic are apt to regard Sherlock Holmes' "deductions" as misnamed. Rather, they say, Holmes was reasoning inductively or abductively when he constructed explanations for his observations. In my view, if we are willing to grant that the inference of a man's death is deductive, given his unimpeded fall to the pavement from the top of a skyscraper, then some of Holmes' inferences are equally deductive. If the former is not deductive, then no inferences based on world knowledge are deductive, whether directed forward or backward in time.

where the known effect axioms characterize *all* the ways the changes of interest can come about (Generalized Completeness Assumption). For such cases, he shows how EC axioms can be derived from effect axioms, and combined with them into biconditionals ("successor state axioms"); e.g., a wall changes color *if and only if* it is painted or wallpapered. This mechanical derivation should allay some of the above qualms about the lack of invariance of EC axioms when new actions are added. Reiter further shows how to use such axioms for sound and complete goal regression.[4]

However, I will keep effect axioms and EC axioms separate for the sake of generality, since I believe that the GCA is valid only to the extent that the "blanket closure" assumptions implicit in nonmonotonic approaches are valid. It breaks down for realistically complex domains, and even for some simple worlds of interest. For instance, we may know that a robot's $Goto(x)$ action brings about $nextto(Robot, x)$. But it would be wrong to biconditionalize this to say that $nextto(Robot, x)$ becomes true *if and only if* the robot moves next to x. After all, there may be objects near x which the robot may also end up next to (and these "side effects" may depend more or less unpredictably on low-level path planning). Yet we can state an EC axiom that $nextto(Robot, x)$ becomes true *only if* the robot goes to some y and $x = y$ or x is near y; this may be quite sufficient for the persistence reasoning needed for practical purposes (see further details in Sch90). Sandewall's test suite provides additional illustrations [29], [30] (henceforth San91, San92).[5] For instance, in the "stuffy room" problem (discussed at length later on), various EC axioms are possible (without change to the effect axioms), depending on how much freedom to "flit about" we want to allow objects when a vent is blocked or unblocked (creating drafts, one imagines). In general, we cannot characterize changes in terms of conditions that are both necessary and sufficient for those changes to occur. When we abstract away details in high-level axiomatizations (e.g., by using predicates like *nextto*), or have only partial knowledge of the behavior of a domain (because of its lack of familiarity, complexity, or inherent nondeterminism), then the best we can do is to provide some (practically certain) postulates about sufficient conditions for change, and others about necessary ones.

The test suite provides an unprecedented opportunity to examine the strengths and shortcomings of various methods for reasoning about change in a systematic way. I will show that the approach based on EC-reasoning fares very well indeed. Moreover, the proffered solutions are monotonic except in the case of one variant of McCarthy's "potato in the tailpipe" problem (where I suggest a probabilistic approach). This seems to me to call for a reassessment of the proper roles of monotonic and nonmonotonic (or probabilistic) methods in reasoning about change. While nonmonotonic methods still retain an important role in reasoning about an uncertain, incompletely known world (as the "potato in the tailpipe" problem and other instances of the qualification problem show), monotonic methods can deal straightforwardly

---

[4] The usefulness of EC axioms in planning has also become apparent in more recent work on SAT-planning (e.g., [14]).

[5] These publications were precursors of the monograph *Features and Fluents* [31].

with many of the scenarios viewed as motivating examples for nonmonotonic methods.

The examples will also serve to illustrate a version of EC-based reasoning within a temporal calculus loosely modelled on Sandewall's DFL (discrete fluent logic). They will further illustrate the form and importance of *action closure* (AC) axioms in the temporal calculus, and allow us to probe the limits of the monotonic approach.

2 DFL, TC, and the test scenarios

Sandewall's *discrete fluent logic* (DFL), outlined in a preliminary way in San91 and developed into several variants in San92 (see also [31]), offers a concise notation for time-dependent descriptions of dynamic worlds. A very interesting aspect of DFL is the theory of entailment, whose central idea is that an agent can view the world as inert, with all fluents retaining their values *except* when forced to undergo change by the agent's actions. (In the model theory, each action has associated with it certain "trajectories" of change for a finite number of fluents, for each state in which the action may be initiated. I will have some further remarks about this semantics later on.) Another idea Sandewall pursues is that actions can "occlude" the fluents they may affect, for the duration of the action; i.e., the values of occluded fluents cannot be presumed to persist. A model preference criterion may then be employed according to which less occluded models and those that postpone *transparent* (non-occluded) change are preferred.

Of particular interest for my present purposes is Sandewall's effort to identify and catalogue many of the defects of extant nonmonotonic logics, and provide old and new test problems which bring these defects to light. Sandewall's preliminary assessment in 1991 was that his study "... provides reasons for renewed disappointment. The situation in 1991 is only marginally different from the one in 1986 [the year of the Hanks & McDermott paper]... most of the 'most popular' approaches actually fail on the test scenarios." (*ibid.*: sec. 7). In the more recent work (San92), however, the emphasis is on viewing various NM logics as "tools", whose utility for various purposes can be assessed via Sandewall's inertia-world semantics.

The "temporal calculus" (TC) notation I will use emulates Sandewall's DFL syntax to facilitate comparisons. Thus it consists of the usual first-order syntax plus the following DFL-like temporal notation (but without involvement of occlusion): Truth of a formula φ at (moment of) time τ is written $[\tau]\varphi$, and truth at all times in $[\tau_1, \tau_2]$ is written $[\tau_1, \tau_2]\varphi$. Also $[\tau_1, \tau_2]\varphi := v$ means that $[\tau_1]\neg(\varphi = v)$ and $[\tau_2]\varphi = v$, i.e., the value of φ becomes v somewhere in the interval $[\tau_1, \tau_2]$. If φ is a formula, we use $\varphi = T$ and $\varphi = F$ equivalently with φ and $\neg\varphi$ respectively (as in DFL). As a semantic basis for the notation so far, an interpretation of the fluent predicates and functions is assumed to provide their extensions at each moment of time. (The time line could be taken to be discrete or the real line.) We will also use an *action* predicate do, where $[\tau_1, \tau_2]do(\alpha, \beta)$ is true or false of an agent α, action β and time *interval* $[\tau_1, \tau_2]$, viz., the interval over which the action takes place. An interpretation of TC is assumed to specify the extension of do at all time *intervals*, rather than at all times. A useful abbreviation will be

$$[\tau_1..\tau_2]do(\alpha,\beta),$$

which stands for

$$(\exists\tau_1')(\exists\tau_2')[\tau_1 \le \tau_1' \le \tau_2' \le \tau_2] \wedge [\tau_1',\tau_2']do(\alpha,\beta),$$

i.e., $do(\alpha,\beta)$ happens *somewhere* between τ_1 and τ_2. Though I will mostly use temporally annotated formulas of the types described, "timeless" formulas (e.g., specifying entity types) are also useful. These can be equivalently thought of as true at all times, and clearly the following is a sound rule of inference, for ϕ any formula:

$$\frac{\phi}{[t_1,t_2]\phi}.$$

TC solutions to the test problems are generally more perspicuous and concise than solutions in the situation calculus (SC). (See examples of the latter in Sch90.) This is mainly because the TC notation allows us to index states of affairs directly via time variables, instead of requiring us to index them via sequences of actions. However, the most interesting difference lies in the way the action closure (AC) assumption – that all relevant actions are known – is encoded. In SC versions, the assumption is implicit in the functional dependence of situations on actions. In TC versions, times (and hence fluent values at those times) are introduced independently of actions, and so the assumption of complete knowledge of relevant actions needs to be stated separately. It will typically (though not always) be represented by the "only if" part of an equivalence of form, "x did y from time t_1 to time t_2 iff (x,y,t_1,t_2) is one of the following tuples...". Such axioms will be called "action chronicles" (with apologies to those, including Sandewall, who have employed the term differently).

An important question here is whether AC assumptions are by their nature excessively strong. Does it not require God-like omniscience to know what all the relevant actions are that could have affected the fluents of interest? The answer is no, provided that we are only looking for *practical* certainty rather than *absolute* certainty. The *relevant* actions are often ones which occur in a very limited spatiotemporal domain, for instance in a certain room during a short time interval. We often have good reasons to believe that we know all the relevant actions within such confines. For example, if we are physically "on the scene of the action", we can often be sure that we are aware of all the relevant physical actions (e.g., which objects were painted or moved about) thanks to our perceptual and cognitive abilities; and when there are possible relevant actions beyond our purview, we are often well *aware* of just what those gaps in our knowledge are.

If we are simply being told a *story*, we can rely on the narrator to withhold nothing of relevance from us. The narrator will not neglect to mention that Joe unloaded the gun before pulling the trigger on Fred. As Amsterdam [1] argues, narrators are expected to tell their story in a way that puts the hearer/ reader on the scene (vicariously, through the narrator's perceptions), and this entails reporting everything of relevance that happened. To be sure, there are many qualifications to be made and subtleties to be explored here. But my point is that the source of closure in narration is the narrator, not the hearer or God. (Formally, Amsterdam assumes that no actions occurred other than those deducible from the narrative, or that could have transpired during explicitly reported lapses in the narrator's awareness. I will have further comments on Amsterdam's proposals later.)

If instead a scenario represents a plan of action, whose consequences are yet to be observed (once the plan is carried out), then clearly it is the planner's *intention* to shield the fluents of interest from capricious disturbances. If you *plan* to kill Fred by loading the gun, aiming at Fred, and pulling the trigger, you surely plan *not* to unload the gun before pulling the trigger. And if you plan to repaint the walls a certain color, you surely do not intend to let others meddle at will. Thus it is the planner who is the source of action closure. He may ensure closure, for instance, by arranging to be the only agent on the scene, or to have only co-agents who will do his bidding, or who at least can be relied on not to interfere. That is all that is needed to justify AC axioms. Moreover, it is an important advantage of the explicit AC approach that we can arbitrarily *delimit* the spatial and temporal locations, the agents, and the kinds of actions for which our action chronicles are complete. By contrast, NM logics generally have much stronger, universal completeness assumptions built into their semantics, and this can lead to bizarre and unexpected inferences for larger, non-transparent examples.

Of course, if we demand *absolute* reliability of our axioms, then God-like omniscience is indeed required; after all, even the most carefully insulated and controlled setting is subject to freak occurrences. But that is not an observation about EC or AC axioms in particular, but about *all* nonlogical axioms. Moreover, a monotonic approach to the inference of change and persistence does not preclude the addition of *belief revision* mechanisms, capable of retracting, amending, or adding to the beliefs which form the basis for these monotonic inferences. When I discover that the wall I painted blue turned green when it dried, I'll revise my effect axioms; and if I find that while my back was turned, a prankster who had been hiding in the closet repainted the wall red, I'll revise my action chronicle. But unless and until that happens, I may well be best off reasoning monotonically with "practically certain" axioms.

The test scenarios which follow adhere closely to Sandewall's formulations. Each scenario is described very briefly, the intended conclusions are indicated, and then the TC formalization is shown. Although detailed proofs exist in all cases, the style of reasoning used to reach the desired conclusions should be clear enough from just a few sample proofs here and there. (The reader might in particular look at the reasoning given for the Hiding Turkey Scenario (HTS).) I hope that the axiomatizations are sufficiently transparent to allow the reader to reconstruct the rest. The headers are worth paying close attention to; they encapsulate essential dimensions of variation among test cases, largely as identified by Sandewall – dimensions often difficult for any one nonmonotonic logic to measure up to simultaneously.

In all of the axiomatizations, names beginning with obs, chr, eff, exp, and ineq respectively are used for axioms describing observations at particular situations or times, action chronicles, effect axioms, explanation closure axioms, and inequality axioms. These names serve no theoretical purpose, only a mnemonic one (unlike DFL conventions). As in Sch90, constants and functions will start with an upper case letter and variables and predicates will be lower case. Top-level free variables are implicitly universally quantified (with maximal quantifier scope). The predicate u ("unequal") takes any number of arguments and asserts that they are pairwise distinct.

Prediction: Yale Turkey Shoot (YTS)

There are two truth-valued fluents, a (alive) and l (loaded). Initially the turkey is alive and the gun not loaded. The agent loads, waits and fires. Loading brings about l (from prior state $\neg l$ or l), and firing brings about $\neg a$ and $\neg l$ provided that l held prior to it. We wish to conclude that at the end of firing, $\neg a$ holds (the turkey is not alive).

I will slightly embellish the usual action repertoire to include Unload, Spin, and Chopneck, for illustration and for consistency with later variants. For simplicity Chopneck has been given no preconditions and the effect axiom for Unload has been omitted, since these actions play no role here.

obs1	$[0]a \wedge \neg l$
chr1	$[t_1, t_2]do(Joe, y) \Leftrightarrow (t_1, t_2, y) \in \{(4, 6, Load), (10, 12, Fire)\}$
eff1	$[t_1, t_2]do(Joe, Load) \Rightarrow [t_2]l$
eff2	$[t_1]l \wedge [t_1, t_2]do(Joe, Fire) \Rightarrow [t_2](\neg a \wedge \neg l)$
eff3	$[t_1, t_2]do(Joe, Chopneck) \Rightarrow [t_2]\neg a$
exp1	$[t_1, t_2]l := T \Rightarrow [t_1..t_2]do(Joe, Load) \vee [t_1..t_2]do(Joe, Spin)$
exp2	$[t_1, t_2]l := F \Rightarrow (\exists y \in \{Fire, Unload, Spin\})[t_1..t_2]do(Joe, y)$
exp3	$[t_1, t_2]a := F \Rightarrow (\exists t_1')[t_1 \leq t_1' \leq t_2 \wedge [t_1']l \wedge [t_1'..t_2]do(Joe, Fire)]$
	$\vee [t_1..t_2]do(Joe, Chopneck)$
ineq1	$u(Load, Unload, Fire, Spin, Chopneck)$

Reasoning: We infer $[4]\neg l$ by noting $[0]\neg l$ and that if $[4]l$ were true, a Load or Spin action would have had to occur between times 0 and 4, by exp1. But this is ruled out by chr1. Hence by chr1 and eff1, $[6]l$. Similarly $[10]a$ since if this were false we would have had a Fire or Chopneck action between times 0 and 10 by exp3, contrary to chr1 and ineq1.

Now we infer $[10]l$ in much the same way, using the fact that its falsity would imply a Fire, Unload, or Spin action by exp2, which can be ruled out by chr1. Hence by chr1 & eff2, $[12](\neg a \wedge \neg l)$. $\neg l$ is easily shown to persist. $\neg a$ will persist if we add $[t](\neg a \wedge d \geq 0) \Rightarrow [t + d]\neg a$. □

Though superficially close to Sandewall's axiomatization, the TC version makes significantly stronger assumptions at the outset. For instance, chr1 leaves Joe inactive between loading and firing, and this together with exp2 ensures that the gun remains loaded. But in the DFL version, this is a defeasible chronicle completion inference. Should it be? Suppose the problem specification included the statement, "Between loading and firing, another action either did or did not take place". Intuitively, this blocks the inference that the gun remained loaded – despite the fact that the added statement is logically vacuous (a *tautology*)!

Clearly, it is a mistake to simply render the given English sentences as directly as possible in some logic, and then make it a matter of the semantics of that logic to deliver the intuitively required conclusions. How could *any* reasonable logic have entailments defeasible by tautologies? This once again raises the important question of "what's in a problem statement". As noted earlier, Amsterdam [1] drew attention to the role of narrative conventions in story-like problem statements, in particular the requirement that the author relate everything his audience would have observed under the reported circumstances – except perhaps events that transpired during explicitly reported lapses of attention (e.g., where the author indicates that some time passed, or says "I blacked out for a moment", etc.). This is formally written as UA_t, i.e., it is unknown whether action A occurred.

It is interesting to note that Amsterdam's assumption about what actions did and did not occur is closely related to the AC assumption. Stated a little more fully than before, his assumption is that an action A occurred at t if A_t is provable, and did not occur if neither A_t nor UA_t is provable. My AC assumption is computationally less problematic: it says that all the actions that bear on the fluents of interest are explicitly known, without invoking provability. Also, Amsterdam makes an assumption closely related to EC: roughly speaking, changes that are provable effects of provable actions (according to some theory of what constitutes an "effect") definitely occurred, and no change occurred unless it is the effect of some A_t, where A_t or UA_t is provable. (For the exact formulation, see [1].) Amsterdam notes that his approach fails to allow for actions which people regard as "obvious" inferences from certain state changes. His example is one where a character is sitting by the fireplace in one sentence and standing by the door in the next. These are precisely the action inferences supplied by EC!

Amsterdam's attempt to capture narrative conventions by nonmonotonic action and effect closure and the modal U operator is interesting, but it remains to be seen how far it can be taken. Besides computational intractability and the problem about action inference noted by Amsterdam, there is also the problem that real stories allow for many actions and events that are neither entailed by the story nor occluded by lapses in the narrator's attention. For instance, it certainly seems possible in a story like *Little Red Riding Hood* that the heroine hopped over a small creek, or glanced at some birds overhead on her way to Grandmother's house, even though nothing in the story entails this or even suggests that this *may* have occurred. The narrator simply did not judge such events relevant, and therefore, abiding by the Gricean maxim, omitted them. The view taken here is that narrative implicatures and domain reasoning are separable phenomena, and that it is therefore worthwhile to study domain reasoning methods as far as possible independently of story understanding. This means that we begin by extracting *all* of the information intuitively conveyed by a narrative – the positive as well as the negative, the asserted as well as the "conversationally implicated" information – while setting aside the question of exactly *how* the narration managed to convey that information. Only then do we ask what *follows* from what we have been told.

Regardless of strategy, however, what is important about Amsterdam's work is its recognition of the importance of narrative conventions and maxims in shaping what we take a story to imply. Much of the heated debate about which nonmonotonic logic is the right one for chronicle completion seems attributable to the neglect of information implicitly conveyed through these conventions and maxims, or misguided attempts to make this information fall out of the logic.

Retrodiction: the Stanford Murder Mystery (SMM)

The world is the same as for the YTS, but the gun is initially loaded, firing and waiting are performed in succession, and then the turkey is not alive. We are to infer that the gun was initially loaded, and the turkey was not alive after firing (prior to the wait).

obs1 $[0]a$
chr1 $[t_1, t_2]do(Joe, y) \Leftrightarrow (t_1, t_2, y) = (10, 12, Fire)$
obs2 $[14]\neg a$
eff1–ineq1 as above (YTS)

Ambiguous prediction: the Ferryboat Connection Problem (FCP)

A motorcycle M goes from F, some location on island Fyen, to the ferry landing L, and gets there between times 99 and 101. If it gets there before time 100, it will catch the ferry and be in Jutland (J) as of time 110, otherwise it stays at L. We are to infer that at time 110, M is either on L or on J (but should not infer one or the other).

Actually, Sandewall's DFL formalization makes the problem a little harder by saying, in effect:

At time 0, the bike is on Fyen. At some time T between 99 and 101, the bike arrives at the landing. If its arrival T is before time 100, then the bike gets on board the ferry at time 100. If the bike is on board at time 105, it arrives on Jutland at time 110.

I will use a similar encoding for the TC version. The TC version assumes more, but, I will argue, rightly so.

obs1 $[0]on(M, F)$
chr1 $[t_1, t_2]do(M, y) \Leftrightarrow [(t_1, t_2, y) = (0, T, GotoL)]$
        $\vee\, [T < 100 \wedge (t_1, t_2, y) = (100, 101, Board)]$
        $\vee\, [T \geq 100 \wedge T \leq t_1 \leq t_2 \leq 110 \wedge y = Wait]$
obs2 $99 \leq T \leq 101$
eff1 $[t_1, t_2]do(M, GotoL) \Rightarrow [t_2]on(M, L)$
eff2 $[t_1, t_2]do(M, Board) \Rightarrow [t_2]on(M, B)$
eff3 $[105]on(M, B) \Rightarrow [110]on(M, J)$
eff4 $[t_1, t_2]do(M, Wait) \wedge [t_1]on(M, y) \Rightarrow [t_1, t_2]on(M, y)$

exp1 $[t_1, t_2]on(M, B) := F \Rightarrow [t_1..t_2]do(M, Unboard)$
ineq1 $u(GotoL, Board, Unboard, Wait)$

Reasoning: Suppose $T < 100$. Then by chr1, $[100, 101]do(M, Board)$ and hence by eff2, $[101]on(M, B)$. By exp1, if $[105]\neg on(M, B)$ then $[101..105]do(M, Unboard)$, which can be seen to be false from chr1(according to which there are no actions beginning at time 101 or later if $T < 100$). Hence $[105]on(M, B)$, and so by eff3, M gets to Jutland at time 110.

Now suppose $T \geq 100$. Then by chr1 & obs2, $[T, 110]do(M, Wait)$. (Likewise for all subintervals of $[T, 110]$, but that doesn't interest us.) Also by chr1, $[0, T]do(M, GotoL)$ and hence by eff1, $[T]on(M, L)$. So by eff4, $[T, 110]on(M, L)$.

Clearly there is no basis for supposing either $T < 100$ or $T \geq 100$, and no unequivocal final location for M can be obtained. □

I said above that the TC version assumes more than Sandewall's DFL version. I was referring to the use of Unboard in the reasoning. The explanation closure axiom giving Unboarding as an explanation for $on(M, B)$ becoming false (i.e., exp1) is not just an embellishment but is essential to the inference that once M in on board, it stays on board till time 110.

I claim that this is an entirely reasonable assumption – in fact, that the desired conclusion about ending up at L or J should *not* be reached based on the information assumed by Sandewall. To illustrate the point, I will give a few syntactic variants of the "story" which lead to different conclusions.

At time 0, Albert is at home and hungry (state F).
At some time T between 99 and 101, Albert arrives (hungry) in the foyer of his favorite restaurant (state L).
If his arrival time T is before time 100, then he gets seated (still hungry) at time 100 (state B). (Maybe the manager usually holds a table for him till then, or maybe with the random comings and goings of customers it just happens to work out that way).
If he is seated and still hungry at time 105, he is seated and not hungry at time 110 (state J).

Can we conclude Albert is either in state L (hungry and in the foyer) or state J (seated and not hungry) at time 110? Clearly not: he could equally well be in state B (seated and still hungry), after having waited for a while in the foyer and finally gotten a table. He could even be in state F again (home and hungry) after stalking out of the restaurant, or even home and not hungry, having ordered and consumed a pizza (this is none of F, L, J). The point is that the conclusions we draw from even the simplest story about persistence of states are subtly dependent on world knowledge and narrative conventions, so we should not expect them to follow simply from superficial logical translations of the story sentences.

Here are two more variants:

At time 0, the subway is at station F.
At some time T between 99 and 101, it arrives at station L.

If its arrival T is before time 100, then it gets to station B at time 100.
If it is at station B at time 105, then it gets to station J at time 110.

In this case the alternative of still being at L seems quite unlikely. Moreover, the subway is not likely to be at J much past time 110.

At time 0, the house is on fire (but not wet) (state F).
At some time T between 99 and 101, the fire truck arrives and at that point the house becomes wet and on fire (state L).
If this happened before time 100 (say, the "flare point" of the fire), it will stop being on fire (while still being wet) at time 100 (state B).
If it was wet and not on fire at time 105, it'll be dry and not on fire at time 110. (state J).

Here again state B (wet and not on fire) can't be ruled out – the fire may have been doused by then anyway.

Prediction from disjunction: the Russian Turkey Shoot (RTS)

The problem differs from YTS only in that a Spin action (spinning the chamber of the gun) is inserted between the Wait and the Fire. The inference that the turkey dies should be disabled.

chr1 $[t_1, t_2]do(Joe, y) \Leftrightarrow (t_1, t_2, y) \in \{(4, 6, Load), (7, 9, Spin),$
$(10, 12, Fire)\}$

obs1, eff1–3, exp1–3, ineq1   as in YTS

Ambiguous retrodiction: Stolen Car Problem (SCP)

At the beginning of the first night, the car is in my possession (expressed by predicate p). I perform the action of "leaving the car overnight in my garage" on two successive nights. On the following evening, the car is not in my possession.

I cannot lose possession of the car during the day. Once I've lost possession of it, I can't regain it. The intended conclusion is that I lost possession of the car during one of the two nights (with no conclusion about which night it was).

To illustrate that complete action closure is in general unnecessary, I will merely assume that the only Leave-car-overnight actions were those on the given nights ($[0, 2]$ and $[4, 6]$), so (given that only these can lead to car loss) the car couldn't have been lost during the day. The even weaker assumption that there were no Leave-car-overnight actions on the given days would have been sufficient, as well.

obs1 $[0]p$
chr1 $[t_1, t_2]do(I, Leave\text{-}car\text{-}overnight) \Rightarrow (t_1, t_2) \in \{(0, 2), (4, 6)\}$
obs2 $[8]\neg p$
eff1 $[t_1]\neg p \Rightarrow [t_1, t_2]\neg p$
exp1 $[t_1, t_2]p := F \Rightarrow [t_1..t_2]do(I, Leave\text{-}car\text{-}overnight)$

Random, but probable events: Ticketed Car Problem (TCP)

In some nonmonotonic approaches to the SCP above, the theft of the car would be treated as an exceptional event, and this will affect the axiomatization. Therefore San92 also offers a variant in which a car left overnight in a certain spot is *quite likely* to be ticketed. Other than that, the scenario and the desired conclusions are just as in the SCP. In the EC/AC approach, the distinction makes no difference so I omit the specifics.

Logically related fluents: Dead Xor Alive Problem (DXA)

This is a slight reformulation of the YTS, with "becoming not alive" replaced by "becoming dead", and the equivalence axiom $[t]\neg a \Leftrightarrow [t]d$ added (where d means "dead"). Such logical connections lead to "autoramifications" (in Sandewall's terminology). In our monotonic approach the reformulation leads unproblematically to the conclusion that the turkey is d (and hence $\neg a$) after firing, and no sooner, much as before.

Logically related fluents: Walking Turkey Problem (WTP)

This is another slight variant of the YTS, in which the turkey is initially known to be walking (w) (but it is not explicitly given that he is alive), and the conditional $[t]w \Rightarrow [t]a$ is known. We are to conclude that the turkey is not walking after the firing. We easily infer $[0]a$ from $[0]w$ and reason as in YTS, concluding $[10]a$ and $[12]\neg a$ and hence $[12]\neg w$ by the contrapositive of the new conditional.

Prediction from disjunction: Hiding Turkey Scenario (HTS)

In this variant of Sandewall's, the turkey may or may not be deaf, and if it is not, it goes into hiding when the gun is loaded (where it is initially unhidden). Gun-loading, waiting, and firing take place in succession as in the YTS, but firing only kills the turkey if it is not hiding.

The intended conclusion is that at the end of firing, the turkey is either deaf and not alive, or nondeaf and alive. Sandewall points out that this problem confutes methods like Kautz's [19] which unconditionally prefer later changes to earlier ones (and so leave the turkey unhidden and hence deaf and doomed). In an EC-based approach, this variant is quite analogous to the RTS. We add an effect axiom that Hide brings about h (eff4), and EC axioms that *only* Hide and Unhide can bring about h and $\neg h$ respectively (exp3, exp4). We further add an assumption stating that if Fred is ever deaf, then he always was and always will be deaf (eff5).[6]

I will represent the gunman's (Joe's) and the turkey's (Fred's) actions by separate chronicles for clarity (chr1 and chr2).

---

[6] On a more careful analysis, the events causing or remedying deafness are like those causing or remedying a plugged car exhaust (see "Improbable disturbances" below). However, for the purposes of the present scenario it seems reasonable to treat deafness and nondeafness as permanent.

obs1	$[0]a \wedge \neg l \wedge \neg h$
chr1	$[t_1, t_2]do(Joe, y) \Leftrightarrow (t_1, t_2, y) \in \{(4, 6, Load), (10, 12, Fire)\}$
chr2	$[t_1, t_2]do(Fred, y) \Leftrightarrow [[5]\neg d \wedge (t_1, t_2, y) = (7, 9, Hide)]$
eff1	$[t_1, t_2]do(Joe, Load) \Rightarrow [t_2]l$
eff2	$[t_1](l \wedge \neg h) \wedge [t_1, t_2]do(Joe, Fire) \Rightarrow [t_2](\neg a \wedge \neg l)$
eff3	$[t_1, t_2]do(Joe, Chopneck) \Rightarrow [t_2]\neg a$
eff4	$[t_1, t_2]do(Fred, Hide) \Rightarrow [t_2]h$
eff5	$[t_1]d \Rightarrow [t_2]d$
exp1	$[t_1, t_2]l := F \Rightarrow (\exists y \in \{Fire, Unload, Spin\})[t_1..t_2]do(Joe, y)$
exp2	$[t_1, t_2]a := F \Rightarrow (\exists t_1')[t_1 \leq t_1' \leq t_2 \wedge [t_1'](l \wedge \neg h)$
	$\wedge [t_1'..t_2]do(Joe, Fire)]$
	$\vee [t_1..t_2]do(Joe, Chopneck)$
exp3	$[t_1, t_2]h := T \Rightarrow [t_1..t_2]do(Fred, Hide)$
exp4	$[t_1, t_2]h := F \Rightarrow [t_1..t_2]do(Fred, Unhide)$
ineq1	$u(Load, Unload, Fire, Spin, Chopneck, Hide, Unhide)$

Reasoning: Suppose the turkey is initially deaf, $[0]d$. Then he is still deaf after the Load by eff5, hence he fails to Hide after the Load (or indeed, at any time) by chr2. Since he is initially unhidden according to obs1, he remains unhidden by exp3(etc.), so that in particular $[10]\neg h$. Likewise the l property inferrable at time 6 from the Load action and eff1 persists by EC-reasoning to time 10. Hence the Fire action is fatal by eff2, and so $[12]\neg a$ and $[12]d$ (after another application of eff5).

On the other hand, if the turkey is not initially deaf, he is still nondeaf at time 5 during the Load (by the contrapositive of eff5). Hence Fred Hides during $[7, 9]$ by chr2. He remains hidden through the subsequent actions by EC-reasoning based on exp4, in particular $[10]h$. After proving persistence of a from the initial state to time 10 in the usual way, we can also prove its persistence through the Fire action from exp2. Thus $[12]a$ and $[12]\neg d$ in this case (after another application of eff5). The assumption of initial deafness or non-deafness can each be made consistently, so that we can only infer the disjunction of the corresponding conclusions. $\square$

Improbable disturbances: Potato in the Tailpipe (TPP)

Initially the car engine is not running ($\neg r$). The action of attempting to start the car is performed. On the assumption that there is usually no potato in the tailpipe (predicate p is usually false), and that the car will start if there isn't, we are to conclude that the car will start.

Sandewall approximates the premise that there is usually no potato in the tailpipe by saying that there is no potato in the tailpipe at time 0, by default. Default axioms are used only for final ranking of the most preferred models of the remaining axioms, and thus may be violated.

Ordinary first-order logic cannot express explicitly uncertain premises (such as ones involving "usually") and so cannot accurately model reasoning based upon them. To my mind the most attractive approach to uncertain reasoning is one based

on *direct inference* of epistemic probabilities (i.e., probabilities for particular propositions) from "statistical" generalizations (e.g., [2], [3], [6], [17], [18]). The advantage of a probabilistic approach to nonmonotonicity is that it allows systematically
for degrees of belief, and that it can provide a coherent basis for decision-making
by an intelligent agent. The proof theory for direct inference is not yet fully developed, though the known techniques nicely handle many standard examples in
nonmonotonic inheritance [6]. The present version of the TPP seems beyond the
scope of current syntactic proof techniques, but can be analysed directly in terms of
the model theory. For illustrations of direct inference for other variants of TPP, see
[33], [34] and [3, 180-2].

The following, then, is a TC-like axiomatization of TPP based on a statistical
interpretation of the statement about potatoes in the tailpipe. $[[t]\neg p]_t$ denotes the
proportion of times t at which $[t]\neg p$ holds.

stat1 $[[t]\neg p]_t > .99$
obs1 $[0]\neg r$
chr1 $[t_1, t_2]do(Joe, y) \Leftrightarrow (t_1, t_2, y) = (6, 8, Start)$
eff1 $[t_1]\neg p \wedge [t_1, t_2]do(Joe, Start) \Rightarrow [t_2]r$

Reasoning: We appeal directly to the model-theoretic definition of epistemic prob-
abilities [6]. Essentially $Prob([8]r|KB)$, where $KB =$ stat1$\wedge$ obs1$\wedge$ chr1
$\wedge$ eff1, is the proportion of models of KB in which $[8]r$ holds. (More exactly, one
considers the limit ratio as the number of time points comprising the time domain
approaches ∞.)

Let the number of interpretations of p satisfying stat1 be M (for some fixed
finite time domain). Since in each of these interpretations the proportion of time
points at which $\neg p$ holds is $> 99\%$, it is clear that more than 99% of these interpretations, say M' of them where $M' > .99M$, will have $[6]\neg p$ true in them.

Now each of these M' interpretations can be extended to a model of KB by
using any interpretations of r and *do* that satisfy obs1$\wedge$ chr1$\wedge$ $[8]r$. (Note that
this entire conjunction does not involve p.) So if there are N such interpretations of
obs1$\wedge$ chr1$\wedge$ $[8]r$, we obtain $M'N$ models of KB in which $[6]\neg p$ and $[8]r$ hold.

This leaves the remaining $M - M'$ $(< .01M)$ interpretations of stat1 to be
considered for which $[6]p$ holds. Each of these interpretations can be extended to
a model of KB using any interpretations of r and *do* that satisfy obs1$\wedge$ chr1
(since the antecedent of eff1 will always be false when $[6]p$ holds, so that eff1
will be satisfied regardless of the truth or falsity of $[8]r$). Of these interpretations of
r and *do*, N satisfy obs1$\wedge$ chr1$\wedge$ $[8]r$, and (as is not hard to see) an equal number
satisfy obs1$\wedge$ chr1$\wedge$ $[8]\neg r$.

Thus,

$$Prob([8]r|KB) = \frac{M'N+(M-M')N}{M'N+2(M-M')N} = \frac{1}{1+(M-M')/M} > \frac{1}{1.01} > .99. \quad \Box^7$$

---

[7] It is interesting to note that since we haven't said $[t_1]p \wedge [t_1, t_2]do(Joe, Start) \Rightarrow [t_2]\neg r$,
the model counting technique predicts that with the tailpipe plugged, the car still has a
50% chance of starting, and that's why the lower bound on overall success probability is
expected to be slightly better than 99/100, namely 100/101.

As long as we demand that very improbable, but nevertheless possible, events be explicitly allowed for, the TPP cannot be monotonically represented. Still, the following trivial monotonic approximation is worth noting. Here the tailpipe is assumed to be initially clear, and the assumed chronicle and EC axiom for tailpipe plugging-up rule out any mischief.

obs1 $[0]\neg r$

obs2 $[0]\neg p$

chr1 $[t_1, t_2]do(x, y) \Leftrightarrow (t_1, t_2, x, y) = (6, 8, Joe, Start)$

eff1 $[t_1]\neg p \wedge [t_1, t_2]do(Joe, Start) \Rightarrow [t_2]r$

exp1 $[t_1, t_2]p := T \Rightarrow (\exists x)[t_1..t_2]do(x, Plug)$

Event-time ambiguity: the Tailpipe Marauder (TPM)

This variant of Sandewall's assumes that a potato *is* put in the tailpipe somewhere between 8am and 5pm, but it is not known when. The attempt to start the car takes place at 1:30pm, and the aim is *not* to reach a conclusion about whether the car starts or not.

TPM is much less problematic for a monotonic approach than the original TPP, since it merely involves *incomplete* knowledge (about the time of a known event), rather than "defeasible" knowledge (where one of the possibilities consistent with our incomplete knowledge is much more probable than the others). I'll arbitrarily call the protagonist Joe and the antagonist Moe, and assign a duration of 2 (two hundredths of an hour) to the Plug and Start actions.

obs1 $[0]\neg r$

obs2 $[0]\neg p$

chr1 $[t_1, t_2]do(x, y) \Leftrightarrow (t_1, t_2, x, y) \in \{(1350, 1352, Joe, Start),$
$(T, T{+}2, Moe, Plug)\},$

$800 \leq T \leq 1699$

eff1 $[t_1, t_2]do(x, Plug) \Rightarrow [t_2]p$

eff2 $[t_1, t_2]do(x, Start) \Rightarrow [[[t_1]p \Rightarrow [t_2]\neg r] \wedge [[t_1]\neg p \Rightarrow [t_2]r]]$

exp1 $[t_1, t_2]p := T \Rightarrow [t_1..t_2]do(Moe, Plug)$

exp2 $[t_1, t_2]p := F \Rightarrow [t_1..t_2]do(Joe, Unplug)$

exp3 $[t_1, t_2]r := T \Rightarrow [t_1..t_2]do(Joe, Start)$

ineq1 $u(Start, Plug, Unplug)$

It is straightforward to show that neither $[1352]\neg r$ nor $[1352]r$ can be inferred. Note that if we are *given* $[1352]\neg r$, we can infer $T < 1350$ and if we are *given* $[1352]r$, we can infer $T \geq 1350$.

Event-order ambiguity: Tailpipe Repairman Scenario (TPR)

In this variant, Sandewall assumes that the tailpipe is initially blocked, and the actions of unplugging the tailpipe and trying to start the car are done in arbitrary order.

No action ordering should be inferrable, but it should follow that the car starts iff the unplugging is done first.

I include the gratuitous assumption that the tailpipe was unobstructed prior to 8am, for conformity with Sandewall's axiomatization.

obs1	$[800]\neg r$	(not running at 8am)
obs2	$[0]\neg p$	(no potato previous midnight)
obs3	$[800]p$	(potato in tailpipe at 8am)
chr1	$[[t_1, t_2]do(x, y) \wedge (800 \le t_1 \le 1698) \wedge (800 \le t_2 \le 1698)]$	
	$\Leftrightarrow (t_1, t_2, x, y) \in \{(T_1, T_1+2, Joe, Start), (T_2, T_2+2, Joe, Unplug)\}$	
eff1	$[[t_1]\neg p \wedge [t_1, t_2]do(Joe, Start)] \Rightarrow [t_2]r$	
eff2	$[t_1, t_2]do(Joe, Unplug) \Rightarrow [t_2]\neg p$	
exp1	$[t_1, t_2]p := F \Rightarrow (\exists x)[t_1..t_2]do(x, Unplug)$	
exp2	$[t_1, t_2]p := T \Rightarrow (\exists x)[t_1..t_2]do(x, Plug)$	
exp3	$[t_1, t_2]r := T \Rightarrow (\exists t_1' \ge t_1)(\exists x)[t_1']\neg p \wedge [t_1'..t_2]do(x, Start)$	
ineq1	$u(Start, Plug, Unplug)$	

Neither $[T_1 + 2]r$ nor $[T_1 + 2]\neg r$ can be inferred. With assumption $T_2 + 2 \le T_1$, we would get $[T_1 + 2]r$, and for the contrary assumption we find the car will never run. Given the extra premise `obs2`, it is also possible to deduce a `Plug` action prior to 8am.

Conditional durations: Furniture Assembly Scenario

A furniture kit is initially unassembled. It is not known whether assembly instructions are included or not. (i or $\neg i$ may hold.) The `Assemble` action is performed, and this requires 20 minutes for completion if the instructions were included and 60 minutes otherwise. The desired conclusion is just that if the instructions were included, the kit is assembled within 20 minutes, and if not, within 60 minutes.

In the following axiomatization, T is the unknown assembly time. The chronicle says that the `Assemble` action is my only action, and could easily be refined to say it is my only action between times 0 and T. (In fact, we could just have asserted $[0, T]do(I, Assemble)$ – completeness is irrelevant here.) Inclusion of the instructions is treated as an atemporal (or permanent) property, though it could also be treated as a fluent. No EC axioms are needed and so none are shown.

obs1	$[0]\neg a$	
chr1	$[t_1, t_2]do(I, x) \Leftrightarrow (t_1, t_2, x) = (0, T, Assemble)$	
eff1	$i \wedge [t_1, t_2]do(I, Assemble) \Rightarrow [t_2]a \wedge t_2 = t_1 + 20$	
eff2	$\neg i \wedge [t_1, t_2]do(I, Assemble) \Rightarrow [t_2]a \wedge t_2 = t_1 + 60$	

Reasoning: We easily reach the conclusion that $[T]a$ and that $T = 20$ if i and $T = 60$ if $\neg i$, using reasoning by cases.□

A stable world: Lifschitz's N blocks

Lifschitz's N-blocks world [21] provides one example of a slightly more complex world than the previous ones. The world in the immediately following example (the "stuffy room" scenario) is likewise more complex, and also less sedate than the N-blocks world. As far as the EC/AC-based approach is concerned, neither presents any unusual challenge (and indeed at least equally complicated cases were treated in Sch90).

The N-blocks world allows movement of one block onto another (with the usual clear-top conditions, formulated in a slightly unusual way in terms of a *top* function) or onto the table, and painting of a block with one of three colors. There are axioms about uniqueness of destinations and resultant colors, and so on. The aim is to come up with the same state-transition characterization of this world as Lifschitz obtains circumscriptively, i.e., (roughly) nothing else changes when a block is moved or painted.

I will not spell out the details of the EC/AC-approach here, as this would be rather pointless. In essence, one just adds EC axioms about the 5 fluent predicates employed ($at, color, true, false, clear$): blocks change at properties only when moved, and change color only when painted, etc. In fact, Reiter's technique for automatic biconditionalization of effect axioms would work well here. The reason it would be pointless to spell all this out is that in doing so, one would assert precisely what Lifschitz sets out to *prove* by circumscribing *causes* and *precond*!

The circumscriptive approach would be preferable if it could be depended on to give the desired persistence properties independently of the domain, without any need to specify EC axioms. Let me reiterate that this is not in general true, because we do not in general have complete knowledge of effects (recall *nextto*).

An unstable world: Agatha's stuffy room

Lifschitz's world is as stable as one would expect a blocks world to be, whereas in Ginsberg & Smith's "stuffy room" world [12] there is considerable latitude for objects to "flit about" unpredictably. They are apt to do so when Tyro, Aunt Agatha's robot, moves an object onto or away from one of the two ventilation ducts on the floor. This is claimed to be consistent with the intuition that light objects like newspapers may be shifted when the airflow in the room changes.

I will consider Winslett's variants of the original scenarios ([35]). These scenarios are designed to illustrate the advantages of Winslett's "possible models approach" (PMA) over Ginsberg & Smith's In essence, the advantage is insensitivity of nonmonotonic inferences to the syntactic form in which information is supplied.

Sandewall [30, 205-6] discusses the scenarios briefly, but does not attempt to duplicate the results in his DFL-2 logic. His reason is that he does not think the conclusions drawn are convincing. In particular, he questions the assumption that an object placed on a vent will stay put, while at the same time this blockage of air flow can cause motion of an object at *another* vent. More generally, he suggests that we should not equivocate about the causal model ("abstraction"): either things remain inert when we block a vent, or we should model the way in which blockage increases pressure, and the way in which this pressure in turn shifts lightweight objects.

Sandewall has a point. In fact, close scrutiny of the examples reveals that the minimization of net change in the PMA (as in the PWA) has some peculiar effects. For instance, the two vents act as "object attractors" when both are initially blocked and one of the blocking objects is removed. The reason turns out to be that by attracting a blocking object, the unblocked duct can maintain its own "blocked" property and the "stuffy" property of the room! On the other hand, when one vent is already blocked and a blocking object is placed on the second vent, the first vent is apt to blow off its obstructing object, so as to maintain its "unblocked" property – and the room's nonstuffy property. More generally, it seems quite odd to claim that the PMA (or the PWA) somehow allows precisely for the physically plausible sets of alternative side effects induced by an action. There surely is no limit to the number and type of potential side effects. Once we have opened the door to drafts, why should we not also admit effects transmitted through attached strings, magnets, electrical conductors, etc.? These may have been cleverly hidden, and be no more apparent to the eye than the ventilator drafts; and their relative *improbability* is surely not something that can magically pop out of the *logic* we use.

Notwithstanding all that, it is of interest to encode this slightly bizarre world monotonically, as a test of the flexibility of the EC-based approach. After all, one *can* make up a physics story about why the vents attract and repel objects as predicted by the PMA. Once one makes these physical assumptions explicit through EC axioms, the charge of arbitrariness will no longer stick. With regard to Sandewall's specific objection, one can imagine that Tyro holds on to an object after moving it, until the gusts caused by the changes in duct blockage have settled down. Encoding a more causally coherent world such as Sandewall envisages would certainly be possible as well, indeed easier. EC axioms allow us to tailor the persistence knowledge to fit the physics, relieving us from trying to make the physics fall out of the semantics.

We begin with a set of timeless "laws" constraining Aunt Agatha's living room R. Everything is either a location or is *on* something. For one thing to be on another, the latter must be a location while the former must not. There are exactly two floor ducts D_1 and D_2. These and the floor (*Floor*) are the only *locations*. A thing can be *on* only one location and only the floor can have more than one thing on it. A duct is *blocked* iff something is on it. The room is stuffy iff both ducts are blocked. Symbolically,

law1 $location(x) \lor (\exists y)on(x, y)$
law2 $on(x, y) \Rightarrow [location(y) \land \neg location(x)]$
law3 $duct(x) \Leftrightarrow x \in \{D_1, D_2\}$
law4 $location(x) \Leftrightarrow [duct(x) \lor x = Floor]$
law5 $[on(x, y) \land on(x, z)] \Rightarrow y = z$
law6 $[on(x, y) \land on(z, y)] \Rightarrow [z = x \lor y = Floor]$
law7 $[duct(d) \land (\exists x)on(x, d)] \Leftrightarrow blocked(d)$

law8 $[blocked(D_1) \wedge blocked(D_2)] \Leftrightarrow stuffy(R)$

Winslett's first scenario was designed to illustrate the difficulties that the PWA encounters with the frame problem when some properties of the initial state are entailed by the axioms but not explicitly asserted. Agatha's TV, birdcage C and magazine M are on D_1, D_2, and $Floor$ respectively. There is also a newspaper N but nothing is specified about it (except, one presumes, that it is distinct from the other things in this world). Note that if N is not a location it must be on a location (law1), and since the ducts are occupied, it must in fact be on the floor. The only available action (performable by Tyro) is $Move(x, y)$, for which it is necessary that y is the floor, or nothing is *on* y, or x is already *on* y. Under these conditions the effect is that x is *on* y.

By a careful consideration of the possible models in all 3 of Winslett's scenarios, we find that the EC laws of this world should say the following. First, an object can flit spontaneously to a duct only if that duct is initially blocked and Tyro moves away a blocking object from either one of the ducts. (Model-theoretically, this can avert changes in "blocked" and "stuffy" properties.) Second, an object can flit spontaneously to the floor only if it is on a duct initially, the other duct is not blocked, and Tyro moves an object from the floor onto the other duct. (Model-theoretically, this can avert a change from a nonstuffy room to a stuffy one.)

Agatha now asks Tyro to move the TV to the floor. The desired conclusion is that the TV will be on the floor, while other objects may flit to one or the other duct, in conformity with one of Winslett's 6 models.

obs1 $[0](on(TV, D_1) \wedge on(C, D_2) \wedge on(M, Floor))$

chr1 $[t_1, t_2]do(Tyro, x) \Leftrightarrow (t_1, t_2, x) = (1, 2, Move(TV, Floor))$

eff1 $[[y = Floor \vee [t_1]\neg on(z, y) \vee [t_1]on(x, y)]$
$\wedge [t_1, t_2]do(Tyro, Move(x, y)] \Rightarrow [t_2]on(x, y)$

exp1 $[[t_1, t_2]on(x, y) := T \wedge y \in \{D_1, D_2\}] \Rightarrow$
$\qquad [t_1 .. t_2]do(Tyro, Move(x, y))$
$\qquad \vee [(\exists t \geq t_1)[t]blocked(y)$
$\qquad\qquad \wedge (\exists x')[t][(on(x', D_1) \vee on(x', D_2))$
$\qquad\qquad\qquad \wedge (\exists y')[t][\neg on(x', y') \wedge$
$\qquad\qquad\qquad [t .. t_2]do(Tyro, Move(x', y'))]]]]$

exp2 $[[t_1, t_2]on(x, Floor) := T \Rightarrow [t_1 .. t_2]do(Tyro, Move(x, Floor))$
$\qquad \vee [(\exists t \geq t_1)(\exists d_1, d_2 \in \{D_1, D_2\})[t]on(x, d_1) \wedge \neg on(z, d_2)$
$\qquad\qquad \wedge [t .. t_2]do(Tyro, Move(x', d_2))]]$

ineq1 $u(D_1, D_2, Floor, M, N, TV)$

ineq2 $(x \neq x' \vee y \neq y') \Rightarrow u(Move(x, y), Move(x', y'))$

Reasoning: First, we obviously get $[2]on(TV, Floor)$ from chr1 and eff1. This persists to time 3 since if it became false, $on(TV, x)$ would have to become true for some x and so by exp1 there would have to be an additional Move between times 2 and 3, contrary to chr1. The additional conclusions desired can be reformulated as follows:

1. If nothing flits to duct D_1, then only the TV moves to a new location;

2. If one of M, N flits to D_1, then only that object and the TV move to a new location;

3. If C flits to D_1, then there are no constraints on further flitting except those dictated by the " laws". (So there is nothing further to be proved in this case.)

First we note that when Tyro moves the TV to the *Floor*, he makes no other concurrent move. This follows from chr1, ineq1 and ineq2. So any additional shifts are due to "flitting". So to prove (1), assume nothing flits to D_1, so that $[2]\neg on(x, D_1)$. To complete the analysis for time 2, we need only show that cage C does not flit to the *Floor*. If C did flit to the *Floor*, the second disjunct of exp2 would apply, and chr1 would force the identification $t = 1$; at that time, if $d_1 = D_1$ then $on(x, d_1)$ would be false, and if $d_2 = D_1$ then $\neg on(z, d_2)$ would be false. In either case exp1 would be violated and so C does not flit to the *Floor*.

To prove (2), assume first that magazine M flits to D_1. Once again we need only show that C does not flit to the floor. As before we find that if it did, the second disjunct of exp2 would be violated. The argument for the case that newspaper N flits to D_1 is completely analogous. It remains to show that there is no change from time 2 to time 3, and this follows from the fact that exp1 would require a further action by Tyro between those times for any *on* relationship to change, and this is ruled out by chr1. (The " laws" then also prevent change of *blocked* and *stuffy*.) □

Second and third " stuffy room" variants

Winslett's second variant was designed to show that the PWA can generate unwarranted conclusions in the presence of a logically redundant disjunction, and the third to show that it makes a difference for inferences under the PWA whether or not an entailed negative literal is explicitly present. I will not go into these except to say that for the effect and EC axioms above one obtains the same alternative outcomes for these scenarios as are obtained by Winslett's PMA.

Concurrent actions

In Sch90 I suggested that many of the alleged deficiencies of the situation calculus, as a general calculus for action and change, were due simply to neglect of the possibilities inherent in *functions* of situations and actions. In particular, I suggested that (1) external change could be accommodated by letting the usual $Result(a, s)$ function predict such change (for instance, $Result(Wait\text{-}a\text{-}minute, s)$ might differ significantly from s if s is a dynamic situation such as one where the sun is about to rise); (2) continuous time and continuous change could be accommodated with functions like $Clock\text{-}time(s)$ and $Trunc(a, t)$, where the latter supplies an initial segment of duration t of action a (so that $s' = Result(Trunc(a, t), s)$ is a situation t seconds after situation s and $Clock\text{-}time(s') = Clock\text{-}time(s) + t$); and (3) most importantly composite actions, including concurrent ones, could be accommodated through action composition functions such as $Seq(a, b)$ and $Costart(a, b)$.

I worked through an example involving a man, a robot, and a cat, where the man walks from one place to another while the robot concurrently picks up and carries a box containing the (inactive) cat. I showed how persistence reasoning based on EC could be extended to such a setting. For instance, the EC axiom for color change says that if the color of an object is changed in the result state of a composite action, then that action must have had a primitive part in which the object is painted or dyed. I also showed how the usual effects of independent actions executed concurrently could be predicted if actions are provably *compatible*. In the example, compatibility of the concurrent actions was taken to be a consequence of disjointness of the "action corridors" within which they happen to occur. Rather than repeating such an example here, let me just reiterate that the solution was entirely monotonic, and that it can easily be recast in TC form.

Gelfond *et al.*[25] independently made some proposals similar to my own concerning the use of action combinators for dealing with concurrent actions in the SC. Just as I was concerned with showing certain concurrent actions to be *compatible*, they are concerned with showing that composite actions (with concurrent components) are *free of conflict*. For them, this means that the concurrent components do not have effects leading to different values for the same fluent, and they employ circumscription of conflict to minimize this sort of adverse interaction. While I considered only the case where compatibility (lack of conflict) is due to action disjointness, Gelfond *et al.* also allow for *constructive* interference, whereby certain effects of individual actions (like spilling of soup in one-handed lifting of a soup bowl) are "cancelled" in the concurrent case. These are interesting ideas, though their formulation is limited by the need to specify causal relations in a situation-independent way (*a la* [Lifschitz 1987]; *cf.,* [Baker 1991]). More importantly from the present perspective, the "blanket closure" assumptions implemented through circumscription are too strong, for much the same reasons that closure of effects is in general too strong. (Interference is, after all, due to the *effects* of actions on each other.)

Lin and Shoham [24] provide a third, and also closely related, preliminary proposal for allowing concurrency in SC. Their main concurrent combinator is written with set brackets, i.e., $\{Action_1, ..., Action_n\}$. Much as in the earlier attempts, a central concern is encoding noninterference between concurrent actions. They make the apt observation that this problem is analogous to the frame problem, i.e., by and large, actions don't interfere; accordingly, they tackle the problem by circumscriptively minimizing pairwise "cancellation" of given concurrent actions in given situations. In keeping with my approach to the frame problem, I would instead suggest the use of EC-like axioms to rule out interference; i.e., we specify various *necessary* conditions for various kinds of actions to interfere, and rule these out *where our observations and world knowledge allow us to do so*. In fact, the reasoning about "action corridors" in Sch90 I mentioned above involves just such an EC-like axiom, viz., one that states that for the physical motion of two objects to interfere, their paths must intersect – a reasonable postulate in many settings. If their paths are known *not* to intersect within a given time frame then we can infer noninterference.

In this way we can avoid the extreme requirement of "epistemic completeness" which Lin and Shoham propose as a desideratum for action formalizations.

The ramification problem

The ramification problem arises from the fact that the changes directly produced by an action can entail additional changes, which can entail still further changes, and so on. The problem is to avoid exhaustively enumerating all the resultant changes in describing the effects of an action, yet be able to infer those changes (as needed).

Ginsberg [11] apparently regards the ramification problem as a difficulty for EC. However, a rather elaborate "robot's world" example in Sch90 [section 3] showed how well EC works even in the presence of ramifications. The ramifications in that example are ones resulting from arbitrarily stacked containment and *on*-relations. For instance, the robot may carry a box which contains a cup which in turn contains an egg. The inference that the cup and egg are transported along with the box is easily made. We merely need to be sure that the *in* and *on* relations persist, and for this we apply straightforward EC axioms stating, for example, what needs to happen in order for an *in*-relation to change. (In the axiomatization in Sch90, the robot needed to take the object out of the container, or take something else out of the container that "carries" the object along with it; "carries" was in turn axiomatized to allow for stacked in/on/part-of relations.)

I see no particular difficulties in extending these techniques to arbitrarily complex worlds. We neither need to *directly* axiomatize the cascaded effects of an action (rather we need only axiomatize "one link at a time" of the causal chains), nor retrace these cascades explicitly in the EC axioms.

3   Coda: The Metaphysics of Change

A recent trend in NMR research has been the development of criteria of *correctness* for nonmonotonic theories of action, based on "inertial" models. Sandewall [30], [31] is a prime example of this, and Gelfond & Lifschitz [8] is in a similar spirit. In particular, these correctness criteria assume that (A) the world is totally inert except for changes wrought by the agent (Sandewall's "ego"), and (B) we have total knowledge of actions and action laws (and state constraints, if allowed).

I do not accept (A), i.e., "commonsense inertia". I think that while this is a reasonable working assumption for some highly restricted, tightly controlled domains, it is an untenable metaphysical position *vis a vis* "the world at large".

We do certainly *seek* invariants in the way we conceptualize the world; and to the extent we succeed, we reduce its perceived complexity and can cope more readily with it. But to extrapolate from our partial success in this quest for invariance to a metaphysics according to which the world really *is* a set of passive objects with static features, altered only by the intervention of one or a few agents, seems to me utterly implausible. Does anyone, naive or sophisticated, actually *believe* this? Ought not semantics accord with our intuitions about the world?

Perhaps academic researchers contemplating these matters are more apt than others to be impressed with the stability of the world, as their gaze wanders over tranquil office furnishings and inert papers and books, and their obedient workstation passively awaits the next keystroke. If in addition they understand computation, they may find the notion that the world is like a computer's internal store, modified only at the behest of the CPU, nearly irresistible (and indeed this analogy has been alluded to by the father of "commonsense inertia" – McCarthy [26]). I suspect that cab drivers, factory workers, weather reporters, stockbrokers, firemen or fishermen may have less affinity for such a metaphysics.

In my own metaphysics only the *laws* that govern fluents are constant, not the fluents themselves. The world is a chaotic place teeming with activity and change at all time scales and "granularities"; and only careful choice of vocabulary and coordinate frame and calculated neglect of many obvious variables imposes a semblance of order and stability on some patches of this hubbub.

But aren't these patches of stability enough to afford the proponents of inertia worlds a foothold? Yes, but note that this amounts to a pretense: it's not that the world *is* inert, just that for some purposes we can do business *as if* it were. And this pretense, like any other, is brittle: it lacks the robustness of truth, breaking down at the edges as we shift out of the narrow domain for which the pretense was contrived.

Of course, to the inertia adherents this simply indicates the need for a certain nimbleness in switching from one pretense to another – shifting to a new coordinate frame, a new vocabulary of fluents and actions reflecting new criteria of relevance, and semantically, to a new make-believe world of inert objects (see [22]). So there is one frame for physical action within the confines of the office, another for coping with rush-hour traffic, another for maintaining the lawn, others for functioning as part of various social groups and organizations (teaching, administering, parenting, choir singing, etc.), and so on.

But such nimbleness will be hard to achieve, since (i) the number of special domains is large, (ii) the boundaries are blurry, (iii) they can merge rather arbitrarily (a business meeting in an office, parenting while driving in rush-hour traffic, etc.), (iv) they don't always admit a static view of the relevant aspects, however adroitly we choose our frame, and (v) worst of all – at least from a logicist perspective – the *knowledge* of what frame is appropriate under what circumstances can have no coherent semantics, since there is no comprehensive inertia world in which the various make-believe micro-worlds can be embedded. The real world just *isn't* inert!

Wouldn't it be preferable to view the world realistically in the first place – exploiting the stabilities and regularities we find, of course, but treating these as contingent knowledge, for instance as knowledge about (commonsense) physics, or (commonsense) psychology, etc., rather than as a matter or *meta*physics? I claim that this can and should be done, through appropriate, limited explanation closure axioms (in conjunction with effect axioms).

Nor do I accept the epistemic assumption (B). Briefly, (i) We don't know all the actions in the world that have taken place or will take place. (ii) We don't know all the effects of all the actions we know about on all the fluents we care about. (iii) We don't know all relevant state constraints. (iv) We live in a world where there are many other agents as well as spontaneous change. (v) Effects may ramify unboundedly and affect unboundedly many fluents. (E.g., consider an object's distance from all others, when that object is moved.)

In short, it seems to me that methods based on inertia-world semantics are forever doomed to be applicable only to narrow, largely passive, insulated, thoroughly "predigested" domains, where moreover we have more or less complete knowledge of relevant actions and fluents, and the laws that govern them. The EC/AC approach lacks these metaphysical and epistemic overcommitments, yet can deal with the frame problem and has the flexibility to encompass multiple domains without inconsistency.

4   Conclusion

I have tried in this report to explore the scope of a particular technique, EC/AC-based reasoning in dynamic worlds, more fully than is the standard practice. I hope to have provided enough of the technical gist of the proposed EC/AC-based solutions to Sandewall's test suite to support my contention that much of the reasoning commonly thought to require nonmonotonic methods can in fact be done monotonically.

I should reiterate that in saying this, I am not suggesting that monotonic reasoning is all you really need. A monotonic theory of any realistically complex, dynamic world is bound to be an approximation, in the sense that it ignores both improbable qualifications on the effects of actions, and far-fetched explanations for change. We simply cannot *express* in ordinary FOL that certain kinds of events are very unlikely but may nonetheless occur. For this, we need to go beyond FOL, as has been done in nonmonotonic and probabilistic logics.

But I think the literature on nonmonotonic logics has put *too much* of the burden of commonsense reasoning (especially too much of the task of inferring persistence and change) on nonmonotonic methods. An adverse effect has been a confusion between narrative principles and logic, and between physics and logic. The very terms "persistence" and "inertia" used as *model-theoretic* notions are suspect, since objects stay put, or keep moving, for physical rather than model-theoretic reasons. As well, the over-deployment of nonmonotonic methods has created computational intractability problems, where relatively simple monotonic methods would have sufficed.

The EC/AC-based approach seems to deal with most of the issues addressed by Sandewall's test suite rather handily. It does not render things quiescent (or nonexistent) merely because nothing is known about them, it does not spawn spurious events to minimize change, it does not fail when aimed backward in time, and it does not arbitrarily choose between disjuncts. Plausible EC and AC axioms are not hard to conjure up (and as Reiter showed, the former can sometimes be obtained mechanically), they do not work in mysterious ways, and they work computably and even efficiently (in STRIPS-like settings). It therefore seems well worthwhile to further investigate EC/AC-based methods, e.g., for planning applications. One of the most interesting directions for further work is to use probabilistically qualified EC and AC axioms in a probabilistic logic setting (cf. the earlier citations of work by Bacchus and Tenenberg & Weber); i.e., we would say such-and-such a change is *very likely* due to this or that kind of action, and such-and-such actions are *very probably* the only relevant ones that occurred in a certain setting. At that point we would be ready to address the qualification problem in full, while still exploiting the power of EC and AC to infer (probable) persistence or change.

Acknowledgements

Conversations with Ray Reiter and Andy Haas have clarified my understanding of the relation between effect axioms, EC axioms, and the qualification problem. The paper also benefited from Michael Georgeff's, Chung Hee Hwang's, Vladimir Lifschitz', and anonymous JLC referees' helpful comments. Support was provided by ONR/DARPA research contract no. N00014-82-K-0193 and Rome Lab contract F30602-91-C-0010. This paper previously appeared in *Journal of Logic and Computation* 4(5), pp. 679-799, 1994, and is reprinted with the permission of Springer-Verlag.

References

1. J. Amsterdam. Temporal reasoning and narrative conventions. In *Proc. of the 2nd RInt. Conf. on Principles of Knowledge Representation and Reasoning (KR'91)*, pages 15–21, Cambridge, MA, April 22-25 1991.
2. F. Bacchus. Statistically founded degrees of belief. In *Proc. of the 7th Bienn. Conf. of the Can. Soc. for Computational Studies of Intelligence (CSCSI '88)*, pages 59–66, Edmonton, Alberta, June 6-10 1988.
3. F. Bacchus. *Representing and Reasoning with Probabilistic Knowledge.* MIT Press, Cambridge, MA, 1990.
4. F.M. Brown, editor. *The Frame Problem in Artificial Intelligence. Proc. of the 1987 Workshop*, Lawrence, KS, Apr. 12-15 1987. Morgan Kaufmann Publishers, Los Altos, CA.
5. E. Davis. Axiomatizing qualitative process theory. In *Proc. of the 3rd Int. Conf. on Principles of Knowledge Representation and Reasoning (KR'92)*, pages 177–188, 1992.
6. J.Y. Halpern F. Bacchus, A.J. Grove and D. Koller. Statistical foundations for default reasoning. In *Proc. of the Int. Joint Conf. on Artificial Intelligence (IJCAI-93)*, pages 563–9, 1993.
7. G. Ferguson and J.F. Allen. Actions and events in interval temporal logic. *Journal of Logic and Computation*, 4(5) (Special Issue on Actions and Processes):531–579, 1994.
8. M. Gelfond and V. Lifschitz. Representing actions in extended logic programming. In K. Apt, editor, *Proc. of the Joint Int. Conf. and Symp. on Logic Programming*, pages 558–573. MIT Press, 1992.
9. M.P. Georgeff. Actions, processes, causality. In M.P. Georgeff and A.L. Lansky (1987), pages 99–122, 1987.

10. M.P. Georgeff and A.L. Lansky, editors. *Reasoning about Actions and Plans: Proc. of the 1986 Workshop*, Timberline, OR, June 30-July 2, 1987. Morgan Kaufmann Publ., Los Altos, CA.

11. M. Ginsberg. *Essentials of Artificial Intelligence*. Morgan Kaufmann, Los Altos, CA, 1993.

12. M. Ginsberg and D.E. Smith. Reasoning about actions i: A possible worlds approach. In F. M. Brown (1987), pages 233–258. 1987. Also in *Artificial Intelligence 35* (1988): 165–195.

13. M. Ginsberg and D.E. Smith. Reasoning about actions ii: The qualification problem. In F. M. Brown (1987), pages 259–287. 1987. Also in *Artificial Intelligence 35* (1988): 311–342.

14. D. McAllester H. Kautz and B. Selman. Encoding plans in propositional logic. In *Proc. of the 5th Int. Conf. on Principles of Knowledge Representation and Reasoning (KR'96)*, pages 374–384, Cambridge, MA, November 5-8 1996.

15. A.R. Haas. The case for domain-specific frame axioms. In F. M. Brown (1987), pages 343–348. 1987.

16. A.R. Haas. A reactive planner that uses explanation closure. In *Proc. of the 3rd Int. Conf. on Principles of Knowledge Representation and Reasoning (KR'92)*, pages 93–102, Cambridge, MA, 1992.

17. J.Y. Halpern. An analysis of first-order logics of probability. *Artificial Intelligence*, 46:311–350, 1990.

18. Jr. H.E. Kyburg. Probabilistic inference and probabilistic reasoning. In Shachter and Levitt, editors, *The Fourth Workshop on Uncertainty in Artif. Intell.*, pages 237–244. 1988.

19. H. Kautz. The logic of persistence. In *Proc. of the 5th Nat. Conf. on AI (AAAI 86)*, pages 401–405, Philadelphia, PA, August 11-15 1986.

20. A.L. Lansky. A representation of parallel activity based on events, structure, and causality. In M.P. Georgeff and A.L. Lansky (1987), pages 123–159. 1987.

21. V. Lifschitz. Formal theories of action. In F. M. Brown (1987), pages 35–57. 1987.

22. V. Lifschitz. Frames in the space of situations. *Artificial Intelligence*, 46:365–376, 1990.

23. F. Lin and Y. Shoham. Provably correct theories of action (preliminary report). In *Proc. of the 9th Nat. Conf. on AI (AAAI-91)*, pages 349–354, Anaheim, CA, 1991.

24. F. Lin and Y. Shoham. Concurrent actions in the situation calculus. In *Proc. of the 10th Nat. Conf. on AI (AAAI-92)*, pages 580–585, San Jose, CA, 1992.

25. V. Lifschitz M. Gelfond and A. Rabinov. What are the limitations of the situation calculus? In *Working Notes, AAAI Symp. on Logical Formalizations of Commonsense Reasoning*, pages 59–69, Stanford Univ., Stanford, CA, 1991.

26. J. McCarthy. The frame problem today. In F. M. Brown (1987), page 3. 1987. (abstract).

27. L. Morgenstern and L.A. Stein. Why things go wrong: a formal theory of causal reasoning. In *Proc. of the 7th Nat. Conf. on AI (AAAI-88)*, pages 518–523, Saint Paul, MN, August 21-26 1988.

28. R. Reiter. The frame problem in the situation calculus: a simple solution (sometimes) and a completeness result for goal regression. In V. Lifschitz, editor, *Artificial Intelligence and Mathematical Theory of Computation*, pages 359–380. Academic Press, 1991.

29. E. Sandewall. Features and fluents. Technical Report Res. Rep. LiTH-IDA-R-91-29, Dept. of Computer and Information Science, Linköping University, Linköping, Sweden, 1991. Review version of parts of a book.

30. E. Sandewall. Features and fluents. Technical Report Res. Rep. LiTH-IDA-R-92-30, Dept. of Computer and Information Science, Linköping University, Linköping, Sweden, 1992. Second review version of parts of a book.

31. E. Sandewall. *Features and Fluents. The Representation of Knowledge about Dynamical Systems. Volume I.* Oxford University Press, 1994.

32. L.K. Schubert. Monotonic solution of the frame problem in the situation calculus: an efficient method for worlds with fully specified actions. In R. Loui H. Kyburg and G. Carlson, editors, *Knowledge Representation and Defeasible Reasoning*, pages 23–67. 1990.

33. J. Tenenberg. Abandoning the completeness assumption: a statistical approach to solving the frame problem. *Int. J. of Expert Systems*, 3(4):383–408, 1990.

34. J. Tenenberg and J. Weber. A statistical approach to the qualification problem. Technical Report Technical Report 397, Dept. of Computer Science, Univ. of Rochester, Rochester, NY, 1992.

35. M. Winslett. Reasoning about action using a possible models approach. In *Proc. of the 7th Nat. Conf. on AI (AAAI-88)*, pages 89–93, Saint Paul, MN, August 21-26, 1988.

What Sort of Computation Mediates Best Between Perception and Action?

Murray Shanahan

Department of Electrical and Electronic Engineering
Imperial College
Exhibition Road
London SW7 2BT
England.
m.shanahan@ic.ac.uk

Summary. This paper addresses the question of what sort of computation should be used to mediate between perception and action in a mobile robot. Drawing on recent work in the area of cognitive robotics, the paper argues for the viability of an answer based on a rigorous, logical account of interleaved perception, planning and action. In addition, a number of common criticisms of approaches in this style are reviewed, and an implemented logic-based robot controller is described.

1 Introduction

This paper addresses a fundamental methodological question in Artificial Intelligence, namely what sort of computation mediates best between perception and action in a mobile robot. The answer offered here has a traditional feel, and appeals to a conceptual framework which is inherited from the time of the field's inception in the Fifties. In particular, the paper supports the view that robot design can be based on the manipulation of sentences of logic [Lespérance, et al., 1994]. One of the paper's main concerns is to defend this idea from criticisms which rest on a faulty view of logic and its limitations. Rather than relying on abstract argument to justify its views, the paper appeals to work that has been carried out with a real robot based on a rigorously logical formalism for representing and reasoning about action, continuous change, shape and space.

Before proceeding with the argument of the paper, the question under discussion requires clarification. Two points come to mind.

First, determining the sort of computation that mediates best between perception and action is, as far as this paper is concerned, an engineering matter, albeit one which is made more interesting by its deployment of terms which have currency in contemporary philosophy of mind. That is to say, while we are free to seek inspiration from biology, psychology or philosophy, the only criteria on which an answer to the question should be judged are engineering criteria, such as whether our choice of computational medium facilitates the construction of more capable, more robust, easier-to-modify, and easier-to-maintain machines.

Second, if we accept the Church-Turing thesis then, from a reductionist point of view, all computation is the same. But this observation is unhelpful in the present context. Our interest here is in the choice of the fundamental units of computation and (if relevant) of representation, and how these fit in to the architecture of an embodied agent situated in a world like our own.

We could opt for neural computation, in which the unit of computation is the artificial neuron, and representation corresponds to a pattern of activation over a network of such units. Or we could opt for the logical formula as the unit of representation, with rewrites of these formulae as the units of computation.

We could altogether reject the assumption that representation needs to play a part in what mediates between perception and action. Indeed, we might even reject the whole idea of computation. According to the argument for this position, the human brain (on which the argument assumes our research should be based) is a complex dynamical system, and it is misleading to suggest that it performs computation.

In sum, the methodological possibilities are numerous, and the choices we make will impinge on the engineering concerns voiced above.

The paper opens by rehearsing some of the arguments for the importance of embodiment in AI research, and then moves on to survey some of the advantages of a logic-based approach. This leads to the presentation of the paper's main argument, which is that a logical approach to embodied cognition is feasible. A rigorously logical account of perception, reason and action is presented which confronts the usual criticisms levelled at representational approaches to AI.

2 Embodiment

In the late Eighties and early Nineties, traditional approaches to Artificial Intelligence were frequently lambasted for working in disembodied, abstract domains.

> The only input to most AI programs is a restricted set of simple assertions deduced from the real data by humans. The problems of recognition, spatial understanding, dealing with sensor noise, partial models, etc. are all ignored.

> [Brooks, 1991a, p. 143]

> Traditional Artificial Intelligence has adopted a style of research where the agents that are built to test theories of intelligence are essentially problem solvers that work in a symbolic abstract domain.

> [Brooks, 1991b, p. 583]

Taking their cue from Rodney Brooks, many researchers placed a new emphasis on *autonomous* systems *situated* in dynamic environments.

Rather than working on computer programs that appear to mimic some limited aspect of high-level human intelligence ... the new approach concentrates instead on studying complete autonomous agents.

[Cliff, 1994, p. 800]

For many researchers, the issue of embodiment is also crucial because,

... unless you saddle yourself with all the problems of making a concrete agent take care of itself in the real world, you will tend to overlook, underestimate or misconstrue the deepest problems of design.

[Dennett, 1994, p. 143]

The "deepest" problems of design often concern incompleteness and uncertainty.

Sensors deliver very uncertain values even in a stable world ... The data delivered by sensors are not direct descriptions of the world as objects and their relationships ... Commands to actuators have very uncertain effects.

[Brooks, 1991c, p. 5]

Underlying the concerns of these researchers is the belief that,

- the isolated study of different aspects of intelligence leads to the development of incompatible sub-theories, and
- the temptation to idealise away the imperfection of a robot's connection to the world leads to the development of theories which are useless in practice.

These concerns can be addressed without appealing to embodiment, by conducting research using complete agents situated in some artificial environment, such as the Internet. But a further argument mitigates against this approach, insofar as we are interested in one day achieving human level intelligence in a machine as well as in designing useful products for today.

This is how the argument goes. The primary purpose of cognition is to enhance an agent's ability to interact with a world of spatio-temporally located objects given only incomplete and uncertain information. The incompleteness arises because of an agent's limited window on the world, and the uncertainty arises because of sensor and motor noise. In other words, the agent is confronted with what McCarthy calls the common sense informatic situation [McCarthy, 1989]. A capacity to deal with the common sense informatic situation is the substrate on which other cognitive skills rest. Only by building on such a substrate will we be able to duplicate human-level cognitive ability in a machine.[1]

---

[1] This includes linguistic ability. Perhaps natural language will even turn out to be a relatively straightforward phenomenon once we understand how to cope with the common sense informatic situation. After all, linguistic skills are evolution's most recent innovation, the final fold in the human cortex.

This doesn't set a limit, in principle, to the kind of research that can be carried out with agents situated in a simulated world. The argument is methodological rather than metaphysical. The suggestion is only that human-level cognitive skills are tied to the human epistemic predicament - the need to act in the presence of incomplete and uncertain information about a world of spatio-temporally located objects - and that this predicament is a consequence of our embodiment.

3 Logic

The logicist agenda in AI dates back to the Fifties [McCarthy, 1959]. According to the logical approach to AI, knowledge is represented by sentences in a formal language, and intelligent behaviour is mediated by the proof of theorems with those sentences. In the late Sixties, Green presented his classical account of planning [Green, 1969], and in the early Seventies the logical approach was applied to robotics in the form of the Shakey project [Nilsson, 1984]. Sadly, further progress was slow, and the popularity of logic in the robotics community subsequently declined. However, in spite of frequent premature obituaries, the logicist research programme has been vigorously pursued by a substantial minority of enthusiasts in the AI community ever since McCarthy's 1959 proposal. Today, as we shall see, the logic-based approach to high-level robot control, usually called *cognitive robotics* [Lespérance, et al., 1994], is enjoying a renaissance.

It's important to note that the logicist prescription does not demand a one-to-one correspondence between the data structures in the machine and the sentences of the chosen formal language. In other words, representations in the machine do not have to be explicitly stored sentences of logic. Similarly, the logicist prescription does not demand the use of algorithms whose state transitions correspond exactly to the steps of a proof. In other words, the machine does not have to implement a theorem prover directly. Between the abstract description of a logic-based AI program and the actual implementation can come many steps of transformation, compilation, and optimisation. The final product's functional equivalence to the abstract specification counts more than anything else.[2]

The chief advantages of a logic-based approach to AI in general, and to robotics in particular, are threefold. First, if a robot's design is logic-based, we can supply a rigorous, mathematical account of its success or otherwise at achieving its goals. Second, because its knowledge and goals are expressed in a universal declarative language, such a robot is easily modified and maintained. Third, it is relatively clear how to incorporate high-level cognitive skills in a logic-based robot, such as the ability to plan, to reason about other agents, or to reason about its own knowledge.

The remaining two sections of this paper report recent logic-based work on perception and action in robots. In doing so, they address some common criticisms

---

[2] However, it should be emphasised that, for many researchers, the explicit storage of sentences of logic and their manipulation by theorem proving techniques is an important ideal. The argument here often appeals to the idea that declaratively represented knowledge can be used in many different ways.

levelled at traditional approaches to AI. The first of these criticisms concerns the supposed inability of the traditional symbolic approach to AI to deal with incomplete information, the hallmark of the common sense informatic situation. Here are two representative quotes.

> The key problem I see with [all work in the style of Shakey] is that it relied on the assumption that a complete world model could be built internally and then manipulated.

> [Brooks, 1991b, p. 577]

> [In traditional AI] the key issue on which emphasis is laid is a complete, correct internal model, a perfect copy of the world (with all its object and relationships) inside the system, which the system can rely on to predict how the problem can be solved.

> [Maes, 1993, p. 4]

Unfortunately, these claims are based on an incorrect view of the nature of the representational approach to AI. More specifically, they betray a lack of understanding of the nature of predicate logic, which underpins the symbolic paradigm. At a foundational level, the problem of incomplete information was solved by Frege and Tarski, who gave us a good formal account of disjunction and existential quantification. Naturally, a good deal of work is still required to translate their mathematical insights into robotics practice. Section 3 reports an attempt to do this for robot perception.

The second criticism to be addressed relates to robot action. According to Brooks, because it worked in carefully engineered, static domain, the planner in Shakey,

> ... could ignore the actual world, and operate in the model to produce a plan of action for the robot to achieve whatever goal it had been given.

> [Brooks, 1991b, p. 570]

Maes makes a similar point.

> [In traditional AI] the central system evaluates the current situation (as represented in the internal model) and uses a search process to systematically explore the different ways in which this situation can be affected so as to achieve the desired goal.

> [Maes, 1993, p. 4]

Accordingly, a robot like Shakey is slow and intolerant to changes in its environment. This view is now uncontroversial, and robot architectures which incorporate a degree of *reactivity* are the norm. However, although systems which combine reactive and deliberative elements have been around for some time [Georgeff & Lansky, 1987], [Mitchell, 1990], [Gat, 1992], a rigorous logical account of interleaved planning, sensing and acting has only recently been achieved. This account is outlined in Section 6.

4   Action and Change

The next couple of sections provide a semi-technical overview of the abductive account of robot perception developed in a series of papers, [Shanahan, 1996a], [Shanahan, 1996b], [Shanahan, 1997b] and [Shanahan, 1998]. In all of this work, a mobile robot's sensor data is abductively explained by postulating the existence of objects with suitable locations and shapes. The account is the product of the following three steps.

1. Design a logic-based formalism for reasoning about action and space.
2. Using this formalism, construct a theory which captures the robot's relationship to the world, that is to say the effect of the robot's actions on the world, and the impact of changes in the world on the robot's sensors.
3. Consider the task of sensor data assimilation as a form of abduction with respect to this theory.

The logic-based formalism for reasoning about action and change must be able to cope with a variety of phenomena. First, because the robot's motion is continuous, it must be able to represent *continuous change*. Second, since events in the world can occur at any time, it must be able to represent *concurrent events*. Third, in order to deal with noise it needs to be able to handle actions and events with *non-deterministic effects*.

The action formalism employed in this work is adapted from [Shanahan, 1995b]. It is based on the *event calculus* of Kowalski and Sergot [1986], but is expressed in full predicate calculus augmented with circumscription [McCarthy, 1986]. Circumscription is deployed to overcome the frame problem, using a technique whereby the *circumscription* is split into parts inspired by [Kartha & Lifschitz, 1995] and explored more fully in [Shanahan, 1997a]. Figure 1 gives the flavour of the formalism, and shows a (slightly simplified) sample axiom. The basic predicates are Happens, which is used to describe a narrative of events, Initiates and Terminates, which are used to describe the effects of actions, and HoldsAt, which says what fluents hold at what times.

5   Space, Noise and Perception

In [Shanahan, 1996a] and [Shanahan, 1996b], space is represented as the plane $\Re^2$. Objects, including both the robot and the obstacles in its workspace, occupy open, path-connected subsets of $\Re^2$. A fluent for spatial occupancy is employed, and for reasons set out in [Shanahan, 1995a], spatial occupancy is minimised using circumscription. The formalism includes a number of axioms about continuous change and spatial occupancy, which are gathered together in the theory Σ_B.

The theory Σ_E, which is expressed in the language of this formalism (using Initiates and Terminates formulae), describes the effects of the robot's motor activity on the world, and the consequent effect the world has on the robot's sensors. For example, consider a wheeled mobile robot equipped with bump sensors. Σ_E captures

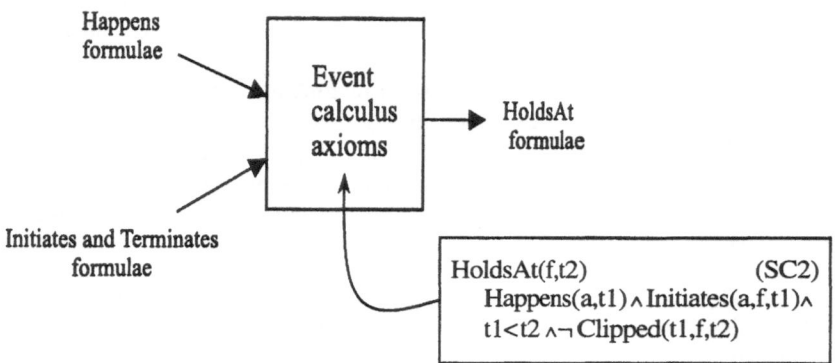

Fig. 1. The Event Calculus

the fact that, if the robot executes a move forward command, its location in $\Re^2$ will start to vary continuously. Σ_E also captures the fact that this continuous variation in location will cease if the robot collides with an obstacle, and that the robot's bump sensors will be tripped as a result of the collision.

Given Σ_E, sensor data assimilation can be considered as abduction. Let's examine the deterministic, noise-free case first. If a stream of sensor data is represented as the conjunction Ψ of a set of observation sentences (Happens and/or HoldsAt formulae), the task is to find an explanation of Ψ in the form of a logical description (a map) Δ_M of the initial locations and shapes of a number of objects, such that,

$$\Sigma_B \wedge \Sigma_E \wedge \Delta_N \wedge \Delta_M \models \Psi$$

where,

- Σ_B is a background theory, comprising axioms about action and change (the axioms of the event calculus), plus axioms about space and shape,
- Σ_E is a theory comprising Initiates and Terminates formulae that relates the shapes and movements of objects, in particular the robot itself, to the robot's sensor data, as described above, and
- Δ_N is a logical description of the movements of objects, in particular the robot itself, expressed as Happens formulae.

The incompleteness of the robot's knowledge due to its limited window on the world is reflected in the fact that there are always many explanations Δ_M for any given collection Ψ of sensor data.

The uncertainty in the robot's knowledge that arises from the inevitable presence of sensor and motor noise can be considered as non-determinism [Shanahan, 1996b]. Let's take motor noise first. Instead of including in Σ_E a formula describing an exact trajectory for the robot as it moves, we can include a formula which describes an ever-increasing circle of uncertainty within which the robot's location is known to fall.

This modification motivates the deployment of a new form of abduction. The non-monotonic nature of Δ_M, which uses circumscription to minimise spatial occupancy, entitles us to use a consistency-based form of abduction similar to that described by Reiter [1987]. Given a stream of sensor data Ψ, the task is now to find conjunctions Δ_M such that,

$$\Sigma_B \wedge \Sigma_E \wedge \Delta_N \wedge \Delta_M \not\models \neg\Psi.$$

This, in essence is the account of robot perception provided in [Shanahan, 1996a] and [Shanahan, 199b]. Unlike the symbols used in disembodied systems, the symbols appearing in Δ_M are *grounded* in the robot's interaction with the world (see [Harnad, 1990]), as well as acquiring *meaning* via Tarski-style model theory. The dual notions of groundedness and meaning allow us to appeal to the *correctness* of the robot's representations and reasoning processes in explaining the success or failure of its behaviour.

In [Shanahan, 1997b], this account of perception is extended to deal with sensor noise, and a different spatial ontology is adopted that permits a simplification of the abductive treatment of non-determinism and cuts down on multiple explanations. The central idea is to consider object *boundaries* as ontologically primitive, and to emphasise the *transitions* that occur in the robot's stream of sensor data as a result of its interactions with those boundaries. For example, suppose the robot is following a wall on its left using infra-red proximity sensors. Then, when a doorway is encountered, the signal from the left-hand infra-red sensor suddenly drops.

An even simpler spatial ontology is deployed in [Shanahan, 1998], where an implemented robot controller is presented. The robot in question uses wall-following to navigate a miniaturised office-like environment comprising a number of rooms connected by doorways. The simplicity of the environment means that the spatial representation comprises only rooms, corners and doorways, and the abductive interpretation of sensor data is consequently very straightforward. The chief attraction of this system, though, is that it interleaves sensor data assimilation with planning and plan execution. Planning in this system, just like sensor data assimilation, is carried out via abduction with the event calculus. Section 6 covers this topic in more detail.

6   Planning, Sensing and Computation

Following [Eshghi, 1988], planning in the event calculus can be considered as an abductive process in a way which closely resembles the above treatment of sensor data assimilation. Given a goal Γ (a conjunction of HoldsAt formulae), the task is to find a sequence of robot actions Δ_N such that,

$$\Sigma_B \wedge \Sigma_E \wedge \Delta_N \wedge \Delta_M \models \Gamma$$

where, Σ_B, Σ_E and Δ_M are defined as in the previous section. More detail can be found in [Shanahan, 1997c]. The details are very similar to those of the logical account of perception outlined above.

The logical accounts of robot perception and planning we now have should be considered as specifications for processes that are to be implemented on actual robot hardware. But, bearing in mind our underlying engineering concerns, how should we render these specifications into code?

A range of methodological options is available. At one extreme, the fundamental units of representation are taken to be sentences of formal logic, and the fundamental unit of computation is the proof step. Ideally, the path from specification to implementation is then a very short one, involving simply the application of a general purpose theorem prover to the theories Σ_B and Σ_E.

This approach preserves all the advantages of declarative representation. The same sentences of logic and the same theorem prover can perform both abductive sensor data assimilation and planning, as well as other reasoning tasks involving Σ_B and Σ_E. Unfortunately, no general purpose theorem prover exists which is up to the job, and it doesn't seem likely, at the time of writing, that future research will come up with one.

At the other end of the methodological spectrum, algorithms for interleaved planning, perception and action can be hand-designed, and proved correct with respect to formal specifications derived from the logical accounts supplied above. This approach has several drawbacks. First, there is no systematic process by which the implementation is derived from the specification. Second, the opaqueness of the implementation with respect to the specification renders it difficult to maintain and modify.

Perhaps the most attractive option is a logic programming approach, which lies in the middle of the methodological spectrum. Following Kowalski's slogan, "Algorithm = Logic + Control", the idea here is to preserve as much as possible of the logic of the specification, while rendering it into a computationally feasible form [Kowalski, 1979]. This is achieved by isolating a clausal fragment, subjecting the result to transformation employing the many tricks of the logic programmer's trade, and submitting the final product to a resolution-based theorem prover.

As well as maintaining a close relationship between specification and implementation, and retaining many of the advantages of declarative representation, the logic programming option has the attraction that it renders computation *transparent* in the following sense. Intermediate computational states are representationally meaningful, since they correspond to sentences of logic expressed in the same language as Σ_B and Σ_E. One consequence of this is that the computational process can be interrupted at any time and still produce useful results, a feature taken advantage of in the approach to interleaved planning, sensing and acting presented in the next section.

The logic programming approach is taken in [Shanahan, 1997c], where an implemented event calculus planner is described. The heart of the system is an abductive meta-interpreter with the event calculus axioms compiled in. The computation carried out by the planner is very close to that carried out by hand-coded partial order planning algorithms, such as UCPOP [Penberthy & Weld, 1992], with direct counterparts to the concepts of protected links, promotion, and demotion found in the planning literature. The planner also incorporates facilities for hierarchical plan decomposition, which, as we'll see shortly, is crucial to interleaving planning, sensing and plan execution.

A version of this system is used for planning in the logic-based robot controller reported in [Shanahan, 1998], where, thanks to the similarity between abductive sensor data assimilation and abductive planning, it also forms the basis of the system's perceptual processing capabilities. The interaction between these two processes – perception and planning – is the topic of the next section.

7   The Sense-Plan-Act Cycle

The main components of a logic-based robot control system have now been presented. We have logical accounts of planning and perception, based on the event calculus, along with a logic programming implementation. It remains only to see how to fit these elements together in a working system.

Work in the logicist tradition on planning and acting dates back to Green's seminal contribution in the late Sixties [Green, 1969]. However, it's only recently that the spirit of this work has been revived in the form of the cognitive robotics research programme [Lespèrance, et al., 1994]. This section reports recent efforts to supply a more up-to-date logical account of the interplay between planning, sensing and acting.

One approach to this issue is to preserve as much of classical Green-style planning as possible. This is the policy adopted by Levesque [1996]. Green's characterisation of the planning task, which is based on the situation calculus [McCarthy & Hayes, 1969] is as follows: given a description Σ of the effects of actions, an initial situation S_0, and a goal Γ, find a sequence of actions σ, such that Σ logically entails that Γ is the case after the execution of σ in S_0.

Green's account, in which planning and execution are sharply delineated, assumes that it's appropriate to attempt to plan an entire course of actions given only knowledge of the initial situation (see the two quotes at the end of Section 2). Levesque's modified account goes beyond this by allowing plans which incorporate sensing actions, as well as familiar programming constructs such as repetition and conditional action.

A potential drawback to Levesque's work is that it still assumes a sharp division between planning and execution. A complete plan for achieving the goal has to be constructed before any action is executed. In this respect, the logic programming account offered by Kowalski [1995] departs more radically from classical planning. Kowalski's presentation interleaves planning, sensing and execution.

To achieve this interleaving, the planner has to be able to recommend an action to be executed using only a bounded amount of computation, because the planning process is subject to constant suspension while actions are performed and new sensor data is acquired. This is a potential problem, as most logic-based planners generate plans in *regression* order, in other words latest action first (see [Eshghi, 1988], [Shanahan, 1989], or [Missiaen, et al., 1995], for example). By contrast, we require a planner that generates actions in progression order, that is to say earliest action first.

In Kowalski's proposal, the need to generate actions in progression order motivates the introduction of a style of formula for describing the effects of actions which is incompatible with that used in Σ_E in the previous section. This style of formula also undermines one of the most pleasing properties of classical planning, namely the tight logical relationship between the goal and the plan.

In [Shanahan, 1998], Kowalski's account of interleaved planning, sensing and execution is adopted, but the event calculus style of effect formula used in Σ_E is largely preserved. This is achieved through the use of *hierarchical* planning, in which a plan is generated by the successive decomposition of high-level actions into lower level actions until primitive, executable actions are reached. If the hierarchy of actions is designed in the right way, a plan is very quickly found comprising an executable first action followed by a high-level abstract action yet to be decomposed. The first action can then be executed right away, before the rest of the plan has been fully decomposed.[3] This results in a degree of reactivity.

Fortunately, as shown in [Shanahan, 1997c], the event calculus can easily be used to represent high-level, compound actions, using " Happens if Happens" clauses (see also [Jung, 1998]). The abductive logic programming techniques that are deployed in [Shanahan, 1997c] for partial order planning carry over with little modification to hierarchical decomposition.

Using hierarchical planning in this way enables a robot to start carrying out a plan before that plan is complete, and to work on finishing the plan while it is already executing it.

It also gives the robot the opportunity to sense and react to its environment during plan execution. The mechanism by which this is achieved in [Shanahan, 1998] exploits the fact that the planner has to record " protected links"/, (negated Clipped formulae, in event calculus terms), while the abductive interpretation of incoming sensor data yields Happens formula, which can violate those protected links. Conflicts of this sort are detected and precipitate immediate replanning.

In the system described in [Shanahan, 1998] replanning is carried out from scratch, rather than any attempt being made to repair the old plan. This strategy

---

[3] The viability of this technique depends on the downward solution property [Russell & Norvig, 1995, Ch. 12], which insists that every abstract plan can be decomposed into a plan comprising only executable actions.

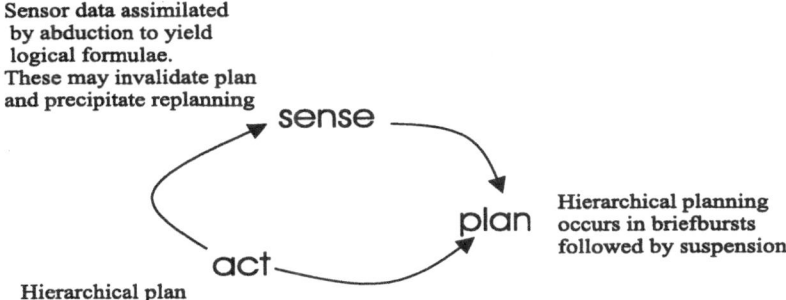

Sensor data assimilated
 by abduction to yield
 logical formulae.
 These may invalidate plan
 and precipitate replanning

sense

plan

Hierarchical planning
occurs in briefbursts
followed by suspension

act

Hierarchical plan

Fig. 2. The Event Calculus

is justified if the hierarchy of compound actions has been designed in such a way as to ensure that a first action is always quickly found in any situation.[4]

```
axiom(happens(go_to_room(R,R),T,T),[]).

axiom(happens(go_to_room(R1,R3),T1,T4),
[towards(R2,R3,R1), connects(D,R1,R2),
holds_at(door_open(D),T1),
happens(go_through(D),T1,T2),
happens(go_to_room(R2,R3),T3,T4),before(T2,T3),
not(clipped(T2,in(robot,R2),T3))]).

axiom(initiates(go_to_room(R1,R2),in(robot,R2),T),
[holds_at(in(robot,R1),T)]).
```

Fig.3. A Fragment of Code from the Robot Controller

8   An Implemented Robot Controller

The foregoing ideas have been realised in a high-level control system for a miniature two-wheeled Khepera robot with a suite of eight infra-red proximity sensors. The

---

[4] In [De Giacomo, et al., 1998], by contrast, the system commits to a course of actions, and when things go wrong, it attempts to reach a state from which its old plan can still be executed.

robot inhabits a model office-like environment of rooms, walls and doorways. It has a small repertoire of low-level actions, such as wall-following, turning corners, and turning into doorways.

The robot has successfully been programmed to perform navigation tasks using the techniques described in the preceding sections. Owing to the simplicity of the proximity sensors, a closed door is indistinguishable to the robot from a wall, and since the robot doesn't maintain any distance information while wall-following, this offers a challenge for abductive sensor data assimilation. When, unknown to the robot, a door is closed during its plan execution, the control system correctly abduces a " close door" event to explain the sensor data it receives when it unexpectedly arrives at a corner. If this new event conflicts with a protected link in the current plan, then replanning is triggered.

Figure 3 shows a sample of the event calculus formulae used to control this robot. The have been cut and pasted directly from the program, so they are still in the form used by the abductive meta-interpreter. $\mathtt{axiom}$ ($\lambda, [\lambda_1 \ldots \lambda_n]$) means $\lambda \leftarrow \lambda_1 \wedge \ldots \wedge \lambda_n$. This particular fragment describes the effects and decomposition of a high-level action $\mathtt{go\_to\_room(R1,R2)}$, which comprises a number of $\mathtt{go\_through(D)}$ actions. Note that the definition is recursive.

Preliminary experiments using the same system for map-building have also been carried out. Instead of being given a map initially, as in the navigation task, the robot performs an exploratory routine, during which it abduces the existence of walls, doorways and corners to explain incoming sensor data.

9   Conclusion

In response to McCarthy's appeal to Dreyfus to supply a well defined problem which he believes the logical approach to Artificial Intelligence will have difficulty solving [McCarthy, 1996], Dreyfus submits the following challenge to "McCarthy and his followers" .

> How does one spell out propositionally our know-how concerning the body and its motor and perceptual systems?
>
> [Dreyfus, 1996]

The thrust of the work reported here is not *our* know-how, nor *human* motor and perceptual systems, but rather those of a mobile robot. However, I believe that Dreyfus is right to place emphasis on the issues of embodiment, perception and action. The present paper and the developments it reports offer a reply to Dreyfus's challenge.

The paper's main purpose, though, is to address the question posed in the title. What sort of computation mediates best between perception and action? Naturally, this paper cannot offer a clear answer, and it seems safe to assume that no definitive answer is possible. However, the work reviewed in this paper suggests that an approach based on logic is feasible, with all the advantages of rigour and flexibility that brings.

Moreover, the work presented here is just one of several attempts to tackle high-level robotic control via logic. The research programme being carried out by Ray Reiter and his colleagues at the University of Toronto is another example [Lespèrance, et al., 1994]. Based on the situation calculus [McCarthy & Hayes, 1969] rather than the event calculus, and employing Reiter's " monotonic" solution to the frame problem [Reiter, 1991], [Reiter, 1998] rather than circumscription, their methodology places emphasis on robot *programming* rather than robot *planning*, using the GOLOG language [Levesque, et al., 1997], which permits the description of compound actions using operators such as sequence, choice and iteration.

However, these differences with the present work are not significant - the circumscriptive solution to the frame problem used here usually reduces to monotonic predicate completion, and the use of action hierarchies in hierarchical planning also amounts to a pragmatic rejection of traditional search-based planning as the main mechanism for deciding on a course of actions. (These hierarchies also allow compound actions using programming constructs such as recursion, sequence and choice.)

A more marked difference between the two strands of work is in their treatment of sensor data. In the Toronto approach, an execution monitor oversees the execution of a GOLOG program [De Giacomo, et al., 1998].[5] It's assumed that the execution monitor knows about every exogenous action - that is to say every action performed by an agent other than the robot itself - that could be relevant to the execution of the program. Conflicts that arise from such actions trigger a recovery procedure that attempts to get the world back into a state from which execution of the original program can continue. In the present paper, by contrast, there is no assumption of knowledge of exogenous action occurrences. Instead, the robot has to *reason* abductively that such actions have occurred on the basis of its raw sensor data.

Furthermore, the Toronto approach requires the provision of extra information to the execution monitor to represent the relevance of exogenous actions to the program being executed. In the form in which high-level actions are represented in the event calculus, this information is already present in the shape of protected links (such as the not clipped conjunct in Figure 3). This is because of the need, in the event calculus approach, to represent explicitly the effects of high-level actions as well as primitive ones. This need calls for the inclusion of not clipped conjuncts (protected links) in the description of a high-level action. Without these, thanks to the possibility of intervening exogenous actions, its effects would not be guaranteed to follow from the effects of its sub-actions. In GOLOG there is no notion of a goal, while in the event calculus approach, " programs" (that is to say high-level, compound actions used in hierarchical planning) are also plans, whose purpose is to bring about some goal state, and whose effects therefore need to be made explicit. If

---

[5] Reiter [1998] describes a reactive version of GOLOG, to which this discussion doesn't apply.

high-level actions are just programs, as they are in GOLOG, their effects don't have to be specified directly.

Finally, the two approaches differ in their method for dealing with actions with knowledge producing effects. For the Toronto team, this is a separate issue from that of execution monitoring. Levesque's situation calculus account of knowledge producing actions has no obvious connection with execution monitoring [Levesque, 1996]. In the present approach, on the other hand, all physical actions are potentially knowledge producing, as any action the robot performs can lead to incoming sensor data whose abductive processing results in the acquisition of knowledge. (The knowledge producing effects of an action can be described with the same style of effect axiom as their physical effects, only using newly introduced epistemic fluents). So the same mechanism, namely abductive processing of sensor data within the sense-plan-act cycle, is the basis for handling both knowledge producing actions and " execution monitoring" .

The extent to which the differences between the Toronto work and that described here represent a fundamental incompatibility is not yet clear. For both communities, much work remains to be done, especially to find out the extent to which the ideas scale up to robots with richer sensor and motor capabilities. Perhaps this will reveal that we've all got it wrong.

10   Acknowledgments

The work reported in this paper was carried out as part of the EPSRC-funded project GR/L20023 " Cognitive Robotics" . Thanks to Mark Witkowski for doing all the low-level robot programming.

References

[Brooks, 1991a] .A.Brooks, Intelligence Without Representation, Artificial Intelligence, vol 47 (1991), pp. 139-159.

[Brooks, 1991b] .A.Brooks, Intelligence Without Reason, Proceedings IJCAI 91, pp. 569-595.

[Brooks, 1991c] .A.Brooks, Artificial Life and Real Robots, Proceedings of the First European Conference on Artificial Life (1991), pp. 3-10.

[Cliff, 1994] .Cliff, AI and A-Life: Never Mind the Blocksworld, Proceedings ECAI 94, pp. 799-804.

[De Giacomo, et al., 1998] .De Giacomo, R.Reiter and M.Soutchanski, Execution Monitoring of High-Level Robot Programs, Proceedings KR 98, pp. 453-464.

[Dennett, 1994] .Dennett, The Practical Requirements for Making a Conscious Robot, Philosophical Transactions of the Royal Society of London, Series A, vol 349 (1994), pp. 133-146.

[Dreyfus, 1996] .L.Dreyfus, Response to My Critics, Artificial Intelligence, vol 80 (1996), pp. 171-191.

[Eshghi, 1988] .Eshghi, Abductive Planning with Event Calculus, Proceedings of the Fifth International Conference on Logic Programming (1988), pp. 562-579.

[Gat, 1992] .Gat, Integrating Planning and Reacting in a Heterogeneous Asynchronous Architecture for Controlling Real-World Mobile Robots, Proceedings AAAI 92, pp. 809-815.

[Georgeff & Lansky, 1987] .Georgeff and A.Lansky, Reactive Reasoning and Planning, Proceedings AAAI 87, pp. 677-682.

[Green, 1969] .Green, Applications of Theorem Proving to Problem Solving, Proceedings IJCAI 69, pp. 219-240.

[Harnad, 1990] .Harnad, The Symbol Grounding Problem, Physica D, vol 42 (1990), pp. 335-346.

[Jung, 1998] .G.Jung, Situated Abstraction Planning by Abductive Temporal Reasoning, Proceedings ECAI 98, pp. 383-387.

[Kartha & Lifschitz, 1995] .N.Kartha and V.Lifschitz, A Simple Formalization of Actions Using Circumscription, Proceedings IJCAI 95, pp. 1970-1975.

[Kowalski, 1979] .A.Kowalski, Algorithm = Logic + Control, Communications of the ACM, vol 22 (1979), pp. 424-436.

[Kowalski, 1995] .A.Kowalski, Using Meta-Logic to Reconcile Reactive with Rational Agents, in Meta-Logics and Logic Programming, ed. K.R.Apt and F.Turini, MIT Press (1995), pp. 227-242.

[Kowalski & Sergot, 1986] .A.Kowalski and M.J.Sergot, A Logic-Based Calculus of Events, New Generation Computing, vol 4 (1986), pp. 67-95.

[Lespèrance, et al., 1994] .Lespèrance, H.J.Levesque, F.Lin, D.Marcu, R.Reiter, and R.B.Scherl, A Logical Approach to High-Level Robot Programming: A Progress Report, in Control of the Physical World by Intelligent Systems: Papers from the 1994 AAAI Fall Symposium, ed. B.Kuipers, New Orleans (1994), pp. 79-85.

[Levesque, 1996] .Levesque, What Is Planning in the Presence of Sensing? Proceedings AAAI 96, pp. 1139-1146.

[Levesque, et al., 1997] .Levesque, R.Reiter, Y.Lespèrance, F.Lin and R.B.Scherl, GOLOG: A Logic Programming Language for Dynamic Domains, The Journal of Logic Programming, vol 31 (1997), pp. 59-83.

[Maes, 1993] .Maes, Behavior-Based Artificial Intelligence, Proceedings of the Second International Conference on the Simulation of Adaptive Behavior (SAB 93), pp. 2-10.

[McCarthy, 1959] .McCarthy, Programs with Common Sense, Proceedings of the Teddington Conference on the Mechanization of Thought Processes, Her Majesty's Stationery Office, London (1959), pp. 75-91.

[McCarthy, 1986] .McCarthy, Applications of Circumscription to Formalizing Common Sense Knowledge, Artificial Intelligence, vol 26 (1986), pp. 89-116.

[McCarthy, 1989] .McCarthy, Artificial Intelligence, Logic and Formalizing Common Sense, in Philosophical Logic and Artificial Intelligence, ed. R.Thomason, Kluwer Academic (1989), pp. 161-190.

[McCarthy, 1996] .McCarthy, Book Review of "What Computers Still Can't Do" by Hubert Dreyfus, Artificial Intelligence, vol 80 (1996), pp. 143-150.

[McCarthy & Hayes, 1969] .McCarthy and P.J.Hayes, Some Philosophical Problems from the Standpoint of Artificial Intelligence, in Machine Intelligence 4, ed. D.Michie and B.Meltzer, Edinburgh University Press (1969), pp. 463-502.

[Missiaen, et al., 1995] .Missiaen, M.Bruynooghe and M.Denecker, CHICA, A Planning System Based on Event Calculus, The Journal of Logic and Computation, vol. 5, no. 5 (1995), pp. 579-602.

[Mitchell, 1990] .Mitchell, Becoming Increasingly Reactive, Proceedings AAAI 90, pp. 1051-1058.

[Nilsson, 1984] .J.Nilsson, ed., Shakey the Robot, SRI Technical Note no. 323 (1984), SRI, Menlo Park, California.

[Penberthy & Weld, 1992] .S.Penberthy and D.S.Weld, UCPOP: A Sound, Complete, Partial Order Planner for ADL, Proceedings KR 92, pp. 103-114.

[Reiter, 1987] .Reiter, A Theory of Diagnosis from First Principles, Artificial Intelligence, vol 32 (1987), pp. 57-95.

[Reiter, 1991] .Reiter, The Frame Problem in the Situation Calculus: A Simple Solution (Sometimes) and a Completeness Result for Goal Regression, in Artificial Intelligence and Mathematical Theory of Computation: Papers in Honor of John McCarthy, ed. V.Lifschitz, Academic Press (1991), pp. 359-380.

[Reiter, 1998] .Reiter, Knowledge in Action: Logical Foundations for Describing and Implementing Dynamical Systems, draft book manuscript (1998).

[Russell & Norvig, 1995] .Russell and P.Norvig, Artificial Intelligence: A Modern Approach, Prentice Hall International (1995).

[Shanahan, 1995a] .P.Shanahan, Default Reasoning about Spatial Occupancy, Artificial Intelligence, vol 74 (1995), pp. 147-163.

[Shanahan, 1995b] .P.Shanahan, A Circumscriptive Calculus of Events, Artificial Intelligence, vol 77 (1995), pp. 249-284.

[Shanahan, 1996a] .P.Shanahan, Robotics and the Common Sense Informatic Situation, Proceedings ECAI 96, pp. 684-688.

[Shanahan, 1996b] .P.Shanahan, Noise and the Common Sense Informatic Situation for a Mobile Robot, Proceedings AAAI 96, pp. 1098-1103.

[Shanahan, 1997a] .P.Shanahan, Solving the Frame Problem: A Mathematical Investigation of the Common Sense Law of Inertia, MIT Press (1997).

[Shanahan, 1997b] .P.Shanahan, Noise, Non-Determinism and Spatial Uncertainty, Proceedings AAAI 97, pp. 153-158.

[Shanahan, 1997c] .P.Shanahan, Event Calculus Planning Revisited, Proceedings 4th European Conference on Planning (ECP 97), Springer Lecture Notes in Artificial Intelligence no. 1348 (1997), pp. 390-402.

[Shanahan, 1998] .P.Shanahan, Reinventing Shakey, Working Notes 1998 AAAI Fall Symposium on Cognitive Robotics, pp. 125-135.

11   Postscript: The Influence of Ray Reiter

The presence of a handful of Ray Reiter's publications in the bibliography completely fails to convey the magnitude of his influence on the present paper. Indeed, without Ray's work, it's unlikely that my research would have taken on a robotics flavour at all. At the IJCAI 93 conference, at a time when I was feeling sceptical about the relevance of logic-based theoretical work to the long-term vision of AI, and was in danger of moving in an altogether different direction, Ray Reiter presented his Research Excellence Award lecture. Ray's presentation was unashamedly logicist. He argued for the contemporary relevance of the situation calculus, a thirty year old formalism, and praised the ideals and achievements of the Shakey project from twenty five years past. His aim was to revive those ideals, and to apply logic to robotics once again, armed with a modern solution to the frame problem.

Sitting in the audience, I knew that this was the kind of thing I wanted to hear. In spite of my reservations, I remained very attached to logic, its rigour and beauty and potential role in AI as an engineering discipline. Ray had sent up a rallying cry. Logic could form the basis of a viable methodology for building intelligent machines, in particular for building robots. It took a couple of year's to come to fruition, but an idea germinated in my mind that day, the idea that the theoretical work I had done on the frame problem could also be applied to robotics. At the ECAI 96 conference, my first paper along these lines won the Best Paper Award, a thing I would never have achieved without Ray's original inspiration. That inspiration continues to this day.

Modeling and Analysis of Hybrid Control Systems
An Elevator Case Study

Ying Zhang[1] and Alan K. Mackworth[2]

[1] Xerox Palo Alto Research Center,
    3333 Coyote Hill Road,
    Palo Alto, CA, USA,
    yzhang@parc.xerox.com
[2] Department of Computer Science,
    University of British Columbia,
    Vancouver, B.C., Canada,
    mack@cs.ubc.ca

Summary. We propose a formal approach to the modeling and analysis of hybrid control systems. The approach consists of the interleaved phases of hybrid dynamic system modeling, requirements specification, hybrid control design and overall behavior verification. We have developed Constraint Nets as a semantic model for hybrid dynamic systems. Using this model, continuous, discrete and event-driven components of a dynamic system can be represented uniformly. We have developed timed ∀-automata as a requirements specification language for dynamic behaviors. Using this language, many important properties of a dynamic system, such as safety, stability, reachability and real-time response can be formally stated. We have also proposed a verification method for checking whether a constraint net model satisfies a timed ∀-automaton specification. The method uses the induction principle and generalizes both Liapunov stability analysis for dynamic systems and monotonicity of well-foundedness in discrete-event systems. The power of these techniques is demonstrated with a simple elevator system. An elevator system is a typical hybrid system with continuous motion following Newtonian dynamics and discrete event control responding to users' request. We model the complete elevator system using Constraint Nets and verify the overall behavior of the system against the requirements specification in timed ∀-automata.

1 Introduction

Hybrid dynamic systems are systems that consist of coupled discrete and continuous components. Any electromechanical system with a computerized controller is a hybrid system in general. In the past, the modeling and analysis of a hybrid system have been done separately for its discrete and continuous components. The overall system is designed in a rather empirical fashion. Since computer-aided control is becoming more and more significant in modern system design practice, we face a major challenge: the development of intelligent, reliable, robust and safe computer-controlled systems. The foundation for modeling and analysis of hybrid systems must be established.

1.1 Our approach

We advocate a formal modeling and analysis framework for the development of hybrid control systems. The framework consists of the interleaved phases of hybrid system modeling, requirements specification, hybrid control design and behavior verification. System modeling can also be called system specification, which precisely defines how a system is structured in terms of its components and interconnections, and how each of its component works. Requirements specification expresses global properties of a system such as safety (i.e., a bad state will never be reached), stability (i.e., a final state will be reached), reachability (i.e., a state is reachable) and real-time response (i.e., a state will be reached in bounded time). Control system design takes a system model (including the plant and its envoronment), which in general is continuous or hybrid, and its requirements specification, produces a control system, which in general is discrete or hybrid. Behavior verification ensures that the behavior of the coupled overall system (i.e. plant+environment +control) satisfies the specified requirements.

For hybrid system modeling, we have developed the Constraint Net (CN) model [37], [34]. CN provides a formal syntax and semantics of a hybrid dynamic system so that its continuous and discrete (fixed sampling time or event-driven) components can be modeled uniformly. CN provides aggregation operators so that a complex system can be modeled hierarchically. Therefore, a system and its control can be modeled individually in CN and then be composed to generate an overall system model. The overall hybrid system is defined mathematically so that it can be analyzed without ambiguity. CN also supports multiple levels of abstraction, based on abstract algebra and topology; therefore, a system can be modeled and analyzed at different levels of detail.

For requirements specification, we have developed timed $\forall$-automata [36], [35]. Timed $\forall$-automata are essentially finite automata modified from $\forall$-automata [24] by generalizing to continuous time and adding timing constraints; yet they are powerful enough to specify global system properties of sequential and timed behaviors of hybrid dynamic systems, such as safety, stability, reachability and real-time response.

For control design, we have proposed constraint-based techniques [38], [39]. In this paper, we advocate a two-level design architecture. A hybrid control system can be developed in a two-level structure, with the lower level as a continuous component guided by an analog control law and the higher level as a discrete logic component driven by events from the lower level or from the environment.

For behavior verification, we have proposed a formal model checking method [35], [34]. The method uses the induction principle and combines Liapunov stability analysis for dynamic systems and monotonicity of well-foundedness in discrete-event systems.

1.2 Related work

Much work has been done on the modeling and analysis of hybrid control systems [10]. Various formal models/languages have been developed for this purpose.

Roughly speaking, formal models/languages for modeling and analysis can be characterized as belonging to one of the four categories: (1) state transition models, (2) algebraic processes, (3) block diagrams/nets/dataflow/equations, and (4) temporal logics/ω-languages.

For example, Phase Transition Systems [23], [26] are a typical state transition model, where computation consist of alternating phases of discrete transitions and continuous activities. Similarly, Nerode and Kohn's Hybrid Automata [25] consist of a digital control automaton and a plant automaton. The plant automaton can be modeled as a state transition system over intervals. Alur et al. [1] develop a model for hybrid systems which generalizes timed automata [3], [2]. Lynch's group at MIT has been using Timed Automata [22] for modeling and verification of automated transit systems [21].

In the formalism of algebraic processes, Gupta et al. [11] proposes Hybrid cc, a generalization (or a new member) of the cc family [28], for modeling systems with timers and continuous activities.

The Constraint Net model we have developed belongs to the category of block diagrams / nets / dataflow / equations. In this formalism, a system is a network of processes, processes are represented by blocks, and interactions between processes are represented by connections between blocks. All processes are considered to be running in parallel and data flow through these processes where operations on the data are performed. An equational representation of the model can also be obtained, with each process corresponding to an equation. For example, SIGNAL [6] is a typical language in this formalism. Conventional continuous or discrete control structures are best modeled in block diagrams. Krogh [17] develops condition/event signal interfaces for block diagrams, which extends block diagrams with logic operators. Brockett [8] studies motion control systems with an equational event-driven model. Some commercial products for control simulation are also belong to this category, such as Simulink [14] and SystemBuild [13].

Timed $\forall$-automata we have developed are closely related to temporal logics/ω-languages. Many extensions of temporal logics/regular languages to timed and/or hybrid systems have been proposed. For example, Lamport [18] develops TLA$^+$, an extension to TLA (Temporal Logic of Action), for modeling hybrid systems. Alur and Henzinger [4] propose a really temporal logic with metric time. Duration Calculus [33], a generalization of interval temporal logics, has also been applied to the formal design of hybrid systems. From the formal automata/language point of view, TBA (Timed Buchi Automata) [2] are developed to generate timed ω-languages, and its properties and decision procedures are also studied.

Related work is also been done in software engineering for safety issues. For example, the Requirements State Machine (RSM) [16] provides semantic analysis of real-time process-control software requirements, and the Requirements State Machine Language (RSML) [19] is then designed to write requirements specifications for an industrial aircraft collision avoidance system. Time Petri Nets are applied to modeling and verification of time dependent systems [7].

From a methodological point of view, there are two schools: one uses a single general model/language (e.g. state transition systems or temporal logics) for both system modeling and requirements specification; the other proposes different formalisms for modeling and specification: e.g., state transition systems for modeling and temporal logics for specification. Our approach here belongs to the second school.

1.3 Our contributions

Our contributions to the modeling and analysis of hybrid control systems are three-fold.

First of all, we develop Constraint Nets for modeling complex structured hybrid dynamic systems. Like all the net-based models, CN is modular and hierarchical with a formal syntax and semantics for simulation. Unlike all the other existing models, CN is developed on abstract time and domain structures. Instead of combining models of discrete and continuous dynamic systems, we start from a general model of dynamic systems, of which both discrete and continuous systems are special cases. As a result, CN provides a powerful and formal model for complex hybrid dynamic systems.

Secondly, we develop timed $\forall$-automata for specifying simple global properties of hybrid dynamic systems. Unlike most existing timed automata, timed $\forall$-automata have finite state without clock or data variables. Timed $\forall$-automata are similar to propositional temporal logics. However, timed $\forall$-automata are defined on abstract time structures which can be either discrete or dense. With notions of both local and global timeout, timed $\forall$-automata can represent real-time responses as well. As a result, timed $\forall$-automata provide a simple and useful specification language for timed behaviors.

Thirdly, we develop a formal behavior verification method that uses the induction principle and generalizes both Liapunov stability analysis for dynamic systems and monotonicity of well-foundedness in discrete-event systems.

1.4 A simple elevator system

We use a simple elevator system here to demonstrate the use of our approach. Elevator systems have been used in various communities as benchmark examples of methodologies for software engineering and real-time systems [15], [27], [12], [5], [9].

However, most previous examples of elevator systems focus on discrete-event structures. In this paper, we model an elevator system as a hybrid system with continuous motion following Newtonian dynamics and discrete control responding to users' request. In particular, we show how the coupling of the discrete and continuous components can affect the behavior of the overall system.

A simple elevator system for an n-floor building consists of one elevator. Inside the elevator there is a board with n *floor buttons*, each associated with one floor. Outside the elevator there are two *direction buttons* for service call on each floor, except the first floor and the top floor where only one button is needed (see Figure 1.4).

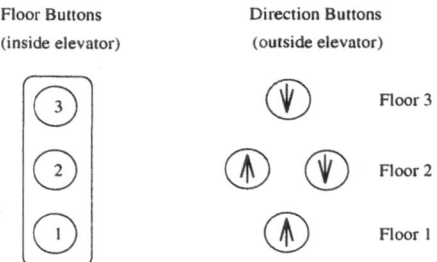

Floor Buttons
(inside elevator)

Direction Buttons
(outside elevator)

Floor 3

Floor 2

Floor 1

Fig. 1. A simple 3-floor elevator

Any button can be pushed at any time. After being pushed, a floor button will be on until the elevator stops at the floor, and a direction button will be on until the elevator stops at the floor and is going to move in the same direction. (Note that a more complex elevator would have open and close door buttons, and alarm or emergency buttons which, for simplicity, we do not model in this paper.)

Using constraint nets, we first model this elevator system at two levels of detail. At the lower level, the dynamics of continuous motion is modeled; and at the higher level, the abstraction of the desired discrete system is represented. Then we model the overall hybrid system, the elevator with control in black boxes, in Constraint Nets.

A well-designed elevator system should satisfy the property of real-time response, i.e., any request will be served within some bounded time. Using timed ∀-automata, we explore the meaning of such requirements and specify them precisely.

Then we design a continuous controller and a discrete controller by analyzing the requirements.

Finally, we explore approaches to verifying the behavior of a dynamic system against its requirements specification. For the elevator case study, we propose and answer the question: is the elevator system well-designed?

1.5   Outline of this paper

The rest of this paper is organized as follows. Section 2 briefly introduces Constraint Nets, and demonstrates constraint net modeling using the elevator example. Section 3 presents timed ∀-automata and gives the requirements specification of the real-time behavior of the elevator system. Section 4 designs the control system for the elevator by analyzing the requirements specification. Section 5 develops a model checking method which determines whether the constraint net model of the elevator system satisfies the timed $\forall$-automaton specification of the desired behavior. Section 6 draws some conclusions.

2 System Modeling in Constraint Nets

A complex dynamic system should be modeled in terms of its components and interconnections, at multiple levels of abstraction. We model the elevator system this way using Constraint Nets.

2.1 Concepts of dynamic systems

We start by introducing some general concepts of dynamic systems which we use later. For formal definitions of these concepts, the reader is referred to [34].

- *Time $\mathcal{T}$* is a linearly ordered set with a minimal element as the start time point, for example, the set of natural numbers or the set of non-negative real numbers with the arithmetic ordering. The former is called *discrete time* and the latter is called *continuous time*.
- A *trace $v_{\mathcal{T}}^A : \mathcal{T} \to A$* is a function from time $\mathcal{T}$ to a domain A of values.
- An *event trace $e_{\mathcal{T}} : \mathcal{T} \to B$* is a special type of trace whose domain B is boolean. An *event* in an event trace is a transition from 0 to 1 or from 1 to 0. An event trace characterizes some *event-based time* where the set of events in the trace is the time set. The time domain of an event trace is called the *reference time* of the event-based time.
- A *transduction $F : \mathcal{V}_I \to \mathcal{V}_O$* is a mapping from a tuple of input traces to a tuple of output traces, which satisfies the causality constraint between the inputs and the outputs, i.e., the output values at any time depend only on the inputs up to that time. For instance, a state automaton with an initial state defines a transduction on discrete time; a temporal integration with a given initial value is a typical transduction on continuous time. Just as nullary functions represent constants, nullary transductions represent traces.
- A transduction F is called a *transliteration* if it is a pointwise extension of a function f, i.e. $F(v)(t) = f(v(t))$. Intuitively, a transliteration is a transformational process without memory (internal state), such as a combinational circuit.
- States or memories are introduced by delays. A *delay* transduction is a sequential process where the output value at any time is the input value at a previous time, i.e., $\delta(v)(t) = v(t - \delta)$. Normally, *unit* delays are for discrete time and *transport* delays are for continuous time.
- The linkages between discrete and continuous components are modeled by event-driven transductions. An *event-driven transduction* is a transduction augmented with an extra input which is an event trace; the event-driven transduction operates at every event and its output value holds from each event to the next.

2.2 Constraint net model

The Constraint Net model is built upon these general concepts of dynamic systems. It is a net/dataflow- or equation-based model with formal syntax and semantics. It also provides modular structures with composition hierarchy.

Syntax and semantics A constraint net consists of a finite set of locations, a finite set of transductions and a finite set of connections. Intuitively, locations represent states, memories, variables or communication channels; transductions represent processes, operating according to a global reference time or activated by external events; and connections represent the interaction structures or data flows of the modeled system. Formally, a *constraint net* is a triple $CN = \langle Lc, Td, Cn \rangle$, where Lc is a finite set of *locations*, Td is a finite set of labels of *transductions*, each with an *output port* and a set of *input ports*, Cn is a set of *connections* between locations and ports, with the following restrictions: (1) there is at most one output port connected to each location, (2) each port of a transduction connects to a unique location and (3) no location is isolated.

A location is an *output* of the constraint net if it is connected to the output of some transduction; otherwise it is an *input*. A constraint net is *open* if there is an input location; otherwise it is *closed*.

A constraint net represents a set of equations, with locations as variables and transductions as functions. The *semantics* of the constraint net, with each location denoting a trace, is the least solution of the set of equations. The semantics is defined on abstract data types and abstract reference time which can be discrete or continuous. For detailed formal semantics, the reader is referred to [37], [34].

Graphically, a constraint net is depicted by a bipartite graph where locations are depicted by circles, transductions by rectangular blocks and connections by arcs.

For example, the graph in Figure 2, where f is a transliteration and δ is a unit delay, depicts a state transition system. This open net, with discrete time, can be represented by the equation: $s(n) = f(u(n-1), s(n-1)), s(0) = s_0$. We can also simply write $s' = f(u, s), s(0) = s_0$ where s' denotes the next state of s.

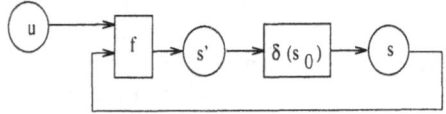

Fig. 2. The constraint net representing a state transition system

Similarly, the net depicted by the graph in Figure 3, with continuous time, models the differential equation $\dot{s} = f(u, s), s(0) = s_0$.

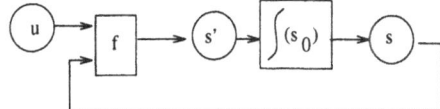

Fig. 3. The constraint net representing a differential equation

Modules and composition A complex system is generally composed of multiple components. We define a *module* as a constraint net with a set of locations as its interface. A module is depicted by a box with rounded corners. For example, the state transition system in Figure 2 is grouped to a module in Figure 4 where u and s are the input and output interface, respectively. The double circles depict its interface.

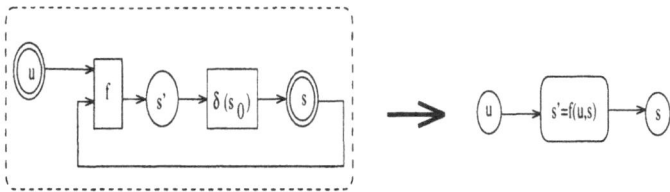

Fig. 4. A module and its representation

A constraint net can be composed hierarchically using modular and aggregation operators on modules. There are three basic operations — union, coalescence and hiding — that can be applied to obtain a new module from existing ones. The *union* operation generates a new module by putting two modules side by side. The *coalescence* operation coalesces two locations in the interface of a module into one, with the restriction that at least one of these two locations is an input location. The *hiding* operation deletes a location from the interface. We can model non-determinism with hidden inputs and model internal states with hidden outputs.

A hybrid system can be modeled by a composition of continuous components with event-driven discrete components. For example, Figure 5 depicts a hybrid system consisting of a state transition system and a differential equation. The event input port of the event-driven state transition is illustrated by a small circle. Events can be generated by clock signals from outside of the net as shown in (a) or by outputs within the net itself as shown in (b) (where events at e are generated by module event with input from y).

2.3   Models of the elevator system

An elevator system is a typical hybrid system with continuous motion following Newton's dynamics and event-based control responding to users' request. In the rest of this section, we will model the elevator system in Constraint Nets. The controllers are left as black boxes and will be designed and modeled in Section 4, after the section on requirements specification.

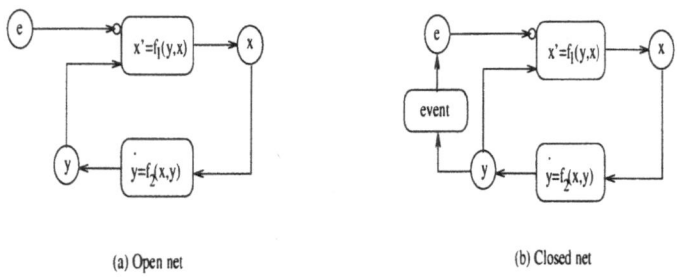

(a) Open net   (b) Closed net

Fig. 5. Hybrid systems in Constraint Nets

Continuous model We model the elevator body by a second order differential equation following Newton's Second Law

$$F - K\dot{h} = \ddot{h} \tag{1}$$

where F is the motor force, K is the coefficient of friction and h is the height of the elevator (Figure 6). Here we assume that the mass of the body is 1 since it can be

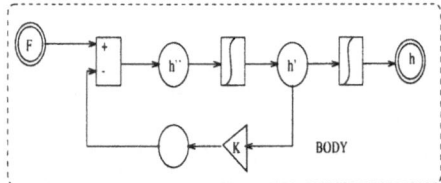

Fig. 6. The BODY module

scaled by F and K. We also ignore gravity since it can be added to F to compensate its effect.

Let the separation between floors be H. Given the current height h, the current floor number can be obtained as

$$f = [h/H] + 1 \tag{2}$$

where $[x]$ denotes the integer closest to x, and the distance to the nearest floor is

$$d_s = h - (f - 1)H. \tag{3}$$

Furthermore, we say that the elevator is in a home position if

$$e_h : |d_s| \leq \epsilon \tag{4}$$

for some $\epsilon > 0$. In real life, f, d_s and the home event e_h can be either sensed directly or calculated from h.

The continuous component of the elevator system is depicted in Figure 7 where Com is the higher level command which can be 1, −1 and 0, denoting up, down and stop, respectively, and CONTROL0 is an analog controller that generates the signal that determines the force to drive the elevator body, BODY is the module in Figure 6, FLOOR is Equation 2 and HOME is Condition 4.

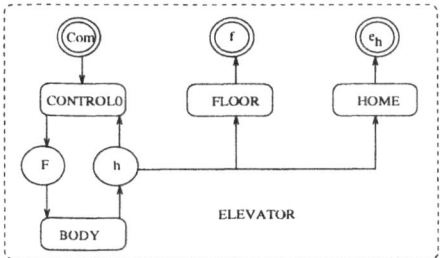

Fig. 7. The continuous components of the elevator system

Discrete model For an n-floor elevator system, the relationship between the command Com and the current floor f modeled in the continuous component of the elevator system can be abstracted by a state transition system with state transition function: $f' = NextFloor(f, Com)$ where

$$NextFloor(f, Com) = \begin{cases} \min(f+1, n) & \text{if } Com = 1 \\ \max(f-1, 1) & \text{if } Com = -1 \\ f & \text{if } Com = 0 \end{cases}$$

provided that CONTROL0 works correctly.

We model push buttons as an array of flip-flops. A button will be set to 1 when it is pushed by a user and be reset to 0 when the request is served.

Formally, let Ub, Db and Fb denote up, down and floor buttons, respectively. For an n-floor elevator, $Ub, Db, Fb \in \{0,1\}^n$ are boolean vectors of n-elements with $Ub(n) = 0$ and $Db(1) = 0$. The request state of a push button is determined by two factors: the users' input and the reset signal when the request has been served. Let the users' inputs, the states of the push buttons and the reset signals be $\langle Ub_i, Db_i, Fb_i \rangle$, $\langle Ub_s, Db_s, Fb_s \rangle$ and $\langle Ub_r, Db_r, Fb_r \rangle$, respectively. If a button is pushed by a user, the state of the button is set (to 1); or, if the request has been served, the state is reset (to 0); otherwise, the state is unchanged. The state transition function of the flip-flop is a logical expression:

$$b'_s = FlipFlop(b_i, b_r, b_s)$$
$$= b_i \vee (\neg b_r \wedge b_s)$$

Note that this flip-flop has higher priority for set than for reset, i.e., if a user pushes a button for service while the elevator is just finishing the service, the elevator should still consider the new request at that time.

Let s be the serving state of the elevator which can be up, down or idle. The reset signal b_r indicates which requests have been served, formally:

$$b_r = ResetSignal(f, Com, s)$$

such that $\forall 1 \leq k \leq n$,

$$Ub_r(k) = (f = k) \wedge (Com = 0) \wedge (s = up)$$
$$Db_r(k) = (f = k) \wedge (Com = 0) \wedge (s = down)$$
$$Fb_r(k) = (f = k) \wedge (Com = 0) \wedge (s \neq idle).$$

Summarizing, the push button module is shown in Figure 8.

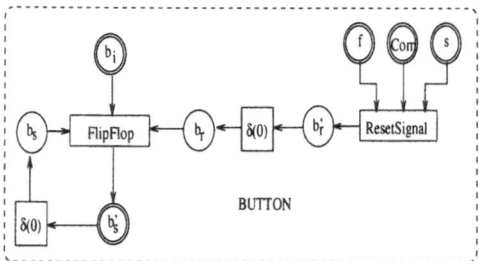

Fig. 8. The BUTTON module

For a continuous reference time, we assume that the event-driven time (which could be generated by a built-in clock) of the push button module has much higher frequency than users' inputs.

Hybrid model Let CONTROL1 be a discrete control with the current floor number f and the current button states b_r as inputs and with command Com and serving state s as outputs. CONTROL1 is driven by event e which is the "event-or" (for the concepts of event logics, the reader is referred to [30]) of the following three events: (1) a user pushes a button at the elevator's idle state ($s = idle \wedge \bigvee_{1 \leq k \leq n}(Ub_s(k) \vee Db_s(k) \vee Fb_s(k))$), (2) the elevator comes to a home position ($|d_s| \leq \epsilon$) and (3) a request has been served ($Com = 0 \wedge \neg(Fb_s(f) \vee (Ub_s(f) \wedge s = up) \vee (Db_s(f) \wedge s = down))$) for certain time. If it takes τ seconds to serve a request, a transport delay of τ will be used.

Finally, the hybrid model of the elevator and its control is shown in Figure 9, where EVENT implements the event logic that produces the events for triggering the discrete control, BUTTON is Figure 8 and ELEVATOR is Figure 7.

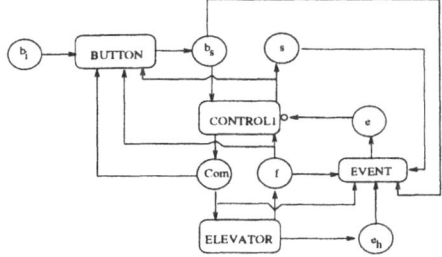

Fig. 9. The hybrid model of the elevator system

3   Requirements Specification in Timed ∀-Automata

While modeling focuses on the underlying structure of a system, the organization and coordination of components or subsystems, the overall behavior of the modeled system is not explicitly expressed. However, for many situations, it is important to specify some global properties and guarantee that these properties hold in the system being designed. For example, a well-designed elevator system should service any request within some bounded waiting time. Requirements specification in timed ∀-automata provides a formal method for this purpose.

3.1   Timed ∀-automata

Discrete ∀-automata are non-deterministic finite state automata over infinite sequences. These automata were originally proposed as a formalism for the specification and verification of temporal properties of concurrent programs [24]. We augment discrete ∀-automata to timed ∀-automata by generalizing time from discrete to continuous and by specifying time constraints on automaton-states.

A ∀-*automaton* $\mathcal{A}$ is a quintuple $\langle Q, R, S, e, c \rangle$ where Q is a finite set of *automaton-states*, $R \subseteq Q$ is a set of *recurrent states* and $S \subseteq Q$ is a set of *stable states*. With each $q \in Q$, we associate an assertion $e(q)$, which characterizes the *entry condition* under which the automaton may start its activity in q. With each pair $q, q' \in Q$, we associate an assertion $c(q, q')$, which characterizes the *transition condition* under which the automaton may move from q to q'. R and S are the generalization of *accepting* states to the case of infinite inputs. We denote by $B = Q - (R \cup S)$ the set of *non-accepting (bad)* states.

A ∀-automaton can be depicted by a labeled directed graph where automaton-states are depicted by nodes and transition relations by arcs. Furthermore, some automaton-states are marked by a small arrow, an *entry arc*, pointing to it. Each recurrent state is depicted by a diamond inscribed within a circle. Each stable state is depicted by a square inscribed within a circle. Nodes and arcs are labeled by assertions. A node or an arc that is left unlabeled is considered to be labeled with *true*. The labels define the entry conditions and the transition conditions of the associated automaton as follows.

- Let $q \in Q$ be a node in the diagram corresponding to an automaton-state. If q is labeled by ψ and the entry arc is labeled by φ, the entry condition $e(q)$ is given by $e(q) = \varphi \wedge \psi$. If there is no entry arc, then $e(q) = false$.
- Let q, q' be two nodes in the diagram corresponding to automaton-states. If q' is labeled by ψ, and arcs from q to q' are labeled by $\varphi_i, i = 1 \cdots n$, the transition condition $c(q, q')$ is given by $c(q, q') = (\varphi_1 \vee \cdots \vee \varphi_n) \wedge \psi$. If there is no arc from q to q', $c(q, q') = false$.

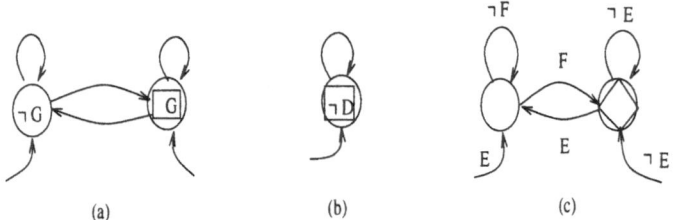

(a)   (b)   (c)

Fig. 10. $\forall$-automata: (a) reachability (b) safety (c) bounded response

Some examples of $\forall$-automata are shown in Figure 10: (a) states that the system should finally satisfy G; (b) states that the system should never satisfy D and (c) states that whenever the system satisfies E, it will satisfy F in some bounded time.

The formal semantics of $\forall$-automata is defined as follows. Let A be a domain of values. An assertion α on A corresponds to a subset $V(\alpha)$ of A. A value $a \in A$ satisfies an assertion α on A, written $a \models \alpha$, iff $a \in V(\alpha)$. Let $\mathcal{T}$ be the time domain and $v : \mathcal{T} \to A$ be a trace. Given a $\forall$-automaton $\mathcal{A}$ as $\langle Q, R, S, e, c \rangle$, a *run* of $\mathcal{A}$ over v is a mapping $r : \mathcal{T} \to Q$ such that the following two conditions are satisfied:

1. *Initiality*: Let $0 \in \mathcal{T}$ be the start time point, $v(0) \models e(r(0))$; and
2. *Consecution*: If $\mathcal{T}$ is discrete time, then for all $t > 0, v(t) \models c(r(pre(t)), r(t))$, where $pre(t)$ is the previous time point of t. If $\mathcal{T}$ is continuous time, the following two conditions must be satisfied:
   (a) *Inductivity*: $\forall t > 0, \exists q \in Q, t' < t, \forall t'', t' \leq t'' < t, r(t'') = q$ and $v(t) \models c(r(t''), r(t))$ and
   (b) *Continuity*: $\forall t, \exists q \in Q, t' > t, \forall t'', t < t'' < t', r(t'') = q$ and $v(t'') \models c(r(t), r(t''))$.
   These two conditions are derived from continuous induction principle. Condition (a) corresponds to a right-closed transition and condition (b) corresponds to a left-closed transition.

If r is a run, let $Inf(r)$ be the set of automaton-states appearing infinitely many times in r, i.e., $Inf(r) = \{q | \forall t \exists t_0 \geq t, r(t_0) = q\}$. A run r is defined to be *accepting* iff:

1. $Inf(r) \cap R \neq \emptyset$, i.e., *some* of the states appearing infinitely many times in r belong to R, or

2. $Inf(r) \subseteq S$, i.e., *all* the states appearing infinitely many times in r belong to S.

A $\forall$-automaton $\mathcal{A}$ *accepts* a trace v, written $v \models \mathcal{A}$, iff *all* possible runs of $\mathcal{A}$ over v are accepting.

For example, Figure 10(a) accepts the trace $x(t) = Ce^{-t}$ for $G =_{def} |x| < \epsilon$. Figure 10(b) accepts the trace $x(t) = \sin(t)$ for $D =_{def} |x| > 1$. Figure 10(c) accepts the trace $x(t) = sin(t)$ for $E =_{def} x \geq 0$ and $F =_{def} x < 0$.

Timed $\forall$-automata are $\forall$-automata augmented with timed automaton-states and time bounds. Let $\mathcal{R}^+$ be the set of non-negative real numbers. A *timed $\forall$-automaton* $\mathcal{TA}$ is a triple $\langle \mathcal{A}, T, \tau \rangle$ where $\mathcal{A} = \langle Q, R, S, e, c \rangle$ is a $\forall$-automaton, $T \subseteq Q$ is a set of *timed* automaton-states and $\tau : T \cup \{bad\} \to \mathcal{R}^+ \cup \{\infty\}$ is a *time function*. A $\forall$-automaton is a special timed $\forall$-automaton with $T = \emptyset$ and $\tau(bad) = \infty$. Graphically, a T-state is denoted by a nonnegative real number indicating its time bound. Figure 11 shows two examples of timed $\forall$-automata.

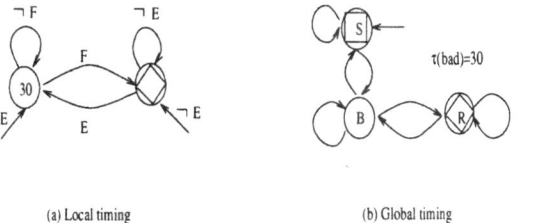

(a) Local timing   (b) Global timing

Fig. 11. Examples of timed automata

The formal semantics of timed $\forall$-automata is defined as follows. Let $I \subseteq \mathcal{T}$ be a time interval and $\mu(I) \in \mathcal{R}^+$ be the real-time measurement of I. Let $r : \mathcal{T} \to Q$ be a run of $\mathcal{A}$. For any $P \subset Q$, let $Sg(P)$ be the set of consecutive P-state segments of r, i.e., $r_{|I} \in Sg(P)$ for some interval I iff $\forall t \in I, r(t) \in P$. A run r *satisfies the time constraints* iff

1. (local time constraint) for any $q \in T$ and any interval I of $\mathcal{T}$, if $r_{|I} \in Sg(\{q\})$ then $\mu(I) \leq \tau(q)$ and
2. (global time constraint) let $B = Q - (R \cup S)$ and $\chi_B : Q \to \{0, 1\}$ be the characteristic function for set B; for any interval I of $\mathcal{T}$, if $r_{|I} \in Sg(B \cup S)$ then $\int_I \chi_B(r(t))dt \leq \tau(bad)$.

Condition 1 says if τ is the local timing bound for q, then the time duration of any segment of q's must be no more than τ. Condition 2 says if τ is the global timing bound, then the total time on bad states must be no more than τ.

Let $v : \mathcal{T} \to A$ be a trace. A *run* r of $\mathcal{TA}$ over v is a run of $\mathcal{A}$ over v; r is *accepting* for $\mathcal{TA}$ iff

1. r is accepting for $\mathcal{A}$ and
2. r satisfies the time constraints.

A timed ∀-automaton $\mathcal{T}\!A$ *accepts* a trace v, written $v \models \mathcal{T}\!A$, iff *all* possible runs of $\mathcal{T}\!A$ over v are accepting. For example, Figure 11 (a) specifies a real-time response property meaning that any event (E) will be responded to (F) within 30 time units. Figure 11 (b) specifies a conditioned real-time response property meaning that R will always be reached within 30 time units of B, i.e., the time in S is not counted.

It has been shown [24] that discrete ∀-automata have the same expressive power as Buchi automata [31] and the extended temporal logic (ETL) [32], which are strictly more powerful than the propositional linear temporal logic (PLTL) [31], [32]. However, timed ∀-automata are not directly comparable with TBA [2]. A local timing constraint in (discrete) timed ∀-automata can also be specified in TBA. However, global timing constraints cannot be specified within TBA, since it is not possible to stop a clock except by resetting it. On the other hand, there are properties of timed behaviors that can be specified by TBA but cannot be specified by timed ∀-automata.

3.2   Requirements specification for elevator systems

A well-designed elevator system should guarantee that any request will be served within some bounded time. We can specify such requirements in timed ∀-automata.

There are three kinds of requests: to go to a particular floor after entering the elevator, or to go up or down when waiting for the elevator. The following are some examples of assertions.

- $R2 : Fb_s(2) = 1$ denotes that "there is a request to go to the second floor."
- $R2S : Fb_s(2) = 0$ denotes that "the request to go to the second floor is served."
- $RU2 : Ub_s(2) = 1$ denotes that "there is a request to go up at the second floor."
- $RU2S : Ub_s(2) = 0$ denotes that "the request to go up at the second floor is served."

Notice that an elevator may stop at a floor forever given that a request button in that floor is pushed continuously. The real-time response specification in Figure 11 (a) will not be satisfied in general (e.g. given $E =_{def} R2$ and $F =_{def} R2S$). Instead, a conditioned real-time response specification in Figure 11 (b) should be adopted. For example, let $S =_{def} Com = 0$, $B =_{def} RU2 \wedge (Com \neq 0)$ and $R =_{def} RU2S$, the specification in Figure 11 (b) means that a request to go up in the second floor will be served within 30 time units of elevator's motion time.

4   Hybrid Control System Design

Given a model of the system and its requirements specification, it is always a challenge to design a "correct" control system so that the overall system satisfies the specification. There is no automatic method in general. However, there are design principles that can be applied case by case.

In section 2, we have modeled the complete elevator system with modules CONTROL0 and CONTROL1 as the black boxes. CONTROL0 is an analog controller that generates force to drive the elevator body, and CONTROL1 is a discrete controller with the current floor number f and the current button states b_s as inputs and with command Com and serving state s as outputs. In this section, we will design these two control modules as a case study.

4.1 Continuous control design

The simplest analog controller is a linear proportional and derivative (PD) controller. Let

$$F = \begin{cases} F_0 & \text{if } Com = 1 \\ -F_0 & \text{if } Com = -1 \\ -K_p d_s - K_v \dot{d}_s & \text{if } Com = 0 \end{cases} \tag{5}$$

where d_s is the distance to the closest floor, F_0 is a positive constant, K_p is a proportional gain and K_v is a derivative gain. The rest of design is to choose F_0, K_p, K_v so that the following two conditions are satisfied:

1. Continuous Stability: the continuous control is stable;
2. Hybrid Consistency: the interface to the discrete control is consistent.

- **Stability**
  Notice that $\dot{h} = \dot{d}_s$, so for $Com = 0$, we have

$$\ddot{d}_s + (K + K_v)\dot{d}_s + K_p d_s = 0. \tag{6}$$

by combining Equation 1 with Equation 5.

The stability theorem for the second order differential equation tells us that if both $K + K_v$ and K_p are positive, the continuous system is stable, because the roots of the characteristic equation have negative real parts.

- **Consistency**
  The home position is defined as $|d_s| \leq \epsilon$ for some $\epsilon > 0$. In order to have hybrid consistency, we have to make sure that if $Com = 0$, then $Com = 0$ implies $|d_s| \leq \epsilon$ for all time (this formal requirement is represented by a $\forall$-automaton in Figure 12), i.e., no overshoot above ϵ should happen after a stop command is issued.

Let

$$K'_v = K + K_v \tag{7}$$

and

$$(K'_v)^2 = 4K_p. \tag{8}$$

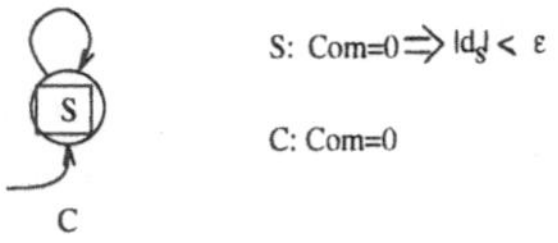

S: Com=0 $\Rightarrow$ |d$_s$| < ϵ

C: Com=0

Fig. 12. The specification of the stop control

The solution to the differential equation Equation 6 is

$$d_s = (C_0 + C_1 t)e^{-\frac{1}{2}K'_v t} \qquad (9)$$

where $C_0 = -\epsilon$ and $C_1 = V_0 - \frac{1}{2}K'_v\epsilon$, where V_0 is the initial speed at distance ϵ.

The maximum distance to the floor D can be computed as follows: The maximum distance is achieved at $\dot{d}_s = 0$. Derived from Equation 9, we have $C_1 = \frac{1}{2}K'_v(C_0 + C_1 t)$, i.e., the maximum distance happens at time $t = 2/K'_v - C_0/C_1$. Putting this t back to Equation 9, this control will overshoot at most D where

$$D = (2C_1/K'_v)e^{-1-C_0/(2C_1/K'_v)}$$
$$= (2V_0/K'_v - \epsilon)e^{-1-\epsilon/(2V_0/K'_v-\epsilon)}.$$

If we choose

$$K'_v = V_0/\epsilon, \qquad (10)$$

we have $D = \epsilon e^{-2}$. Therefore, $\max_t |d_s(t)| \leq \epsilon$.

Given F_0, the maximum speed of the elevator is then

$$V_0 = F_0/K \qquad (11)$$

since the solution to Equation 1 is $\dot{h} = (F_0/K)(1 - e^{-Kt})$ given that F is F_0. If the maximum acceleration of the motor is a, we have to choose

$$F_0 \leq a \qquad (12)$$

since $F = \ddot{h} + K\dot{h}$. On the other hand, if the maximum deceleration of the motor is d, V_0 must satisfy

$$K'_v V_0 - K_p\epsilon \leq d \qquad (13)$$

according to Equation 6. Combining Equation 13 with Equation 10 and Equation 8, we have

$$V_0 \leq \sqrt{4\epsilon d/3}. \qquad (14)$$

Combining Equation 12 and Equation 14,

$$F_0 \leq \min(K\sqrt{4\epsilon d/3}, a).\tag{15}$$

For the rest of parameters, we can have

$$K'_v = F_0/(K\epsilon)\tag{16}$$

using Equation 10 and Equation 11. Then

$$K_v = K_{v'} - K\tag{17}$$

using Equation 7 and

$$K_p = K_v^2/4\tag{18}$$

using Equation 8.

Suppose $K = 1.0$, $a = d = 0.5$, and $\epsilon = 0.15$, we can set $F_0 = 0.33$ that satisfies Equation 15, $K_v = 1.2$ and $K_p = 1.21$.

4.2 Discrete control design

The discrete controller CONTROL1 is a logical component. It receives the current request from the state of push buttons b_s, according to the current floor number f and the current serving state s, determines the next motion of the elevator Com and the next serving state s. We assume here three kinds of serving states: up, down and idle. Furthermore, we assume that the elevator is always parked (idle) at the first floor.

There are three types of request to the elevator: UpRequest, DownRequest and StopRequest.

- Let Ur indicate whether or not there is a request for the elevator to go up:

$$Ur = UpRequest(f, Ub, Db, Fb)$$
$$= Ub(f) \vee \bigvee_{n \geq k > f} (Ub(k) \vee Db(k) \vee Fb(k)).$$

- Let Dr indicate whether or not there is a request for the elevator to go down:

$$Dr = DownRequest(f, Ub, Db, Fb)$$
$$= Db(f) \vee \bigvee_{1 \leq k < f} (Ub(k) \vee Db(k) \vee Fb(k)).$$

- Let Sr indicate whether or not there is a request for the elevator to stop and the request is consistent to the current serving state s:

$$Sr = StopRequest(f, Ub, Db, Fb, s)$$
$$= \begin{cases} Db(f) \vee Fb(f) & \text{if } s = down \\ Ub(f) \vee Fb(f) & \text{otherwise.} \end{cases}$$

In order to satisfy the requirements specification that any request is served in some bounded time, we have to make sure that there is no dead-lock or live-lock for all possible situations. For example, a control that always serves the nearest floor may get stuck and never respond to a request further away. Here is a simple strategy we use: the elevator will move persistently in one direction until there is no request in that direction. More specifically, if there is a request for the elevator to go up and either the last serving state is up or there is no request to go down, the current serving state will be up; if there is no request to go up and the elevator is not at the first floor, or the last serving state is down and there is a request to go down, then the the current serving state will be down; otherwise the current serving state will be idle, that is, the elevator will be parked at the first floor if there are no more requests. Formally,

$$s' = ServingState(f, s, Ur, Dr)$$
$$= \begin{cases} up & \text{if } Ur \wedge (s \neq down \vee \neg Dr) \\ down & \text{if } (\neg Ur \wedge f > 1) \vee (Dr \wedge s = down) \\ idle & \text{otherwise} \end{cases}$$

Then, the current command can be determined as follows:

$$Com = Command(Sr, s)$$
$$= \begin{cases} 0 & \text{if } Sr \text{ or } s = idle \\ 1 & \text{if } \neg Sr \text{ and } s = up \\ -1 & \text{otherwise.} \end{cases}$$

Putting all the functions together, the constraint net model of the discrete controller is shown in Figure 13.

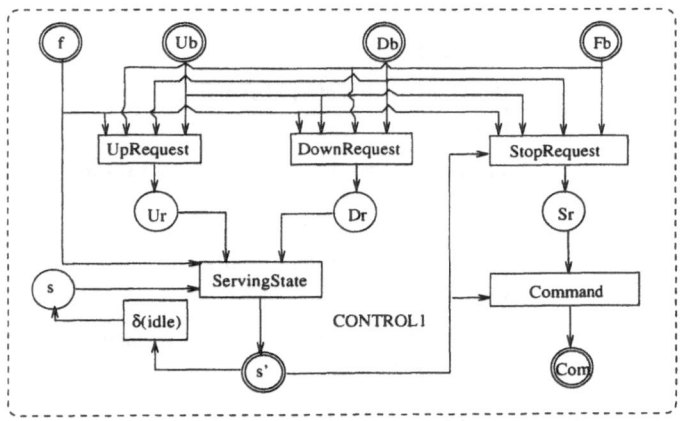

Fig. 13. The discrete control module

5 Behavior Verification Using Model Checking

Given a constraint net model of a system and a timed $\forall$-automaton specification of a behavior, the behavior of the system satisfies the requirements specification if and only if the (behavior) traces of the system are accepting for the timed $\forall$-automaton.

We develop a model checking method that uses the induction principle and generalizes both Liapunov stability analysis for dynamic systems and monotonicity of well-foundedness in discrete-event systems.

A representation between constraint nets and timed $\forall$-automata is a state-based transition system, such as a Kripke structure. The verification rules are applied to the Kripke structure.

5.1 Generalized Kripke structure

A useful and important type of behavior is state-based and time-invariant. Intuitively, a state-based and time-invariant behavior is a behavior whose traces after any time are totally dependent on the current snapshot. State-based and time-invariant behaviors can be defined using generalized Kripke structures [35].

A *generalized Kripke structure* $\mathcal{K}$ as a triple $\langle \mathcal{S}, \rightarrow, \Theta \rangle$ where $\mathcal{S}$ is a set of states, $\rightarrow \subset \mathcal{S} \times \mathcal{R}^+ \times \mathcal{S}$ is a state transition relation, and $\Theta \subseteq \mathcal{S}$ is a set of initial states. We denote $\langle s_1, t, s_2 \rangle \in \rightarrow$ as $s_1 \overset{t}{\rightarrow} s_2$. The state transition relation $\rightarrow$ satisfies the following conditions:

- *initiality*: $s \overset{0}{\rightarrow} s$;
- *transitivity*: if $s_1 \overset{t_1}{\rightarrow} s_2$ and $s_2 \overset{t_2}{\rightarrow} s_3$, then $s_1 \overset{t_1+t_2}{\rightarrow} s_3$;
- *infinity*: $\forall s \in \mathcal{S}, \exists t > 0, s' \in \mathcal{S}, s \overset{t}{\rightarrow} s'$.

For example, the behavior of $\dot{x} = -x$ can be represented by a generalized Kripke structure $\langle \mathcal{R}, \rightarrow, \Theta \rangle$ with $s_1 \overset{t}{\rightarrow} s_2$ iff $s_2 = s_1 e^{-t}$.

Let φ and ψ be assertions on states and time durations. For a generalized Kripke structure $\mathcal{K} = \langle \mathcal{S}, \rightarrow, \Theta \rangle$, let $\{\varphi\}\mathcal{K}\{\psi\}$ denote the validity of the following two consecution conditions:

- *Inductivity* $\{\varphi\}\mathcal{K}^-\{\psi\}$: $\exists \delta > 0, \forall 0 < t \leq \delta, \forall s, (\varphi(s) \wedge (s \overset{t}{\rightarrow} s')) \Rightarrow \psi(s', t))$.
- *Continuity* $\{\varphi\}\mathcal{K}^+\{\psi\}$: $\varphi(s) \Rightarrow \exists \delta > 0, \forall 0 < t < \delta, \forall s', ((s \overset{t}{\rightarrow} s') \Rightarrow \psi(s', t))$.

These two conditions are derived from the continuous induction principle. If time is discrete, these two conditions reduce to one, i.e., $\varphi(s) \wedge (s \overset{\delta}{\rightarrow} s') \Rightarrow \psi(s', \delta)$ where δ is the time duration between s and s'.

5.2 Verification rules

The formal method for behavior verification consists of a set of model-checking rules, which is a generalization of the model-checking rules developed for concurrent programs [24].

There are three types of rules: invariance rules (I), stability or eventuality rules (L) and timeliness rules (T). Let $\mathcal{A}$ be a $\forall$-automaton $\langle Q, R, S, e, c \rangle$ and $\mathcal{K}$ be a generalized Kripke structure $\langle \mathcal{S}, \rightarrow, \Theta \rangle$. The invariance rules check to see if a set of assertions $\{\alpha\}_{q \in Q}$ is a set of invariants for $\mathcal{A}$ and $\mathcal{K}$, i.e., for any trace v of $\mathcal{K}$ and any run r of $\mathcal{A}$ over v, $\forall t \in \mathcal{T}, v(t) \models \alpha_{r(t)}$. Given $B = Q - (R \cup S)$, the stability or eventuality rules check if the B-states in any run of $\mathcal{A}$ over any trace of $\mathcal{K}$ will be terminated eventually. Given $\mathcal{TA}$ as a timed $\forall$-automaton $\langle \mathcal{A}, T, \tau \rangle$, the timeliness rules check if the T-states and the B-states in any run of $\mathcal{A}$ over any trace of $\mathcal{K}$ are bounded by the time function τ. The set of model-checking rules can be represented in first-order logic, some of which are in the form of $\{\varphi\}\mathcal{K}\{\psi\}$.

Here are the model-checking rules for a behavior represented by $\mathcal{K} = \langle \mathcal{S}, \rightarrow, \Theta \rangle$ and a specification represented by $\mathcal{TA} = \langle \mathcal{A}, T, \tau \rangle$ where $\mathcal{A} = \langle Q, R, S, e, c \rangle$:

Invariance Rules (I): A set of assertions $\{\alpha_q\}_{q \in Q}$ is called a set of *invariants* for $\mathcal{K}$ and $\mathcal{A}$ iff

(I1) *Initiality*: $\forall q \in Q, \Theta \wedge e(q) \Rightarrow \alpha_q$.
(I2) *Consecution*: $\forall q, q' \in Q, \{\alpha_q\}\mathcal{K}\{c(q, q')\} \Rightarrow \alpha_{q'}\}$.

The Invariance Rules are the same as those in [24] except that the condition for consecution is generalized.

Stability or Eventuality Rules (L): Given that $\{\alpha_q\}_{q \in Q}$ is a set of invariants for $\mathcal{K}$ and $\mathcal{A}$, a set of partial functions $\{\rho_q\}_{q \in Q} : \mathcal{S} \rightarrow \mathcal{R}^+$ is called a set of *Liapunov functions* for $\mathcal{K}$ and $\mathcal{A}$ iff the following conditions are satisfied:

(L1) *Definedness*: $\forall q \in Q, \ \alpha_q \Rightarrow \exists w, \rho_q = w$.
(L2) *Non-increase*: $\forall q \in S, q' \in Q, \{\alpha_q \wedge \rho_q = w\}\mathcal{K}^-\{c(q, q')\} \Rightarrow \rho_{q'} \leq w\}$ and $\forall q \in Q, q' \in S, \{\alpha_q \wedge \rho_q = w\}\mathcal{K}^+\{c(q, q')\} \Rightarrow \rho_{q'} \leq w\}$.
(L3) *Decrease*: $\exists \epsilon > 0, \forall q \in B, q' \in Q, \{\alpha_q \wedge \rho_q = w\}\mathcal{K}^-\{c(q, q')\} \Rightarrow \frac{\rho_{q'} - w}{t} \leq -\epsilon\}$ and $\forall q \in Q, q' \in B, \{\alpha_q \wedge \rho_q = w\}\mathcal{K}^+\{c(q, q')\} \Rightarrow \frac{\rho_{q'} - w}{t} \leq -\epsilon\}$.

If time is discrete, $\mathcal{K}^+$ rules will not be used. The Stability or Eventuality Rules generalize both stability analysis of discrete or continuous dynamic systems [20] and well-foundedness for finite termination in concurrent systems [24].

Timeliness Rules (T): Corresponding to two types of time bound, we define two timing functions. Let $\{\alpha_q\}_{q \in Q}$ be invariants for $\mathcal{K}$ and $\mathcal{A}$. A set of partial functions $\{\gamma_q\}_{q \in T}$ is called a set of *local timing functions* for $\mathcal{K}$ and $\mathcal{TA}$ iff $\gamma_q : \mathcal{S} \rightarrow \mathcal{R}^+$ satisfies the following conditions:

(T1) *Boundedness*: $\forall q \in T, \alpha_q \Rightarrow \gamma_q \leq \tau(q)$ and $\forall q \in T, q' \in Q, \{\alpha_q \wedge \gamma_q = w\}\mathcal{K}^-\{c(q, q')\} \Rightarrow w \geq t\}$.
(T2) *Decrease*: $\forall q \in T, \{\alpha_q \wedge \gamma_q = w\}\mathcal{K}\{c(q, q)\} \Rightarrow \frac{\gamma_q - w}{t} \leq -1\}$.

A set of partial functions $\{\eta_q\}_{q \in Q}$ is called a set of *global timing functions* for $\mathcal{K}$ and $\mathcal{TA}$ iff $\eta_q : \mathcal{S} \rightarrow \mathcal{R}^+$ satisfies the following conditions:

(T3) *Definedness*: $\forall q \in Q, \alpha_q \Rightarrow \exists w, \eta_q = w$.
(T4) *Boundedness*: $\forall q \in B, \alpha_q \Rightarrow \eta_q \leq \tau(bad)$.

(T5) *Non-increase*: $\forall q \in S, q' \in Q, \{\alpha_q \wedge \eta_q = w\} \mathcal{K}^- \{c(q,q') \Rightarrow \eta_{q'} \leq w\}$ and $\forall q \in Q, q' \in S, \{\alpha_q \wedge \eta_q = w\} \mathcal{K}^+ \{c(q,q') \Rightarrow \eta_{q'} \leq w\}$.

(T6) *Decrease*: $\forall q \in B, q' \in Q, \{\alpha_q \wedge \eta_q = w\} \mathcal{K}^- \{c(q,q') \Rightarrow \frac{\eta_{q'} - w}{t} \leq -1\}$ and $\forall q \in Q, q' \in B, \{\alpha_q \wedge \eta_q = w\} \mathcal{K}^+ \{c(q,q') \Rightarrow \frac{\eta_{q'} - w}{t} \leq -1\}$.

If time is discrete, $\mathcal{K}^+$ rules will not be used. The Timeliness Rules are modifications of the Eventuality Rules; they enforce real-time boundedness, in addition to termination.

A set of model-checking rules is *sound* if verification by the rules guarantees the correctness of the behavior against the specification; it is *complete* if the correctness of the behavior against the specification guarantees verification by the rules. The soundness and completeness of the rules are discussed in [35].

Also notice that the set of verification rules are formal guidance for the proof of a system; it is by no means automatic. However, if the system is discrete and there are computer-aided proof systems available, semi-automatic proofs can be applied. If the system is discrete and finite, automatic algorithms can be derived from the rules [34].

5.3 Is the elevator system well-designed?

This question can be answered only if we can answer the following three questions: (1) Does the model reflect the real system at the appropriate level of abstraction? (2) Is the set of requirements complete? and (3) Does the model satisfy the requirements? The method that we propose here will answer the third question.

Given a 3-floor elevator system, we can check whether or not a request to go up at the second floor will be served within some bounded time units of elevator's motion time, i.e., if the constraint net model in Figure 9 satisfies the timed ∀-automaton specification in Figure 11 (b).

Our method involves the following steps:

First, find a Kripke structure of the constraint nets corresponding to the level of specification. A continuous system can have a discrete Kripke structure if the specification is on high level. In this case, if there is a request to go up in the second floor, a Kripke structure of the behavior of the system can be obtained (Figure 14). Each state is of form (f, s, Com) indicating the current floor, the serving state, and the command of the elevator, with $(1, up, 0)$ as the initial state. The dashed transitions indicate the transitions inhibited by the control given the current request state, and self loops indicate the stationary transitions.

Since the Kripke Structure we use is finite for this case, automatic proof can be applied. The algorithm is directly obtained from the rules, so we will illustrate how the rules are applied here.

1. Find a set of invariants for the timed ∀-automaton in Figure 11 (b) that satisfies the Invariance Rules. In this case, let q_1, q_2, q_3 be states with B, S and R assertions respectively. If $B =_{def} RU2 \wedge (Com \neq 0)$, $S =_{def} Com = 0$, and $R =_{def} RU2S$, we can see that B, S and R are invariants for q_1, q_2 and q_3 respectively.

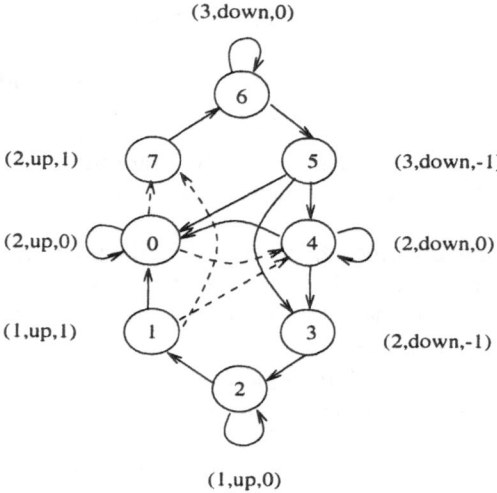

(3,down,0)

(2,up,1)   (3,down,-1)

(2,up,0)   (2,down,0)

(1,up,1)   (2,down,-1)

(1,up,0)

S=(f,s,Com)

Fig. 14. Generalized Kripke structure of the behavior

2. Find a set of Liapunov functions. Let ρ be a function whose value for each state is the longest possible path from that state to the desired shaded state, without self loops (see the number in each state in Figure 14). It is easy to see that rho (for all q_1, q_2 and q_3), together with the invariants, satisfy the Stability or Eventuality Rules. That is, the request will be served within bounded time of motion.

3. Find a set of global timing functions. Let the maximum traversal time of the elevator from one floor to another be T, we define a global timing function as follows:

$$\eta(2, up, 0) = 0;$$
$$\eta(1, up, 1) = T;$$
$$\eta(1, up, 0) = T;$$
$$\eta(2, down, -1) = 2 * T;$$
$$\eta(2, down, 0) = 2 * T;$$
$$\eta(3, down, -1) = 3 * T;$$
$$\eta(3, down, 0) = 3 * T;$$
$$\eta(2, up, 1) = 4 * T;$$

It is easy to see that η (for all q_1, q_2 and q_3), is a global timing function iff $4 * T \leq 30$.

we can calculate the maximum traversal time T of the elevator from one floor to another as follows. Since $\dot{h} = (F/K)(1 - e^{-Kt})$ and $h(0) = 0$, we have $h = (F/K)(t + (1/K)e^{-Kt}) - F/(K^2)$. Suppose the distance between floors is H. The time to traverse one level from stationary state will be $H \geq (F/K)(t) - F/(K^2)$, i.e., $t \leq HK/F + 1/K$. If $H = 2.0$, $K = 1.0$ and $F = 0.33$, we have $t \leq 7.061$. Therefore $T = 7.061$.

6 Conclusions

We have presented a formal approach to the modeling and analysis of hybrid control systems. The approach consists of four interleaving or concurrent phases: hybrid dynamic system modeling, requirements specification, hybrid control design and behavior verification. We have used Constraint Nets for modeling hybrid dynamic systems, timed $\forall$-automata for specifying requirements, coupled continuous control with event-driven logic, and the rule-based model checking for verifying behaviors of the overall systems.

Our approach is demonstrated by an elevator case study. The hybrid control system of the elevator couples a quite complex discrete control logic with an analog control law. The inconsistent behavior between the discrete and continuous components, as well as the incorrect behavior at each level, may cause the malfunction of the overall system. We have simulated the elevator system modeled by Constraint Nets in both Simulink and SystemBuild.

As we reviewed in this paper, much work has been done in modeling and analysis of hybrid systems [10]. Which method to use is largely dependent on the problems at hand.

Like most dataflow/net formalisms, Constraint Nets are modular and hierarchical so as to model complex hybrid systems. Developed from abstract algebra, CN provides a uniform representation for hybrid systems with a formal syntax and semantics for simulation. However, since there are no closed-form solutions for most problems, the analysis is, in general, hard.

Timed $\forall$-automata are simple for requirements specification, However, they are not powerful enough to represent all the possible behaviors.

Our verification method provides a set of formal model checking rules, which can be used to guide a formal proof procedure. However, the invariants, Liapunov and timing functions are not automatically created and the verification of the rules is not automatic, in general.

Nevertheless, the approach we developed here can be used to solve many problems in hybrid systems modeling, designing, specification and verification. With this approach, we have also developed controllers for robot soccer players [39] and hydraulically controlled robot arms [34]. The same approach can be applied to the modeling and control of most complex electromechanical systems.

Acknowledgements

Thanks to Peter Caines, Hector Levesque, Ray Reiter and Maarten van Emden for stimulating us to work on the elevator system. This work was supported in part by the Natural Sciences and Engineering Research Council, the Institute for Robotics and Intelligent Systems and the Canadian Institute for Advanced Research.

References

1. R. Alur, C. Courcoubetis, T. A. Henzinger, and P. Ho. Hybrid automata: An algorithmic approach to the specification and verification of hybrid systems. In R. L. Grossman, A. Nerode, A. P. Ravn, and H. Rischel, editors, *Hybrid Systems*, number 736 in Lecture Notes on Computer Science, pages 209 – 229. Springer-Verlag, 1993.
2. R. Alur and D. Dill. Automata for modeling real-time systems. In M. S. Paterson, editor, *ICALP90: Automata, Languages and Programming*, number 443 in Lecture Notes on Computer Science, pages 322 – 335. Springer-Verlag, 1990.
3. R. Alur and D. Dill. The theory of timed automata. In J.W. deBakker, C. Huizing, W.P. dePoever, and G. Rozenberg, editors, *Real-Time: Theory in Practice*, number 600 in Lecture Notes on Computer Science, pages 45 – 73. Springer-Verlag, 1991.
4. R. Alur and T. A. Henzinger. A really temporal logic. In *30th Annual Symposium on Foundations of Computer Science*, pages 164 – 169, 1989.
5. H. Barringer. Up and down the temporal way. Technical report, Computer Science, University of Manchester, England, September 1985.
6. A. Benveniste and P. LeGuernic. Hybrid dynamical systems theory and the SIGNAL language. *IEEE Transactions on Automatic Control*, 35(5):535 – 546, May 1990.
7. B. Berthomieu and M. Diaz. Modeling and verification of time dependent systems using Time Petri Nets. *IEEE Transactions on Software Engineering*, 17(3):259 – 273, March 1991.
8. R. Brockett. Hybrid models for motion control systems. In H. Trentelman and J. C. Willems, editors, *Perspectives in Control*, pages 29 – 54. 1993.
9. D.N. Dyck and P.E. Caines. The logical control of an elevator. *IEEE Trans. Automatic Control*, (3):480 – 486, March 1995.
10. R. L. Grossman, A. Nerode, A. P. Ravn, and H. Rischel, editors. *Hybrid Systems*. Number 736 in Lecture Notes on Computer Science. Springer-Verlag, 1993.
11. V. Gupta, R. Jagadeesan, V. Saraswat, and D. Bobrow. Programming in hybrid constraint languages. In P. Antsaklis, W. Kohn, A. Nerode, and S. Sastry, editors, *Hybrid Systems II*, number 999 in Lecture Notes on Computer Science. Springer-Verlag, 1995.
12. R. Hale. Using temporal logic for prototyping: The design of a lift controller. In H.S.M. Zedan, editor, *Real-Time Systems, Theory and Applications*. Elsevier Science Publishers B.V. (North-Holland), 1990.
13. Integrated Systems Inc. SystemBuild User's Guide.
14. The MathWorks Inc. Simulink User's Guide.
15. M.A. Jackson. *System Development*. Prentice-Hall, Englewood Cliffs, 1983.
16. M. S. Jaffe, N. G. Leveson, M. P. E. Heimdahl, and B. E. Melhart. Software requirements analysis for real-time process-control systems. *IEEE Transactions on Software Engineering*, 17(3):241 – 257, March 1991.
17. B. H. Krogh. Condition/event signal interfaces for block diagram modeling and analysis of hybrid systems. In *8th IEEE International Symposium on Intelligent Control*, Chicago, IL, August 25–27 1993.

18. L. Lamport. Hybrid systems in TLA+. In R. L. Grossman, A. Nerode, A. P. Ravn, and H. Rischel, editors, *Hybrid Systems*, number 736 in Lecture Notes on Computer Science, pages 77 – 102. Springer-Verlag, 1993.

19. N. G. Leveson, M. P. E. Heimdahl, H. Hildreth, and J. D. Reese. Requirements specification for process-control systems. *IEEE Transactions on Software Engineering*, 20(9):684 – 707, September 1994.

20. D. G. Luenberger. *Introduction to Dynamic Systems: Theory, Models and Applications.* John Wiley & Sons, 1979.

21. N. Lynch. Modeling and verification of automated transit systems, using timed automata, invariants and simulations. In *Hybrid Systems III*. in Verification and Control of Hybrid Systems, October, 1995, to appear in LNCS.

22. N. Lynch. Simulation techniques for proving properties of real-time systems. In J.W. deBakker, W.P. dePoever, and G. Rozenberg, editors, *A Decade of Concurrency*, number 803 in Lecture Notes on Computer Science, pages 376 – 424. Springer-Verlag, 1993.

23. O. Maler, Z. Manna, and A. Pnueli. From timed to hybrid systems. In J.W. deBakker, C. Huizing, W.P. dePoever, and G. Rozenberg, editors, *Real-Time: Theory in Practice*, number 600 in Lecture Notes on Computer Science, pages 448 – 484. Springer-Verlag, 1991.

24. Z. Manna and A. Pnueli. Specification and verification of concurrent programs by ∀-automata. In *Proc. 14th Ann. ACM Symp. on Principles of Programming Languages*, pages 1–12, 1987.

25. A. Nerode and W. Kohn. Models for hybrid systems: Automata, topologies, controllability, observability. In R. L. Grossman, A. Nerode, A. P. Ravn, and H. Rischel, editors, *Hybrid Systems*, number 736 in Lecture Notes on Computer Science, pages 317 – 356. Springer-Verlag, 1993.

26. X. Nicollin and J. Sifakis. From ATP to timed graphs and hybrid systems. In J.W. deBakker, C. Huizing, W.P. dePoever, and G. Rozenberg, editors, *Real-Time: Theory in Practice*, number 600 in Lecture Notes on Computer Science, pages 549 – 572. Springer-Verlag, 1991.

27. B. Sanden. An entity-life modeling approach to the design of concurrent software. *Communication of the ACM*, 32(3):230 – 243, March 1989.

28. V. Saraswat. Concurrent constraint programming languages. Technical report, Computer Science Department, Carnegie–Mellon University, 1989. Ph. D. thesis.

29. M. Shaw. Comparing architectural design styles. *IEEE Software*, pages 27 – 41, November 1995.

30. I. E. Sutherland. Micropipeline. *Communication of ACM*, 32(6):720 – 738, June 1989.

31. W. Thomas. Automata on infinite objects. In Jan Van Leeuwen, editor, *Handbook of Theoretical Computer Science*. MIT Press, 1990.

32. P. Wolper. Temporal logic can be more expressive. *Information and Control*, 56:72 – 99, 1983.

33. X. Yu, J. Wang, C. Zhou, and P. Pandya. Formal design of hybrid systems. In H. Langmaack, W.P. deRoever, and J. Vytopil, editors, *Formal Techniques in Real-Time and Fault-Tolerant Systems*, number 863 in Lecture Notes on Computer Science. Springer-Verlag, 1994.

34. Y. Zhang. A foundation for the design and analysis of robotic systems and behaviors. Technical Report 94-26, Department of Computer Science, University of British Columbia, 1994. Ph.D. thesis.

35. Y. Zhang and A. K. Mackworth. Specification and verification of hybrid dynamic systems using timed ∀-automata. In *Hybrid Systems III*. in Verification and Control of Hybrid Systems, October, 1995, to appear in LNCS.

36. Y. Zhang and A. K. Mackworth. Specification and verification of constraint-based dynamic systems. In A. Borning, editor, *Principles and Practice of Constraint Programming*, Lecture Notes in Computer Science 874, pages 229 – 242. Springer Verlag, 1994.
37. Y. Zhang and A. K. Mackworth. Constraint Nets: A semantic model for hybrid dynamic systems. *Theoretical Computer Science*, 138(1):211 – 239, 1995. Special Issue on Hybrid Systems.
38. Y. Zhang and A. K. Mackworth. Constraint programming in constraint nets. In V. Saraswat and P. Van Hentenryck, editors, *Principles and Practice of Constraint Programming*, pages 49 – 68. MIT Press, 1995.
39. Y. Zhang and A. K. Mackworth. Synthesis of hybrid constraint-based controllers. In P. Antsaklis, W. Kohn, A. Nerode, and S. Sastry, editors, *Hybrid Systems II*, Lecture Notes in Computer Science 999, pages 552 – 567. Springer Verlag, 1995.

Index

N. J. Nilsson: Principles of Artificial Intelligence. XV, 476 pages, 139 figs., 1982

J. H. Siekmann, G. Wrightson (Eds.): Automation of Reasoning 2. Classical Papers on Computational Logic 1967–1970. XXII, 638 pages, 1983

R. S. Michalski, J. G. Carbonell, T. M. Mitchell (Eds.): Machine Learning. An Artificial Intelligence Approach. XI, 572 pages, 1984

J. W. Lloyd: Foundations of Logic Programming. Second, extended edition. XII, 212 pages, 1987

N. Cercone, G. McCalla (Eds.): The Knowledge Frontier. Essays in the Representation of Knowledge. XXXV, 512 pages, 93 figs., 1987

G. Rayna: REDUCE. Software for Algebraic Computation. IX, 329 pages, 1987

L. Kanal, V. Kumar (Eds.): Search in Artificial Intelligence. X, 482 pages, 67 figs., 1988

H. Abramson, V. Dahl: Logic Grammars. XIV, 234 pages, 40 figs., 1989

P. Besnard: An Introduction to Default Logic. XI, 201 pages, 1989

A. Kobsa, W. Wahlster (Eds.): User Models in Dialog Systems. XI, 471 pages, 113 figs., 1989

Y. Peng, J. A. Reggia: Abductive Inference Models for Diagnostic Problem-Solving. XII, 284 pages, 25 figs., 1990

A. Bundy (Ed.): Catalogue of Artificial Intelligence Techniques. Fourth revised edition. XVI, 141 pages, 1997 (first three editions published in the series)

R. Kruse, E. Schwecke, J. Heinsohn: Uncertainty and Vagueness in Knowledge Based Systems. Numerical Methods. XI, 491 pages, 59 figs., 1991

Z. Michalewicz: Genetic Algorithms + Data Structures = Evolution Programs. Third, revised and extended edition. XX, 387 pages, 68 figs., 1996 (first edition published in the series)

V. W. Marek, M. Truszczyński: Nonmonotonic Logic. Context-Dependent Reasoning. XIII, 417 pages, 14 figs., 1993

V. S. Subrahmanian, S. Jajodia (Eds.): Multimedia Database Systems. XVI, 323 pages, 104 figs., 1996

Q. Yang: Intelligent Planning. XXII, 252 pages, 76 figs., 1997

J. Debenham: Knowledge Engineering. Unifying Knowledge Base and Database Design. XIV, 465 pages, 288 figs., 1998

H. J. Levesque, F. Pirri: Logical Foundations for Cognitive Agents. Contributions in Honor of Ray Reiter. XII, 405 pages, 32 figs., 1999

K. R. Apt, V. W. Marek, M. Truszczynski, D. S. Warren (Eds.): The Logic Programming Paradigm. A 25-Year Perspective. XVI, 456 pages, 57 figs., 1999